INTERNATIONAL HANDBOOK OF NETWORK INDUSTRIES

International Handbook of Network Industries

The Liberalization of Infrastructure

Edited by

Matthias Finger

College of Management and Technology, ÉPFL – École Polytechnique Fédérale Lausanne, Switzerland

Rolf W. Künneke

Faculty of Technology, Policy and Management, Delft University of Technology, The Netherlands

Edward Elgar

Cheltenham, UK • Northampton, MA, USA

Published by
Edward Elgar Publishing Limited
The Lypiatts
15 Lansdown Road
Cheltenham
Glos GL50 2JA
UK

Edward Elgar Publishing, Inc.
William Pratt House
9 Dewey Court
Northampton
Massachusetts 01060
USA

A catalogue record for this book
is available from the British Library

Library of Congress Control Number: 2011922860

ISBN 978 1 84720 642 8

Typeset by Servis Filmsetting Ltd, Stockport, Cheshire
Printed and bound by MPG Books Group, UK

Contents

Contributors

Maarten J. Arentsen, University of Twente, The Netherlands

Demi Chung, University of Sydney and The University of New South Wales, Australia

Aad Correljé, Delft University of Technology, The Netherlands

Michael A. Crew, Rutgers University (Newark), USA

Hella Engerer, German Institute for Economic Research (Berlin), Germany

Catarina Figueira, Cranfield University, United Kingdom

Matthias Finger, École Polytechnique Fédérale Lausanne, Switzerland

Dominique Finon, French National Centre of Scientific Research (CNRS), Centre CIRED, Paris, France

Jean-Michel Glachant, European University Institute (Florence), Italy

Marianela Gonzalez, Universidad de las Palmas (Gran Canaria), Spain

John Groenewegen, Delft University of Technology, The Netherlands

Sveinn Vidar Gudmundsson, Toulouse Business School, France

Shahid Hasan, The Energy and Resources Institute (TERI) (New Delhi), India

David A. Hensher, University of Sydney, Australia

Christian Jaag, University of St. Gallen, Switzerland

Mark A. Jamison, University of Florida (Gainesville), USA

Paul R. Kleindorfer, Wharton Business School (Philadelphia), USA and INSEAD (Paris), France

Rolf Künneke, Delft University of Technology, The Netherlands

Pun-Lee Lam, Hong Kong Polytechnic University, China

Marc Laperrouza, École Polytechnique Fédérale Lausanne, Switzerland

Wolter Lemstra, Delft University of Technology, The Netherlands

Gary Madden, Curtin University of Technology (Perth), Australia

William H. Melody, Technical University of Denmark (Lyngby), Denmark

Claude Ménard, Université Paris 1 Sorbonne-Pantheon, France

Robert Millward, University of Manchester, United Kingdom

Miguel A. Montoya, Technologico de Monterrey (Guadalajara), Mexico

Aaron Morey, Curtin University of Technology (Perth), Australia

Milton L. Mueller, Syracuse University, USA

David Parker, Cranfield University, United Kingdom

Aleksandra Peeroo, Université Paris 1 Sorbonne-Pantheon, France

Yannick Perez, Université Paris-Sud 11, France

Jeffrey Petchey, Curtin University of Technology (Perth), Australia

Alice Shiu, Hong Kong Polytechnic University, China

Pablo T. Spiller, University of California (Berkeley), USA

Leena Srivastava, The Energy and Resources Institute (TERI) (New Delhi), India

John Stanley, University of Sydney, Australia

Francesc Trillas, Technologico de Monterrey (Guadalajara), Mexico

Urs Trinkner, University of Zürich, Switzerland

Lourdes Trujillo, Universidad de las Palmas (Gran Canaria), Spain

Lucia van Geuns, Clingendael International Energy Programme, The Netherlands

1 Introduction

Matthias Finger and Rolf Künneke

LIBERALIZATION OF INFRASTRUCTURES

Infrastructures have been significantly restructured during the past three decades. Formerly these sectors have been perceived as utilities that received special governmental supervision and support in order to provide essential services to society under politically acceptable conditions. Public ownership and regulated monopolies have been very common in order to guarantee adequate service levels and pricing schemes under close political supervision. Nowadays, infrastructures are increasingly positioned as commercial economic sectors that need to efficiently satisfy customers' needs and expectations. Accordingly, competition is introduced, private interests are accentuated for instance by private ownership and the roles and responsibilities of governments are redefined. This liberalization[1] defines a fundamental change in the governance of infrastructures with significant consequences for their operations and performance, something that is amply demonstrated throughout this book. The importance of this phenomenon is emphasized by the fact that liberalization occurred in all infrastructure sectors in one way or another and in almost all countries in the world, as illustrated in Table 1.1. Apparently liberalization is under way even under very different technological conditions of the sectors involved, different socio-political preferences and needs, and different political ideologies.

The scope and magnitude of the liberalization of infrastructures is even more remarkable if we consider that these sectors are of fundamental importance for the socio-economic development of countries. Typically, infrastructures provide services in the fields of energy (electricity, gas, oil), communication (telephony, Internet, postal services), transportation (aviation, railways, maritime transport, public transportation, roads), as well as drinking water and sewerage. These are essential services to citizens and industry that are indispensible to modern societies. Generally, liberalization is introduced as a means to improve the quality of services, increase efficiency, lower prices and stimulate technological innovations. But there are also deeply rooted worries, for instance, when it comes to guaranteeing public service obligations, national interests, security of supply, or the stimulation of environmentally sustainable developments which are not necessarily supported in a liberalized environment. However, despite these concerns, liberalization seems to be inevitable. The process of restructuring of network industries is evolving in different sectors and countries according to its own dynamics and in quite different ways with unknown outcomes.

Infrastructure liberalization is unprecedented not only with respect to its sectoral and geographical scope, but also with respect to the fundamental changes of governance structures, to political ambitions and objectives, as well as with respect to technological innovations. Many publications focus on specific sectors or countries. This handbook is one of the few attempts to provide a general overview across sectors and countries.

Table 1.1 Some milestones for the restructuring of infrastructures in different sectors and countries

Country/sector	Year
USA	
Airline deregulation Act	1978
Staggers Act (rail deregulation)	1980
AT&T divestiture (telecommunication deregulation)	1984
FERC order 636 (gas deregulation)	1992
FERC order 888 (electricity deregulation)	1996
Telecommunication Act	1996
Europe	
UK privatization British Telecom	1984
UK privatization British Gas	1986
UK privatization British Airways	1987
EU liberalization directive Telecom	1990
EU liberalization directive Railways	1991
EU liberalization directive Electricity	1996
EU liberalization directive Gas	1998
Norway Energy Act	1990
Finland regulatory reform power sector	1997
Other parts of the world	
South America: Chile, regulatory reform of the power sector	1978
Australia power sector reform	1994
New Zealand power sector reform	1984
China: Formal establishment of the China United Communications Corporation	1994
India: Announcement of the New Economic Policy	1991
Russia: Transformation of Gazprom into a corporate group	1992
South Africa: Electricity Act Amendment	1995

Source: Based on Kessides (2004, p. 32), various chapters in Sioshansi and Pfaffenberger (2006), Gaunt (2008) and various chapters in this handbook.

AMBITIONS OF THIS HANDBOOK

This handbook provides an overview of the features and outcomes of liberalization in different sectors and parts of the world from a governance perspective. Since each sector and region is specific in its own way, and since the different authors of this book have their own interpretation of the sector's reforms, it is not possible to use a strict format for their description and analysis. However, certain general aspects can be identified.

The sector studies roughly deal with five issues. First, they provide insights into the specificities of various network industries with respect to technology, economics, institutions or policy. For example: which parts of the industry are open for competition? Which are regulated? What is the importance of this sector with respect to public services or national interests? Are there any geographic specificities, such as particular regions from any given sector's liberalization emerged or evolved? Which technological innovations facilitated liberalization? A second aspect is the historical overview of the sector's restructuring. How was the sector organized traditionally and what were important drivers for change? Possibly, different periods of liberalization can be identified, pointing to different drivers of this process. As a third point, sector performance is reviewed. What are the results of the liberalization process from a socio-economic, a political or technical perspective? Important aspects include prices, costs, services, quality, public service, national interests and technological reliability. Fourthly, an overview is provided of regulatory models and approaches. Among others, regulatory governance and regulatory bodies are specified. Specific issues of regulation are addressed, such as sustainability, security of supply, innovation, risk, geopolitical considerations or the nature of competition (for instance, competition in the market, competition for the market or network competition). A final (fifth) point addresses the future challenges and problems of the sector.

The choice of the infrastructure sectors that are discussed in this handbook is somewhat arbitrary and certainly incomplete. First, the question arises, what to consider as a network industry or an infrastructure. In our perception infrastructures are related to large-scale physical networks that provide essential goods and services to society. Governments are concerned with the provision of such goods because they are considered to be a public service which needs to be accessible for citizens and industry under socially acceptable conditions. Often, the provision of these goods or services is associated with market failures, which requires governmental intervention or regulation (Firth et al., 1998).

Against this background, this handbook covers a broad range of different infrastructures in the fields of energy, transportation, communication and the provision of drinking water, as mentioned earlier. We are aware that the definitions and perceptions of infrastructures are quite different and, accordingly, the delineation of these sectors is debatable. However, we are sure that this handbook covers the most relevant infrastructure sectors.

The country studies touch on similar aspects as the sector studies. First, the technological, economic, institutional, political and geographical specificities of the region or country are highlighted. How have the infrastructures evolved over time? Which infrastructures were reformed? Which ones are still traditionally organized? What are important cultural aspects, norms or values that influence infrastructure development? As a second point, the regulatory models and approaches are summarized. This provides an overview of the different approaches or models of restructuring in different sectors in any given country or region. What are the differences or similarities between different sectors and what are the possible reasons for this? What are specific issues of regulation? Are there country-specific approaches to liberalization? Thirdly, the current challenges and opportunities of restructuring are discussed. For instance: what is the overall economic performance of the liberalized sectors in the region? What is the technical performance (reliability, security of supply), the social acceptance, or the institutional capacity for restructuring? In some countries international institutions might play an important role (for instance, the European Union or the World Bank), as well as multinational corporations or the emergence of global markets. Finally, an outlook will be offered on the future opportunities and challenges of the liberalization of infrastructures in this national or regional context. The selection of the countries or regions is certainly not exhaustive, and primarily aims to illustrate different national or regional approaches to the liberalization of infrastructures. The range of national cases is, of course, much broader than what we are able to cover in this handbook.

In addition to the focus on specific sectors and countries, some generic issues of the liberalization of infrastructures are addressed. This includes the basic economic principles of liberalization, the scope and safeguarding of public service obligations, the institutional design and the institutional economic history of liberalized infrastructures.

STRUCTURE OF THE BOOK

This handbook consists of three parts, that is, generic aspects of infrastructure liberalization, sector studies and country studies.

Part I: Generic Aspects of Infrastructure Liberalization

Pablo Spiller reviews, in Chapter 2, basic economic principles that govern the relationships between government and the private sector in infrastructure provision, namely, the different types of contracts and the various problems related to such contracts, in particular, opportunism. These theoretical elements will be found throughout the book in situations of privatization and private sector participation, namely, in the areas of water, airports, ports, toll roads and others.

Christian Jaag and Urs Trinkner outline, in Chapter 3, some theoretical (economic) foundations of infrastructure regulation, in particular for the regulation of so-called monopolistic bottlenecks. The authors define, from a neo-liberal perspective, where such regulation is justified, as well as the criteria for assessing its effectiveness. These theoretical elements will be found throughout the book whenever competitive elements interface with monopolistic bottlenecks, such as in the case of railway, electricity or gas liberalization.

Matthias Finger and Dominique Finon address, in Chapter 4, the relevance of public service obligations in the liberalization of infrastructures. The authors crystallize the political dimensions of infrastructure services provision, which, traditionally, was its public service dimension. They show, that, in the case of Europe, this traditional idea of public service has given rise to 'universal service', a concept that is increasingly used around the world. The concept denotes basically the protection of consumers who do not automatically benefit from the liberalization of infrastructures.

John Groenewegen extends, in Chapter 5, the basic principles of infrastructure liberalization of Chapter 2 (for example, contracts) to transactions and the institutional arrangements (for example, the governance structures) that accompany successful infrastructure liberalization. He also builds on the considerations of Chapter 3 (regulation), inasmuch as such institutions – which generally include but are not limited to regulatory bodies – need to be efficient from an economic point view. Such considerations about designing economically efficient governance structures for the liberalized infrastructures are increasingly prevalent worldwide and can be found throughout this book.

In Chapter 6, Robert Millward approaches the evolution of the infrastructures from a historical perspective. He recalls the evolution of the main infrastructures mainly in Europe and the USA prior to their liberalization, and highlights, once again, the important role of the public policies in their shaping.

Part II: Sector Studies

The sector studies are clustered into four categories: electronic communication (Chapters 7, 8, 9), energy (Chapters 10, 11, 12), transport (Chapters 13 – 17), and 'others' (Chapters 18, 19).

In Chapter 7 William Melody recalls the main features and outcomes of telecommunications' liberalization in the USA, Europe and the developing countries. He focuses in particular on the pioneering role played by the Federal Communications Commission (FCC) in liberalizing and re-regulating the sector, as well as on the importance of the US telecom liberalization and regulation for the development of regulatory economic theory.

In Chapter 8 William Melody and Wolter Lemstra present the liberalization in radio spectrum management, which has been at the heart of the telecommunication reform process and is closely related to the increasing importance of mobile communication. They review spectrum characteristics and describe the evolution and structure of spectrum management and regulation and their performance. Different proposals to improve spectrum efficiency and management are reviewed, with special attention being given to economic criteria, such as spectrum markets or auctions. The authors are sceptical as to whether spectrum management institutions are able to adapt in time, and effectively, to the rapidly changing technologies and service markets.

The liberalization of the Internet is elaborated in Chapter 9 by Milton Mueller and Wolter Lemstra by defining and presenting the key characteristics of the global Internet. The authors show in particular how the development of the Internet is related to the liberalization of telecommunications' infrastructures and how their co-evolution leads to interesting new challenges, both for telecommunications and Internet regulation (for example, net neutrality).

The next three chapters deal with the liberalization of energy sectors, that is, electricity, gas and oil. In Chapter 10 Jean-Michel Glachant and Yannick Perez present the main economic principles of the liberalization of the electricity sector. Among others, they highlight different approaches to this sector's reform in the USA and Europe, which are rooted in the diversity of the various historical and institutional conditions.

Chapter 11, which is authored by Maarten Arentsen, highlights the two currently dominant models in the gas sector, namely, the model striving towards gas market liberalization (USA, Europe, Latin America, Australia) on the one hand, and the model stressing state ownership and resource nationalism, a model not astonishingly favoured by large gas producers (for example, Russia, Norway, Middle Eastern countries). New challenges emerge, however, from technological changes such as LNG (liquid natural gas) as well as from climate change.

The restructuring of the oil sector is addressed in Chapter 12 by Aad Correljé and Lucia van Geuns. They present the oil value chain, the power relationships among the major actors along this value chain – namely, the producer countries, the oil companies and the consumer countries – as well as their subsequent struggle for rents. Based on a historical approach, the authors conclude to a certain cyclicality in the evolution of the oil industry, whereby the intervention of governments has less to do with markets and more with geopolitics and rents.

The liberalization of different transport sectors is presented in Chapters 13 to 17. Chapter 13, by Marc Laperrouza, deals with the railway sector. The author recalls the specificities of this network industry, which explain the comparatively slow pace of railway reforms across the world. Many challenges remain, notably technical (interoperability), institutional (appropriate regulatory framework; unbundling), and economic challenges (third-party access and corresponding pricing). The author doubts whether the unbundled railway operator will be the model for the future.

Chapter 14, on air transport, is authored by Sveinn Gudmundsson. The chapter describes air transport liberalization in the USA and Europe mainly. It covers not only airlines, but also the liberalization of the supporting industries, such as ground handling, airport slots, ticket advertising and computer reservation systems. In addition, problems with state subsidies are an issue. Overall, the author deems airline liberalization to be a success, especially when it comes to efficiency, growth, prices and competition.

Chapter 15 is about road infrastructures. David Hensher and Demi Chung present and discuss the growing and worldwide involvement of the private sector in road construction and maintenance, focusing in particular on questions of risk-sharing between governments and private sector in so-called public–private partnerships. The authors consider that toll charges may well be a precursor of much broader congestion charging.

Chapter 16, by John Stanley, focuses on the liberalization of public bus transport. The main approach to liberalizing in this sector has been competitive tendering. The author identifies and reviews the critical success factors. He argues that tendering has delivered on its promises and has led to cost reductions, as well as to growing patronage.

Chapter 17 on seaports is co-authored by Lourdes Trujillo and Marianela Gonzalez. The authors present the various activities and components, as well as the different types of seaports around the world. They show that the liberalization of seaports mainly takes the form of concessions, with the so-called 'landlord' model becoming the dominant way of managing seaports.

The last two sector chapters are about infrastructures in the sectors of drinking water and postal services, which we earlier categorized as 'others'. In Chapter 18, Claude Ménard and Aleksandra Peeroo elaborate the particular characteristics of the liberalization of the drinking water sector. The particular features of this sector explain its rather slow liberalization, which takes the form of changes in decision rights, rather than property rights. By way of three significant country cases – UK and Wales, France and Germany – the authors show that the liberalization euphoria seems to be exhausted and that corporatization of water companies seems to become the model.

Chapter 19, by Michael Crew and Paul Kleindorfer, describes the recent transformation of the postal sector in the industrialized countries and highlights in particular the examples of corporatization (Sweden Post and New Zealand Post) and privatization (in the cases of Germany and the Netherlands), as well as the unique case of the USA. Overall, the authors conclude that the liberalization of the postal sector has led to efficiency and innovation but they highlight the particular challenge to balance universal service obligation, commercial freedom of the historical operator, and market opening.

Part III: Country and Regional Studies

This part covers major countries and regions in the world in order to illustrate different approaches to liberalization across network industries. As mentioned earlier, this overview is by no means complete or exhaustive.

The case of Europe is presented in Chapter 20 by Matthias Finger and Marc Laperrouza. In this chapter, the authors highlight the very particular European approach to infrastructure liberalization, which is embedded in broader policy objectives, whereby infrastructure markets are ultimately means for enhancing the competitiveness of a unified Europe. They also show how, in part as a result of such liberalization, the broader public policy objectives are currently evolving.

Chapter 21, on the USA, has been written by Mark Jamison. He highlights the quite specific nature of utility liberalization in the USA, shaped as it is by both national sector-specific and state comprehensive utility regulation. More precisely, the chapter highlights the evolution of regulation in the telecommunications (fixed and wireless), the gas and the electricity sectors, and concludes that policy-makers (and thus regulators) often lag behind the technological changes in these sectors.

Chapter 22 is about Latin America and authored by Francesc Trillas and Miguel Montoya. They present the infrastructure reforms that have taken place in this part of the world with particular attention being given to the case of Chile, which is often considered to be Latin America's success story. They also explain the special conditions of reform in Latin America, as compared to the other regions of the world, in particular the prevalent inequalities. The balance sheet of infrastructure reforms in Latin America is not, at least in light of public opinion, overall positive, and the authors conclude that such reforms should not be conducted independently of broader macroeconomic policies focusing especially on distributive questions.

China is presented in Chapter 23 by Pun-Lee Lam and Alice Shiu. They focus specifically on the reform of China's electricity and telecommunications sectors. The authors crystallize the policy steps that were taken for such successful reforms, namely, the creation of competition and industry restructure before corporatization and partial privatization. They also highlight the challenges that remain, in particular the challenges of governance of the firms, the challenges of regulatory governance of the sectors, not to mention the environmental challenges in the energy sector, along with the challenge of rural telecommunications penetration.

Chapter 24 by Leena Srivastava and Shahid Hasan discusses the liberalization of infrastructures in India. The authors highlight the, so far, profound transformations of India's infrastructure sectors, as well as the significant challenges that still lay ahead, especially in terms of accessing the rural poor, of investment needs, but also in terms of regulatory institutions and governance of the often decentralized and still state-owned operators.

The Russian approach to infrastructure liberalization is presented by Hella Engerer in Chapter 25. She describes in much detail the reform in the main Russian infrastructure sectors, namely, electricity, gas, railways, telecommunications, postal services, and water and sanitation services. The chapter highlights the difficulties and slow progress of such reform in international comparison, given the communist past of monopolistic state-owned enterprises. Not surprisingly, reform appears to be slowest where former state-owned enterprises remain dominant, that is, especially in gas, railways, and telecommunications.

Chapter 26 is about Australia and authored by Gary Madden, Jeffrey Petchey and Aaron Morey. They highlight the reforms in the different sectors, covering railways (characterized by unbundling), roads (network extension), telecommunications (privatization), ports (minimal reforms), airports (privatization), electricity and gas (unbundling), and water (unbundling and trading). Australia being a federalist country, most infrastructure services are provided at the state level, which leads the authors to focus in particular on questions of funding. Overall, the balance sheet of the reforms is, according to the authors, not yet satisfactory, which has to do with the regulatory environment as well as with the (limited) size of the Australian market.

Finally Chapter 27 addresses infrastructure liberalization in developing countries overall. Catarina Figueira and David Parker focus on the relationship between investments in infrastructures and the regulatory environment in developing countries. In the context of such countries private sector investments into infrastructures are particularly important. The authors are able to show, however, that it is less the sector-specific regulation that determines such investments than rather broader conditions of (good) governance.

Overall, this book offers a comprehensive perspective of the main infrastructure sectors across the globe, focusing in particular on their transformation of the past 20 or more years.

NOTE

1. We use the term 'liberalization' as the all-encompassing term. Deregulation is defined as the introduction of competition, whereas privatization is either the transfer of ownership or the participation of the private sector in the management of public assets. The redefinition of the roles and responsibilities of government in this context is termed as 're-regulation'.

REFERENCES

Firth, Lucy, Karin Boersma and William M. Melody (1998), 'Infrastructure concepts and classifications. A framework for scenario analysis of infrastructures in an economic perspective', in Margot P.C. Weijnen and Ernst F. ten Heuvelhof (eds), *The Infrastructure Playing Filed in 2030*, Proceedings of the First Annual Symposium, Delft, pp. 21–40.

Gaunt, C.T. (2008), 'Electricity distribution industry restructuring in South Africa: a case study', *Energy Policy*, **36** (11), 3448–59.

Kessides, Ioanis N. (2004), *Reforming Infrastructure: Privatization, Regulation, and Competition*, Washington, DC: World Bank and Oxford University Press.

Sioshansi, Fereidoon P. and Wolfgang Pfaffenberger (eds) (2006), *Electricity Market Reform: An International Perspective*, Amsterdam: Elsevier.

PART I

GENERIC ASPECTS OF INFRASTRUCTURE LIBERALIZATION

2 Basic economic principles of infrastructure liberalization: a transaction cost perspective

Pablo T. Spiller

LIBERALIZATION AS AN INSTITUTIONAL CHOICE

Infrastructure regulations take many forms, and their appearance is not random, but rather reflect the complex interplay of a nation's public decision-making. Thus, to understand the nature of infrastructure regulation, and, in particular, the move towards liberalization, it is fundamental to understand how public policy is undertaken in a particular environment, and how such a process interacts with the features of the infrastructure sector.

In this chapter I approach infrastructure regulatory choice as an institutional choice, where by institutional choice I follow Williamson's path-breaking work which, starting with *Markets and Hierarchies*, opened a new approach to institutions. In *Markets and Hierarchies* (Williamson, 1975) institutions are not seen simply as organigrams or chains of command whose layers are determined by the workings of communications channels. Instead, institutions are now seen as designed to deal with the basic time inconsistency inherent to most transactions, namely, opportunism coupled with the inability to write fully contingent contracts. Indeed, in explaining the differences with the prior literature, Williamson writes (1975, p. 7):

> (2) I expressly introduce the notion of opportunism and am interested in the ways that opportunistic behavior is influenced by economic organization. (3) I emphasize that it is not uncertainty or small numbers, individually or together that occasion market failure, but it is rather the joining of these factors with bounded rationality on the one hand and opportunism on the other that gives rise to exchange difficulties.

The main point of this chapter is that to understand the institutional choice in the infrastructure sector, a deeper understanding of opportunism as it relates to public contracting in the infrastructure is necessary. Paraphrasing Williamson (1975, p. 7), 'opportunism, in a rich variety of forms, is made to play a central role in the analysis of [institutions] herein'. In this chapter I focus on the role of opportunism in its various forms for the design of public contracting in the infrastructure sector and, in particular, on how regulatory policy adjusts, or not, to provide the necessary commitment for investments in the infrastructure sector.

A fundamental divergence from the previous literature on regulation that also considered infrastructure regulation as a contract between the state and the operators, is that, following Spiller (2009), I look at regulation as a public contract. A fundamental difference between private and public contracts is that public contracts are in the public sphere and, thus, although politics is normally not necessary to understand private contracting, it becomes fundamental to understanding public contracting. Furthermore, the analysis of public contracting – of which liberalization is a form – must be done within the proper institutional comparison (Coase, 1964) and with a heavy micro-analytic dose (Williamson, 1979). In this sense, the supposed inefficiency of public contracting must be assessed in reference to all relevant alternatives (Williamson, 1996). Thus, when considering infrastructure liberalization, all relevant alternatives must be considered within an appropriate institutional comparison. Infrastructure regulation faces peculiar hazards arising from both the nature of infrastructure sectors, and its public contracting feature. The combination of these features creates a highly complex set of hazards that require subtle political management. We conclude that not paying attention to such hazards will necessarily bring about policy reversals, and conflict.

KEY FEATURES OF INFRASTRUCTURE PROJECTS

Infrastructure projects, by involving contracts with the government, have three fundamental types of contractual hazards: (1) what I will call here, 'vanilla opportunism', which reflects the hazards associated with any large-scale contract involving substantial amounts of sunk cost investments. In these cases, special governance is required to limit opportunism by the contracting parties. Sunk investments here play a large role in determining the type of long-term contract and its governance structure. (2) Because the purchasing party is the government, the investing party (which could be a private or public entity) faces potential unilateral changes in the rules of the game, or what I elsewhere called 'governmental' opportunism, which it would not be exposed to in a normal contract with another private purchaser. Finally, (3) because the purchasing party is the government, the contract becomes a public contract, and is therefore exposed to what I have called elsewhere 'third-party opportunism' (Spiller, 2009), which reflects the risks that interested third parties may have incentives to challenge the 'probity' of the public agent involved in the transaction, even if the transaction is being undertaken in an honest way.

What Defines Infrastructure Projects

Three features define infrastructure projects: first, in general, their technologies are characterized by large specific, sunk, investments; second, their technologies are characterized by very large economies of scale and potentially of scope; and third, their products are massively consumed. Consider, for example, a water distribution company. Its assets (mostly water mains and pumping stations) have very little value in alternative uses (it is extremely expensive, if not impossible, to dig out the mains and move them to an alternative location/use, and so on); network externalities and economies of density imply that it may not be economical to have multiple water mains and distribution networks deployed on the same street or even in the same neighbourhood; furthermore, the variable cost of service provision is nil compared with the overall economic cost of the service; and, finally, its product is consumed by a large proportion of the city's population. Compare this situation with that of another industry characterized by large sunk investments: steel. While steel mills have very little value in alternative uses, the economies of scale and scope are trivial compared with the size of the market, and furthermore, while everybody indirectly consume steel products, very few individuals in society pay any attention to the price of steel. Thus, it is not simply specific investments that characterize infrastructure. Nor is it simply economies of scale. Nor is it widespread consumption. What separates the infrastructure sector from the rest of the economy is the combination of the three features: specific investments, economies of scale and widespread domestic consumption. These features are at the core of contracting problems that have traditionally raised the need for governmental ownership and regulation of public infrastructure.

In turn, they make the pricing of infrastructure sectors inherently political. The reason for the politicization of infrastructure pricing is threefold. First, the fact that a large component of infrastructure investments is sunk, implies that once the investment is undertaken the operator will be willing to continue operating as long as operating revenues exceed operating costs. Since operating costs do not include a return on sunk investments (but only on the alternative value of these assets), the operating company will be willing to operate even if prices are below total average costs. Second, economies of scale imply that in most infrastructure services, there will be few suppliers in each locality. Thus, the whiff of monopoly will always surround infrastructure operations. Finally, the fact that infrastructure services tend to be massively consumed, and thus that the set of consumers closely approximates the set of voters, implies that politicians and interest groups will care about the level of infrastructure pricing. Thus, massive consumption, economies of scale and sunk investments provide governments (either national or local) with the opportunity to behave opportunistically vis-à-vis the investing company. For example, after the investment is sunk, the government may try to restrict the operating company's pricing flexibility, may require the company to undertake special investments, purchasing or employment patterns or may try to restrict the movement of capital. All these are attempts to expropriate the company's sunk costs by administrative measures. Thus, expropriation may be indirect and undertaken by subtle means.

Expropriation of the firm's sunk assets, however, does not mean that the government takes over the operation of the company, but rather that it sets operating conditions that just compensate for the firm's operating costs and the return on its non-specific assets. Such returns will provide sufficient *ex post* incentives for the firm to operate, but not to invest. Indeed, sunk assets' expropriation has been more prevalent in the developing world than direct utility takeovers or expropriation without compensation. While the government may uphold and protect traditionally conceived property rights, it may still capture the companies' quasi-rents via regulatory decision-making, what is commonly called 'indirect expropriation' in international law. By setting prices, investment or quality requirements, limiting promised subsidies, imposing taxes and the like, the state may limit the ability of the investor to recover its sunk investments, while still granting it enough cash flow to cover its variable operating costs. The government, in this way, obtains political support from relatively low prices *and* the maintenance of service, albeit with diminished incentives to invest and expand and in continuous conflicts with the operator.

Governmental Opportunism

Sunk assets' expropriation may be profitable for a government if the direct costs (reputation loss vis-à-vis other potential investors, lack of future investments by utilities) are small compared to the (short-term) benefits of such action (achieving re-election by reducing utilities' prices, by attacking the monopoly, and so on), and if the indirect institutional costs (for example, disregarding the judiciary, not following the proper, or traditional, administrative procedures, and so on) are not too large. Thus, incentives for expropriation of sunk assets should be expected to be largest in environments where indirect institutional costs are low, direct costs are also small and, perhaps more importantly, the government's horizon is relatively short. Forecasting such expropriation, investors, whether private or public, will not undertake investments in the first place. Thus, government direct intervention may become the default mode of operation.

The performance implications of government opportunism

If, in the presence of such incentives a government wants to motivate investment in the infrastructure sector, then, it will have to design institutional arrangements that will limit its own ability to behave opportunistically once the investment programme is undertaken. Such institutional arrangements are nothing but the design of regulatory governance. Regulatory governance, if credible, solves a key contracting problem between the government and the investor – whether public or private (Savedoff and Spiller, 1999) – by restraining the government from opportunistically expropriating the utilities' sunk investments. This, however, does not mean that the operator has to receive assurances of a rate of return nature, or that it has to receive exclusive licences. In some countries, however, where the incentives for governmental opportunism are high, exclusive licences and well-specified assurances on rates of return may be the only way to grant investors sufficient incentives to invest. Unless regulatory governance is credible, then, investments will not be undertaken or, if undertaken, will not be efficient. Investment inefficiencies may arise in several fronts. A first-order effect is underinvestment. Although the operator may invest, it will do so exclusively in areas whose market return is very high and where the payback period is relatively short. Second, maintenance expenditures may be kept to the minimum, thus degrading quality. Third, investment may be undertaken with technologies that have a lower degree of specificity, even at the cost of, again, degrading quality. Fourth, upfront rents may be achieved by very high prices which although they may provide incentives for some investment, may be politically unsustainable. Regulatory governance schemes that do not limit the potential for governmental opportunism, then, create strong inefficiencies and poor sector performance.

Contrasting operator with governmental opportunism

Standard approaches to infrastructure regulation (whether of the Chicago School or the Incentives approach) involve a normative or passive view of the regulatory process. In the Chicago School, regulation can be perceived as an arena where conflicting private interests are accommodated, while in the incentives regulation literature, regulatory rules are optimally designed to placate the firm's profit motive. The former approaches political actors as essentially passive, while the latter approaches political actors as benevolent. We see political actors differently. They are neither passive nor benevolent. They are no different to any of us. They are opportunistic – willing to lie and deceive and to pursue 'self-interest with guile' (Williamson, 1975, p. 26). There is, though, a fundamental difference between governmental opportunism and the opportunism or exercise of market power that is perceived to be at the root of the regulatory problem by most neoclassical economists. If what drove the design of regulatory policy-making is the potential for private firm opportunism or exercise of market power, then that could be undertaken by the application of general antitrust (and common law) provisions.

There would be no need for industry-specific regulation. Indeed, granting the power to limit the exercise of market power by regulatory fiat – say by setting maximum prices, conducting cost reviews, requiring specific investments, and the like – grants the political and administrative power that is behind governmental opportunism. Troesken (1996, 1997) has convincingly argued that the early 1900 movement away from municipal regulation towards state regulation was a way to reduce the incentives to behave opportunistically by the municipal regulators. Although some will argue that the complexities of modern regulatory issues (for example, interconnection, prices and standards) tilts the balance towards regulatory agencies, regulatory agencies were created more than 100 years ago, at times when the pressing regulatory issues were less interconnection and other complex regulatory matters, and more investment incentives in the presence of strong pressure to limit prices. Thus, although on a day-to-day basis regulators' main concerns are indeed firm opportunism and the restrain of market power, rather than thinking how to restrain themselves from expropriating the firms' quasi-rents, the origins of regulatory governance is rooted in restraining governmental opportunism.

Although in some environments regulatory governance may have been designed to facilitate private capture, such design exposes the regulatory process to *political capture* following a turn of the political wheel. Investors fearing such events will be cautious on long-term investments, and more interested in short-term gains. Thus, regulatory design that limits the potential for governmental opportunism not only facilitates investment, but also serves to credibly enhance the political restraint over operators' opportunism.

The Public Contracting Feature of Infrastructure Projects

The need to control for governmental opportunism, drives infrastructure projects into the public sphere. Although one could think of a motorway being 100 per cent organized, funded and operated by private investors, the need to prevent the government from setting its prices below cost, drives the government into designing governance structures for those projects that move them beyond the purely private sphere, and into a highly explicit contract with the government. The essence of public contracting, however, is its publicity. Public contracting involves, directly or indirectly, the use of public monies, and thus it affects, albeit indirectly, the lives of all citizens. Reasonably working societies, then, will naturally develop ways for public contracts to be subject to public scrutiny so as to avoid corruption and bribery. Public contract scrutiny is normally undertaken by designated agencies in charge of contract supervision. In the USA, while individual departments have agencies in charge of auditing their procurement, the Office of Management and Budget, the Government Accountability Office and the Congressional Budget Office routinely examine the procurement performance of government agencies and of the auditing agencies themselves. Apart from official agencies, a substantial amount of supervision and control is done by interested third parties (McCubbins and Schwartz, 1984; McCubbins et al., 1987, 1989). In fact, McCubbins and Schwartz (1984) in their seminal article argue that politicians can manage the bureaucracy via 'fire alarms', whereby interest groups (interested third parties) will 'pull the alarm' when agencies stray from the politician's preferred policy path. They further make the point that 'a predominantly fire-alarm oversight policy is likely to be more effective . . . than a predominantly police-patrol policy'. Thus, third-party supervision is fundamental in a democratic society.

A fundamental feature of interest groups as monitors, though, is that they are interested. In other words, they are biased. They provide information only when it is to their advantage. That is, the third-party (or parties) may behave opportunistically. As it relates to public contracts, interested third parties may have incentives to challenge the 'probity' of a particular public agent when by such action they may benefit. Such incentives may exist when third parties compete with the public agent in another (political) market. Benefits may arise in the political and the economic sphere. In both it may involve the displacement of the incumbent (and competing) public agent. In the political sphere, the challenge may be deemed successful if because of the challenge the public agent is, eventually, replaced by an agent related to, or more to the liking of, the interested third-party. As it relates to the economic sphere, the challenge may be deemed successful if the private party is replaced or the terms of the contract are changed in ways that benefit the third-party. But it is precisely because of competition in the political market that such challenges are particularly dangerous to the public agent. In a competitive political market environment third-party opportunism, depending on the challenge's credibility, may entail significant costs to the public agent. The public agent may have to incur significant time and expense to defend its actions, may have to leave its public position or, in the extreme, if the challenge is fully successful, may be prosecuted.

The potential for successful challenges has different types of implications. Dal Bó and Di Tella (2006) show, for example, that the potential for threats of the type discussed here – that is, actions by third parties that may imply costs to public agents – has an impact on the selection of public agents. While the potential for positive transfers (that is, bribes) increases the competition to become a public agent, the potential for costly and credible threats reduces it, thus reducing the quality of public agents.

Here I will explore the impact that potential challenges of this sort have on the nature of public contracting in infrastructure projects. The exposure to third-party opportunism increases the risk to both the public agent and the private party contracting with the state. In response, both will have incentives to increase the specificity of these contracts as compared to equivalent contracts among private parties. Moreover, to mitigate the risk of third-party opportunism, these contracts are likely to demand more rigid procedural processes, including formal procedures for renegotiation. Nevertheless, these adjustments are unlikely to mitigate third-party opportunism altogether, and the implementation of public contracts is more likely to experience a higher degree of conflict than contracts among private parties. In other words, the risk of third-party opportunism means that 'relational' contracting is less likely to evolve in infrastructure projects. 'Relational contracting', though, is precisely the type of contracting required for infrastructure projects. Infrastructure projects are long lasting, although they have large upfront investments, maintenance is fundamental, and improvements or modifications over time are normally required as demand and costs change. Those additional activities, and their value, are however, uncertain at the time of project selection and execution, and adaptations will therefore be required. All contracts, however, are incomplete, and to a large extent the degree of incompleteness is chosen by the parties. In the extreme, and in the absence of a third-party enforcer, contracts, whether explicit or implicit, may be sustained exclusively via implicit or explicit threats of actions, such as discontinuation of a relationship, executing implicit or explicit guarantees (Williamson, 1983), decentralized enforcement by market participants (Milgrom et al., 1990) or expelling the breaching party from membership in a clan, sect or family.

In fact, even in the face of third-party enforcement, companies normally rely on inter-firm relationships to support contracting (Macaulay, 1963) and, in particular, on private ordering in a discriminatory way (Williamson, 2002). Inter-firm relationships help manage conflicts that would otherwise be resolved via formal mechanisms (courts), with the corresponding uncertainties, delay and costs (Williamson, 1975). MacNeil's (1978) three-way categorization of contracts into classical, neoclassical and relational properly highlights that as relations become long term and complex, the reference point for adaptations becomes less the original contract as specified, and more the entire relation as it developed, the 'relational web' as MacNeil (1974, p. 595) puts it. In fact, 'adaptation is now taken to be the chief mission of economic organization', with cooperative adaptation being achieved via 'complex contractual modes' (Williamson, 2002).

Although informal and continuous adaptations may be enough to sustain a private contract over time, thereby reducing haggling and negotiating costs, infrastructure projects with their very large upfront sunk investments fit the 'hazardous transactions' description of Williamson, requiring, then 'more complex modes of governance [typically] reserved for more hazardous transactions' (Williamson, 2002). Relational contracting, then, appears in highly complex transactions, transactions where parties have sunk highly specific investments (Williamson, 1985), and where parties have highly particular, detailed and tacit knowledge of their specific situations, which they can use to 'adapt to new information as it becomes available' (Baker et al., 2002). In other words, infrastructure would be prone to be governed by 'relational contracting'. In sum, discriminating alignment is the foundation of private contract adaptability. The main risk facing the parties in complex transactions is opportunistic behaviour by the trading party. Contractual governance is precisely designed to limit such behavior so as to successfully implement the transaction, or as Williamson (2002, p. 439) puts it, to 'infuse *order*, thereby to relieve *conflict*, and realize *mutual gain*'. Figure 2.1 reproduces Williamson's famous 'simple contractual schema', whereby simple transactions ($k = 0$) get implemented via simple contractual methods such as unassisted market transactions, while complex transactions ($k > 0$) exposing the parties to transaction hazards require the design of complex governance structures. Figure 2.1 shows the existence of multiple governance structures – credible contracting at node (C) and hierarchy at node (D) – which counter entering into the transaction with unrelieved hazards (node B).

Infrastructure projects, then fall in the C (Hybrid) or D (Hierarchy) node, where the investment is safeguarded via either credible contracting or administrative operation. Different from with private contracts, though, infrastructure contracts have the added third-party opportunism that limits the ability of the parties to engage in 'relational' contracting, and instead need to enter into highly complex and formal contracting hazard of third-party opportunism that limits the ability of the parties to engage in 'relational' contracting, and instead need to enter into highly complex and formal contracting.

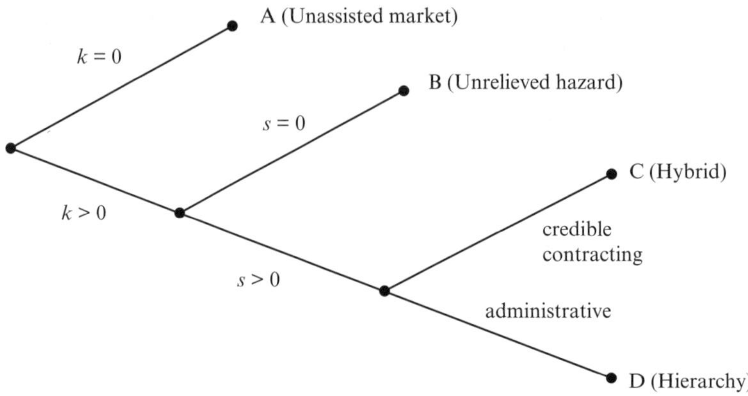

Source: Williamson (2005).

Figure 2.1 Williamson's simple contractual schema

Basic Framework of Third-party Opportunism

The following provides a basic framework to understand the workings of third-party opportunism and its impact on infrastructure projects. Consider a political environment with four players:

1. Incumbent public agent – who is of uncertain quality (his or her predisposition to corruption being a good example), is responsible for choosing the nature of the contract with the private party (which for simplicity can be considered as high power or low power), and which may or not involve a payment to him or her.
2. Contractor – who could be a supplier or employee. His or her job is to accept or reject the contract offered by the incumbent public agent, and who will then share or not share the proceeds with the incumbent.
3. Political competitor – the potential third-party opportunist. This competitor's decision is whether to challenge the contract implementation. Challenges are costly, however, so not all contract types will be challenged. The political competitor also is able to observe some signals concerning the realization of the contract which are not verifiable.
4. The public (which could be considered also to be the court), who will, based on the actual realization of the contract and the challenge of the political competitor, decide to retain the incumbent or to replace him or her with the political competitor (if the challenge took place).

In this environment it is straightforward to show that low-power incentives will be preferred by public incumbents. By relating payments to verifiable incurred costs, low-power contracts limit the potential for successful political challenges. Contracts with low power incentives, however, may generate fewer social benefits, or be more costly to the public, but avoid the potentially negative political consequences of high-power incentive contracts. Similarly, high-power incentives do not provide the same benefits they could provide in private transactions. Challenges have also risks for the private party. In fact, challenges may impede, *ex post*, the payment of high power compensation when it is due. Thus, third-party opportunism also increases the risk to the private agent of transacting with the state. The response to third-party opportunism for both the incumbent public agent and the private transacting party is to provide low-power contracts, to increase the degree of contract specificity, to limit the potential discretion granted to the public agent and to increase procedural rigidity. Since the potential for third-party opportunism increases with complexity, complex public contracting projects would then have more contractual rigidities than simpler public contracts. Rigidity, however, when combined with complexity, will increase conflict. The potential for conflict, together with the potential for third-party opportunism, then, favours low-power incentives. High-power incentives simply may not be implementable given the potential for third-party opportunism and the expected level of contractual conflict. Thus, third-party opportunism limits the potential for 'relational public contracting'. Public agencies will have difficulty entering into a close relationship with an infrastructure investor, in which contract adaptation takes place without formal renegotiations, and/or litigation. Furthermore, public contracting will not only be more complex, involving multiple rules and procedures, but will also be more subject to litigation.

The added complexity required to limit the potential for third-party opportunism will make public contracting look 'inefficient'. This inefficiency, however, does not pass Williamson's (1999) remediableness test. In other words, the perceived inefficiency of public contracting is an equilibrium response to its hazards and, in particular, to the hazard of third-party opportunism, a defining feature of public contracting. In fact, to eliminate the alleged inefficiency, one could consider two options: one, to move the transaction within the public sphere completely, that is, to vertically integrate (node (D) in Figure 2.1); second, to drive it off the public and into the private sector. Consider, first, moving a complex infrastructure project within the public sphere completely. Public vertical integration does not solve the 'within the bureaucracy' contracting problems. In fact, similar types of concerns arise with the implementation of high-power incentives within the bureaucracy itself. For the same reason that high-power incentives are not appropriate for public contracting, high-power incentives are not often appropriate for within the bureaucracy relations as high transfers to public employees will naturally raise probity questions, and will thus increase the risk of third-party opportunism.

Complete privatization of the government activity may not be feasible either. As discussed, commitment to not interfere with infrastructure pricing may simply not exist in democratic societies, and thus investors would not dare invest without appropriate safeguards. Public contracts provide one such, albeit imperfect, safeguard. Thus, the undertaking and delivery of public infrastructure projects and services will have to be undertaken in coordination with the government, that is, under public contracting. Whether public contracting is, or is not, inefficient cannot be inferred from the observation that it does not replicate private contracting. A proper comparison must be institutionally consistent and pass the Williamson's remediableness test, and that can only be assessed on a case-by-case basis. The potential for third-party opportunism, furthermore, may be exacerbated by the nature of the institutional environment in which the infrastructure project takes place. To thrive, third-party opportunism requires some political contestability and fragmentation. Although internal party politics could provide the environment for fragmentation, and for the type of political displacement required for third-party opportunists to prosper, centralized party power limits the upward mobility of political mavericks, and thus the potential for internal third-party opportunists. On the other hand, political instability, the cauldron where governmental opportunism thrives, is also conducive for third-party opportunism as the cost of removing incumbent politicians falls. In the middle, between stable centralized party control and rampant political instability, is where most of the world democracies fall. 'Open access' states, following the nomenclature introduced by North et al., (2006), naturally facilitate the development and organization of third-party interest groups. In these societies, public policies become depersonalized, and governments are constrained in their ability to limit – whether by withdrawing funding, political harassment or direct violence – the development and organization of such groups. It is in these societies where the threat of third-party opportunism becomes more credible, as such challenges may not be easily covered up by side payments or the direct threat of recourse to violence. In 'natural' states, following again North et al. (2006) nomenclature, the public agent may have more instruments at her disposal to quash such challenges, and thus, it could be argued that her ability to overcome a third-party challenge is increased. This discussion, then, suggests that, as it relates to the infrastructure sector, the relevance of third-party opportunism and governmental opportunism may vary across countries. First, governmental opportunism requires the existence of important sunk investments, and infrastructure projects are characterized by them. Thus, the hazard of governmental opportunism is ever present in the sector. The potential for governmental opportunism is aggravated in institutional environments with few institutional limitations to governmental discretion, the essence of 'natural' states. Third-party opportunism, which may appear even when the contractor may move its assets costlessly elsewhere, is limited in 'natural' states precisely by the same discretionary ability of governments. Third-party opportunism, then, is aggravated in open access states, thus suggesting that public agents would adapt public contracting further away from relational contracting in 'open access' than in 'natural' states. In sum, the infrastructure sector is plagued by third-party and governmental opportunism. While the institutional environment most propitious for the development of both types of opportunism differ, as do the governance adaptations required for successful provision of infrastructure services, the framework provided here suggests that both types of hazards interact in increasing the specificity and rigidity of infrastructure contracts causing difficulties in adapting to shocks and leading to low-powered incentives. Similarly, both types of hazards, and their combination, would lead to more 'inefficient' termination of public infrastructure and private infrastructure contracts. These 'inefficiencies' persist in equilibrium, and thus public contracting cannot be simply compared to private contracting, nor can the standard measure of performance from private contracting be applied to public contracts. With this in mind, I proceed to analyse the implications of the framework to regulatory reform in the infrastructure sector, and in particular to the use of concession contracts as privatization instruments.

CONCESSION CONTRACTS AS PRIVATIZATION INSTRUMENTS

Concession contracts are part of the general set of licences and permits through which states grant the right to a private organization to undertake a particular public-service activity. Concession contracts differ from other types of legal instruments in that while taking the form of operating licences, they tend to embed the basic regulatory framework that will guide their evolution as it relates to basic features such as prices, quality, penalties, termination, and the like. In that sense, the state that chooses to grant a concession contract is in essence embedding the regulatory framework within a formal contract. Concession contracts, then, differ from simple operating licences, such as those common in the USA, as the latter are silent about the regulatory framework, leaving the management of the terms of operation of the licensee to other instruments, such as administrative procedures or specific legislation. Since including the regulatory framework within the licence makes it more rigid, significant change in regulatory public contracting is subject to governmental opportunism requiring its own types of adaptations.

Concession contracts, then, arise as a mode for organizing provision of public services precisely because regulation by contracts sets limits to unilateral regulatory changes, and by doing that, it mitigates the potential for governmental opportunism (Levy and Spiller, 1994). In particular, regulation by contract is preferred in sectors with a high level of sunk assets (for example, water, transport, natural gas and electricity distribution), in politically unstable environments and when regulation by law does not provide enough credibility to protect the investments of the service providers. In these cases the providers are vulnerable to governmental opportunism, and so private service providers will not enter the market or demand too high a premium on services under direct regulation. Concession contracts, by creating an individualized regulatory framework for the investments at hand, limit such opportunism. The choice of regulatory contracts shows the inexorable trade-off between commitment and flexibility (Spiller, 1996). To be credible, a regulatory contract must be specific – that is, it must lay out in detail how the parties will deal with each eventuality. However, adjustments in the relationship require flexibility in the contract, which in turn requires that the contract be complex. Complexity is limited by administrative capability and negotiation costs (which depend on differences among potential investors and political incentives). Concession contracts are rigid by 'origin'. A flexible concession contract that grants substantial discretion to the state would fail to provide the required commitment, and thus would not be seen by the investor as mitigating the risk of governmental opportunism. Thus, concession contracts are born with less flexibility than normal private contracts. They are also not 'relational contracts'. In fact, adaptations to shocks seldom involve serious contractual deviations without triggering formal contractual adaptations. Figure 2.2 shows the distribution across sectors, showing that most transport and water concession contracts were renegotiated.

The large percentage of formally renegotiated contracts is only surprising if we look at it from the perspective of 'relational contracts'. But regulatory adaptations should not be surprising. Most concession contracts should be adapted over time as, by nature, these contracts are long term and incomplete. As a consequence, it is unreasonable to expect that through the long life of these concessions, there will be no economic (for example, an unexpected increase in external or internal costs) or political (for example, new political

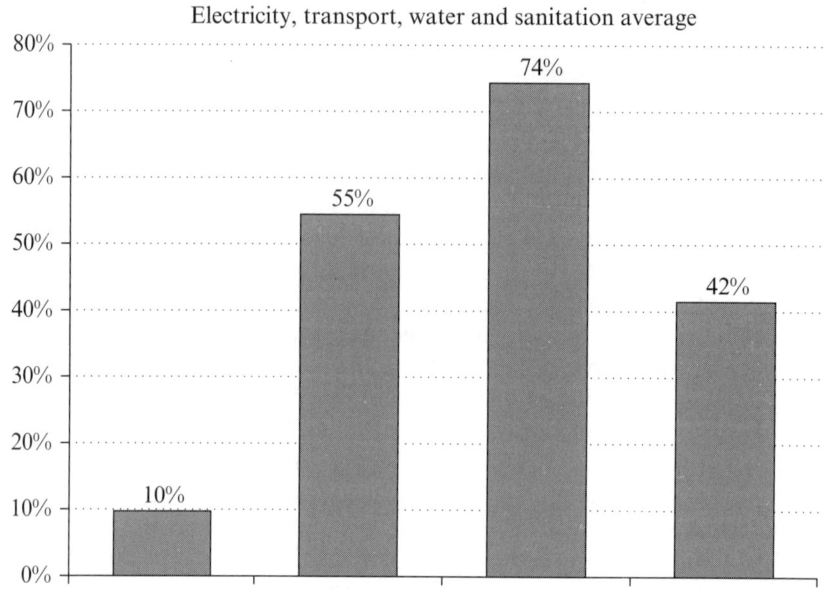

Source: Guasch (2004).

Figure 2.2 Latin America concession contracts renegotiated 1985–2000

needs) shocks that would call for a change in the agreed terms of the contract. An unadjusted contract may lead to unproductive actions or even default, generating unnecessary social losses.

Lack of adaptation following a shock that threatens the viability of the concession may, in fact, trigger contract termination. While termination may be 'socially inefficient', it may be politically preferable for the public agent than bearing the third-party opportunism risks associated with renegotiating the contract. All other things being equal, it is reasonable to expect that the larger the investments undertaken in relation to future network expansion needs, the higher the government's incentive to terminate the concession (Troesken, 1997), both because of governmental opportunism and because of the difficulties of renegotiating given third-party opportunism hazards.

Given the risk of early termination, concession contracts normally provide for very specific and explicit compensation at termination, whether early or normal. The degree of specificity of these clauses is obvious: the greater the ability of the state to manipulate the compensation, the higher the government's incentive to terminate. Although there are inherent and institutional limits to governmental opportunism, third-party opportunism may not be subject to these limitations. Even in contractual circumstances where the potential for governmental opportunism has been limited by expectations of future investments or institutional limitations, third-party opportunism may still trigger 'inefficient' contract termination. Consider, for example, waterworks in the USA. A large number of municipalities across the USA use concession contracts to operate and manage their waterworks.

Not all contracts, though, are for city-wide service, with most covering only partial operations, such as operation of a particular pumping or treatment plant, or the provision of billing services (Water Science and Technology Board, 2002). These contracts are long term, often dealing with highly complex physical environments. These contracts assign risks to the contractor and the city in highly complex and sophisticated fashions, providing for cost escalation clauses, and for renegotiation facing a large-scale shock, such as a natural disaster. The transfer of existing operations to a private contractor is, though, as complex in the USA as elsewhere. Consider, for example, the breakdown of the Atlanta waterworks contract. In January 2003, the City of Atlanta terminated what was then a 22-year management and operation contract of the whole waterworks system of Atlanta, only four years after its transfer. Immediately after the transfer the parties were in conflict over the costs required to meet the standards agreed upon in the contract. United Water, the operator, claimed that much of the baseline data were inaccurate and thus, the targets and annual fees agreed upon were unrealistic, forcing it to lose $10 million a year out of a $20.8 million/year contract.

Renegotiation, then, would have to involve a substantially higher per year fee, as well as a modification of the service targets. The city, however, preferred to take back the contract at a cost of $40 million/year – against the $22 million/year of the contract, rather than having to handle the public criticism of its management of the contract. Following the reasoning in this chapter, renegotiation would have been 'in neither party's interest'. It would not have been in the city's interest as it would be extremely difficult to explain that a contract drafted and crafted by the city just a few years ago was in such seeming disarray that it requires a substantial fee increase. For the city it would be much easier to blame all problems on the operator's bad performance. Similarly, it would not have been in the operator's interest, as, given the public perception of wrongful operator performance, it would be difficult to expect that the renegotiated contract would fully compensate it. Eventually, the parties ended the contract 'amicably'. The contract termination, however, was preceded by a series of corruption allegations involving an agreement to grant a $4 million/year for 17 years payment increase following the operator's request for $80 million compensation for work not included in the contract, and allegations that the Mayor had business relations with a competing water operator.

The Atlanta case is an example of a contract falling apart not because of governmental opportunism, but rather by the inability of the parties to draft and implement contracts with sufficient flexibility that can adapt to uncertain operating circumstances. Instead, the parties entered into a highly inflexible contract, forcing them to renegotiate or terminate. The city, however, was unwilling to pay the political price of renegotiation. The main thrust of this chapter is that contract inflexibility is inherent to public contracts, and that, facing the potential for third-party opportunism, large shocks may trigger termination and/or litigation rather than renegotiation.

IMPLICATIONS FOR LIBERALIZATION OF INFRASTRUCTURE PROJECTS

This analysis has some implications for the liberalization of infrastructure sectors. Moving infrastructure sectors away from public ownership towards private operation must take into account the hazards of governmental and third-party opportunism. I have postulated that infrastructure contracts that require high cash-flow fluctuations, such as those involving price cap regulatory frameworks or minimum investment requirements (rather than service targets), increases the potential of third-party opportunistic behaviour. On the other hand, regulatory systems, such as rate of return regulation, which generate steadier cash flows, tend to limit third-party opportunism and, thus, reduce the overall cost of the service. The evidence that Guasch and his colleagues collected of formal renegotiations in Latin America is very instructive. In Guasch et al. (2003), they show that formal renegotiations are more likely to occur in infrastructure projects involving price caps (as against rate of return regulation), specific investment requirements (as against service performance requirements) and purely private financing (as against public sector shoring up the upfront burden). These three features of contracts require large swings in cash flow, which, as seen above, are not easily made compatible with public contracting. Since governments are not interested in triggering service dislocation, governments facing important economic shocks will tend to favour renegotiation over cancelling the concession and re-licensing. The Williamsonian fundamental transformation (from one buyer with many suppliers to one buyer to one supplier) is also present in infrastructure projects (see Williamson, 1976).

Thus, renegotiation is going to be a fact of life in infrastructure contracts. That renegotiation is the norm rather than the exception in public infrastructure contracts should be taken into account when granting the original contract. Concession contracts are normally granted via some type of bidding process. The bidding process may be for the lowest average price, for the highest annual – or upfront – payment, for the highest investment level or faster deployment or for any combination of these (see Abdala and Spiller, 2000). Not all concession contracts will be renegotiated rapidly, though. The better designed the contract (that is, the lower the uneconomic investments, the stronger the collecting rights of the operator, or annual payments to the government) the lower the potential for renegotiation. If, however, the regulatory framework is designed with social policy in mind, then the infrastructure contract, as well as the bidding procedures must be adjusted. Bidding for lowest price or highest annual payment may not be optimal if the constraints on the company are such that contracts will have to be renegotiated readily after the concession was granted. In those environments, governments ought to consider undertaking substantial part of the investments, limiting both third-party opportunism – via a reduction in cash flow volatility – and the potential for governmental opportunism – via limiting the ability to reduce prices below costs.

REFERENCES

Abdala, M.A. and Spiller, P.T. (2000) *Institutions, Contracts and Regulation in Argentina* (in Spanish), Buenos Aires: Temas.

Baker, G., R. Gibbons and K.J. Murphy (2002). 'Relational contracts and the theory of the firm', *The Quarterly Journal of Economics*, **117**(1), 39.

Baron, D. and R. Myerson (1982), 'Regulating a monopolist with unknown costs,' *Econometrica*, **50**, 911–30.

Coase, R. (1964), 'the regulated industries: discussion', *American Economic Review*, **54**, 194–7.

Dal Bó, E. and R. Di Tella (2003), 'Capture by threat', *Journal of Political Economy*, **111**(5), 1123–54.

Guasch, J.L. (2004), *Granting and Renegotiating Infrastructure Contracts: Doing It Right*, Washington, DC: World Bank Institute.

Guasch, J.L., J.-J. Laffont and S. Straub (2003), 'Renegotiation of concession contracts in Latin America', World Bank Policy Research Working Paper 3011.
Levy, B. and Spiller, P.T. (1994), 'The institutional foundations of regulatory commitment: a comparative analysis of telecommunications regulation', *Journal of Law, Economics, and Organization*, **10**, 201–46.
Macaulay, S. (1963), 'Non-contractual relations in business: a preliminary study,' *American Sociological Review*, **28**, 55–67.
MacNeil, I.R. (1974), 'Restatement (second) of contracts and presentation', *Virginia Law Review*, **60**, 589–610.
MacNeil, I.R. (1978) 'Contracts: adjustment of long term economic relations under classical, neoclassical and relational contract law', *Northwestern University Law Review*, **72**, 854–905.
McCubbins, M.D. and Schwartz, T. (1984), 'Congressional oversight overlooked: police patrol vs fire alarms', *American Journal of Political Sciences*, **28**, 165–79.
McCubbins, M., R. Noll and B. Weingast (1987), 'Administrative procedures as instruments of political control', *Journal of Law, Economics and Organization*, **3**, 243–77.
McCubbins, M., R. Noll and B. Weingast (1989), 'Structure and process, politics and policy: administrative arrangements and the political control of agencies', *Virginia Law Review*, **75**, 431–82.
Milgrom, P., D. North and B. Weingast (1990), 'The role of institutions in the revival of trade: the law merchant, private judges and the champagne fairs', *Economics and Politics*, **2**, 1–23.
North, D., J.J. Wallis and B.R. Weingast (2006), 'A conceptual framework for understanding recorded human history', NBER Working Paper 12795.
Savedoff, W. and P.T. Spiller (1999), *Spilled Water: Institutional Commitment in the Provision of Water Services in Latin America*, Washington, DC: Interamerican Development Bank.
Spiller, P.T. (1996), *Regulations, Institutions and Commitment: Comparative Studies of Telecommunications*, Cambridge: Cambridge University Press.
Spiller, P.T. (2009), 'An institutional theory of public contracts: regulatory implications', in C. Ménard and M. Ghertman (eds), *Regulation, Deregulation, Reregulation – Institutional Perspectives*, Cheltenham, UK and Northampton, MA, USA: Edward Elgar, pp. 45–66.
Troesken, W. (1996), *Why Regulate Utilities? The New Institutional Economics and the Chicago Gas Industry, 1849–1924*, Ann Arbor, MI: University of Michigan Press.
Troesken, W. (1997), 'The sources of public ownership: historical evidence from the gas industry', *Journal of Law, Economics and Organization*, **13**(1), 1–25.
Troesken, W. and R. Geddes (2002), 'Municipalizing American waterworks: 1897–1915', *Journal of Law, Economics and Organization*, **19**(2), 373–400.
Water Science and Technology Board (2002), *Privatization of Water Services in the United States: An Assessment of Issues and Experience*, Washington, DC: National Academy Press.
Williamson, O.E. (1975), *Markets and Hierarchies: Analysis and Antitrust Implications*, New York: Free Press.
Williamson, O.E. (1976), 'Franchise bidding for natural monopolies: in general and with respect to CATV', *Bell Journal of Economics*, **7**(1), 73–104.
Williamson, O.E. (1979), 'Transaction-cost economics: the governance of contractual relations', *Journal of Law and Economics*, **22**(2), 233–61.
Williamson, O.E. (1983), 'Credible commitments: using hostages to support exchange', *American Economic Review*, **73**, 519–40.
Williamson, O.E. (1985), *The Economic Institutions of Capitalism*, New York: Free Press.
Williamson, O.E. (1996), *The Mechanisms of Governance*, Oxford: Oxford University Press.
Williamson, O.E. (1999), 'Public and private bureaucracies: a transaction cost economics perspective', *Journal of Law Economics and Organization*, **15**, 306–42.
Williamson, O.E. (2002), 'The lens of contract: private ordering', *American Economic Review*, **92**, 438–43.
Williamson, O.E. (2005), 'The economics of governance', *American Economic Review*, **95**(2), 1–18.

3 A general framework for regulation and liberalization in network industries[1]

Christian Jaag and Urs Trinkner

INTRODUCTION

Traditionally, most network industries used to be dominated by state-owned regulated monopolies. During the past 20 years, governments in many parts of the world have started liberalizing their network industries, for example, telecommunications, postal services, electricity and transport. This liberalization process started in the USA in the late 1970s and in the UK in the early 1980s. Since then, sectors such as telecommunications and air transport have become fully liberalized in the European Union (EU) and are becoming increasingly competitive. The electricity sector, postal services and railways are not yet fully liberalized.

In parallel with liberalization, sector-specific regulation in network industries has become a widely discussed topic among academics, policy-makers, industry economists and regulators themselves. The issue of these debates has usually been on whether such regulation is necessary and if so what its optimal design should be. Some argue for complete deregulation (that is, the complete abolishment of sector-specific regulations), whereas others propose re-regulations, that is, the replacement of pre-existing (monopoly) regulations by new regulations aiming at safeguarding service levels and competition. The resulting compromise is often somewhere in between; the liberalization process usually entails the partial replacement and realignment of sector-specific regulatory intervention by the disciplining forces of competition protected by competition law. Consequently, competition law and sector-specific regulation play complementary roles.[2]

In this chapter, we present a general economic framework to assess regulatory remedies in network industries. Therein, liberalization, or more generally market access regulation, can be assessed as well. It provides a foundation and explanation of current sector-specific regulations in place. Our analytical starting point is the observation of industry-specific deviations from the concept of perfect markets. Where markets are perfect, efficient outcomes can be expected. In such a world, the only rationale for sector-specific regulation would be redistribution. This standard result of economic theory is based on quite rigid assumptions. In real markets, there are important deviations which give rise to market failures and inefficient market equilibria. We find that in the network industries, most regulatory interventions can be explained by market failures. In particular, natural market power, incomplete markets and asymmetric information are sound rationales for sector-specific market interventions. However, political processes often fail to produce consistent regulatory models,[3] and moreover, there are no benevolent regulatory authorities able to implement these models in practice perfectly.[4] Hence, there is the risk of potential regulatory failure too. Therefore, market failures are necessary but not sufficient preconditions for market interventions. In particular, market failure and regulatory failure must be balanced against each other. In the second part of the chapter, we propose a framework on how to perform this assessment. Applying the framework to market power regulation, we find that pre-existing regulations should not be fully replaced by competition law in the presence of infrastructures or network layers which are monopolistic by nature.

Market failures as a starting point for regulations have been proposed by many authors. For example, Baldwin and Cave (1999) provide rationales for regulation that are closely linked to deviations from perfect markets.[5] While others take a different starting point,[6] our approach follows these lines. It is new to the extent that it provides a comprehensive framework for understanding and analysing the need for sector-specific regulations in the network industries (including the role of market access and universal service obligations). Moreover, it enables directly deriving an appropriate and consistent regulatory model to cope with prevalent or potentially harmful market failures.

The chapter proceeds as follows: the next section presents the analytical framework to assess the need for regulatory intervention. The third section applies this framework to bottleneck regulation in network industries. We conclude in the fourth section.

ANALYTICAL FRAMEWORK

Our theoretical framework starts from a free market primacy assumption: markets, if they function properly, provide firms with the right incentives to enter markets, set prices, and invest in innovation at a socially optimal level. Of course, there are a number of obstacles to markets functioning well in that sense. So-called market failures give rise to potentially beneficial regulation – at the risk of these regulations failing as well. Generally, those market failures should be tackled by regulatory interventions that result in a clear welfare improvement in light of possible regulatory failures.

Figure 3.1 illustrates the basic framework. Starting from a free market situation, if there is no market failure or no harmful potential market failure, there is no need to intervene. If the market fails persistently and government chooses to intervene, it establishes regulations.[7] If the regulated situation still fails when confronted with the socially desirable outcome – either because the selected regulation is not appropriate to cope with the initial market failure or due to other political or social goals – there is scope for re-regulation. However, if regulation in general is deemed not appropriate or detrimental to attain social goals, this should result in deregulation, that is, a lighter regulatory framework leaving market players more leeway to act in the market. The right-hand side of Figure 3.1 summarizes the main criteria that should be considered when assessing the need for regulatory remedies. We will develop the rationale behind these criteria in the remainder of the chapter.

Economic Foundations of Sector-specific Regulation

In perfectly competitive markets, the resulting competitive market equilibrium is Pareto optimal (first fundamental theorem of welfare economics) and there are no

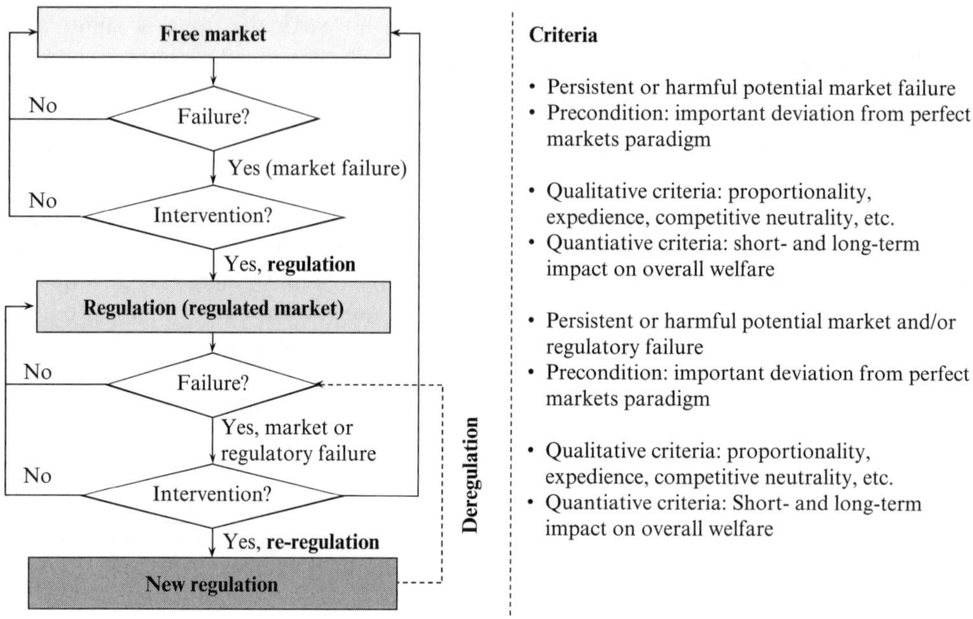

Figure 3.1 Illustration of general framework

market failures. The market equilibrium is efficient in the sense that one cannot make anyone better off without making at least somebody else worse off. This rather descriptive welfare criterion is not very ambitious; depending on the initial endowment of the economy, a Pareto efficient equilibrium may constitute an uneven distribution of wealth among individuals.

The second theorem of welfare economics establishes that by use of appropriate lump sum transfers, one can achieve different Pareto efficient market equilibria with different wealth distributions. This second theorem gives rise to a first fundamental source of market interventions: redistribution. For example, if a society agreed on a social goal like an even income distribution, it could do so by appropriately defining lump sum transfers.

From this viewpoint, rent-seeking activities are compliant with the second fundamental welfare theorem – rent-seekers try to establish a different Pareto equilibrium. In fact, regulatory economists such as Crew and Kleindorfer (2002) conclude that in practice, liberalization of network industries might largely be motivated by redistributions of wealth (in particular from small customers to large customers): 'Rent seekers will devote considerable resources to obtaining a share of the monopoly rents of which government now has taken control through the regulatory process' (p. 8), and 'Deregulation's likely primary driver is the division and redistribution of the monopoly rents. Large customers are attempting to gain at the expense of small customers' (p. 12).

In summary, the only market interventions that are economically justifiable in perfect markets are lump sum transfers aimed at redistribution. In practice, however, markets are hardly ever perfect and redistribution is hardly ever lump sum.

Attempts for redistribution are present in any industry and market. Although redistribution is at the heart of any regulatory discussion and political process in network industries, the second theorem of welfare economics cannot explain why topics like access regulation or universal service obligations are specific to network industries.

In the remainder of the chapter we will present a theoretical framework that explains why sector-specific regulations implemented in network industries do and should differ both from other industries as well as among the various network industries.

Our analytical starting point is deviations of real markets from the above paradigm of 'perfect markets' (or 'perfectly competitive markets'). Recall that the perfect market assumption lies at the basis of the two welfare theorems. Important assumptions are:

A1: Any company or consumer in the economy acts as price-taker, that is, there is *no bargaining or market power*.
A2: Markets are complete, that is, there exists a true price for any good – there are *no externalities* (or they are readily traded and thus internalized correctly).[8]
A3: Information is symmetrically distributed (*no asymmetric information*) and there are *no transaction costs*.

These assumptions are very rigid and are hardly ever met in practice. This gives room for market failures. Market failures are 'situations in which some of the assumptions of the welfare theorems do not hold and in which, as a consequence, market equilibria cannot be relied on to yield Pareto optimal outcomes' (Mas-Colell et al., 1995, p. 350). As a consequence, sector-specific regulation tackling the relevant market imperfections can potentially lead to Pareto improvements (that is, Pareto efficient equilibria).

However, many deviations from perfect markets can be observed in most if not all markets. It makes sense to tackle them uniformly and *equally among all sectors* of an economy. First, and in light of important information asymmetries, any society must find ways to allocate and enforce property rights, and to secure commercial freedom and the free movement of prices. Consequently, an important share of civil and commercial law deals with securing (intellectual) property rights and making contracts better enforceable in a world of incomplete contracts and asymmetric information. Further important sources of general regulations are competition or antitrust laws to limit harmful abuse of market power.

In principle, the area of application of these laws and acts covers all markets of an economy. However, they may not be sufficient to cope with sector-specific market failures. Hence, potential sector-specific regulations must be assessed against these prevailing laws. Non-sector-specific market failures (or regulatory failures) should be dealt with in general laws to avoid uneven and inconsistent treatment of market players among various industries.[9] In network industries, the most important source of prevailing regulations is competition law.

We summarize as follows: (1) any new sector-specific regulation in question should be based on deviations of the perfect markets assumptions leading to (potential) market failures with harmful effects on overall welfare; (2) new regulation should be introduced only if these failures are not sufficiently tackled by general regulations or prevailing sector specific regulations, and (3) moreover, the new regulations should clearly improve the situation with respect to potential regulatory failures.

Sources for Sector-specific Regulation in Network Industries

In most industries and in network industries in particular, there are important deviations from the perfect markets assumptions A1 to A3 that have been introduced above. On the grounds of these deviations we will be able to explain most elements of sector-specific regulations in network industries.[10] Figure 3.2 provides an overview of our approach and illustrates that monopolistic bottlenecks (A1), externalities (A2) and further market imperfections such as information asymmetries (A3) provide sound reasons for sector-specific regulations that typically are encountered in industrialized countries. We will now briefly discuss these three dimensions of potential market failures and their regulatory remedies.

A1: Market power – regulation of bottlenecks

Economic agents do not act as price-takers if they have persistent market power. In such a case they are able to profitably alter prices away from competitive levels without provoking new market entries or other counteractions that force prices back to efficient levels within a reasonable time limit. Because of the specific characteristics of network industries, market power is an important issue in these industries.

Our starting point for the localization of persistent market power is the concept of barriers to market entry dating back from Bain (1956) and Stigler (1968). Stigler defines barriers to entry as follows: 'A barrier to entry may be defined as a cost of producing (at some or every rate of output) which must be borne by a firm which seeks to enter an industry but is not borne by firms already in the industry' (1968, p. 67). The more persistent such barriers to entry are the more likely harmful price-setting by incumbent operators is. However, barriers to entry are not per se a rationale for regulation. Virtually any industry exhibits some kind of barriers to entry and one can expect new entrants to overcome some barriers by other advantages such as new processes, technologies and innovation. Consequently, we must take a closer look at barriers to entry, of which there are various classifications. For example, Church and Ware (1999) distinguish structural and strategic barriers. As De Bjil et al. (2006) and others, we differentiate between 'legal barriers to entry', 'strategic barriers to entry' and 'natural barriers to entry'.

Legal barriers prohibit new entrants to access markets ('market access regulations'). Usually, some kind of exclusive rights are granted to a governmental or private undertaking. If such legal barriers to entry are politically desired (for example, a state monopoly on military forces, the US monopoly for selected postal services), there is no reason to grant access to new entrants. However, safeguards should be established that limit the abuse of market power based on the granted exclusive rights. Price regulations and measures to avoid cross-subsidization into adjacent and other markets are very popular. The resultant regulations are sector-specific by nature.

Strategic barriers to entry arise by *conduct*. The behaviour of one or several market players, for example a cartel, can lead to market power that in turn can be used for anti-competitive and welfare-decreasing practices. Such harmful conduct is usually tackled by competition law and should therefore not be duplicated by sector-specific regulations.[11]

Natural barriers (also called structural barriers) arise where the *cost structure* of an industry leads to stable market power ('natural market power'). Recall that the concept of perfect markets assumes constant returns to scale. Increasing returns to scale – which

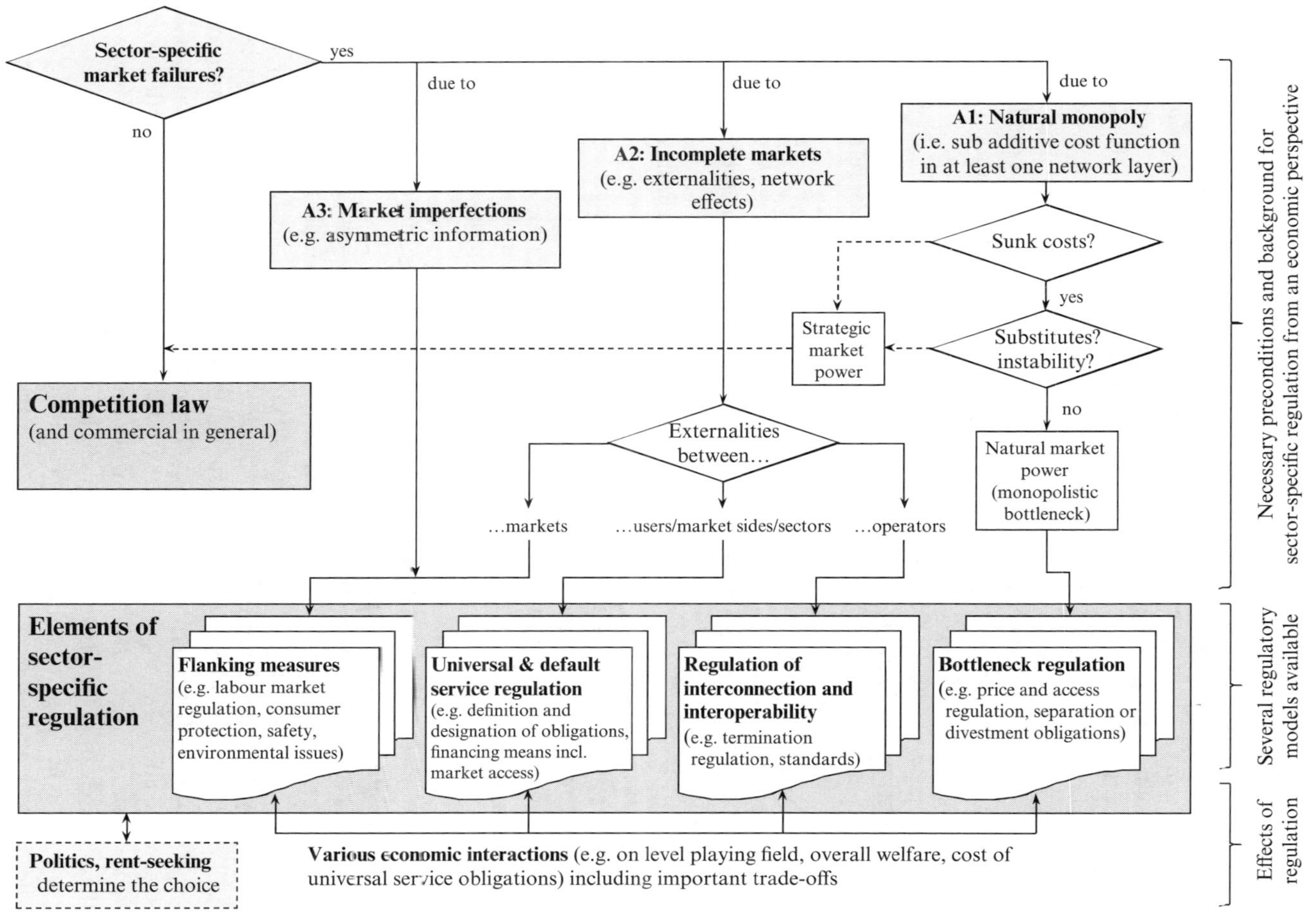

Figure 3.2 Application to network industries

are usually present in network industries – might give rise for regulation from an economic point of view. However, neither scale economies nor a natural monopoly (defined as a subadditive cost structure[12]) are a sufficient condition for regulation as shown by the theory of contestable markets. The theory has its origins in the work of Baumol et al. (1982) and essentially extends the economic efficiency properties of perfect markets to all markets which are contestable, that is, markets without relevant sunk and transaction costs.[13] If such sunk costs are present together with a natural monopoly (cost subadditivity), we speak of a 'monopolistic bottleneck'. If this bottleneck cannot be duplicated nor substituted by other means, it is called a 'stable monopolistic bottleneck'[14] (in US antitrust law this is referred to as an 'essential facility'). Such infrastructures give the owner natural market power and potential entrants will not be able to enter the market, even if the incumbent charged excess prices. Therefore, it will be crucial to ensure in liberalized markets that new market players get timely access to bottleneck facilities at reasonable terms and conditions that prevent the abuse of market power. On the other hand, any sector-specific access regulation is to be seen as a massive intervention into the bottleneck owner's property rights. Recall that securing property rights stays at the heart of making markets function and plays a major role for providing appropriate innovation and investment incentives.

In line with Knieps (2000a) and others we conclude that the goal of any regulation of a stable monopolistic bottleneck is to *enable non-discriminatory access to these bottlenecks at reasonable conditions while minimizing the infringement of property rights on the bottleneck resource.*

In practice, competition law is often not sufficient to tackle harmful behaviour of companies with natural market power. In the network industries, one network layer often has the property of a stable monopolistic bottleneck. Moreover, the issues are quite technical and sector-specific knowledge is needed. Hence, sector-specific regulation dealing with natural market power might be particularly appropriate in network industries. Later on we explore the issue in more detail and provide a framework to select the appropriate regulatory intervention to cope with monopolistic bottlenecks.

A2: Externalities – universal and default service regulations, interconnection and interoperability regulations, flanking measures

External effects are present when one economic agent's actions affect the actions of other agents in the economy. One agent's action can have positive or negative externalities on other agents. The classic case for a negative externality is a chemical plant which pollutes a river, thereby reducing the prospects of the fishery located downstream. Analytically, externalities are closely linked with incomplete markets; when an economic agent's own action has a positive effect on others but is not rewarded in return, there exists no market for this external effect. In general, when external effects are present, market equilibria are not efficient, as these effects are not taken into account in individual decisions. However, potentially, they can be 'internalized', for example, by public obligations, taxes, quotas or the allocation of property rights (where these are in fact enforceable).[15]

In the network industries, many important externalities exist and, thus, there is a potential for market failure. The following types of externalities are present which give room for universal service or default obligations, regulations of interconnection and interoperability, market access regulations, or flanking measures.

Network externalities among users The utility of a user increases with the number of users connected to the network. For example, a phone subscription is much more valuable if others are connected to the network too.[16] Similarly, letters as a media are much more attractive if they can reach anyone. Positive network externalities give rise to universal service regulations that aim to connect everyone to the telecommunications or postal network. Such obligations can be found in virtually any industrialized communications and postal market.

Externalities between market sides Closely linked to network externalities are externalities between different market sides of a platform. Where lump sum price redistributions between market sides affect overall demand, markets are said to be two- or multi-sided.[17] Many network industries, such as telecommunication, postal or cable networks can be understood as being such platforms. These platforms link senders and receivers, as well as sellers and buyers. An example for a positive externality would be the mere existence of a broad installed base, which makes entering a market attractive if interconnection is available. Negative externalities usually arise in the context of advertisement. Platforms are convenient means to reach people. Their operators have incentives to sell as much direct mail, telemarketing and television advertisements as possible. The receivers perceive these as 'bads', and it is crucial to find the right balance between content (good) and advertisement (bad). If platforms fail to tackle these issues, regulatory answers are Robinson lists or the acceptance of 'no advertisement' stickers on mailboxes.[18]

Network externalities between various operators By analogy to network externalities, demand in a given mobile network is higher if the network is interconnected with other competing networks (a subscriber's network is extended to a much larger subscriber base). Termination issues in telecommunications should primarily be decided against this background (and only secondary against market power[19]). Similarly, in liberalized postal markets, post office (PO) box access is essentially a termination issue.[20] In the railways industry, integrated and interconnected systems provide the passengers much more utility than isolated solutions. Hence, from a user's perspective, it makes sense to interconnect competing networks. Often, an industry-specific self-regulation arises. Some of the oldest international organizations deal with interconnection issues. Examples are the Universal Postal Union (UPU) or the International Telecommunications Union (ITU). However, under some circumstances, negotiations might fail in light of the commercial interests of the involved operators. Again, regulation can be a reasonable means to internalize these positive externalities.

The issue of interconnection is closely linked with interoperability. In most technical systems, interoperability is a precondition for interconnection. Consequently, sector-specific regulation in the network industries usually includes interoperability (for example, in railways, telecommunications or electricity markets). Industry standards generally can be seen as platforms.[21] Where self-regulation fails, government intervention might be reasonable from an economic perspective.

Externalities onto other sectors of the economy Network industries usually provide an economy nationwide with essential services such as water, power, communications, transport, and delivery. A new train service will boost a region's economic activity in many aspects and attract new businesses and residents. Among others, beneficiaries include people who will not use the train connections (for example, property owners). Hence, the passengers' willingness to pay for the future service is too low. As a consequence, such projects often need government support. Similarly, postal services can be seen as 'economic enablers'; the interplay between letters, parcels and payment services is manifold and various external effects are present. Universal service regulations might be needed to ensure the necessary minimal standards. In the electricity industry, blackouts cause serious harm (large negative externalities on other suppliers as well as customers). However, the market will fail to provide the necessary overcapacities and quality standards, as these extra costs are not compensated in the marketplace. Default service regulations might be the only solution to mitigate the problem. Other negative externalities include pollution. For example, full market opening in postal markets will lead to overlapping and parallel networks, thereby potentially increasing pollution (that is, a negative externality). Hence, one might think of flanking measures to internalize these effects (if not already accounted for in road traffic regulations).

A3: Other important imperfections – other flanking measures

The perfect market hypothesis assumes symmetric information, that is, everybody knows about everything at every point in time. This is hardly ever the case. However, many information asymmetries have found market solutions over time (for example, private certification bodies) or are dealt with in general law.[22] Still, there might be sector-specific failures caused by asymmetric information.

A first issue related with asymmetric information is safeguarding demand. The issue is closely linked with customer protection measures (protecting against asymmetric information). In postal markets, operators must comply with a number of obligations which aim at ensuring the integrity of mail (for example, the postal secrecy). Similarly, operators are obliged to publish figures on their quality of service which is not observable by individuals. With regards to safety, airlines or railways must ensure that their employees meet certain professional standards. Such 'flanking measures' are usually implemented by application of individual or general authorization procedures (the former needing *ex ante* approval by the regulatory authority before being active in the market, the latter – if at all – *ex post* approvals, subject to a declaration of fulfilling the respective obligations).

Further flanking measures might relate to working conditions. Examples can be found in the railways, telecommunications and postal industries. These might be motivated mainly by redistributive (social) goals. In industrial countries, the following sector-specific issues are relevant. First, in light of the establishment of a level playing field, the regulation of labour regulations (for example, minimal wages) might assure even conditions for old and new operators where historically grown labour costs are rigid (the issue is comparable with stranded costs in the electricity industry). Note that high minimal standards might raise important barriers to entry. Second, once parallel networks (and overcapacities) evolve, operators tend towards marginal cost pricing. This is not sustainable given the decreasing average costs in network industries and leads to underinvestment or downward pressure on labour conditions (for example, in the postal industry, where an important fraction of total cost is labour cost). From an economic point of view, this would be a concern if it had negative consequences on service quality (and the externalities provided to other industries). As a consequence, governments might impose flanking measures to guide market forces away from harmful 'price competition' towards 'quality competition', where operators shall, given common lower boundaries for working conditions, compete for quality and innovation.

Finally, sector-specific regulations need to be implemented in the market. In a world of asymmetric information, the law alone will not be sufficient. Where necessary, independent bodies must be established and empowered to enforce compliance with the rules. Consequently, sector-specific acts usually contain provisions that deal with the establishment of regulatory bodies and their powers, with processes and with information and accounting requirements of market participants. Thereby, it is a challenge to establish 'independence', that is, the compliance of the competent authorities themselves (ministries, regulatory authorities, others).[23] A second challenge is the establishment of appropriate processes that empower the authorities to ensure compliance while protecting market actors from regulatory arbitrariness.

The Role of Liberalization in the Framework

In the previous section we discussed various sources of market failures and their potential regulatory remedies. What is the role of liberalization here? In essence, liberalization is a change of market access regulations towards a more entry-friendly regime. Note that for the discussed three sources of market failures, limiting market access/restricting competition (that is, the contrary of liberalization) is a potential regulatory remedy. Consequently, we understand market access regulations in general (and liberalization in particular) as ordinary regulatory remedies to deal with market imperfections.

Thereby, establishing a legal monopoly would be the most rigid form of (market access) regulation; it precludes competitive forces to a large extent.[24] In our framework which starts from the 'free market primacy', a monopoly should be introduced as a 'remedy of last resort' only. Recall that monopolizing a market will usually not provide the right signals for pricing, efficiency, innovation and customer friendliness. A priori, regulations preventing market access are harmful. In any case, they should be combined with appropriate regulations that prevent the exploitation of market power and that aim to give the right incentives for efficiency and innovation. This is a difficult task and involves many fundamental trade-offs which are increasingly complex to resolve with the degree of asymmetric information.[25]

Initially, most network industries were fully opened and free markets. Over time, virtually all of them got monopolized. Only recently, important sectors like telecommunications, railways, electricity and post have been opened up to competition again.[26] Hence, liberalization is a quite new policy where the starting point are pre-existing sector-specific regulations preventing market entry ('legal barriers to entry') combined with safeguards to prevent the abuse of market power by the legal monopolist. Examples are 'reserved areas' or exclusive concessions granted to state-owned enterprises coupled with price regulations and anti-cross-subsidization measures.

In the past, limiting market access has been a popular regulatory remedy for the following reasons.

- Investments in network industries are high and have a long-term time horizon. In particular, a large sunk cost must be spent in advance. This comes along with investment risks which are higher in fully competitive markets where other market players increase uncertainty and limit pricing flexibility. Moreover, the risks are usually higher for private companies than for public companies as they must be aware of government opportunism (see Spiller, Chapter 2 in this volume). At the same time, private investors usually demand higher risk premia and expect a shorter payback period.[27] Hence, public monopolists might have the best capabilities to undertake the large investments needed to succeed in network industries.

- Related to the point above, monopolies might be a means for financing public obligations (for example, universal service obligations). Compared to competitive markets, pricing must not necessarily be fully cost based (there are no market forces that would drive prices back to costs). This allows for (1) mark-ups over competitive price levels to finance extra obligations, and (2) for redistributions between customer segments (for example, from densely populated low-cost regions to rural high-cost ones). In competitive markets, such pricing policies would cause selective market entries (cherry picking) and would not be sustainable. Third, monopolies prevent inefficient duplication of networks and hence ensure optimal utilization of costly infrastructures.
- Where the industry characteristics inevitably lead to market power (for example, where monopolistic bottlenecks are present), a public monopoly might be superior to a private monopoly.[28] Still today, most bottlenecks remain state-owned or state-dominated (airports and train stations, roads, bridges and tracks, electricity grids, last mile infrastructures in telecommunications).

We conclude that the world we live in is far from the perfect markets paradigm; asymmetric information is the norm. Consequently, there will be both market failures in fully opened markets and regulatory failures in regulated monopolies. The key question regarding market access and liberalization is whether a free market would lead to an outcome less inefficient than a regulated monopoly. In free markets, decision-making is delegated to the various market players and organized by the pricing mechanism (the 'invisible hand'). The more market players there are, the better is the outcome of the decision-making process. Finally, the various market players are much better informed than centralized regulators (or managers). Consequently, economists expect better decisions from free markets under normal circumstances. However, network industries exhibit characteristics where competitive forces might fail (or are not present). Hence, depending on the very situations in individual economies, the optimal answer on how to regulate market access will vary not only from sector to sector but also from country to country.

Normative Criteria to Assess Regulatory Remedies

Above we have introduced our general framework. The framework allows for a broad range of sector-specific regulations and covers most regulations that are encountered in practice. Summarizing, we have developed *three economic preconditions* that should be met when thinking on new sector-specific regulations in network industries. First, one should observe important deviations from the perfect markets paradigm. In network industries, this is certainly the case. Second, there should be none or poor prevailing regulations dealing with these market failures. Third, harmful market or regulatory failures should be prevalent or imminent. If these preconditions are met, various regulatory remedies are available to deal with the detected failures. The crucial normative question is which regulations should be chosen, and which should not. We will now briefly outline our approach. The details are provided below.

For any problem set, there are various *regulatory instruments* available. These may vary in scope and nature. For example, when dealing with monopolistic bottlenecks, there are *ex post* and *ex ante* variations of prices and/or access regulation as well as separation of accounts, functions, structures and/or ownership.

Combinations of such regulatory instruments can be bundled into specific *regulatory models* that one can apply in practice. Of course, these models must be technically and commercially feasible.

From an economic point of view, any such regulatory model must bear up against a thorough *economic analysis and comparison* according to a number of qualitative and quantitative criteria. These are partly contradictory, which mirrors different interests of the various stakeholders in the marketplace: incumbent operators and their residual claimants, new operators, employees, business customers private customers, and taxpayers. The *qualitative criteria* include proportionality, expedience, competitive neutrality (including lightness, symmetry), incentive neutrality, subsidiary, simplicity, transparency and temporality. While these criteria are of qualitative nature and mostly have an indirect impact on welfare, *quantitative criteria* should be included too to compare the economic effects of any model compared to the (non-)regulation in place in the short, medium and long runs (market failure versus regulatory failure).

Thereby, static effects like benefits from regulation, productive efficiency and direct costs of regulation should be considered as well as dynamic aspects such as efficiency incentives, innovation and investment incentives, risks, and the effects of regulatory and organizational dynamics.

The regulatory model with the least net economic cost is then the one which should be chosen – if it sufficiently conforms to the qualitative criteria. Of course, welfare effects are hard to quantify as they often take effect in the long run (for example, innovation incentives) and would therefore have to be explicitly considered in a dynamic setting.

Figure 3.3 summarizes our approach. The following section provides the details by way of application to the most prominent example in network industries, namely the regulation of monopolistic bottlenecks.

APPLICATION TO BOTTLENECK REGULATION IN NETWORK INDUSTRIES

As described above, the existence of a persistent monopolistic bottleneck is an important rationale for regulation. Recall that monopolistic bottlenecks are present if an industry, network layer or value chain element exhibits the properties of (1) natural monopoly (subadditive cost function), (2) considerable sunk costs and (3) no substitutes (that is, no economic possibilities for bypass). Such bottlenecks are present in most

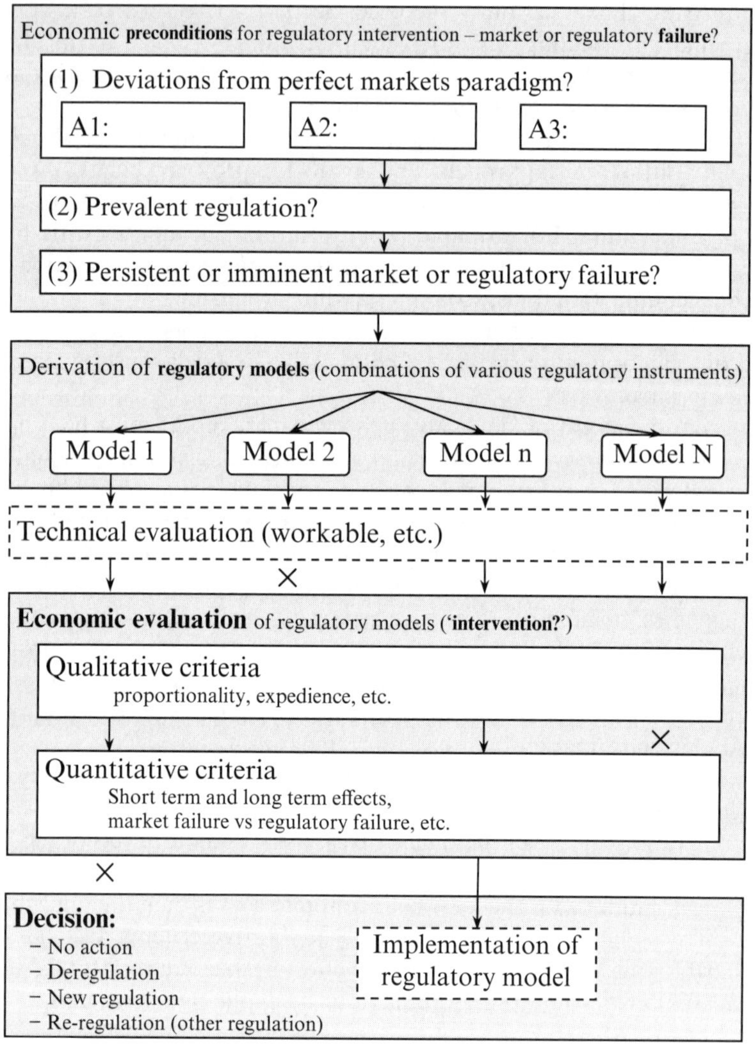

Source: Swiss Economics (2009).

Figure 3.3 Summary of regulatory assessment framework

network industries and raise the issue of natural market power being capable to distort competition in a harmful way.

The presence of a bottleneck does not directly imply the optimal choice of regulatory intervention. To this end, it is important to first exactly locate the source of monopolistic market power (Step 1). Only then is it possible to identify possible regulatory instruments (Step 2) which can be combined to consistent regulatory models (Step 3). In order to make sure that the implementation of regulatory models really improves the market outcome it needs to be assessed against various criteria, which are not only economic, but which may encompass broader goals of market intervention (Step 4 and 5).

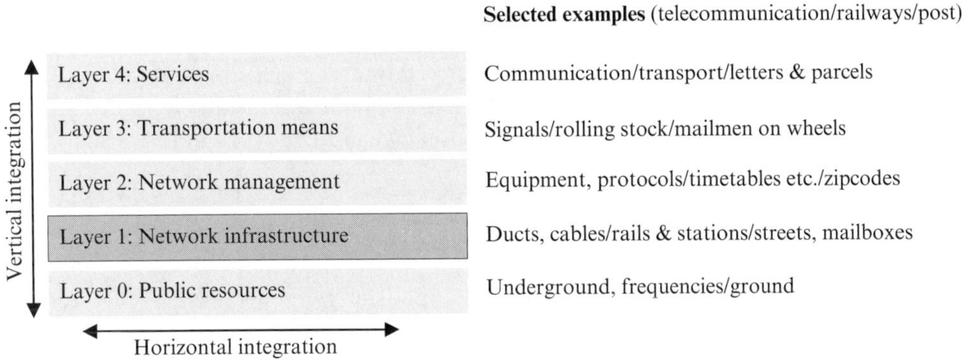

Source: Swiss Economics (2009).

Figure 3.4 Network layers

Step 1: Identification of Monopolistic Bottlenecks

A common characteristic of all network industries is that they form a coherent and interrelated system. The central aspect of networks is their ability to transport goods or information between two geographically diverse locations. At the nodes in the network (which are connected by transportation means), the routing follows specific rules. Networks can be either one- or two-sided. A typical one-sided network is the grid in the electricity sector which is essentially a distribution channel with one side selling electricity to the other.[29] In the telecommunication sector, however, households and businesses are on the originating and terminating side of the network at the same time, and their utility increases with the number of people connected to the network. Generally, two-sided markets increase complexity and call for adaptations of the traditional regulatory recipes (Wright, 2004).

The concept of monopolistic bottlenecks allows for the localization of market power in a disaggregate manner. Among others, Knieps (2000b) proposes to subdivide networks into various network layers. Figure 3.4 illustrates the relevant network layers within our framework. Such a subdivision is applicable both in one- and two-sided network markets. Each layer of a specific network has a specific function in the network. Therefore, some layers may be fully competitive while others constitute persistent monopolistic bottlenecks. The starting point of the disaggregate approach is the differentiation between those network layers or segments in which workable (actual and potential) competition is warranted and those in which there is stable market power. The latter must be expected only in those layers or segments, which are characterized by a subadditive cost function in combination with irreversible costs and the absence of close substitutes. Hence, regulation based on the essential-facilities doctrine should be focused on certain network elements and not the entire network.

This implies that various network layers can and should be analysed separately – notwithstanding the strong connections among the layers. An aggregate analysis would come to the imprecise conclusion that competition is not workable in an entire industry as long as there is at least one bottleneck layer. One single non-competitive layer would bias the entire analysis.

Often, monopolistic bottlenecks are located on layer 1, as shown in Figure 3.4. This layer contains the physical network infrastructure, the establishment of which entails considerable sunk costs. In the process of liberalization, competition will first set in on the higher layers if access to the infrastructure is available at reasonable terms. Hence, regulatory intervention should focus primarily on layer 1 in network industries.

Step 2: Regulation

Instruments for bottleneck regulation

The goal of any regulation of a stable monopolistic bottleneck is to enable non-discriminatory access to these bottlenecks at reasonable conditions while minimizing the infringement of property rights on the bottleneck resource. There are a number of potential regulatory measures to tackle the issue. We distinguish the dimensions *ex post/ex ante*, general competition law versus sector-specific regulation, and regulatory instruments (price regulation, and so on).[30]

Ex post *versus* ex ante *regulation:* Governmental market interventions can either be effective *ex post* or *ex ante*. In the case of *ex post* regulation, the regulatory authority has the right (and often also the duty) to take action only if market participants have not been able or willing to settle access or interconnection disputes by means of self-regulation. This implies that the authority has the competence to correct errors of the second kind which means that it can alter an unexpectedly adverse market outcome.[31] In an *ex ante* regulatory environment, the authority can – based on its legal competences – act directly, without awaiting the outcome of the operators' self-regulation. Hence, this potentially accelerates the regulatory process. Moreover, if the regulator is able to commit to its rulings, it is in principle able to impart proper efficiency incentives as the regulated operators (or their owners) remain the residual claimants of increased profits. The downside of *ex ante* regulations is the creation of errors of the first kind which results in unnecessary regulation with potentially adverse effects on competition.

Competition law versus sector-specific regulation: Competition law and sector-specific regulation differ in several fundamental aspects. Competition law typically contains guidelines that describe acceptable behaviour in broad terms, while regulatory policy specifies detailed rules which often apply to particular firms. Competition law focuses on non-discrimination between comparable customers, with *ex ante* guidelines which are enforced *ex post*. In contrast, sector-specific regulation couples *ex ante* rules with permanent industry control and rule refinement. Moreover, competition law relies on commands to discontinue anti-competitive behaviour, while regulation proscribes certain types of conduct with associated incentives and ongoing intervention.[32]

Elements of competition law to deal with bottlenecks

Banning abusive behaviour by a firm dominating a market, or anti-competitive practices that lead to such a dominant position is one of the three main elements of competition law.[33] Practices controlled in this way include predatory pricing, tying, price gouging and refusal to deal.

Elements of sector-specific regulation to deal with bottlenecks
Regulation potentially has a much broader range of instruments than competition law:

Price regulation: confers the price setting competence to the sector-specific regulatory authority. Such regulation can be applied to retail and/or access prices by means of a price-cap or a cost-plus mechanism (the latter is basically equivalent with a rate of return regulation). In a cost-plus regime, the regulated rate is based on (an estimation) of the regulated firms' costs. This entails serious incentive problems as the regulated operator is no longer the residual claimant of efficiency improvements. This problem is temporarily solved with a price-cap regulation, which adjusts the operator's prices according to the price-cap index that reflects the overall rate of inflation in the economy, the ability of the operator to gain efficiencies (not actual gains in efficiency) and the inflation in the operator's input prices. As long as all these factors influencing the evolution of the price cap are exogenous to the regulated operator, they do not distort his incentives to innovate. However, in practice, both price-setting rules often converge to a rate of return regulation due to the need to re-set the conditions of regulation as the regulator is not able to commit to terms allowing the operator excessively high or low profits.

Access regulation: concerns non-discriminatory prices and conditions of access to a monopolistic bottleneck. This also comprises non-tariff access barriers. Such regulation may either be *ex post* or *ex ante* and based on several possible calculation methods. Among the most prominent methods is the efficient component pricing rule (ECPR, see Baumol and Sidak, 1994; Willig, 1979), long run incremental cost (LRIC) (including variants), Ramsey Pricing, or a global price cap (cf. Crew and Kleindorfer, 1994; Laffont and Tirole, 1993, 1996) (including access or wholesale prices). The details are fairly important but their discussion would be beyond the scope of this chapter.

Separation: of the bottleneck resource from the rest of a vertically integrated operator is an important means to regulate bottleneck resources. Its effect is in principle *ex ante*. Cave (2006) differentiates between eight levels of separation, while other authors define three degrees of separation. There seems to be no consensus in the literature about the demarcation between different types of separation. In the following, we follow the terminology used by the European Group of Regulators (ERG, 2007):

- *Accounting separation* is the least intrusive form of separation. It is a requirement that forces greater transparency in accounts to disclose the real costs involved in the production of regulated bottleneck resources in order to avoid cross-subsidization, margin squeeze or to enable cost-oriented pricing.
- *Functional separation* requires the creation of separate divisions within vertically integrated operators. It does not change ownership of bottleneck resources, however. Nevertheless, it requires operations and management separation and decisions to be made independently by the separated business unit and the rest of the company. While the exact location of separation may be variable, it typically includes a separation of functions, employees and information (systems).

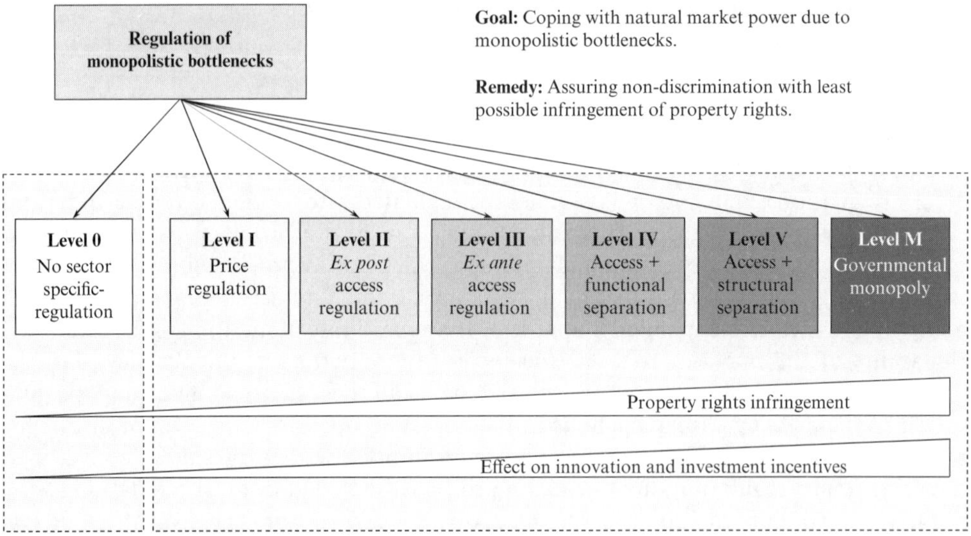

Source: Swiss Economics (2009).

Figure 3.5 Regulatory models for bottleneck regulation

- *Structural separation* implies that the vertically integrated operator is forced by regulation to dissolve the bottleneck away from the rest of the company. The independent unit may still be owned by the incumbent company, be sold to a third party or even be nationalized.

Ownership regulation: of the monopolistic bottleneck. Forced sale to an independent owner or to government are probably the most intrusive regulatory measures to assure non-discriminatory access conditions.

Sector-specific regulation disposes of a much broader spectrum of remedies to bottlenecks than competition law. This difference is due to the latter's generality which does not allow for sector-specificity and is the very reason for the coexistence of competition law and sector-specific regulation.[34]

Step 4: Models for Bottleneck Regulation

The specification of the various dimensions of regulation results in concrete regulatory models which are employed in practice. The models on all regulatory levels must be technically and commercially feasible. Figure 3.5 shows the range of possible regulatory levels in terms of bottleneck regulation. In increasing order, they become more intense and intrusive with respect to property rights infringement.

The different levels of regulation can be characterized as follows:

Level 0: no regulation. This is usually the case if there is no stable monopolistic bottleneck or if there is no (expected) misused market dominance.

Level I: price-cap regulation of existing access prices to a bottleneck resource on the basis of separated accounts.

Level II: *ex post* access regulation to a bottleneck resource based on separated accounts. This implies a subsidiary competence of the regulator to enact regulations if competing operators cannot agree on access conditions (primacy of self-regulation).

Level III: *ex ante* access regulation to a bottleneck resource by a regulator based on separated accounts. The results of such a regulation are the more incisive the larger the regulator's discretion in defining the scope of its regulation. Generally, such competences should not exceed the scope of the regulated stable bottleneck resource.

Level IV: functional separation of the bottleneck resource from the rest of the operator. The separated business unit operating the bottleneck is obligated to grant non-discriminatory access conditions to all operators.

Level V: structural separation by creating an independent legal entity for the operation of the bottleneck resource. The commercial freedom of this operator has to be defined in detail, for example, whether it is allowed to offer own services based on its infrastructure.

Level VI: Nationalization of the bottleneck resource by expropriation of the original owner. This results in a governmental monopoly.

The stronger the property rights of one or all operators are confined, the stronger this will affect investment and innovation incentives in the long run. For example, the abandonment of regulation (Level 0) results in strong incentives to the bottleneck owner to develop it further, as it is the residual claimant of all profits resulting from its development. At the same time, all operators without access to the bottleneck have strong incentives to search for alternative technologies in order to bypass the bottleneck and offer substitute services.

A structural separation of the bottleneck resource (Level V) adds to its stability and therefore creates the need for continuing (follow-up) regulation. If access is granted on a cost basis, the owner's incentives are low to invest in its development as cost savings are automatically passed down to competitors.[35] If there is an (unregulated) bypass opportunity, the bottleneck operator will rather invest there in order to bypass regulation.[36]

Sep 4: Qualitative Assessment

When choosing the optimum level of regulation in case of an abuse of a stable monopolistic bottleneck, from an economic point of view the following criteria should be considered.[37] They are partly contradictory and imply trade-offs which mirrors different interests of the various stakeholders: incumbent operators, new operators, business customers, private customers, taxpayers and employees, among others.

Proportionality: the impact of regulation should be commensurate with the market failure that is to be corrected. Indirect and second-order effects should be taken into account, for example, the long-term influence on industry development. An important prerequisite is a clear understanding of the purpose and the scope of regulation.

Expedience: any regulation should be aimed at mitigating the effects of market failure without compromising the achievement of other objectives. Trade-offs between conflicting goals must be solved in a transparent manner.

Competitive neutrality: regulation should have the least possible impact on working competition and not distort the operators' market behaviour in non-monopolistic market segments/layers. The following two aspects are of greatest importance:

- *Lightness:* regulation should be as weak as possible in order to not create inefficient barriers to entry ('light is right').[38]
- *Symmetry:* if possible, obligations and regulations should be the same for all operators in order to create a playing field as level as possible. Asymmetric regulations induce market distortions which are hard to neutralize (second-order effects of regulation). Universal service obligations that are otherwise not provided in an unregulated market are exempt. In such situations, designating the obligations to just one operator is probably more efficient (and light) than obliging any market player with the same obligations.

Incentive neutrality:

- any regulation should have the least possible effect on investment and innovation incentives.
- at the same time, a risk transfer from corporations to government should be avoided: Regulations should not result in crowding out the operators' direct responsibility, for example, in pricing matters.

Subsidiarity: regulation should come into action rather *ex post* than *ex ante*. While *ex ante* regulation allows a regulator to commit to its regime and set long-term incentives right, *ex post* regulation gives market participants a real chance to organize themselves. This reduces the risk of regulatory failure and the cost of regulation. It is important, however, to design *ex post* regulation such that negotiations between operators are not affected by the expectation of future regulation.

Simplicity: the simpler regulations are and the lower the regulator's information needs are, the easier is the implementation and control of regulation and therefore the lower its direct cost.

Transparency: transparency enables fair processes and strengthens the regulator's and the regulatory system's credibility in the public. Moreover, it also allows competing operators and consumers to indirectly 'regulate' through their informed decisions.

Temporality: technological progress and changing consumer needs put regulation at risk of becoming irrelevant and lapsed.[39] It is therefore important to design regulation as flexible as possible in order to adjust it to a changing environment. Moreover, there should be a sunset clause allowing a regulation to be withdrawn once it becomes obsolete. Temporality is not possible as far as irreversible regulations are concerned and with dynamic developments in which the development of markets and regulation strongly affect each other.

Step 5: Quantitative Assessment

The above criteria for choosing optimum regulation are of a qualitative nature. In the following, we discuss the direct and indirect economic impact of regulation in the short medium and long runs. The instrument with the least net economic cost is then the one which should be chosen – if it sufficiently conforms to the qualitative criteria. However, it must be noted that the welfare effects depend very much on consumer preferences and the operators' cost structure. They are therefore hard to quantify. Moreover, they often take effect in the long run (for example, innovation incentives) and therefore need to be explicitly modeled in a dynamic setting.

Static aspects

Benefits from regulation: what is the direct impact of regulation with respect to consumer prices? Will they decrease due to increased inter-layer competition? Or will they increase because of double marginalization problems in case of a vertical separation? The answers to these questions will strongly affect the overall welfare impact of regulation. Redistribution of profits among operators, however, is welfare-neutral in principle, as they have a distributional effect only.

Productive efficiency: will regulation result in (static) cost savings on the operators' side? In the case of subadditive cost functions – which are prevalent in certain infrastructure layers in network industries – a division of volumes among several operators increases costs per piece. Additionally, will some of the pre-existing vertical synergies be destroyed through the duplication of overhead activities and additional need for coordination?

Direct costs of regulation: regulation does not come free. Regulatory authorities need to be installed and be equipped with resources. Legal disputes resulting from more complex interactions between operators tie up management attention and delay the market development.

Dynamic aspects

Efficiency incentives: strong competition results in a strong pressure to save on costs in order to stay competitive; benchmarking among competing operators allows for mutual learning and further efficiency gains.

Innovation and investment incentives: who is the residual claimant of profits resulting from increased efficiency? Are the operators' innovation incentives protected from

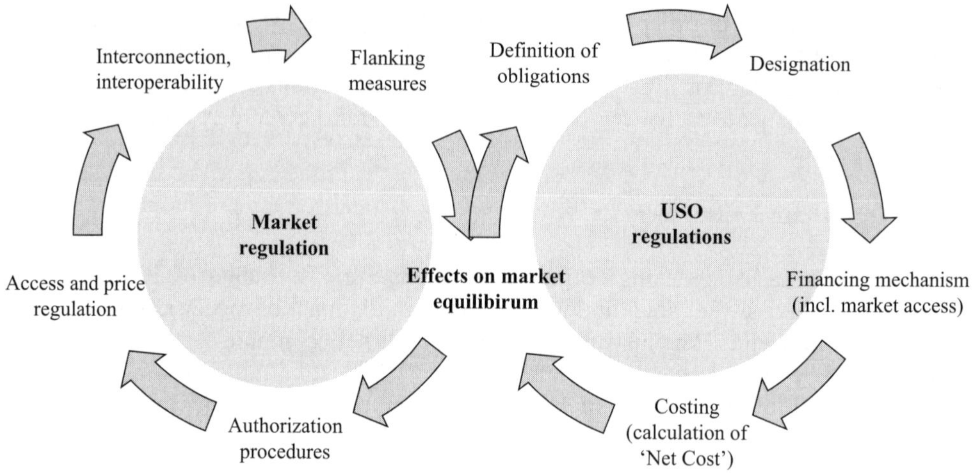

Figure 3.6 Illustration of interactions

regulatory appropriation (for example, through patent protection or 'regulatory holidays')? For example, the regulation of access conditions to a monopolistic bottleneck yields to constant downward pressure on prices on the part of those operators that seek access while access providers tend to try to increase prices, of course. In any case, the incentive to invest and innovate are fundamentally distorted as the owner of a bottleneck resource has little incentives to develop it further and competing operators do not invest in substitutes as long as they have cheap access to existing infrastructures. Abandoning such regulation may therefore yield better dynamic effects (at the cost of static inefficiencies).

Risks: does regulatory discretion directly increase business risk? Does regulation indirectly alter the risk structure in the entire industry? Does it change incentives to deal with such risks? Are there implicit government guarantees?

Regulatory dynamics: what are the costs of a potential regulatory 'encrustation', if new entrants base their business cases on regulatory arbitrage? In such cases, the abandonment of lapsed regulation (due to emerging substitutes) may be delayed. The dynamics of the regulatory framework is also driven by the regulators' (often successful) quest for increased influence on markets to justify their budgets. This results in a dynamic regulatory spiral. Moreover, there is the risk of regulatory capture if lobbyists become increasingly strong over time.

Organizational dynamics: operators which are not regularly re-organized tend to encrustations, just as regulatory systems. If regulations slow down the evolution of entire industries, they curb the elimination of internal inefficiencies.

One fundamental trade-off between static market efficiency and dynamic efficiency incentives is the Williamson trade-off in the context of merger control in antitrust which weighs an increase in market power against cost efficiency.[40] Of course, there are other trade-offs between assessment criteria: for example, high regulatory discretion to cope with time or firm-specific issues comes at the cost of long-term regulatory dynamics.

Interactions

In addition to static and dynamic effects, the assessment of regulatory remedies needs to take into account their interaction with other elements of sector-specific regulation (see Figure 3.2 and 3.6). Often important trade-offs are present. For example, access regulations have an impact on the financing needs of public obligations that are delegated to the market. Similarly, many measures related to universal service obligations (USO) and its costing and financing have an impact on the levelness of the market playing field. Hence, an isolated first-best access solution might have unwanted negative effects on conflicting aims such as investment incentives or universal service provision. Hence, a second-best solution might still prove optimal when tackling the problem more globally. Generally, various elements of market regulations and universal service regulations should not be viewed independently of each other.

The appropriate choice among the various regulatory models ranging from unregulated competition to a government monopoly varies not only with the relevant technological and demand conditions. Also, a regulator's skills and resources, the degree of corruption in a society, the efficiency of the tax system and capital markets, and pre-existing institutions determine which regulatory model might be optimal.

Illustration

In the following, we briefly discuss the above considerations applied to a selection of regulated network industries (see Swiss Economics, 2009, for a detailed analysis).

Postal services

The postal sector is one of the oldest if not the oldest network industry. It is usually not analysed along the layer framework as in the other industries. If so, the only layer exhibiting subadditive costs as well as sunk costs would be the road system positioned on layer 1. It is public and open to anyone on non-discriminatory terms. Rather, a disaggregate analysis focuses on the various parts of the value chain which consist of collection, sorting, transport and delivery of mail and parcel items. Collection and delivery exhibit subadditivity and hence the characteristics of a natural monopoly. As there are no significant sunk costs, there is no bottleneck facility which would justify access regulation or even mandated separation in general. Hence competition law (level 0) should be sufficient. Nevertheless, access to post office boxes or information on change of addresses is often regulated in liberalized markets. Note that these are not monopolistic bottlenecks. In our general framework, regulation can still be justified on the basis of market failures arising from network externalities (between operators, in analogy to termination issues in mobile telecommunication). In such cases, *ex post* access regulation might be justified.

Telecommunication

In the (wire-bound) telecommunications sector, the relevant network layers are the wide-area and last-mile passive infrastructures (ducts and cables), active infrastructures (electronic equipment) and services. While all last-mile infrastructures can be considered a natural monopoly due to subadditive costs, only the last-mile ducts and cables are a monopolistic bottleneck which cannot (sensibly) be duplicated. However, as there is increasing inter-modal competition (for example, by wireless communication, in many developed countries also by local loops established by electricity and cable companies) and fast technological progress (fibre optics), the traditional copper bottleneck in the last mile becomes increasingly contested. Hence, from an economics perspective, access regulation – which can be phased out eventually – to these temporary (copper-based) bottlenecks is more appropriate than a persistent functional or structural separation since the latter are often considered to be irreversible.[41] Once new fibre optics are in place, however, the fibre-optic local loop might be a stable bottleneck in case consumers were to demand capacities that can be delivered by fibre optics-to-the-home infrastructures only.

Railways

Railways in Europe carry both passenger and freight traffic. Despite high/subsidies[42] and their vital economic and social links within countries and across Europe, they have steadily lost market shares to other modes of transportation over past decades. In the railways sector, the relevant network layers are tracks and railway stations, network management (slot allocation, safety measures), rolling stock and services. The tracks and railway stations can be considered a monopolistic bottleneck due to subadditivity and significant sunk costs; hence, there is a potential for regulation. There is only a little technological progress on that layer, meaning that the bottleneck will likely persist. However, railways find themselves in intermodal competition with transportation means on roads, water and in the air. Rail market shares rarely exceed more than 20 per cent in the modal split. Depending on the effectiveness of this intermodal competition, two main scenarios are thinkable. Either price regulation of existing incumbents might be appropriate, or an intense regulation of the monopolistic bottlenecks to enforce competition on the service level. The latter necessitates strong regulatory interventions, such as access regulation combined with functional or structural separation of infrastructures (tracks, train stations) and passenger services in order to assure non-discriminatory access conditions. Thereby, the decision on the intensity of the separation depends on the existing synergies between the various network layers. Caution might be indicated in well-developed railway systems exhibiting tightly synchronized schedules and scarce capacities on rail tracks and in train stations.

CONCLUSIONS

Sector-specific regulation in network industries has become a widely discussed topic among academics, policy-makers, industry economists and regulators themselves. The issue of these debates has usually been whether such regulation is necessary and, if so, what its optimal design should be.

In this chapter we have presented a general economic framework to assess regulatory remedies and have applied it to network industries. It starts from a free market primacy assumption: markets, if they function properly, provide firms with the right incentives to enter markets, set prices and invest in innovation at a socially optimal level. Of course, there are a number of obstacles to markets functioning well in that sense. Such market failures give rise to potentially beneficial regulation – at the risk of these regulations failing as well.

There are *three economic preconditions* that should be met when thinking about new sector-specific regulations in a network industry. First, there need to be significant deviations from the perfect markets paradigm. In network industries, this is certainly the case; often, there are monopolistic cost structures, important externalities and asymmetric information. Second, there should be no or poor prevailing general or sector-specific regulations dealing with these market failures. Third, harmful market or regulatory failures should be prevalent or imminent.

If these preconditions are met, there are various *regulatory instruments* available to deal with the detected failures. These may vary in scope and nature. For example, when dealing with monopolistic bottlenecks, there are *ex post* and *ex ante* variations of price and/or access regulation as well as separation of accounts, functions, structures and/or ownership. Combinations of such regulatory instruments can be bundled into specific *regulatory models* that one can apply in practice. These models must be technically and commercially feasible.

From an economic point of view, any such regulatory model must bear up against a thorough *economic analysis and comparison* according to a number of qualitative and quantitative criteria. The *qualitative criteria* include proportionality, expedience, competitive neutrality (including lightness and symmetry), incentive neutrality, subsidiary, simplicity, transparency and temporality. While these criteria are of qualitative nature and have an indirect impact on welfare, *quantitative criteria* should be included, too, to compare the economic effects of any model compared to the regulation in place (market failure versus regulatory failure). Thereby, static effects like benefits from regulation, productive efficiency and direct costs of regulation should be considered, as well as dynamic aspects such as efficiency incentives, innovation and investment incentives, risks, and the effects of regulatory and organizational dynamics. The regulatory model with the least net economic cost is then the one which should be chosen – if it sufficiently conforms to the qualitative criteria.

Within this framework, the question of *liberalization in network industries* is essentially a decision on market access towards a more liberal regime; new competitors willing to enter a market obtain the right to do so (or fewer duties). In practice, there will be both market failures in fully opened markets and regulatory failures in regulated monopolies. Hence, the key question regarding market access is whether a free market would lead to an outcome *less inefficient* than a regulated monopoly. We have pointed out that sector-specific market access regulations play various roles. On the one hand, monopolies usually impair innovation, cost awareness and customer friendliness. On the other hand, monopolies can provide important investment incentives, are a financing means for policies such as universal service obligations, and might be straightforward where the natural market characteristics tend to monopolize the market. Hence the optimal answer on how to regulate market access will vary not only from sector to sector, but also from country to country.

NOTES

1. For a detailed description of the framework and applications to the postal and telecommunications sector, see Trinkner (2009) and Swiss Economics (2009).
2. The terms 'competition law' and 'antitrust law' are used as synonyms. The former is the technical term in the EU, while the latter is used in the USA.
3. Politics can be interpreted as a rent-seeking or capture process, see Stigler (1971). In this view, consistency would be a product of coincidence.
4. Newbery (1999, p. 134) describes normative and positive theories of regulation. He finds that 'regulation . . . is inevitably inefficient because of problems of information and commitment and, more fundamentally, because of inefficient bargaining between interest groups over potential utility rents'.
5. They also summarize positive theories of regulation, including public interest theories, interest group theories and private interest theories.
6. For example, Spiller (2008) understands regulation as a public contract between the government and operators and develops a theory on liberalization depending on regulatory credibility and opportunism of all involved parties. While Spiller (as many others) treats regulators as opportunistic like any other agent in the economy, the Toulouse school usually assumes benevolent regulators in a world of asymmetric information and aims to provide the right incentives for the companies in the market. More explicitly, Armstrong and Sappington (2006) see the key question of whether or not to liberalize as an unavoidable trade-off between efficiency of operations and rent extraction by operators. They discuss various regulatory policies and identify which ones are suitable and which ones not.
7. There is a broad range of possible regulations including self-provision of certain goods or services by the government itself.
8. For the link between incomplete markets and externalities see Arrow (1969).
9. For example, Panzar (2009) discusses the relationship of competition policy and (sector-specific) regulation in the newly liberalized postal sector.
10. See Trinkner (2009) and Swiss Economics (2009) for applications to the postal and telecommunications sector.
11. Among others, see De Bijl et al. (2006).
12. Subadditivity implies that the cost of producing a set of outputs as a whole are less than the costs of producing the same output subdivided in any combination of subsets. See Baumol (1977) for a precise definition.
13. Postal markets provide empirical support for the theory: despite substantial economies of scale and scope, new entrants have successfully entered newly liberalized markets (among others, see De Bijl et al., 2006).
14. See Knieps (2000a).
15. See Mas-Colell (1995).
16. See Economides (1996) and Farrell and Klemperer (2007).
17. See Rochet and Tirole (2006) for a formal definition.
18. See Tempest (2007) for a discussion of Robinson lists. Uniform pricing constraints can also be explained partly by two-sided markets. In the two-sided postal market, for example, postage is usually charged to the senders. However, the charges are often passed on to the receivers, for example, by banks or distance mail order companies. Thereby, normally single piece prices are charged instead of the effective wholesale prices. Some mail order companies make the bulk of their profits out of this difference. Such behaviour has a negative impact on overall mail volume development as the average price signal is higher than the one actually charged by the platform. Under uniform pricing, the difference is smaller.
19. Market power can play an important a role in private termination negotiations. In such cases, termination regulations should cope with the respective side effects.
20. In particular, post office boxes are not monopolistic bottlenecks. For a discussion of PO box provision in two-sided postal markets, see Jaag and Trinkner (2008).
21. The issue is closely linked with compatibility. Among others, Katz and Shapiro (1985) analyse the conditions under which an industry achieves compatibility by self-regulation. See Economides (1996) for a summary. Essentially, compatibility is a strategic issue from an operator's perspective.
22. Moral, ethical and religious standards are of importance too. For example, most social standards condemn the breach of contracts.
23. In practice, there is no benevolent authority. The institutional design should minimize regulatory capture (from incumbents, new entrants, unions, and so on) and deal with persistent information asymmetries.
24. Unless there is platform competition (facilities-based competition). Similarly, modern incentive regulation has some potential for fostering efficiency.
25. Note that in a world of perfect markets, a benevolent regulator/dictator could also achieve any Pareto

optimal market solution. However, if these assumptions do not hold true, both markets as well as regulators are increasingly likely to fail.

26. Besides the above mentioned reasons (innovations, and so on), a particular reason with regards to efficiency of operations might refer to market power of labour unions in public undertakings. Over time, unions have succeeded in extracting a large part of the incumbents' monopoly rents. Hence, liberalization might be a means to break the unions' bargaining power and to enforce the necessary adjustments on the salary side.
27. In investments with time horizons of, for example, 30 years, a private investor's depreciation rate makes cash flows in the second half of the project almost worthless.
28. In case monopoly rents occur, these remain public. Moreover, pay-offs are symmetric in case of public ownership whereas with private ownership, eventual losses exceeding private equity at stake must be borne by the public.
29. The tendency towards smart grids might increase the fraction of households consuming and producing electricity at the same time. However, the utility of a subscriber depends on the electricity delivered and not on the fraction of consumers connected to the network.
30. Note that some authors do not distinguish between *ex post* versus *ex ante* and competition law versus sector-specific regulation as competition law usually applies *ex post* while regulation applies *ex ante*.
31. Glazer and McMillan (1992) argue that in the liberalization process, regulatory control may sometimes be replaced by the spectre of regulation. A similar effect is at work in cases where self-regulation is used to avert explicit regulation. An example of the latter is Deutsche Post, which provides universal postal services with neither being explicitly mandated nor any form of compensation.
32. See Viscusi et al. (2005) and Carlton and Picker (2007). On the relationship between competition law and regulation see, for example, Bourreau and Dogan (2001), Cave (2004) and Geradin and Sidak (2005).
33. The other two are merger control and prohibiting agreements or practices that restrict free trading and competition between firms. In particular this includes the repression of cartels.
34. See, for example, Shelanski (2002) for a discussion of the relationship between sector-specific regulation and competition law in the telecommunications sector.
35. Price-cap regulations tend to converge to rate-of-return regulations in the long run, as regulators are not able to politically maintain a situation in which the regulated operator generates above-normal profits. Hence, if the regulated firm under a price-cap – owing to good incentives to increase efficiency – is successful in outperforming, future price caps will be renegotiated.
36. For a detailed discussion of the various economic effects of a structural separation in telecommunications see, for example, Crandall and Sidak (2002).
37. Many of these criteria also serve to assess other regulatory remedies apart from bottleneck regulation.
38. See, for example, de Bijl et al. (2006) for an application to the postal sector.
39. Note that only persistent monopolistic bottlenecks without substitutes necessitate the regulation of access conditions. These conditions vary over time.
40. See Williamson (1968).
41. See, for example, OECD (2003).
42. Less than half of the total costs of rail transport in Europe are borne directly by passenger and freight customers.

REFERENCES

Armstrong, Mark and David Sappington (2006), 'Regulation, competition, and liberalization', *Journal of Economic Literature*, **44**, 325–66.

Arrow K. (1969), 'The organisation of economic activity: issues pertinent to market versus non-market allocation', *The Analysis and the Evaluation of Public Expenditure: The PPBS System*, Washington, DC: Congress Joint Economic Committee, pp. 47–64.

Bain, Joe S. (1956), *Barriers to New Competition*, Cambridge, MA: Harvard University Press.

Baldwin, Robert and Martin Cave (1999), *Understanding Regulation: Theory, Strategy, and Practice*, New York: Oxford University Press.

Baumol, W. (1977), 'On the proper cost test for natural monopoly in a multiproduct industry', *American Economic Association*, **65** (5), 809–22.

Baumol, W., J. Panzar and R. Willig (1982), *Contestable Markets and the Theory of Industry Structure*, San Diego, CA: Edward Elgar.

Baumol, William J. and J. Gregory Sidak (1994), *Toward Competition in Local Telephony*, Cambridge, MA: MIT Press; Washington, DC: AEI Press.

Bourreau, Marc and Pinar Dogan (2001), 'Regulation and innovation in the telecommunications industry', *Telecommunications Policy*, **25**(3), 167–84.

Carlton, Dennis W. and Randal C. Picker (2007), *Antitrust and Regulation*, NBER Working Paper Series, vol. w12902.

Cave, M. (2006), 'Six degrees of separation: operational separation as a remedy in European telecommunications regulation', *Communications & Strategies*, **64**(4), 84–104.

Cave, Martin E. (2004), 'Economic aspects of the new regulatory regime for electronic communications services', in Pierre A. Buigues and Patrick Rey (eds), *The Economics of Antitrust and Regulation in Telecommunications: Perspectives for the New European Regulatory Framework*, Cheltenham, UK and Northampton, MA: Edward Elgar Publishing, pp. 27–41.

Church, Jeffrey and Roger Ware (1999), *Industrial Organization: A Strategic Approach*, Boston, MA: Irwin McGraw-Hill.

Crandall, R. and J.G. Sidak (2002), 'Is structural separation of incumbent local exchange carriers necessary for competition?', *Yale Journal on Regulation*, **19**(2), 335–411.

Crew, Michael A. and Paul R. Kleindorfer (1994), 'Pricing, entry, service quality, and innovation under a commercialized postal service', in: J. Gregory Sidak (eds), *Governing the Postal Service*, Washington, DC: AEI Press.

Crew, Michael A. and Paul R. Kleindorfer (2002), 'Regulatory economics: twenty years of progress', *Journal of Regulatory Economics*, **21**(1), 5–22.

De Bijl, Paul, Eric van Damme and Pierre Larouche (2006), 'Regulating access to stimulate competition in postal markets?', in M.A. Crew and P.R. Kleindorfer (eds), *Progress Towards Liberalization of the Postal and Delivery Sector*, New York: Springer, pp. 153–72.

Economides, N. (1996), 'The Economics of Networks', *International Journal of Industrial Organization*, **14**, 673–99.

European Group of Regulators (ERG) (2007), *Opinion on Functional Separation*, ERG (07) 44.

Farrell, Joseph and Paul Klemperer (2007), 'Coordination and lock-in: competition with switching costs and network effects', in M. Armstrong and R. Porter (eds), *Handbook of Industrial Organization*, Vol. 3, Amsterdam: Elsevier.

Geradin, Damien, and J. Gregory Sidak (2005), 'European and American approaches to antitrust and remedies and the institutional design of regulation in telecommunications', in Sumit K. Majumdar, Ingo Vogelsang and Martin E. Cave (eds), *Handbook of Telecommunications Economics, Vol. II: Technology Evolution and the Internet*. Amsterdam: North-Holland, pp. 517–53.

Glazer, Amihai and Henry McMillan (1992), 'Pricing by the firm under regulatory threat', *Quarterly Journal of Economics*, **107**(3), 1089–99.

Jaag, C. and U. Trinkner, (2008), Pricing in competitive two-sided mail markets. In M. Crew and P. Kleindorfer (eds), *Competition and Regulation in the Postal and Delivery Sector*, Cheltenham, UK and Northampton, MA, USA: Edward Elgar Publishing, pp. 136–49.

Katz, M. and C. Shapiro (1985), 'Network externalities, competition and compatibility', *American Economic Review*, **75**(3), 424–40.

Knieps, G. (2000a), 'Interconnection and network access', *Fordham International Law Journal*, **23**, 90–115.

Knieps, G. (2000b), 'Access to networks and interconnection: a disaggregated approach', in C.-D. Ehlermann, L. Gosling (eds), *European Competition Law Annual 1998: Regulating Communications Markets*, Oxford and Portland, OR: Hart, pp. 151–70.

Laffont, Jean-Jacques and Jean Tirole (1993), *A Theory of Incentives in Procurement and Regulation*, Cambridge, MA: MIT Press.

Laffont, Jean-Jacques and Jean Tirole (1996), 'Creating competition through interconnection: theory and practice', *Journal of Regulatory Economics*, **10**(3), 227–56.

Mas-Colell, A., M. Whinston and J. Green (1995), *Microeconomic Theory*, Oxford: Oxford University Press.

Newbery, D.M. (1999), 'Liberalization and regulation of network utilities', in M. Sertel (eds), *Contemporary Economic Issues, Vol. 4, Economic Behaviour and Design*, Basingstoke: International Economic Association, Macmillan Press, pp. 3–29.

OECD (2003), *The Benefits and Costs of Structural Separation of the Local Loop*, Paris: OECD.

Panzar, John C. (2009), 'Interactions between regulatory and antitrust policies in a liberalized postal sector', in Michael A. Crew and Paul R. Kleindorfer (eds), *Handbook of Worldwide Postal Reform*, Cheltenham, UK and Northampton, MA, USA: Edward Elgar, Publishing, pp. 15–31.

Rochet, Jean-Charles and Jean Tirole (2006), 'Two-sided markets: a progress report', *The RAND Journal of Economics*, **35**(3), 645–67.

Shelanski, H.A. (2002), 'From sector-specific regulation to antitrust law for U.S. telecommunications: the prospects for transition', University of California, Berkeley, School of Law Working paper.

Spiller, P.T. (2008), 'An institutional theory of public contracts: regulatory implications', NBER Working Papers 14152, National Bureau of Economic Research, Inc.

Stigler, G.J. (1968), 'Barriers to entry, economies of scale, and firm size', in G.J. Stigler, *The Organization of Industry*, Homewood, IL: Irwin, pp. 67–70.
Stigler, G.J. (1971), 'The theory of economic regulation', *Bell Journal of Economics and Management Science*, **3**, 3–18.
Swiss Economics (2009), *Bestimmung des Regulierungsbedarfs aus ökonomischer Sicht: Angemessenheit und Folgen einer funktionalen oder strukturellen Trennung von Swisscom.* Study on behalf of Sunrise. Authors: Matthias Finger, Christian Jaag, Markus Lang, Martin Lutzenberger and Urs Trinkner, Zurich: Swiss Economics SE AG.
Tempest, Alistair (2007), 'Robinson lists for efficient direct marketing', in Manfred Krafft, Jürgen Hesse, Jürgen Höfling, Kay Peters and Diane Rinas (eds), International Direct Marketing Principles, Best Practices, Marketing Facts, Berlin: Springer.
Trinkner, Urs (2009), 'Neue Postrichtlinie: Spielräume der Mitgliedstaaten bei der Umsetzung, Konsequenzen für den Binnenmarkt und Folgen für die Schweizer Postpolitik', in C. Baudenbacher (ed.), *Aktuelle Entwicklungen des Europäischen und Internationalen Wirtschaftsrechts*, Vol. 11, Basel: Verlag Helbing und Lichtenhahn, pp. 337–424.
Viscusi, W. Kip, Joseph E. Harrington and John M. Vernon (2005), *Economics of Regulation and Antitrust*, 4th edn, MIT Press.
Willig, R. (1979), 'The theory of network access pricing', in H.M. Treibing (ed.), *Issues in Public Utility Regulation*, Michigan State University Public Utilities Papers, East Lansing, MI.
Williamson, O.E. (1968), 'Economies as an antitrust defense: the welfare tradeoffs', *American Economic Review*, **58**, 18–42.
Wright, J. (2004), 'One-sided logic in two-sided markets', *Review of Network Economics*, **3**(1), 42–63.

4 From 'service public' to universal service: the case of the European Union

Matthias Finger and Dominique Finon

INTRODUCTION

This chapter addresses the question of public services in and of the network industries. This is a complex topic at the crossroads of political and economic issues. Indeed, before liberalization, enterprises in the different infrastructure sectors were often equated with a 'public service'. Consequently, the very idea of public service is questioned by infrastructure liberalization. To recall, public services, prior to liberalization, were provided in most of the countries by publicly owned enterprises, themselves the result of public policies. Examples are Post, Telegraph and Telephone Administration (PTT) for postal and telecommunications services, electricity and gas operators, railway companies, water utilities and many others more. Public service in these sectors and for these enterprises meant a combination of objectives, ranging from guaranteeing security of supply, contributing to national and social cohesion by equal treatment of citizens, pursuing industrial policy and recently ensuring environmental protection and many others. Also, from an economic perspective these public enterprises were justified by their character as natural monopolies, by the very nature of their activities, combined with the fact that they provided public and other socially useful goods along with social equity. It was possible for these public enterprises – often also called public utilities – to provide such public services because they were not competing with one another or with private companies, as they were enjoying monopoly protection. Of course, the definition of a public service also varied from country to country, but it generally included services whose cost could not be recovered from the customers who directly benefit from them. This thus implied cross-subsidization from more lucrative activities or subsidies from the public authorities.

This public service model could, however, no longer be sustained in a liberalizing and a liberalized market. Competitors attacked the most lucrative segments of any liberalizing market and were, of course, not interested in providing public services. As a result, the historical enterprises were threatened in two ways. On the one hand, competitors seized the most lucrative market segments, that is, these market segments from which the public enterprise was precisely cross-subsidizing their public service obligations. On the other hand, this very public service obligation did not intrinsically change in the first step of reform, leading to the fact that the historical enterprise, henceforth subject to competition, remained charged with public service obligations it could no longer finance, a development which leads to a so-called 'graveyard spiral'.

However, the public authorities also came under pressure. It appeared that the market did not provide some of the former public services, as they were simply not lucrative enough for the competitors. Also, it became increasingly difficult to impose the traditional public service obligation to the historical enterprises, as this risked seriously disadvantaging them vis-à-vis new entrants, leading even to their bankruptcy. Furthermore, it was inconceivable to simply get rid of the public service, as this would not have been accepted by the population. While this is a general problem for all governments undergoing liberalization processes, the case of Europe appears to be particularly interesting, warranting particular attention in this chapter. Facing this apparently insurmountable dilemma, the European Commission came up with an original solution, namely, by way of distinguishing so-called 'universal services' from 'services of general (economic) interest'. While the European Commission took on the concept of universal service – defined as a means to protect the weakest citizens from market liberalization – it delegated the services of general interest to the nation states.

This chapter is structured as follows. In the next section, we present the public service before and during liberalization, focusing in particular on the European context as well as on the role of the European Commission in this evolution. Thereafter, we discuss the different ways of designing the so-called universal service obligation, in particular the different options to attribute and to fund this obligation. In the final two sections, we first criticize the conception of the universal service and subsequently explore the future of public service in the age of liberalized infrastructures. In particular, we argue that there is a need for public services even after liberalization and that the concept of universal service will not be able to address all public service needs.

THE EVOLUTION FROM 'SERVICE PUBLIC' TO THE UNIVERSAL SERVICE OBLIGATION

In this section, we first present what the public service in the network industries was before the liberalization process and in its different national configurations. We then highlight how liberalization of the infrastructures has affected the public services in the different sectors. Finally, we discuss how, in the particular case of Europe, the European Union has created the concept of universal service, focused on the protection of the most vulnerable consumers, along with the concept of services of general interest.

'Service Public' before Liberalization

There is no sharp definition of what 'public service' (or '*Service Public*' in French) really is. The French legal tradition is probably the most developed in this matter. There, *service public* is an activity of public interest, which the private sector is not capable of guaranteeing completely (Duguit, 1928). More precisely, '*service public*' is defined by the following three principles, namely, (1) equal treatment of all users (equity), (2) continuity of the service and (3) responsiveness of the services with respect to the consumers' changing needs as well as to changing technology. There is also, in the French tradition, the notion of 'General Interest', which is, however, much more loosely defined and thus depends much more upon the political considerations of the day. Furthermore, every country has its own ideas and definition of public service, which therefore also covers different areas, ranging from education to health to infrastructures and beyond.

The very idea of public services delivered by public enterprises is grounded in related public policies. Traditionally, public policies – in the case of the infrastructures – can pertain to social development, such as the reduction of inequities, national and regional development, national security, employment, support of national industries or, more recently, environmental protection, as well as others. Governments want to further such public policy objectives and have, in the past, often used their own, public enterprises in order to do so. More precisely, national governments have ordered their own enterprises, for example, to offer their services at nationally uniform (equity), as well as at affordable, prices (sometimes below production costs). They have also ordered them to offer certain services, which a commercial enterprise would not offer (for example, public transport services in remote areas). Some public enterprises had to offer particular services for national security purposes (for example, military air traffic control, military postal services), which the government generally did not pay for. Other public enterprises had to make particular efforts to employ people above the levels of personnel needed for efficient service provision. Sometimes, they were ordered to source nationally, for example to support the national labour force among the suppliers or to buy their equipment (for example, railways, electricity) from the national suppliers. In extreme cases, and depending on particular political circumstances, public enterprises were forced to locate production facilities in particular regions of their country, so as to create employment there or otherwise support a particular region.

All these different obligations, which public enterprises were asked to do for reasons of public policy, make up for what is called public services. One can say that these are services which the public enterprises offer because they are publicly owned and/or because they were explicitly mandated by politics to do so. Often, however, they were not even mandated, as public service was simply part of their 'DNA', that is, part of their *raison d'être*. It is thus not surprising that, especially in the Latin tradition, public enterprises were often equated, in the past, with public services. In France it often still happens that the railway company (SNCF) and the postal service (La Poste) are called 'a public service'.

However, there are big differences between countries, and especially between politico-administrative traditions, as to what public service really is. We distinguish here between the Latin, the Anglo-Saxon, the German and the Scandinavian traditions.

As a matter of fact, public service is a concept that mainly stems from the so-called *Latin tradition* (France, Spain, Italy, Portugal, Greece, Latin America and former French colonies worldwide). Here, public service is defined at a national level as a service that the government, via its public enterprises and its public administration, provides to all citizens, regardless of geographical location, gender, origin, social status and so on. Public service is indeed an integral part of what governments do for their citizens. Overall, in this tradition, public service is a political science and a legal concept.

In the *Anglo-Saxon tradition* (the UK, the USA and British colonies worldwide) there is no public service per se. Here, the main concept used is the so-called 'public interest', a concept even more fuzzy than that of public service, even though it is grounded in economic theory. In this economic tradition, public interest is equated to public welfare, which means that the public interest is less geared towards citizens as individuals, than rather towards society as a collective. Also, the term is defined negatively: whatever is produced by economic actors is in the general interest (for example, public welfare). However, there may be occurrences, where some particular economic activities are not in the general interest, in which case they will call for regulation, that is, state intervention, so as to stem market failures.

Third, is the *German tradition* of so-called '*soziale Marktwirtschaft*'. While in principle very close to the Anglo-Saxon tradition of a free market working in the public interest at the macro-level, there is a slight difference however at the micro-level. Indeed, the German tradition stresses a much stronger social responsibility of firms: unlike the Anglo-Saxon tradition, *soziale Marktwirtschaft* considers that firms do have a social responsibility, not only vis-à-vis their employees, but also vis-à-vis the society within which they operate. However, unlike in the French tradition, firms have no public service obligations.

Fourth, is the *Scandinavian tradition* of municipal public services. Unlike the other three traditions, where public service is defined in relationship to the nation state, the Scandinavian tradition sees public service mainly at the local level, where municipal companies offer infrastructure services to the inhabitants of the regions they serve. Such infrastructure services comprise energy, public transport, water and sometimes others. This idea of municipal public services can also be found in Germany, Austria and Switzerland.

The challenge – in particular for the European Commission – was to come up with a compromise among these four traditions, namely, a compromise that is not tilted towards one or the other tradition, as neither tradition alone would be acceptable to the other. But before presenting the European Commission's redefinition of public services below, let us first illustrate what happens to the public services as a result of liberalization.

How Does Liberalization Affect the Public Service?

As said above, liberalization challenges the very fact that public enterprises provide public services. As a result of liberalization, firms which previously enjoyed national, regional or local monopolies from where they could cross-subsidize non-lucrative public services, are now faced with competition. Moreover, competitors do what is commonly called cream-skimming or cherry-picking, that is, they attack the most lucrative segments of a market, which are precisely the segments which are vital to the enterprises providing public services. As a result of this evolution, the very financial basis of the public services is being threatened and such public services are no longer affordable, unless the subsidies come directly from the state. Of course, the public services of each of the infrastructure sectors are affected differently by liberalization. We will thus briefly describe what liberalization means for the public service of each of the main sectors (see Bolhöf, 2005; Künneke et al., 2009).

It is in the *postal sector* where the public services have probably been most extensive, liberalization is also having the biggest effect, not only on the public postal services, but on the postal sector as a whole. Traditionally, postal operators had a monopoly protection in order to finance its public service, which basically means the collection and delivery of postal items every day and everywhere throughout a given country. However, the delivery in remote regions is very costly and cannot be financed without some form of subsidy, which currently takes the form of a monopoly protection. As the monopoly in the postal sector is gradually being abolished, the only possibility for postal operators is to reduce the quality of the public service accordingly, generally by reducing the collection points (post offices) but also by reducing everyday delivery of mail and parcels. This is not always easy, as politics seeks to conserve the traditional definition of public postal services while at the same time no longer subsidizing it.

In the *telecommunications sector*, the situation was originally identical as in Post, both being part of the so-called PTT. However, the telecommunications sector being very dynamic, telecommunications services have evolved beyond the traditional public services, which still mainly consist of fixed-line telephone connections. As a result, the new dynamic telecommunications services, that is, the mobile services, are no longer considered to be part of the public service and the market, more generally, takes care of the evolution of the telecommunication sector.

In the *electricity sector*, public service is generally defined as the reliable delivery of electricity to the household at affordable prices and in equal conditions of treatment. Overall, liberalization actually makes the delivery of electricity less reliable than before, for three reasons. First in the countries with 'energy-only' market design, there are weak incentives to invest in peaking capacities because of volatile revenues on the wholesale market (IEA, 2003; Joskow, 2007). Second, price volatility on hourly markets is such that investors prefer to invest in low capital intensive equipments, but high variable costs dependent from volatile fuel price instead of investing in capital-intensive, but low variable costs which are the cheapest in terms of expected complete costs. With insufficient investment in production capacities in capital-intensive equipment, this makes production prices increase after several years, all things being equal (Green, 2007). Third deverticalization of the industry makes technical and economic coordination much more difficult than before reforms with vertical monopolies. It is generally the local electricity distributors who deliver the 'public service', which generally takes the form of continuity and reliability. It is also the historic suppliers who have to deliver electricity to affordable prices. This is somewhat paradoxical, as the local distributors cannot really control production costs or network stability, and the historic suppliers tend to price the electricity for retail sales at the level of the wholesale price with no relation to their own costs, even if they have their own generation assets. Liberalization thus creates a serious problem for public service in the electricity sector.[1]

In the *railway sector*, public service takes the form of national, regional and local transport services provided to a certain timetable, as well as affordably. Most of these transport services, especially the regional transport services, being subsidized anyway, the question simply is whether or not the government continues to subsidize these services and to whom ultimately these subsidies are attributed (for example, owner of the tracks so as to offer cheap access, or the transport company using the tracks).

In the *drinking water sector*, public service means the regular delivery of potable water at a certain sanitary quality at an affordable price. Liberalization in the water sector takes a particular form, namely, the tendering of water services at regular intervals. The consequence of this for public service is that water quality generally goes up, but so also do prices, while the extension of the network to peri-urban areas generally slows down or is even halted (Finger and Allouche, 2002).

Overall, it appears clearly that liberalization negatively affects most of the public services, at least the services as they were defined traditionally. It is also clear that Europe, with its tradition of public service, could not simply accept the reduction of such public services as a result of liberalization.

THE EUROPEAN UNION AND THE EMERGENCE OF THE UNIVERSAL SERVICE OBLIGATION

As a result of this challenge – the pressure on the public services on the one hand and the different European traditions of public service on the other – the European Commission had to clarify the concepts and come up with a definition of its own, namely by way of the new concept called 'universal service'. In this section, we will discuss the different concepts as used by the European Commission, namely the concepts of 'services of general interest', 'services of general economic interest', and 'universal services'. The remainder of this section then focuses on universal services only and outlines its definition by the European Commission.

To recall, the European Union had only become concerned with public services once liberalization was well under way, that is, during the second half of the 1990s. To this effect, it published a so-called *Communication* on services of general interest (SGI) in 1996 (EC, 1996). One must also mention the Treaty of Amsterdam (1997), which allows for certain restrictions on competition in order to accomplish missions of general (public) interest. This in turn enables states to go beyond the universal service protection (see below). However, the discussion on the desirability of a Framework Directive and a charter clarifying the legal status of the services of general interest is still ongoing. In 2009, the Treaty of Lisbon introduced the concept of 'Service of general economic interest'. Ever since, the Commission uses both terms, namely, 'Services of general economic interest' and 'Services of general interest', which are services subject to 'specific obligations imposed by public authorities to a service provider to ensure the achievement of certain objectives of public interest' (EU, 2007, p. 23).

The concept of 'universal service' appears for the first time in 1987 in the Green Paper on the liberalization of telecommunications (EC, 1987). The concept defines a set of principles and common requirements for EU countries by establishing, in several sectors, the need for a basic service available to all consumers throughout a country at a certain quality and at an affordable price. The concept is supposed to solve the tension between market opening and traditional public services. However, the emergence of the concept of universal service does not mean that the concept of public service has disappeared. In some Directives, such as the electricity market Directives (EC, 2004), public service objectives related to social and environmental protection are recognized as such, but they are considered to be part of the broader general interest.

One must also place the concept of universal service within the broader liberalization context and endeavour: it must be stressed that competition remains the main driver of 'universal service', because it leads, in theory, to lower prices, increased supply and innovation. However, market imperfections exist both in practice and in theory: markets end up re-concentrating as can be observed in telecommunications, electricity, gas, air transport and many other industries. In this context, expectations regarding the desired effects of competition are not necessarily met and it is unlikely that the process of liberalization (and privatization) will really provide all or part of the universal services without the use of specific public interventions. Such intervention thus takes the form of consumer protection and concerns the evolution of prices, the quality of services, along with affordable prices.

To sum up, for the European Commission there are universal services which are a form

Table 4.1 Distinction between universal services and services of general interest in different network sectors

Sector	Universal services	Services of general interest
Telecommunications	• Geographical ubiquity of fixed network • Non-discrimination: equality of tariffs, whatever locations in fixed networks	• Social tariffs for vulnerable consumers • Obligatory duties (reserved networks for defences, internet access for education, etc.)
Postal service	• Distribution in remote areas • Non-discrimination: equality of tariffs in the universal service	• Access to banking services
Electricity	• Geographical coverage • Reliability of supply	• Social protection: special tariffs and rules of disconnections • Promotion of clean production and energy efficiency
Gas	• Reliability of supply	• Social protection: special tariffs and rules of disconnections
Railways		• Regional link traffic (funding by grants) • Social tariffs (families, old population)
Air transport		• Regional traffic airways (subsidies after auctioning)
Water	• Local coverage of networks • Water quality	

of consumer protection, on the one hand, and services of general interest (or services of general economic interest) on the other. The latter cover the broader public policy objectives. While universal service covers affordability, quality and accessibility of precisely defined services throughout a country, the general interest covers all other public policy objectives beyond citizen protection, such as long-term energy security, environmental protection, sustainable development and others (EC, 2004).[2] The first task therefore is to clearly distinguish between universal service obligations (USOs) on the one hand and services of general (economic) interest on the other. Table 4.1 presents this distinction for the main infrastructure sectors.

Both, the services of general (economic) interest and the universal service obligations are political in nature. However, the difference is that the former are implemented by way of traditional public policies and corresponding administrative oversight, whereas the latter are regulated by sector specific regulators. This can be done either by way of putting restrictions or obligations on market players, or by way of subsidizing these actors who have to implement the services of general interest out of the state budget (taxpayer). In both cases, obligations and subsidies, the market may not be distorted, something which is to be supervised by the competition authorities. In the second case, the universal service obligations are also politically defined but the attribution of these services, their financing and their regulation will be done differently, as we will show in the next section.

DESIGNING UNIVERSAL SERVICE OBLIGATIONS

While services of general interest are not really problematic in a liberalized environment – they simply need to be financeable by the public authorities and not distort the market – universal services are much more difficult to design, as they are directly market relevant. In this section, we therefore describe how universal service obligations (conceived as consumer protection) are attributed to one of the market actors, how they are financed and how they are regulated.

Attributing the Universal Service Obligation in a Liberalized Market

In the former utility monopoly model, public service obligations attributed to utilities or public firms had their costs funded by cross-subsidies. When private companies were charged with public service obligations, they were generally given a monopoly protection. As an implicit compromise, the costs of public service obligations were passed through via tariffs by means of 'cost plus regulation'. In other words, private firms, and even public firms, were allowed to charge the costs of their services plus a certain margin or they were allowed to cross-subsidize. The advantage of this was its simplicity. But cross-subsidies were quite opaque because cost assessment and cost allocation to different market segments were not codified in the regulation. Moreover, this way of doing it led to wrong incentives.

In the market regime, cross-subsidisation internal to historic operators is no longer possible or desirable. As a first step, the universal service obligation had to be separated from the services of general interest. The responsibility for these services of general interest had to be taken back by the public authorities and who is best suited to provide them or whether they could be imposed upon all the operators had to be examined. As for the universal service obligation (consumer protection), the approach should, in principle, have been the same but, for convenience reasons, the public authorities generally charged their historical operators with this universal service obligation. Nevertheless, in principle the universal service obligation should be open to all market players in order not to give commercial advantages to the historic operator. We identify four main ways of attributing the universal service obligation in a liberalized market (Cave, 1996; Choné et al., 2002; Crew and Kleindorfer, 2000; Lévêque, 2004; Ménard and Ghertman, 2009; NERA, 1998; OXERA, 2007; Valetti et al., 2002):

1. *The 'restricted-entry' system*: in this system, the universal service obligation is attributed and assumed by the historical operator, generally considered best suited for this task. But this implies strict supervision by the regulator in order to avoid undue competitive advantage for the universal service provider, resulting, for example, from an automatic customer base or from an advantageous calculation of the universal service costs. Also, it is not clear if the advantages gained from the responsibility of the universal service obligation (such as for example economies of scope) would have to be subtracted from the universal services costs. This approach thus leads to problems of market distortion on the one hand and of disadvantaging the historical operator on the other, especially if the universal service obligation is not properly retributed (see below).

2. *Imposing the USO to every competitor*: in principle, if the universal service obligation is not too heavy, it is conceivable that it is imposed upon all market players in an equal manner. This solves the problem of market distortion as well as of subsidizing the universal service obligation. In the postal services the generalization of geographical USO would oblige all the operators to be active on the totality of national areas. As the universal service is costly in the postal sector, this basically prevents new entrants. It would be feasible in telecommunications if a universal service obligation were defined for mobile telephony, for example. In electricity, however, it would not be feasible at all.
3. *The 'pay-or-play' system*: in this system the operator can choose whether he or she would like to offer the universal service obligation or pay a tax proportionately to his or her activities. The revenues from this tax will help to compensate US obligations costs. The pay-or-play regulation is applied in Australia in telecommunications and in Finland in the postal sector, but generally entails the problem of limiting market entry.
4. *Auctions*: in this market-based system, the USO is allocated on the basis of the best bid in terms of asked subsidy for cost compensation with respect to the terms of reference and requirements. Universal service obligations are auctioned for the postal service in Germany or for telecommunications in the USA, for example. Furthermore, this system is starting to be used in electricity provision in rural areas in some developing countries. The system can have limitations in situations where there are few potential candidates, as this creates collusion risks on the bid price (Klemperer, 2002).

Funding Mechanisms for the Universal Service Obligation

Funding the USOs is generally a necessity in order to preserve the financial viability of the universal service provider. Others, however, advocate self-funding of universal service obligations (Cohen et al., 2000; Sappington and Sidak, 2001). The basic rationale is the qualitative appreciation that the USO gives as much commercial advantages to the operator as the costs that the obligation makes him or her bear. This perspective concerns telecommunications and postal services. It also concerns electricity and gas suppliers when they have to support an obligation of last resort supplier. But the conclusions from these debates is not unanimous: in the postal services many European operators insist that maintaining a universal service obligation for the incumbent operators without new funding mechanisms forces them to scale down access to services (Sweden) or to compromise their financial equilibrium (UK). However, others retort that incumbents can cover the cost of the USO relatively easily, by increasing the single-piece price of mail delivery, by restructuring to become more commercially oriented, or by taking advantage of economies of scale and scope by using the same network to offer a larger number of more competitive products (such as financial services, and so on).

Nevertheless, public funding is necessary in most of the cases. In principle, and contrary to the services of general interest, funding for the universal service obligation should come out of the sector, that is, it should be borne exclusively by the market players (consumers) and not by the taxpayers. This generally implies a certain role for the regulator who has to assess the real USO costs. The issue of USO cost assessment is not neutral at all. Methods of assessment of USO costs are strategic for those who are or will be in charge of them.[3] Historical operators in telecommunications, who for a long time will benefit from the monopoly of attribution of USO, tend to overestimate its costs in order to gain large compensation. There are several ways of funding the universal service.[4]

1. *Public budget*: public funding from the state budget is in principle not coherent with the universal service idea and should not exist. In the 2008 Postal Directive of the European Commission, however, it is nevertheless tolerated.
2. *Consumers*: here, consumers pay for the universal service costs. This is easy in the case of sectors where there exists an essential facility which every competitor has to use (for example, a distribution network in the case of electricity). In this case, the universal service costs are covered via an additional tax on the usage of the essential facility.[5]
3. *A fund financed by every competitor*: in this system, both the historical operator and the new entrants pay into a fund proportionate to their market shares or proportionate to their profits. The fund then compensates those players in charge of public service obligations. The problem of this system is that, often, the historical operation pays most of the universal service costs which he or she bears at the same time. This is actually the reason why this idea of a fund has never been applied in the postal sector.
4. *A compensation fund financed by new entrants only*: if the historic operator has to assume universal service obligation, it would be sufficient to compensate it by considering contributions to be paid by each entrant. This system is in reality a form of the 'pay-or-play' mechanism mentioned above. Yet, it is difficult to implement. How, for example, to calculate the cost of the USO and the entrants' contribution? Are contributions to be calculated on the basis of entrants' and incumbent's market shares, or on the basis of entrants' benefits, or with a lump sum? Moreover to limit entry barriers, minimal thresholds should have to be defined to let small competitors to enter into the profitable market segments.

It is worth noting that the costs of services of general interest are not funded in the same way as USOs. A first example is regional and suburban links in railways which are financed by regional and local governments, as they are the main stakeholders. A second example pertains to social equity: in some countries, rather than subsidizing the providers of services of electricity, gas, or water, local and national governments subsidize the customers themselves. This is, for example, the case of municipalities in Germany or the USA.

Finally, it should be noted that the regulation of the universal service obligation and the services of general interest is also very different: the latter is basically a matter of public policy and corresponding financing where it must simply be ensured that such funding is not distorting the market. The universal service obligation is much more difficult to regulate, as its financing also comes out of the sector. Market distortion should therefore not occur in collecting the money for the USO or in dispersing it. In principle, this task of regulating the USO is generally attributed to the sector specific regulator.

CRITICISM OF THE UNIVERSAL SERVICE IDEA

Market liberalization and the introduction of the concept of universal service have without doubt helped clarify the vague notion of public service, as well as the responsibilities of the different actors. However, while the 'Universal Service is a dynamic and flexible concept [that can] be redefined periodically in order to adapt it to the social, economic and technological context' (EC, 2004, p. 8), its definition has raised some criticism (Herzog, 2004). Indeed, if the concept of universal service is without doubt a step forward when it comes to precision and responsibilities, it is nevertheless a step backwards when it comes to the scope of the original concept of public service, at least as defined in the Latin tradition. Our criticism has four different dimensions, namely, the intrinsic qualitative difference between universal service and *service public* in the Latin tradition, the weak nature of universal service protection, the issue of long-term efficiency, and the static nature of the Universal Service definition.

To begin with, let us recall that the concept of universal service and the concept of public service in the Latin tradition are qualitatively different. Public Service is the responsibility of the nation state and, as such, analogous to a public policy: a government considers that its public or industrial policy is more effective than competition for the sake of competition; as such a public policy can generate positive externalities in the form of, for example, innovation or job creation. The idea of the universal service instead is totally different: rather than emanating from a public policy – the underlying public policy is market creation in very general terms – the universal service is simply a regulatory tool to correct some market imperfections in the case of particular network industries. In this sense, the universal service is a defensive concept, whereas the public service is a more offensive concept. The universal service is always subordinated to the market: it is the market that provides the main services (not the state) and the universal service is simply a corrective in the case of some equity-related market failures. It is important to note that the European Commission has not disqualified the idea of public service as a public policy but it has redefined it in terms of services of general (economic) interest, which governments can still provide if they can finance them and if they do not distort the market.

The second criticism pertains to the fact that universal service conceived as consumer protection is nevertheless weak. This is basically due to the financing mechanism, namely, the idea that the universal service must be paid out of the sector and cannot receive any subsidies from the government (such subsidies are reserved for services of general interest). This idea leads to the fact that the universal service is automatically limited and constitutes basically a correction mechanism to some of the market imperfections. Indeed, neither do the competitors want to finance the universal service obligation, nor can a heavy universal service obligation be imposed upon the incumbent as this would otherwise threaten his survival.

A third criticism pertains to the long-term consequences of liberalization on public service, something which is not covered by the concept of universal service. Indeed, the universal service is a short-term concept, which ignores the long-term consequences of infrastructure liberalization. We already know about the difficulties to attract and plan heavy private investment over long periods in these infrastructures in many cases (Arts et al., 2009). This is particularly true as market players competing for the provision of universal service do not necessarily guarantee the results. In a number of liberalized network industries – telephone, electricity, gas and railways – which rely on fixed infrastructures, the timely development of infrastructures related to the services' demand growth and improvement of existing installations becomes an issue because market gaming between decentralized players could alter the security of service supply. Incumbents' investment in new infrastructures capacity can furthermore be reduced by bad incentives being built into access-charge regulation. With access charges too low, the incentives of entrants to invest in infrastructures may also be reduced (Cave and Prosperetti, 2001). The unbundling of network services (for instance, passenger transport or electricity transport for consumers) from the infrastructure (the railways system or the transmission distribution networks) also creates a problem of coordination. This separates the function of infrastructure planning from that of the provision of services over the infrastructures. Also, unbundling in capital-intensive industries generates the risk that fragmented responsibility might encourage opportunistic behaviour (Hall, 2008). The inefficiencies arising from unbundling can be seen in industries where the scope for benefits from competition is limited by fixed and sunk costs. It is particularly clear in the railway sector experiences in Britain and in the USA (Pittman, 2005). While this is not really a criticism of the concept of universal service, it might nevertheless question the long-term viability of the infrastructures.

Finally, the universal service is quite static in nature. This is due to the fact that the universal service definition is to be institutionalized and regulated, whereas the market constantly creates new needs for universal service protection. This is very typically the case in the telecommunications sector: here we observe that the definition of universal service has not evolved much except for the inclusion of broadband access into the universal service obligation. Indeed, as the world becomes more connected to information and communication technologies (ICTs), universal service must be technologically neutral and should include Internet access. There are trends in triple play for the use of broadband as a platform for convergence. Universal service relating to voice only becomes less relevant as voice is delivered in combination with other services. Voice over Internet Protocol (VoIP) can help reach more remote areas and Internet Provider (IP) networks are increasingly used to redistribute television and radio services. Universal service must be adapted to reflect a broadband–IP converged world but, of course, consumer protection becomes more complex with many new players and a broader choice of services to address new bottlenecks.

These four criticisms show that, while the concept of universal service has merits, it also has serious limitations. In the long run it may well be that the universal service is protected but for services which the customers no longer use. Instead, new services will not be covered by the Universal service protection because its definition is not dynamic enough, because the firms refuse to finance it or simply because the consumers are too weak to lobby for universal service protection. Finally, while the concept has helped describe and make more precise the notion of public service, it has also reduced its scope and relegated the other public services to the notion of general interest whose responsibility has been handed down to nation states who have no means of addressing the issue, given their financial problems. Overall, the universal service has thus weakened the public service idea.

CONCLUSION: WHICH FUTURE FOR PUBLIC SERVICE IN THE AGE OF LIBERALIZATION?

Universal service, with its geographical and social elements, constitutes an essential element of any service of general interest beyond what is currently established in the 'public service' norm. Its definition has been a step forward in some respects (transparency of roles, costs, funding, responsibilities, and so on), but problems remain. The 'public service norm' gives the public more space for public intervention than the 'universal service norm'. If universal service must be an effective and dynamic concept, it must integrate within its scope the technical and societal developments without which it would, indeed, lose its full meaning as is the case in telecommunications. Moreover, we observe that there is a revival of public service objectives in some network industries for reasons of collective goods preservation in the case of capital-intensive sectors which need long-term planning, strong coordination and risk mitigation for private investors. This is, for example, the case in energy when talking about security of supply, or in railways when referring to investments in infrastructures.

Under the pressures of necessity and changes in political priorities there has been in some cases an adaptation towards a more public-service oriented regulation. In fields like railways, where liberalization by unbundling fails to provide some public goods (for example, an integrated timetable, safety and security of supply) or in electricity and gas, the public service regime has come in through the back door. Recent European Directives, such as the third package on energy and climate (EC, 2009), organize more broadly the possibility for exemptions from market rules and the opportunities for states to assign obligations in terms of 'public services'. They allow for the existence of policies in support of the development of technologies that entail heavy investments.

National and European legislators have identified the problems of *security of supply* and safety created by players' decentralization and unbundling as a new matter of general economic interest. It is a collective good to be guaranteed, and results from the requirement of delivering a service in the best safety conditions as is the case in railways. Subsequently, national laws and even European Directives allow temporary exemptions in the application of competition principles, in particular coordination between market players, when such issues are at stake in a critical way. Furthermore, unbundling does not impede the designation of a technical coordinator who has the legal responsibility to guarantee the safety or the reliability of the system, that is, to offer a collective good the cost of which is being paid by any consumer. But, going further, this wisdom has often been painfully acquired after the industry was unbundled. The problems with the UK rail sector, for example, are now well known, including safety issues, and the economic failure of Railtrack, leading to a de facto renationalization.

Energy network industries are mainly concerned by objectives related to negative externalities and collective goods. Treatment of environmental externalities can rely on conventional instruments of public economics theory, taxation or tradable permits, instead of regulation by standard. Price signals on environmental goods can provoke substitution between generation technologies in a way compatible with the satisfactory functioning of electricity markets. In fact, there are still market barriers to substitution towards renewables-based production and efficient consumption equipments. The internalization of environmental externalities of polluting energy technologies is not sufficient to foster the development of efficient equipments (via electricity price increase) and the replacement of fossil fuels generation with renewable or other non-carbon technologies. There is no guarantee that even a high price of CO_2 emissions will trigger such substitution because of barriers to entry faced by non-carbon electricity technologies (Jaffe et al., 2002, Stern Review, 2007).[6]

In the case of the EU, sustainable development and environmental protection have been recognized as general interest objectives. This means that due to market failures, exemptions to competition policy principles can be allowed for developing instruments which will boost demand for green and efficient electricity technologies. Policies will consist of placing renewable energy sources outside the electricity markets by imposing obligations on electricity suppliers to purchase green electricity at high regulated price over a long period of time (10–15 years), covering the economic lifespan of the equipment (Finon and Perez, 2007). The funding of suppliers' cost is financed directly by consumers by a fund fed via an uplift.

To conclude, in the age of liberalization, there is still a need for public services, and the concept of universal service will never address all the public service needs. Public-service objectives could be served by specific obligations, as well as by adjustment of reforms to complement market games.

NOTES

1. The solutions have been in the consolidation of the regulation at different levels: the adjunction of a capacity credit obligation and increased long-term responsibility of the system operator for solving the former problem, tolerance for vertical integration and long-term contracts between generators and suppliers for the second problem, and protection of the consumers by default tariffs for the third problem; all these measures tempering competition.
2. The chapter 2, article 3, section 2 of the Electricity Market Directive 2003/54/EC (EU, 2004) establishes that: 'Having full regard to the relevant provisions of the Treaty, in particular Article 86 thereof, Member States may impose on undertakings operating in the electricity sector, in the general economic interest, public service obligations which may relate to security, including security of supply, regularity, quality and price of supplies and environmental protection, including energy efficiency and climate protection. Such obligations shall be clearly defined, transparent, non discriminatory, verifiable and shall guarantee equality of access for EU electricity companies to national consumers. (… Consequently), Member States may introduce the implementation of long term planning, taking into account the possibility of third parties seeking access to the system.'
3. Different methods compete for the estimation of universal service cost and their allocation between products in telecommunications and postal services (OXERA, 2007). In the method of incremental costs, the price increment on a service or business line makes no contribution to the business line costs: it is the increase in costs associated with a specified increment of the output relating to the USO. In the 'stand alone costs' method, all the indirect costs are allocated to that service. But, given the large difference between the two options, regulators often develop intermediary methods to allocate USO costs: for example, the so-called equi-proportionate mark-up or 'EPMU' method allocates USO costs to products in the same proportion as that of the products' share of total long run marginal costs.
4. It is worth noting that the methods of compensation (uplift, public fund fed by all competitors, 'pay-or-play' system) are more information costly because they require USO cost assessments. This presents advantages and drawbacks: it sheds light on the cost of the USO, but it exposes the system to recursive disagreement and critics. In this respect the 'pay-or-play' method is the less prone to disagreements because it lets entrants choose.
5. In the postal sector where there is no essential facility, private operators who are obliged to use some infrastructures of the historic operator should pay for access to these infrastructures, this toll being used to compensate for the USO burden.
6. Technologies at an early stage of development and commercial deployment might be expensive now, but

their costs are likely to fall as and when they gain commercial maturity by learning effects in manufacturing and administrative costs. But the prospect of cost decreases by cumulated learning effects and by increasing returns of adoption are difficult to anticipate; meanwhile the dominant technologies continue to progress. Another entry barrier stems from constraints and costs of integrating decentralized, renewable technologies into an existing centralized electricity system infrastructure (comprising network investment cost, balancing cost for intermittent production).

REFERENCES

Arts, G., Dicke, W. and Hancher, L. (eds) (2009), 'New perspectives on investment in infrastructures', report of the Scientific Council for Government Policy, Amsterdam, Amsterdam University Press.

Bölhoff, D. (2005), 'Developments in regulatory regimes: comparison of telecommunications, energy and rail', in D. Coen, and A. Héritier (eds), *Refining Regulatory Regimes: Utilities in Europe*, Cheltenham, UK and Northampton, MA, USA: Edward Elgar Publishing.

Cave, M. (1996), 'Meeting universal service obligations in competitive telecommunications markets: lessons for the postal sector', in U. Stumpf and W. Elsenbast (eds), *Cost of Universal Service: Papers Presented at the 3rd Konigswinter Seminar*, Bad Honnef: WIK.

Cave, M. and Prosperetti, L. (2001), 'European telecommunications infrastructures', *Oxford Review of Economic Policy*, **7**(3), 188–210.

Choné, P., Flochel, L. and Perrot, A. (2002), 'Allocating and funding universal service obligations in a competitive market', *International Journal of Industrial Organization*, **20**(2002), 1247–76.

Cohen, R.H., Ferguson, W, Waller, J., and Xenakis, S. (2000), 'Universal service without a monopoly', in M. Crew and P. Kleindorfer (eds), *Current Directions in Postal Reform*, Boston, MA: Kluwer Academic.

Crew, M. and Kleindorfer, P. (2000), 'Liberalization and the universal service obligation in postal services', in M. Crew and P. Kleindorfer (eds), *Current Directions in Postal Reform*, Boston, MA: Kluwer Academic.

Duguit L. (1928), *Traité de droit constitutionnel*, Ancienne librairie Fontemoing, Paris.

European Commission (EC) (1987), *Towards a Dynamic European Economy, Green Paper on the Development of the Common Market for Telecommunications Services and Equipment*, COM(87) 290, June.

European Commission (EC) (1996), *Services of General Interest in Europe, European Commission, Services of General Interest in Europe*, Brussels 1996, COM (96) 443 final, Brussels, 11 June.

European Commission (EC) (2004), *Livre blanc sur les services d'intérêt général*, COM(2004)/374, Bruxelles.

European Commission (EC) (2009), *European Commission's Energy Package – Energy for a Changing World*, available at: http://www.consilium.europa.eu/uedocs/cms_Data/docs/pressdata/en/misc/107136.pdf

European Commission (EC), *Directive 2003/54/EC of the European Parliament and the Council of 2003-06-26 Concerning Common rules for the Internal Market in Electricity and Repealing Directive 96/92/EC*. COM (2003/54/EC).

European Union (EU) (2007), *Treaty of Lisbon Amending the Treaty on European Union and the Treaty Establishing the European Community, Signed at Lisbon, 13 December 2007*, Eur-Lex, Official Journal 2007/C 306/01.

Finon, D. and Perez, Y. (2007), 'The social efficiency of instruments of promotion of renewable energies: a transaction-cost perspective', *Ecological Economics*, **62**(1), 77–92.

Finger, M. and Allouche, J. (2002), *Water Privatization. Transnational Corporations and the Re-Regulation of the Water Industry*, London: Francis and Taylor.

Green, R. (2007), 'Electricity and markets', in D.R. Helm (ed.), *The New Energy Paradigm*, Oxford: Oxford University Press, pp. 255–88.

Hall, D. (2008), *Private Equity: Financial Investors, Public Services, and Employment. A PSIRU Report*, Public Services International Research Unit (PSIRU), Greenwich University (February), available at: http://www.psiru.org/publicationsindex.asp.

Hall, D. and Lobina, E. (2008), 'The illusions of competition in the water industries, the lessons from other network sectors', Greenwich University (PSIRU report).

Herzog, P. (2004), *Résolution of the European Parliament on the Green Report on the Service of General Interest*, rapporteur P. Herzog, adopted on 14 January 2004.

International Energy Agency (IEA) (2003), *Power Generation Investment in Electricity Markets*, Paris: OECD/IEA.

Jaffe, A., Newell, R. and Stavins, R. (2002), 'Environmental policy and technological change', *Environmental & Resource Economics*, **22**(1–2), 41–69.

Joskow, P. (2007), 'Competitive electricity markets and investment in new generating capacity', in D. Helm (ed.), *The New Energy Paradigm*, Oxford: Oxford University Press.

Klemperer, P.D. (2002) 'What really matters in auction design', *Journal of Economic Perspectives*, **16**, 169–89.

Künneke, R., Groenewegen, J. and Auger, J.-F. (eds) (2009), *The Governance of Network Industries*, Cheltenham, UK and Northampton, MA, USA: Edward Elgar Publishing.

Lévêque, F. (2004), *Economie de la reglementation*, Collection Repères, Paris: Editions La Découverte.

Ménard, C. and Ghertman, M. (eds) (2009), *Regulation, Deregulation, Reregulation*, Cheltenham, UK and Northampton, MA, USA: Edward Elgar Publishing.

National Economic Regulation Associates (NERA) (1998), *Costing and Financing of Universal Service Obligations in the Postal Sector*, Final Report for EC DG XIII, London, November.

Oxford Economic Research Associates (OXERA) (2007), 'Funding universal service obligations in the postal sector', January, report prepared for La Poste.

Pittman, R. (2005), 'Structural separation to create competition? The case of freight railways', *Review of Network Economics*, **4**(3), 181–96.

Sappington, D. and Sidak, J.G. (2001), 'Incentives for anticompetitive behavior by public enterprises', *Review of Industrial Organisation*, **22**(3), 183–206.

Stern Review (2007), *The Economics of Climate change Part IV: Policy Responses for Mitigation*, Cambridge: Cambridge University Press, pp. 347–69.

Valleti, T., Hoerning, S. and Barros, P. (2002), 'Universal service and entry: the role of uniform pricing and coverage constraints', *Journal of Regulatory Economics*, **21**(2), 169–90.

5 Evolution and design of institutions supporting liberalization

John Groenewegen

INTRODUCTION

The transformation in network industries of state-owned enterprises directed by ministries to private firms competing in markets and regulated by public agencies, is a process of institutional change designed with the purpose of increasing efficiency. When network industries are liberalized and competition by entry and exit of firms is made possible, the privatization of previously state-owned firms and a process of deregulation and re-regulation usually come with it.

When, in liberalized markets, competition has become the name of the game, such markets need to be supported by other institutions than was the case under state-controlled coordination. Markets in which consumers can choose between competing suppliers need the support of specific laws and regulations that protect property rights and enforce contracts. Competition in network industries also needs supportive values and norms that are internalized into the behaviour of the actors and that support the competition and corporate laws, the role of regulatory agents, the role of private firms and their way of doing business. Finally, markets in liberalized network industries need the right supportive private governance structures like entrepreneurial firms, (long-term) contracts and strategic alliances that coordinate transactions.

In this chapter we discuss the different institutions that support liberalized network industries; who can and should design the institutions and to what extent are institutions not the result of purposeful design but more or less spontaneously evolve?

In the first section we discuss different types of supportive institutions ranging from values and norms, via laws and regulations to public and private governance structures. In the second section we discuss different theoretical 'lenses' of institutional economics through which we can analyse the different institutions that support the liberalization of markets. We start with transaction cost economics as the design approach of new institutional economics (NIE), which is adequate for explaining public and private governance structures from an efficiency point of view. We then introduce, in the third section, the NIE 'lens' of evolutionary game theory, which is appropriate for explaining the emergence and evolution of values and norms. In both the NIE approaches the concepts of efficiency and equilibrium are central. In the fourth section we introduce another 'lens' of institutional economics: in the so-called original institutional economics (OIE) institutions are first of all perceived as dynamic phenomena of which the change is driven by conflicts between interest groups. Efficiency and equilibrium are not central concepts to understand institutions, but the specific interests of powerful groups are. The OIE approach analyses the institutions in network industries from a more holistic system perspective. It discusses how the public values of the system can be best formulated and identifies the institutions that are instrumental to the realization of those public values. Central is a political valuation process, which should preferably be open and democratic. The supportive institutions that result from an OIE analysis are different from those that an NIE analysis would recommend. In the final section we conclude about how the different theoretical 'lenses' of institutional economics provide insights into the evolution and design of different types of institutions in liberalized network industries.

DIFFERENT TYPES OF INSTITUTIONS

The liberalization of network industries did not come about by itself. There was no evolutionary development that made the transformation from a state-directed industry to a market process happen spontaneously. On the contrary, the decision about the transformation was consciously taken by the political system and the changes in judicial and economic systems were designed and managed by public authorities. Markets were designed where consumers could switch to other suppliers, the entry of competitive suppliers to the markets was taken care off, prices and quality of the services were purposefully regulated, independent regulatory agents were installed, and so on. The purpose of the design of the transformation towards a liberalized market was to influence the behaviour of consumers and producers in such a way that markets would produce more efficient outcomes than in the case of state-directed network industries. Change the structure, then you will have a different behaviour resulting in a more preferred outcome, was the leading idea.

The structure in which actors operate includes different parts, such as the natural resources, the composition of the population, technology, the power structure in society, the division between gender, the income distribution, the values and norms, the laws and regulations, and all kind of public and private organizational structures. The relationship between the actors and the structure is interdependent: actors are (partly) constrained by the structure and can (partly) create the structure as well. Actors and structure are mutually interdependent with varying degrees of autonomy of the actors. There are situations in which, for instance, the technology is a given and actors have no option other than to adapt to the standards of the ruling technology. There are situations when laws leave no other option to the producers in network industries than to unbundle their networks from their production and selling activities. However, there are also situations in which actors have room to manoeuvre, to be creative and to implement their own specific strategies. Investments in research and development (R&D) for instance, can make firms influence the technology, which could open possibilities for a more decentralized production of electricity.

In this chapter we focus on the institutional part of the structural environment of the actors and we apply the perspective of institutional economics to discuss which institutions are supportive of liberalization. In this section we explore how different types of institutions constrain the behaviour of actors and how actors can design institutions to minimize costs or to serve their specific interests. However, not all institutions are designed: large parts of the institutions that coordinate behaviour of actors in network

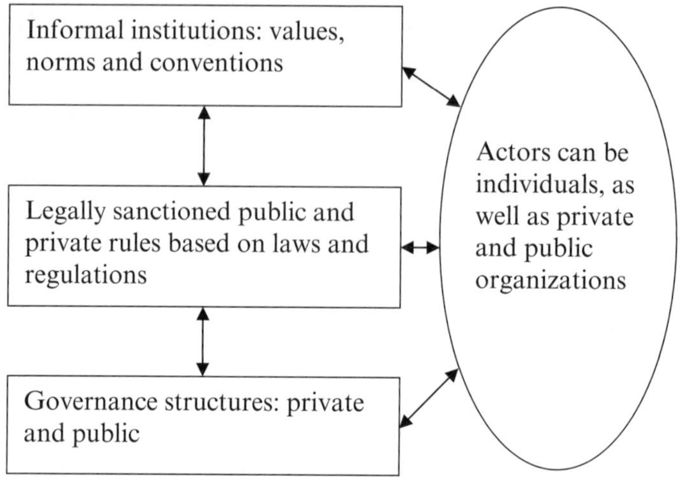

Figure 5.1 A hierarchy of institutions

industries are the result of a spontaneous, evolutionary process, in which design by actors does not play a role.

Institutions

We follow Hodgson (2006) in defining institutions as systems of durable social rules that structure behaviour and social interaction. The 'social' means that the rules are the result of human action, not of technology or nature. The systems of social rules have different sources ranging from very implicit norms concerning how one should behave in negotiations about, for instance, a contract, via organizational relations, such as vertically integrated firms, to explicit laws and regulations with legal sanctions. Language, customs, norms, etiquette, legal rules and standards, such as systems of weights and measures, are thus all considered social rules that structure behaviour and interaction between actors. Also organizations like firms and ministries are considered to be systems of social rules and, as such, they are part of the institutional environment in which actors operate.

Because institutions structure behaviour and social interaction, institutions make behaviour more or less predictable. Because of a specific institutional structure the actors in a network industry can form expectations about the behaviour of the other actors on the basis of which they can make plans for the future and conclude contracts about supply and demand. In other words, institutions make transactions in markets possible because they structure behaviour, and allow for predictions and agreements about future actions.

Below we discuss several types of institutions which support the behaviour of actors in liberalized network industries. In Figure 5.1 we present a hierarchy of institutions, in which we show the relationships between the different levels, the different sanction mechanisms (informal and formal) and whether the institutions are of private or public origin.

Informal Institutions: Values, Norms and Conventions

A distinction often made between institutions refers to the mechanism that makes actors comply with the social rule: is that legally sanctioned or not? In that respect informal institutions are private social rules that have gradually and spontaneously evolved and that do not need any legal enforcement. The rules are sanctioned by the private parties themselves (self regulating), or it is in the self-interest of the actors to adhere to the rules (self-enforcing). The values, norms and conventions that structure behaviour of actors in network industries are examples of such informal institutions. Values are embedded in the culture of a country, sector or firm and are generally held preferences about pursuable goals and embody what the members of the group believe to be good. Values are 'standards of judgement', and valuation would then be the process of applying those standards. Values refer to freedom, justice, security, prosperity and a liveable environment and are not given by nature or God, but are subject to human evaluation and decision-making. As such, values are subject to deliberation and choice (Bush, 2009)

Norms define the ways in which people ought to behave so the values of society are realized, or at least not violated. Examples of norms include competition (to achieve efficiency) and the 'polluter pays principle' (to achieve a liveable environment).

Conventions are practical rules that private actors agree upon and that in general reduce coordination problems. Well-known examples of conventions are the use of a common scale to measure lengths, weights and time and the standards developed by the Institute of Electrical and Electronics Engineers (IEEE). When in a network industry the value is security and the norm is to follow the safety procedures that all actors have agreed upon, then the convention can be that in case something goes wrong one immediately warns the other actors in the system. When in the society the actors agree about the values, norms and conventions and they all consider these institutions to function in their own interest, or all actors have internalized the informal institutions, or the social pressure is such that each actor follows the rules, then the institutions are self-regulating and need no formal sanctions.

Formal Institutions and Governance Structures

In the case of formal institutions the rules are designed by a public or private authority with legislative power and are formally sanctioned by explicit laws and regulations. Another important category of formal institutions is the so-called governance structure, being the framework within which parties coordinate their transactions. Examples include market contracts, private and state-owned enterprises, regulatory public agencies and ministries. A governance structure describes which parties are involved in the transaction, who owns what and who is allowed to make decisions. Governance structures are context-specific and emerge or are designed within the framework of informal and formal institutions. Consequently, when the framework changes and a network industry is, for instance, liberalized, then the actors will change the governance structure to coordinate the transactions. They will switch, for instance, from a vertically integrated firm to a long term contract.

In Figure 5.1 a hierarchy of different types of institutions is shown. Between the institutions a positive feedback mechanism can exist. When individual actors in network industries after liberalization, for instance, appreciate the results of competition and the reward system of differentiated salaries and bonuses that comes with it, they will support the norms and values that belong to such a world of competition. The different institutions in a network industry are complementary and reinforce each other in such a way that a 'logic' to the system emerges. It can also be that the levels in the system do not harmonize and, for instance, the formal rules of the game are changed (liberalization), whereas the values and norms are still legacies of the old state-controlled system. Then the system lacks a specific logic and the hierarchy of institutions consists of conflicting social rules that give contradictory incentives to the actors and, as such, the system of social rules will be less supportive to liberalization.

With the framework shown in Figure 5.1, the institutions that played a role before liberalization of the industry can be identified, as can those that are or should be in place after the liberalization. Also, gaps in the institutional framework or inconsistencies between the different types of institutions can be identified. Furthermore it is important to be aware of the different dynamics of the institutional layers: the informal institutions are mostly of an evolutionary, spontaneous type, whereas the formal institutions and the governance structures are more subject to design. Between the two, a certain degree of coherence should exist and in designing institutions that support liberalization in network industries it is important to take the different dynamics, as well as the degree of coherence, into account. Moreover, next to efficiency, power and vested interest also play an important role in understanding the dynamics of institutions. In the next section we first focus on the role of efficiency.

NEW INSTITUTIONAL ECONOMICS

Institutional economics offers insights into the evolution and design of the institutions that are relevant in the process of liberalization of network industries. In this section we discuss the so-called new institutional economics (NIE), which is complementary to neoclassical economics and, as such, is built along the same methodological lines in which efficiency and equilibrium are central. Following Rutherford (1994) we distinguish two perspectives within the NIE: the first is based on the theory of property rights, agency theory and transaction cost economics and conceptualizes institutions as efficient governance structures that are designed by public and private actors. In this section we focus on transaction cost economics.

The second perspective, which is discussed in the following section 3, is based on evolutionary game theory and conceptualizes institutions as equilibria (Aoki 2001, 2007), that emerge spontaneously out of the behaviour of individual actors. The institutions are then not the result of a collective design, but result from the unintended actions at decentralized level.

Transaction Cost Economics (TCE)

Oliver Williamson introduced in 1975 the approach of new institutional economics: he addressed in his *Markets and Hierarchies* the question of why firms exist and why specific transactions are coordinated inside firms and others through market contracts.

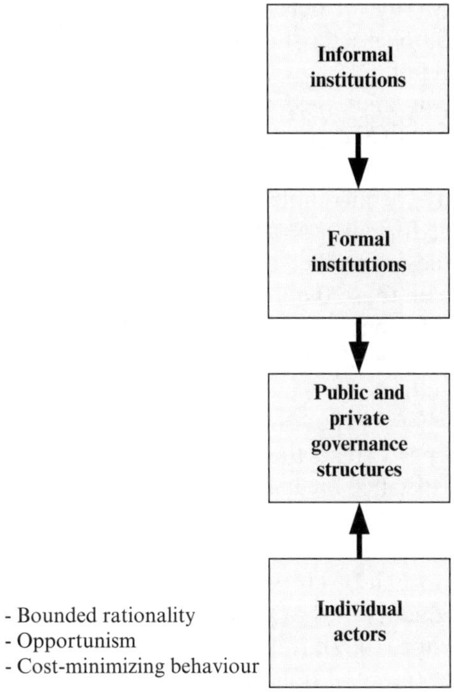

Figure 5.2 A four-layer model of institutions

Transaction cost economics is about institutions as governance structures. The unit of analysis is the transaction to be understood as 'the transfer of rights to use goods and services across a technologically separable interface' (Williamson, 1985, p. 1). Williamson developed TCE to explain different types of governance structures both private, such as contracts and firms, and public, such as regulation and 'public bureaus'. In industries that are liberalized these governance structures are important institutions to support a well functioning market. If, for instance, markets are liberalized in the electricity industry the system only functions well if new firms enter the market, when networks can be accessed by new entrants at the same conditions as applied to incumbent firms and when sufficient transparency exists for consumers to choose.

Figure 5.2 which is based on the three and four level models of Williamson (1996, and 1998) shows the relationships between the governance structures and the (in)formal institutions, as well as the characteristics of the actors.

In Figure 5.2 three categories of institutions are depicted. At level 1 the informal institutions of values, norms and conventions and, at level 2, the formal institutions of property rights, competition and corporate law and regulations. At level 3 are located the private governance structures of contracts, private organizations, like firms, and hybrids such as public private partnerships (PPPs) and networks. Also at level 3 we locate the public governance structures, for example, bureaucracies, political parties, ministries, parliament and public courts. At level 4 the actors are located as legal individual decision-makers characterized by bounded rationality, the possibility to behave opportunistically and the rule of behaviour to minimize production and transaction costs. Below we discuss the characteristics of the actors and the relationships between the different levels in more detail.

In new institutional economics, theorists use the concept of so-called bounded rationality. That is, actors have limited capacity to absorb relevant information and to make *ex ante* calculations. It is important to stress here that bounded rationality implies that contracting will almost always be incomplete. That is, there will be open ends, that parties have not dealt with in the formal contract. When contracts are incomplete, actors have the opportunity to use the unspecified area in their own interest. If one party depends on the other, he or she can abuse the incompleteness of the contract to renegotiate the contract at a later stage. In that respect the 'economics of property rights' (Grossman and Hart, 1986) claims that the allocation of property rights is then of crucial importance (see also Williamson, 2000). After the contract is signed a period of 'post contractual opportunism' begins in which parties can make use of, or abuse, their position and grasp so-called quasi-rents. Property rights then determine who has the legal right to act in that incompletely defined, unspecified area. To better understand the role of property rights, the other characteristic of the actors in the model becomes relevant – opportunism.

It is assumed that the possibility exists that actors will behave opportunistically: they might provide wrong information during the transaction process about, for instance, their capabilities, or the quality of the good or service to be delivered (see, for instance, the case of Enron in Bruijne, 2009). It might even be that actors cheat and simply have no intention to live up to the conditions agreed upon. The fact that opportunism is an attribute of the actors does not imply that it is common practice for actors to try abuse situations or to cheat each other. Most of the time, actors can and do trust each other. The point is that transactions between actors can be such that one of the parties has an opportunity to behave opportunistically and that the other party has to incur costs to find out what actually happens, or to make costs to safeguard against opportunism. The simple fact that costly arrangements should be made to support the coordination of the transactions gives rise to the question, which governance structure would be most efficient to deal with opportunism and to provide the safeguard?

Designing Efficient Institutions

In the NIE approach, of which we consider TCE the core, first efficient institutions should be designed at level 2: 'get the formal institutions right' (Williamson, 1998, 2000). When the political system has designed the efficient rules of the game (for instance, competition law or sector-specific regulation) and also has established efficient public governance structures (bureaucracy, public agencies and courts), then the private actors will 'get the private governance structures right'.

Supporting institutions in liberalized networks are then first about the design of clear property rights and about effective competition and corporate laws. The allocation of property rights (and decision rights) is crucial in the NIE because these rights provide the incentives to the actors to design the efficient private and public governance structures. The competition and corporate laws are equally important because these provide the pressure on actors to use their rights in an efficient way. Competition law addresses the issue of 'effective competition' (aimed at static and dynamic efficiencies) and corporate

Table 5.1 Governance structures, property and decision rights

Governance structures	Property rights (Ownership of assets)	Decision rights (DR)
1. Public governance		
Direct management	Public	Governmental department and/or public authorities
Agencies		Autonomous public authority
Public corporation		Corporatized public entity. DR in the hands of a relatively independent board
2. Public private partnerships		
Service	Public	Delegated limited DR (for some actions only) to private company: service agreements
Management	Public	Private operation and maintenance, but public capital investments
Leasing	Public with respect to core transactions; private investments on peripheral transactions	Operation and management (O&M): private Profits: private
Build, operate, transfer	Delegated private ownership for the contract duration	Operation: private Design: government
Concession	Delegated private ownership for the contract duration	'Framed' private DR with respect to core transactions, but large autonomy of private operator concerning the operation
3. Market		
Regulated market	Private	Competition in the market, subject to sector specific regulation
Competitive market	Private	Competition in the market, subject to general competition policy

Note: This table is based on discussions with Claude Ménard and Rolf Künneke in the period that Claude Ménard was Visiting Professor at Delft University of Technology (Spring 2006)

law addresses the issue of 'good governance' (clear responsibilities and accountability of managers and stakeholders).

Below we present three types of institutional settings in liberalized network industries ranging from public property and public decision rights to private property and private decision rights (see Table 5.1).

Public Governance

Table 5.1 starts with the situation before the liberalization. In the first category the state owns the assets and we distinguish between three types of decision rights. The situation of the 'public governance' would be the one in which the state owns and manages the property; mostly the ministry and the public corporation form one unit then. However, the state has also the possibility to allocate via contracts (part of) the decisions rights to a public agency or corporation. Then decision rights are transferred, for instance, to the board of the corporatized public entity.

In the case of public governance, the state can effectively direct the behaviour of public actors towards the goals of the public interest. This is of special interest in the case of network industries; the idea is that public values, such as universal access to telecommunication services or the quality of drinking water, are only sufficiently guaranteed when public authorities are in full control of the system. The transactions in the network that are crucial for the performance of the network as a whole are called 'critical' transactions (Finger et al., 2005). It can be argued that in the case of liberalized network industries it is important that government maintains control over those parts and transactions in the network that are considered critical for the functioning of the system. However, according to the NIE, there should always be a weighing of the efficiency implication of such public control. In the NIE approach it is pointed out that public governance incurs high transaction costs in terms of organization costs and decrease of 'market incentives' with consequences for the behaviour of the actors.

It would be a mistake to think that the NIE approach always favours private property and private decision rights because such rights would imply 'high powered market incentives'. What is the most appropriate property right regime depends on the type and setting of transactions. In general, private property will create the right incentive for owners to make the required investments to improve resources. However, when the costs of exclusion are very high, private property rights are difficult to assign to individuals. In the case of natural resources, that have, for instance, the characteristics of a common pool, it has been proposed (Ostrom, 1990) to create common ownership together with 'self crafted rules by the users' on how to use the resources.

Public Private Partnerships (PPPs)

The second category, PPPs, indicates a role for both public and private actors. The different types of PPPs concern different combinations of ownership and decision rights. An important reason to form a partnership with private actors is access to private capital and stronger incentives to be cost efficient. In return for taking risks and being responsible, private actors claim rights: rights to supply the service, right to operate and the right to design, build and maintain the asset. On the other hand, government wants to maintain property and decision rights to the extent that it can guarantee a specific minimum performance of the network as a whole, which is in line with the public values of society. So there are different types of PPPs, in some of which the ownership is public and the 'peripheral' parts are privately owned and others in which the building and maintenance is private but government is responsible for the design. There are also PPPs where the rights are sold to private actors as a concession for a specific period, after which the rights are transferred back into public hands (Ricketts, 2002).

Market Regulation

The third category in which the ownership is private, but government wants to guide the private actors in a specific direction consistent with the public interest, is regula-

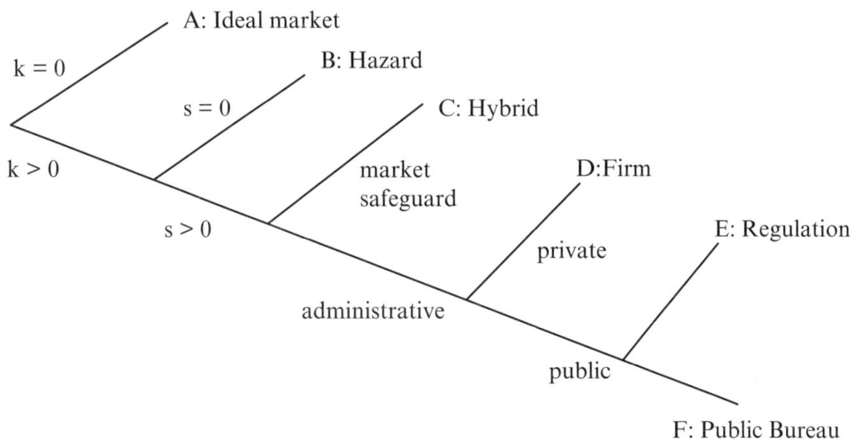

Note: k = level of asset specificity; s = level of safeguards.

Source: Williamson (1998 p. 47).

Figure 5.3 A simple contracting scheme

tion (Ménard and Ghertman, 2009). Sector-specific regulators are then authorized by law to intervene in the structure of the market or directly into the behaviour of the private actors. Regulation of behaviour in the natural monopoly segments can have the form of price, network access and quality regulation, whereas for the 'competitive segments' the rate of return and cost-of-service regulation is more and more replaced by incentive regulation, like price cap regulation (for details and examples, see Ricketts, 2002).

New institutional economics provides insights into why regulation of private actors can incur high transaction costs; the higher the asset specificities involved, the higher the uncertainties and the larger the information asymmetries, the larger the opportunities for private actors to behave opportunistically. Then the rules of regulation and the monitoring and sanction procedures become more strict and detailed; transaction costs rise and forms of public ownership become more efficient.

The last case in Table 5.1 represents the private firm which has property and decisions rights and is regulated by general competition and corporate laws. In the NIE approach it is explained how private actors will then design efficient private governance structures that coordinate transactions efficiently in liberalized markets.

Private Governance Structures

In the case of private actors who own and decide, NIE explains what kind of private governance structures will be designed by the private actors, which range from spot market contracts, via long-term relationships to vertically integrated firms.

In TCE the degree of asset specificity is the central variable to explain which governance structure is most efficient. The continuum in Figure 5.3 shows the type of governance structure related to the degree of asset specificity (k).

The asset specificity concerns the type of investments actors make in order to have a good or service transacted. When the investments are very specific, then the investment is worthless when the transaction is ended. The degree of asset specificity has implications for the possibility of opportunistic behaviour and therefore for the need of safeguards. In Figure 5.3 the asset specificity increases from left to right. When no asset specificity exists, the transaction can be efficiently coordinated with a 'contract on the ideal market' (potential opportunism is absent because of the high level of competition and high substitutability between the competitors). When asset specificity increases, safeguards are built into the contract, for example, through long-term contracting, and so-called 'hybrids', that is, governance structures in which contracts are combined with organizational structures that limit the autonomy of the actors. In a hybrid, actors, who are in a horizontal relationship with each other, are constrained by vertical organizational relations. Their autonomy is limited by the vertical constraints materialized in actions of monitoring, arbitrage and sanctioning (for instance, by arbitrage committees). When asset specificity increases further, the hierarchy (vertical integrated firm) will be the most efficient from a transaction cost point of view. Figure 5.3 also shows how the private governance is complemented by public governance (regulation and 'public bureau') when asset specificity further increases (Williamson, 1998).

INSTITUTIONS: A MATTER OF EFFICIENT EVOLUTION

When NIE discusses institutions this is preferably done in terms of equilibrium and efficiency. In transaction cost economics the central question is about the most efficient governance structures given the characteristics of the transaction, the preferences of the actors, the technology and the informal institutions. In this section we discuss the type of supportive institutions in liberalized network industries known as the informal institutions of values, norms and conventions. We take the evolutionary game theoretical approach of Aoki (2000, 2001, and 2007) as our guide. Aoki (2007 p. 3) has added two interesting problems to the TCE:

> the synchronic problem, whereby the goal is to understand the complexity and diversity of overall institutional arrangements across the economies as an instance of multiple equilibria of some kind, and the diachronic problem, whereby the goal is to understand the mechanism of institutional evolution/change in a framework consistent with an equilibrium view of institutions, but allowing for the possibility of the emergence of novelty.

With his evolutionary analysis, Aoki not only introduces the issue of multiple equilibria, but also the issue of dynamics, of the explanation of the process towards a new equilibrium.

With respect to the synchronic problem quoted above, TCE representatives suggest analysing governance structures is embedded in an institutional environment of specific informal and formal institutions. Figure 5.2 shows that transaction costs minimization does not take place in a vacuum. The institutional environment of values, norms, attitudes and also of the laws and regulations, have an impact on the choice of governance structure as a 'shift parameter'. To coordinate a specific transaction, a firm in a network industry before liberalization will opt, for example, for the governance structure of vertical integration, while after the shift to liberalization the most efficient governance structure is a long-term contract. That choice is not only based on the characteristics of the transaction, such as asset specificity and uncertainty, but the choice is also influenced by the institutional environment. In other words an identical transaction can be differently coordinated efficiently in the USA (for instance, in a vertically integrated firm) than in Japan (for instance, in a network of firms). Below we discuss how the evolutionary approach of Aoki takes the TCE analysis a step further and explains why multiple equilibria can exist and how the different equilibria come about.

With respect to the diachronic problem quoted above, TCE theorists have attempted to include the dynamics of institutions in their analytical framework. The concepts of farsightedness and selection through competition are central. Farsightedness refers to the idea that actors anticipate future (opportunistic) behaviour and create safeguards to protect themselves. In cases where uncertainty and complexity are absent and actors have equal access to resources and similar perceptions of the future, they may indeed be able to anticipate the future correctly. If in addition competition is effective, it can be safely assumed that the most efficient governance structures will be imitated and the predictions of TCE will be confirmed. But if these conditions are not met there is ample reason to assume that actors are not able and will not be forced to select the 'fitter' structures. Without more detailed information about the actors, their perceptions, the complexity and uncertainty about the environment, as well as the selective forces of competition, no prediction about the efficient outcome of the selection process can be made. To put it more precisely, with the introduction of the institutional environment as a 'shift parameter', TCE is more or less able to cope with the synchronic problem, while the diachronic problem is placed into the black box of an assumed selection process of the 'fitter' governance structure. When and how that process will really end in an efficient equilibrium is not part of the TCE analysis (yet).

Efficient Evolution of Institutions

In Figure 5.1 we located informal institutions at the top of the hierarchy, suggesting that they influence the institutions at lower levels. The values, norms and conventions constrain the private and public formal rules and governance structures. In designing supportive institutions in liberalized network industries it is important to be aware of the relationships and interdependencies between the different types of institutions, as well as their different nature. Above, NIE explained institutions as social rules that could be designed. In the other stream of the NIE the emergence and evolution of institutions is not explained as the result of a purposeful collective decision, but as the unintended result of a number of micro decisions. To understand the spontaneous emergence of institutions, the so-called coordinated game is relevant. When the domain of the game is specified (economic, political, judicial, or social) and the choices the agents can make are specified, then it can be shown that boundedly rational actors will over time create stable institutions. The definition of social rules that correspond to this approach is formulated by Aoki (2007, p. 7) as follows: 'An institution is a self-sustaining, salient pattern of social interactions, as represented by meaningful rules that every agent knows and are incorporated as agents' shared beliefs about how the game is played and to be played'.

Aoki conceptualizes institutions as equilibria, which emerge out of the unintended actions of the actors at a decentralized level in the system. Accordingly, Aoki calls his approach the 'institutions-as-an-equilibrium-approach'. This approach in the NIE is relevant for explaining the informal institutions that support the liberalization in network industries. How do these social rules come about? It is argued that their emergence is a matter of spontaneous evolution and not of collective design, and that in designing liberalized markets in network industries it is important to take those institutions into account.

'Institutions are the result of human action, but not of human design', is a well-known saying that captures the essence of the evolutionary approach. The saying refers to actors behaving in a specific way because it is in their own interest to do so and as an unintended outcome an institution like a norm or convention emerges. That behaviour at the individual level of the actors can be intentional (they aim at, for instance, minimizing costs) or habitual (not being aware that actors follow a specific rule). The point in the evolutionary approach is that institutions can come about without individual or collective action intended to create an institution or to change an existing institution. The outcome of all the individual behaviours can be the emergence of an institution that, once in existence, is durable and structures individual behaviour and social interaction.

Under specific conditions independent actors, who behave in their own interest unintentionally create institutions that are durable and that coordinate the behaviour of economic actors. One of the most important conditions is the existence of homogeneous actors. Actors must have the same motivation (to maximize, for instance, their own utility), stable preferences, must operate in the same environment and should be informed by the same signals. All actors consider behaviour in line with the emerging norm to be in their own interest and would like to see others to behave likewise. Each actor discovers that it is costly to ignore the emerging institution and that it is beneficial to copy the behaviour. This insight grows over time when the institution develops and all actors share the same knowledge when the institution is established and is in equilibrium.

Behaviour according to the emerging norm reduces uncertainty, and information costs. For the individual actors it is efficient to behave according to the emerging institution. Under the conditions mentioned above it is possible that over time actors at micro level spontaneously create institutions, which become equilibrium situations. According to the 'institutions-as-an-equilibrium' approach, the regularity has become an institution, that is, a norm, or a convention, when a large majority of the actors in the community have internalized the regularity.

In the literature the institutions-as-an-equilibrium approach is also called 'spontaneous', because it is matter of self-enforcement. No external authority is needed to force actors to behave according to a specific institution. The process emerges spontaneously purely based on the self-interests of the individuals.

Different Supporting Institutions

The way liberalization in network industries takes shape differs among countries and among networks. In evolutionary game theory this is explained by the different historical starting points of the evolution, because of which different evolutionary paths emerge. So history matters. Aoki (2000, p. 13):

we cannot predict endogenously which of those equilibria will be chosen without some other information, such as history, or institutional environments surrounding the domain of the game (institutions existing in surrounding domains). This implies that, in spite of the development of the game theory on which institutional analysis relies, game theory alone cannot provide a complete, closed frame for institutional analysis. The analysis of historical and comparative information must be essentially complementary.

The evolution of the institutions in the economic system is complemented and reinforced by the evolution of the institutions in the political, judicial and social systems. Together the interrelationships between the subsystems make up for a specific 'logic' of the network industry as a whole.

INSTITUTIONS AS PRODUCTS OF COLLECTIVE ACTION

In the institutional economics discussed so far institutions are understood as the result of efficient behaviour of the actors in the economic system. Politicians make efficient laws and regulations, and actors at micro level make efficient governance structures. Or, in an evolutionary process, efficient institutions emerge as equilibrium situations. Both the approaches are relevant to understand the role of supporting institutions in liberalized network industries. However, there are also situations in which institutions are not efficient and exist only because the rules serve the interest of a small elite or a specific interest group of capital or labour suppliers (Spiller and Liao, 2006). Moreover, forces often exist in the system that do not give negative but only positive feedback and, consequently, drive the system ever further away from an equilibrium. Institutions can then not be understood as outcomes of efficient design, nor as products of an evolution to an equilibrium. We are in need of an alternative 'lens' to understand those 'inefficient and disequilibrium institutions'. Original institutional economics (Groenewegen et al., 2010; Tool, 1993) provides an alternative that throws an important light on the question about supportive institutions in liberalized network industries.

Instrumental Valuation

In OIE, institutions are differently conceptualized. John R. Commons, one of the founding fathers of OIE, defined institutions as 'collective action in restraint, liberalisation and expansion of individual action' (Commons, 1934, p. 73). Institutions purposefully restrain individuals (for instance, because they abuse power), liberate individuals (for instance, because they are discriminatory) and expand the possibilities of individuals (for instance, because they have access to education). According to OIE, institutions are always subject to 'instrumental valuation' (Tool and Bush, 2003; Correlje and Groenewegen, 2009), that is, institutions are subject to continuous evaluation in relation to the public values. With respect to the network industries, an OIE analysis would first be about the explicit formulation of the objectives of the network industry (social, technical and economic): the 'ought'. What are the public values the industry should serve and how do these values come about? Then the analysis would explain the actual performance: the 'is'. In that analysis OIE takes a broad multidisciplinary system perspective.

The interactions between the subsystems (economy, social, political, technological and judicial) are explicitly analysed to get an understanding of the dynamics of the system as a whole.

In comparing the 'is' and the 'ought', the required changes in the institutions are determined. The institutions then become the instruments with which to initiate the desired behaviour, which should result in a performance of the system that corresponds with the public values. In the OIE literature this is called the process of instrumental valuation: which institutions are instrumental to societal objectives?

An important question in OIE is about the effectiveness of institutions. Social rules only work efficiently and effectively when they are broadly accepted in society. It is important that they are non-discriminatory, have a certain stability and are at the same time flexible enough to adapt to new situations. Most important in the OIE approach is the nature of valuation and process of change itself. Redesigning institutions implies a redistribution of costs and benefits and, as such, 'instrumental valuation' is always a process of debate, conflict, power play and social pressure (Bromley, 2006). In OIE, great value is attached to the valuation process itself; because the interests of all members of society are at stake, the norm is that the process of institutional change should be open and democratic. Original institutional economics scientists, in many of their writings, demonstrate that they highly value that all members of the society can participate in the political decision-making process. It is extremely important that all actors in society are educated and have the ability to 'reason', to participate in the debate and to convince others with solid arguments. This view of the public values of network industries implies that supporting institutions should not only take care of transparency and competition, but should offer to all participants equal access to information, knowledge and the opportunity to participate in the valuation process.

System Manager

When society has reached conclusions about the public values of network industries and the corresponding institutions are designed and implemented, then government has the responsibility to monitor the developments in the network industry. When actual developments deviate from desired developments and public values are violated, government as a system manager should then intervene. This can imply the supply of additional information to the private actors expecting that additional information will influence their behaviour, but intervention can also imply a change in the incentive structure of the actors through taxes and subsidies. If these indirect interventions of the supply of additional information and the change of the incentives does not result in the desired change of behaviour at decentralized level then government can replace the activities of the private actors and directly manage the system. The system approach in OIE with central attention given to the public values (the 'ought') and the process of instrumental valuation, results in a different view on the supporting institutions in liberalized network industries. The focus is much more on equal participation of all stakeholders in the formulation of the public values and much more on the role of government as a system manager.

CONCLUSION

Markets in liberalized infrastructures need institutional support in order to function properly. Actors in markets never operate in a vacuum, but are always embedded in institutions, which guide and shape behaviour, whereas actors also have options to create institutions. In the institutional structure that make markets in liberalized infrastructures work well, different layers of institutions can be distinguished. The so-called informal institutions of values and norms are more subject to spontaneous evolution then calculative design, whereas the formal institutions, both public (laws, regulations) and private (contract, organizations) are more subject to purposeful (collective) design.

However, it would be a mistake to model and analyse institutions in the different categories separately, each with its own mechanism of operation and change. On the contrary, the institutions supporting markets in infrastructures are better conceptualized as a whole, as a system with a specific logic that manifests itself in all categories and, consequently, also in the mental maps of the actors (North, 2005). This makes the context specificity of the supporting institutions crucial: systems differ in time (their history and phase of development), in place (history and culture) as well as per sector (technology). In designing the institutions that support liberalized markets, the logic of the system and the context specificities are crucial in realizing an effective and efficient institutionalization.

REFERENCES

Aoki, M. (2000), 'Institutional evolution as punctuated equilibria', in C. Ménard (ed.), *Institutions, Contracts and Organizations: Perspectives from New Institutional Economics*, Cheltenham UK and Northampton, MA, USA: Edward Elgar Publishing, pp. 11–36.

Aoki, M. (2001), *Toward a Comparative Institutional Analysis*, Cambridge, MA: MIT Press.

Aoki, M. (2007), 'Endogenizing institutions and institutional changes', *Journal of Institutional Economics*, **3**(1), 1–31.

Bromley, D.W. (2006), *Sufficient Reason: Volitional Pragmatism and the Meaning of Economic Institutions*, Princeton, NJ: Princeton University Press.

Bruijne, Mark de (2009), 'Enron', in E. ten Heuvelhof et al. (eds), *Strategic Behaviour in Network Industries; A Multidisciplinary Approach*, Cheltenham UK and Northampton, MA, USA: Edward Elgar Publishing.

Bush, P.D. (2009), 'The neoinstitutionalist theory of value', *Journal of Economic Issues*, **43**(2), 293–306.

Commons, J.R. (1934), *Institutional Economics*, vol. 1, Madison, WI: University of Wisconsin Press (reprinted 1961).

Correlje, A.F. and Groenewegen, J.P.M (2009), 'Public values in the energy sector: economic perspectives', *International Journal of Public Policy*, **4**(5), 395–413.

Finger, M., Groenewegen, J.P.M. and Künneke, R.W. (2005), 'The quest for coherence between technology and institutions in infrastructures', *Journal of Network Industries*, **6**, 227–59.

Groenewegen, J., Spithoven A and Van den Berg A. (2010), *Institutional Economics; An Introduction*, New York: Palgrave.

Grossman, S. and O. Hart (1986), 'The cost and benefits of ownership: a theory of vertical and lateral integration', *Journal of Political Economy*, **94**(4), 691–719.

Hodgson, G.M. (2006), 'What are institutions?', *Journal of Economic Issues*, **40**(1), 1–25.

Ménard, C. and Ghertman, M. (eds) (2009), *Regulation, Deregulation and Reregulation; Institutional Perspectives*, Cheltenham: UK and Northampton, MA, USA: Edward Elgar Publishing.

North, D.C. (2005), *Understanding the Process of Economic Change*, Princeton: Princeton University Press.

Ostrom, E. (1990), *Governing the Commons: the Evolution of Institutions for Collective Action* (Cambridge: Cambridge University Press).

Ricketts, M. (2002), *The Economics of Business Enterprise*, 3rd edn, Cheltenham, UK and Northampton, MA, USA: Edward Elgar Publishing.

Rutherford, M. (1994), *The Old and New Institutionalism*, Cambridge: Cambridge University Press.
Spiller, P.T. and Liao, S. (2006), 'Buy, lobby or sue: interest groups' participation in policy making. A selective survey', in E. Brousseau and J.-M. Glachant (eds), *New Institutional Economics A Guidebook*, Cambridge: Cambridge University Press, pp. 307–27.
Tool, M.R. (ed.) (1993), *Institutional Economics: Theory, Method, Policy*, Boston, MA, Dordrecht and London: Kluwer.
Tool, M.R. and Bush, P.D. (eds) (2003), *Institutional Analysis and Economic Policy*, Boston, MA, Dordrecht and London: Kluwer Academic.
Williamson, O.E. (1975), *Markets and Hierarchies*, New York: Free Press.
Williamson, O.E. (1985), *The Economic Institutions of Capitalism*, New York: Free Press.
Williamson, O.E. (1996), *The Mechanisms of Governance*, New York and Oxford: Oxford University Press.
Williamson, O.E. (1998), 'Transaction cost economics: how it works; where it is headed', *De Economist*, **146**(1), 23–58.
Williamson, O.E. (2000), 'The new institutional economics: taking stock, looking ahead', *Journal of Economic Literature* **38**(3), 595–613.

6 The institutional economic history of infrastructure industries, 1830–1990: ideology, technology, geopolitics?

Robert Millward

INTRODUCTION

Because the technology and economics of infrastructure industries often involve natural monopoly conditions and the supply of strategic goods and services, the state invariably intervenes in their economic organization and institutional development. That, of course, is a huge subject especially since the intention here is to trace developments over much of the nineteenth and twentieth centuries. However my aim is primarily to explain the differences between countries (Western Europe, Japan, the USA) and this considerably narrows the range of issues. Policy questions about market structure problems are found in every country and often generate similar solutions. What I am going to focus on are the factors which explain differences and see whether there is anything systematic about these differences. The main thrust is to explore how far strategic factors are more important than ideologies about capitalism/socialism/fascism.[1]

Energy, telecommunications and transport are the sectors under investigation. The chronological story starts with the railways, telegraph, coal, gas and water supply in the early nineteenth century which are then joined by electricity, tramways and the telephone in the late nineteenth century, oil and airlines in the early twentieth century and the technological revolution in telecommunications at the end of the twentieth century. Popular accounts see the changes over time and the differences between countries in terms of a strong commitment to laissez-faire in the nineteenth century, especially in the UK and the USA, and state intervention in the twentieth century in Europe culminating in privatization at the end. That approach puts a lot of emphasis on ideological stances and I want here to see how far strategic issues are an equally important element in the story and, indeed, to be able to answer questions like:

1. Why did the governments of France and Prussia intervene in nineteenth century infrastructure industries much more than the British?
2. How do we account for the different responses of Japanese and American governments to the development of railways?
3. Why did Scandinavian countries merge in 1950 into a joint airline, SAS, when most other European countries each had their own airline?
4. Why were telecommunications and airlines (and Britain) the first to privatize in late twentieth century?

Although the recent cross-border activities of Russian enterprises like Gazprom and Western European firms like E.On, Éléctricité de France and RWE have attracted the interest of some scholars (Schröter 2007), it would seem from a recent set of studies on the privatisation experience in Europe that strategic issues have still not yet attracted the interest of economists (Köthenbürger et al, 2006). Nor do some recent econometric studies properly address the issue. Bogart (2006) and Wallsten (2005) have examined nineteenth century telecommunications and railways. Their detailed econometric assessments of network spread in Europe under private and public ownership implicitly view the business–state relationship in the nineteenth century as similar to that in the late twentieth century, with Wallsten even saying his results supported World Bank philosophy. Also Baily (1993) compares labour productivity in airlines and other services in Europe and USA as if the only difference was a commitment to competition.

The historiography on this is fairly limited. Of course much has been written about individual countries and if one were to ask a historian of Japan or Austria-Hungary about early railway developments they would likely say that military factors were important. This would not be the case for the UK and the USA for reasons that will become clear shortly. Moreover, explicit international comparisons have been absent and that is my focus here.

IDEOLOGY, GEOPOLITICS AND TECHNOLOGICAL CHANGE

Comparing countries means, at a minimum, comparing different geographical positions and resource endowments, and secondly different polities (political structure, policies, and so on). For any specific shortish period (say 20 years), four issues are central, namely, (1) ideology, (2) geopolitics, (3) political structures and (4) perceptions of national identity.

Twenty years is proposed because the above issues have to be taken in the context of technological and economic conditions which will change over time. There will now be an explanation of what is meant by each of these categories, then three broad phases will be examined: the nineteenth century, the early twentieth century and, finally, the recent period of privatization.

Ideological differences have coloured much of the traditional literature – contrast, say the mid-twentieth century, between the free enterprise market dominated USA, socialist inspired Britain and France, and fascist Franco Spain. The equation of international differences solely with differing ideologies about economic performance under capitalism, socialism and fascism has carried the implication that big changes in business organization involving the state are strongly associated with a change in political ideology. The literature on privatization and nationalization has often picked up this theme and, in the case of Portugal's rapid shifts from one to the other in the 1970–90 period and Britain's ideological lead on privatization, such a role cannot be denied. Yet, as we shall see, one can argue that nationalization in the 1940s was hardly socialist in its outcome, and privatization was facilitated in no small way by technological changes in sectors like telecommunications, airlines and electricity. Indeed, heavy reliance on conventional distinctions between socialism, fascism and capitalism in the analysis of business–state relations runs into problems with countries like Japan (with its 'developmental state') and Italy where the major state enterprises set up from the 1930s (Istituto per la Ricostrizione Industriale [IRI] and Ente Nazionale Idrocaruri [ENI] held substantial shareholdings in many parts of the economy such that entrepreneurship and finance within the aegis of the state practically eclipsed the purely private sector (Kosai, 1995; Amatori, 2000; Toninelli 2004).

Turning to geopolitical factors, resources and services have strategic significance when (a) they are perceived to be a key input for national unification and/or defence and (b) they have few short-term substitutes. In the nineteenth century, continental Europe's land mass was politically fragmented by nation states, and the form of their strategic concerns contrasts strongly with Britain, Japan and the USA. Many continental European countries had hostile neighbours on their borders (think of France, Prussia and Belgium) which led to strong pressures to exercise controls over communications. As an island economy, economically advanced, well endowed with coal and a massive shipping fleet, Britain had few of these security worries and left its coal and railways in the private sector, electricity, gas and water to local determination and brought only telegraph and telephone under its direct control, reflecting Britain's pervasive interests in imperial and other international business matters (see later). Japan has always been heavily dependant on the Asian mainland and islands for raw materials, including oil. It has often felt geographically isolated and economically vulnerable. In the late nineteenth and early twentieth centuries it intervened decisively in shipping, shipbuilding and railways. The USA had no hostile neighbours on its borders (certainly from the late nineteenth century), foreign trade was important but much less so than for many European countries, and businesses could develop over a huge geographic area with a common language. McCraw's authoritative (1975) survey of the history of business regulation in USA is noticeably empty of any discussion of strategic matters. The internal battle with foreign interests in the south and west in the late eighteenth and early nineteenth centuries, as well as throughout the nineteenth century with Native Americans, was reflected in the Federal government's land grant policy. A rapid development of the West with low prices per acre and small allotments helped the armed forces establish strategic outposts. Once that had been accomplished, the 'internal' strategic concerns which bedevilled Europe were largely absent.

The third general factor is the political structure, which is taken as datum. The powers of the US federal government, on some interpretations, were significantly circumscribed by the constitution. The constituent states had rights and autonomous powers quite different from the constituent regions in the UK, France, Japan, Spain and, even, Germany. Business operations across state boundaries, such as railways and telecommunications, were relatively untouched by regulatory bodies until the 1887 Inter-State Commerce Act. In France the central government was traditionally more powerful, and circumscribed the power of municipalities in economic matters, a factor which largely explains the fact that much of electricity, gas and water supply in the period 1800–1920 was run by private enterprise under concession systems designed by the Conseil d'État (Fernandez, 1999). Regulation of business in Germany and Italy by the late nineteenth century was very much an administrative rather than legal matter (as in the USA). Finally, there is the question of perceptions of national identity. This issue clearly shades in with the others but it is useful to mention it explicitly since, as we shall see, the promotion of social and political unification strongly affected business–state relationships in the nineteenth century, especially for communications, and in the twentieth century is relevant for links with colonies and with the European Union (EU).

It is important, finally, to recognize that the form and relevance of the above factors will change over time, in particular as a result of changes in:

- technology (new forms of transport, information technology in telecoms industry, and so on) and
- economic forces (for example the rise in the size of the public sector 1950–70 in all countries).

THE NINETEENTH CENTURY

The key element in nineteenth-century continental Europe was the emergence of new nation states. Surface transport, the telegraph and coal supplies had potentially strategic significance for them (Belgium, Italy and Germany) and for the older existing states on their borders (France and Holland).

The telecommunications sector was the one area where there were few differences across countries, at least within Europe. The early Chappe flag-waving telegraph which originated in the eighteenth century, was deemed to be secure; not so the electric which frightened many a ruler and official. Thus telegraph development was often financed by the armed forces or other state sources and closely monitored. So also the telephone when it came in the late nineteenth century (and note the absence of these worries in the USA). Local networks gave way to national, and the desire to have a universal service prompted subsidies, reinforced in the case of countries like Sweden by aims for social and political unification. The basic technology had natural monopoly characteristics but the potential mixture of private monopoly and subsidies proved sufficient for arm's length regulation to be rejected, and by 1913 all the trunk networks in Europe were publicly owned. The security dimensions were reflected in the fact that telecommunications were run by government departments, albeit sometimes with separately identified budgets.

In transport, some of the vested interests of state and business arose from these sectors' basic economic characteristics – high fixed costs, natural monopoly conditions and the ability to discriminate in pricing. But since these were common to all countries they cannot, on their own, account for differences across countries. Business and the state were brought together because of rights of way, and everywhere fares and rates were regulated. That would have left the type of loose arm's length regulation found in Britain but how then to account for the very different positions for railway undertakings in continental Europe where relations between business and the state varied from the extensive use of subsidies and capital guarantees in France and Spain, state ownership of trunk lines in Scandinavia, to virtually complete public ownership in Prussia and Italy? The nation states were concerned to enhance social and political unification, ensure key routes were kept open and, in some cases, to directly control operations.

Thus, Belgium was a new state in 1830, Italy in 1860, while Norway and Sweden had scattered communities and fragile political structures. In all cases the state stepped in and, rather than subsidies, it was state ownership which dominated the form of intervention, and this stemmed primarily from the urgency attached to the construction of trunk networks with state operation (as in wartime) effectively bypassing normal market forces. In Italy, the state took over complete responsibility for all the national lines from

1885 and by 1906 the system had effectively been nationalized. They were organized not as companies under private law with government shares (like SNCF in France from 1937) nor as a public corporation (like British Rail in the post-1945 period), but rather as an 'azienda autonoma', an administrative unit under the direction of a government minister. The desire of the state to exercise direct control was so strong, in other words, that in Italy the railways took the form of the Azienda Autonoma delle Ferrovie dell Stato. Similar considerations explain the economic organization of the railway undertakings in Prussia following the unification of Germany in 1870. Faced with France and Russia as neighbours, Prussia not only nationalized its railways in 1879 but through-routes from east to west were sedulously developed in the period 1880–1913. High profits were not initially expected, though they did blossom into an important element of state revenues. Regulation was by administrative means (not by law as in the USA). The Deutsche Reichsbahn which was established by the Federal Constitution of 1919 was supposedly an autonomous enterprise but was strictly under ministerial control. Note finally that the French government planned the whole national rail network, subsidized capital and brought some key lines into state ownership (Dormois, 1999; Schram, 1997; Fremdling and Knieps, 1993; Miercejewski, 1999).

The fortunes of business in Japan in the second half of the nineteenth century were vitally determined by its geopolitical setting and also by the fact that in many key sectors there were few indigenous entrepreneurs and businesses. The geopolitical setting wherein Japan relied on the Asian mainland for many raw materials and viewed it as territory to be developed and controlled, meant that shipping and the shipbuilding industry would be of central importance. They were important in Europe and of course Britain was another island economy, but whereas Japan had only 210 sailing, steam and motor merchant ships in 1880, Italy, for example, had nearly 8000 and the UK 25 0000. The rail network was in no better shape. Only 283 miles had been built by 1883. So if Japan were to expand and dominate economically and militarily, speed in adjustment was essential and this put an onus on the state to push the private sector and/or activate state enterprises. The Sino-Japanese War of 1894–95 and the Russo-Japanese war of 1904–05 vividly drew attention to the need for development in these infrastructure areas. By 1913 the merchant shipping fleet was 1.5 million tons, 29 tons per 1000 population, only slightly less than Italy, Spain and Germany. It had become the third largest ship owner in the world by as early as 1919. At the turn of the century, one-third of all central government subsidies went to railways, a portion exceeded only by the allocations to shipping and shipbuilding. Many railway companies were then taken over and, significantly, the state undertaking, Japanese Railways, was, from 1908, under the control of the Railway Agency, which reported direct to the Prime Minister's office – and included railways in Korea and Manchuria. It became a Railway Ministry in 1920 and took all motorized transport under its wing in 1928 (Crawcour, 1997; Harada, 1993).

By the late nineteenth century the USA had, unlike Japan, already experienced sustained economic growth and needed no dramatic hasty development of its infrastructure to further military ambitions. The main issues were at sub-federal level. Much competition occurred on east–west routes where both states and municipalities were heavily involved. One recent estimate is that by 1871 the states and municipalities owned at least half of US railroad capital, notwithstanding that private investment dominated thereafter. Such competition was rarer in Europe. Parallel routes were for a time encouraged in the Netherlands and Germany, and duplication was allowed in Britain, but by the end of the nineteenth century competition was limited to multiple bidding for contracts. 'Opening up' the territory was not the game in Britain which, by the early nineteenth century, already had a transport network and population that was dense by contemporary standards. In other European countries strategic factors intervened, as we have seen, to generate more route planning and state support.

The American problem was that regulation was initially at state (not federal) level and yet inter-state traffic was the key growth element. For a large part of the nineteenth century (1830s to 1880s) the railway companies argued among themselves and with state courts and legislators about rate regulation. In the 1870s, four states passed the Granger Laws facilitating more intervention in inter-state rates but the railway companies challenged this in the federal courts. It was not until 1886 that the Supreme Court ruled explicitly that states could not regulate inter-state traffic and this prompted Congress to pass the Inter-State Commerce Act in 1887 (Dowd and Dobbin, 2001).

State intervention in infrastructure industries was therefore quite pervasive in the nineteenth century. Note that the state opted for direct ownership rather than subsidies when:

- it wanted the lines and services set up quickly (Belgium 1830s, Sweden 1850s, Italy 1860s) and private firms were proceeding too slowly. This is very similar to a war situation; or
- considerations of national defence and/or socio-political unification required lines which were not only unprofitable but needed such a large subsidy that it would have undermined confidence in the government (France, Italy and also Japan). Modern outsourcing and contracting might mean different possibilities; or
- the state wanted direct operational control of railways for military/security reasons (Prussia, Italy, Japan and all telecommunication).

EARLY TWENTIETH CENTURY

How far did these strategic factors carry over to energy, local utilities and the early twentieth century more generally? Two interesting developments occurred in the UK. The first shows clearly where the UK government's main worries were: the 'City' and the colonies. The telegraph was apparently very suitable for communicating financial transactions. In addition to the Post Office, the government supported the establishment of Cable and Wireless in 1928 to develop overseas networks. Then in 1938 the Treasury acquired £2.6 million. shares. In 1944 there was an agreement with all Commonwealth countries that all companies with international commitments should be publicly owned and in 1946 the Treasury acquired the remaining £27.4 million shares. Second, following the independence of Southern Ireland in the 1920s, what does the new Irish Free state government do? It creates a large number of state enterprises (well before anything comparable in Britain) in electricity, insurance and industrial credit, shipping, steel, aviation, dairies, sugar and surface transport (later). As part of the struggle for a 'self-sufficient and self-supporting state' said the Industry Minister, Lemass, the ailing sugar-processing company was taken over by the state and preferences given to Irish sugar beet producers.

In 1936 the Turf Development Board was established to develop Ireland's key energy source. Historians have seen all this as owing little to socialist theory but rather to a desire for the new nation showing its firm control over these key strategic sectors and declaring it will produce a much more effective economy, than the ailing body it had left. Even by 2006 electricity, gas, railways and water were still public sector monopolies (Barrett, 2006; 2007; O'Grada 1997).

Obviously, local utilities did not generate national security issues and the key point here is that the pattern of development was not much different across countries but again any differences were not because of issues like municipal socialism (on which see Millward, 2005, ch. 3). Many of the utilities were privately owned in the USA – 94 per cent of electricity power generation in 1932. Keller (1990) argues that any impetus for municipal ownership came from a desire to use utility profits to relieve local tax burdens. That seems the explanation in Europe where the more pervasive spread of operations by local government had been traditionally explained in ideological terms (municipal socialism), but since that post-dated the initial rise of municipal enterprise, the fiscal argument is more convincing. That argument does not carry over to water supply where municipal ownership was over 50 per cent in the USA and Europe, noticeably higher than the shares in electricity and gas. The municipal water undertakings were never run for profit and it seems that, faced with the transactions costs of meeting the rising quality standards demanded by ever more public health conscious municipalities and the rising costs of storage and distribution, private companies struggled financially and gave way to municipal ownership in both the USA and Europe (Jacobson, 2000).

One of the key transport developments in the early twentieth century with strategic significance was, of course, airlines. Given the economics of aviation, we might expect a competitive industry to emerge over large land masses and oceanic routes. Elements of such a picture can be seen within the USA in the 1919–39 period, and by the early 1950s it was carrying more passengers than the rest of the world combined. This was because the economic organization of airlines in Europe was very different and stemmed from some key geopolitical factors. Europe was a large land mass but it was spotted with nation states. Since airspace has a strategic value, sovereignty over it was established (in 1919) by each state whether large or small. Since air transport also had socio-political dimensions, at a minimum from servicing the colonies in Africa, Asia and South America and more generally in times of war, each nation state, in the 1919–39 period, protected its airlines by subsidy. Operating in many cases on fairly short domestic routes, these airlines struggled financially so that in the end one airline was allowed to dominate each country and it was publicly owned (see the histories of Air France, KLM, Lufthansa, Lot and British Airways in Dienel and Lyth, 1998). Why SAS? Because these countries had no pretensions to world power status and no colonies. Unable to reap large economies of scale, European airlines were likely to be out-competed by the American airlines. These issues lay at the heart of the inauguration of the fare-fixing body, the International Air Transport Association (IATA) in 1945. The American airlines were given fairly free rein on the Atlantic crossings in exchange for fares fixed elsewhere at levels that would allow the small national carriers to survive.

The first half of the twentieth century. ushered in two general problems for all countries' infrastructure sectors They are not directly relevant to a study of international differences but it is vital to show how they link with the rest of the story:

1. Structural – how to develop national networks (electricity, telephone) sometimes in the face of stubborn resistance from small-scale municipal and private enterprises.
2. Many of the services of all infrastructure industries came to be demanded by politicians, business and residential users as public services. There were pressures for 'low' and 'fair' prices. Often 'universal prices' emerged with similar rates per mile, per kwh, per gas therm, and so on independent of location and cost. Services like railways with such price structures became very vulnerable to the advent of the new competition from road transport.

In principle these problems in the UK and France could have been solved by the establishment of national or regional monopolies, privately owned but regulated by government, as emerged under the Thatcher privatizations of the 1980/90s. In the event, capitalist enterprise did not emerge well from the 1930s and the employers (especially in coal and electricity in the UK, see Hannah, 1979; Supple, 1987) often proved reluctant participants in schemes for reorganization and rationalization. So we find public ownership emerging in the 1940s on a wide national scale in the two countries, France and the UK, where it was least developed. The net result was that by 1950, railways, airlines, coal, electricity, gas and telecommunications were fully or partly government owned everywhere in Western Europe, even though public ownership had emerged in a variety of ideological settings – socialist, fascist, pro market. In Britain and France the rhetoric in the 1940s was certainly socialist but there was little nationalization of manufacturing (outside steel and Renault) and hence neither government could be said to own the 'means of production' (Millward, 1997; Moch, 1953). The nationalizations were concentrated on what the Labour government called the 'basic' industries – energy and transport, plus coal and steel. 'Basic' was never defined but they were central to national reconstruction in the desperate conditions of the late 1940s. So also was cotton and motor cars for exports but as long as the firms met their national targets they were able to escape nationalization (Singleton, 1995). In France the impetus to planning and the key role of state enterprise in those plans stemmed directly from a determination not to suffer a fourth German invasion. The plan gave priority to six 'basic sectors' coal, steel, electricity, transport cement and farm machinery. It demanded control over Germany's coal industry and the Commissariat du Plan was located in the Prime Minister's office (Hackett, and Hackett, 1963; Kuisel, 1981). After 1945 the French government was determined to reduce the country's reliance on the modest coal deposits located near the north-east frontier and hence exposed in the three German invasions since 1870. There was not only a large commitment to nuclear plants for electricity generation but coal prices were allowed to approach world levels and the coal industry to be run down faster than anywhere else in Europe. Chick (2002, 2006) has shown how that determination explains why the principles of marginal cost pricing (including the 'tarif vert' of Électricité) were more readily absorbed in France than in the UK. Finally, we should note that coal was the only industry in the UK where it could be said that the pressure to nationalize from mining unions and Labour Members of Parliament was clearly an expression of socialism and was influential.

So the main conclusions from this section are:

1. Much of the state presence by 1950 in several European countries can be dated back to nineteenth century strategic issues. In the twentieth century, airlines and overseas telecommunication were treated similarly.
2. Some of the state takeovers in Britain and France in the late 1940s were due to the immediate strategic demands of reconstruction and to the perceived failure of arm's length regulation in the inter-war period (you can call this ideological if you like but many writing in the 1950s saw it as 'technical' or 'economic').

PRIVATIZATION

This is a huge topic but here my aim is mainly to link it up with the past experience of the sectors in question. First, it is important to bring out why airlines and telecommunications were the first sectors to change. In the case of airlines, technology change and loss of colonies undermined the case for the national flag carriers. Many opportunities for air travel opened up after the war, especially when the jets came on stream, and the high-cost national carriers were under constant pressure. The emergence of inclusive tours, packaging flight and hotel, eventually, in the 1960s, proved too attractive to resist and the independent operators with charter flights were allowed to blossom. The logic of having one schedule carrier for each country, from the early strategic concerns, lost some force by the 1980s and paved the way for privatization and deregulation.

The telecommunications sector had not only been in public ownership since the late nineteenth century but had been deeply embedded in sections or agencies of government departments, with the employees effectively civil servants. They were, by the 1920s, often integral parts of Ministries of Post, Telegraph and Telephone. We noted earlier the British government's involvement with Cable and Wireless. In both Europe and the USA, the technology available up to the 1970s made for a fairly simple system, a long-distance network with natural monopoly features plus one type of terminal equipment, the telephone. Hence the telecommunications business had strong monopoly power, regulated at a distance in the USA, less concerned at it was by security issues than the European countries where the business was ensconced in government departments. The large element of fixed costs and the low variable costs provided an opportunity for universal service. This meant standard tariffs largely undifferentiated across users but implicitly high (relative to costs) for large business users and urban areas and low for residential customers and rural areas. This appealed to municipalities and central governments in Europe as well as to the Public Utility Commissioners (PUC) in the USA where, in addition, there was a constant battle about the attribution of fixed costs as between local and inter-state charges. Since changes in the volume of inter-state traffic had little impact on local fixed costs, this was effectively an attempt to get the inter-state business to subsidize local telephone exchanges. The PUCs welcomed the benefits to residential users, the simplified administration involved in uniform rates and the fact it seemed to be 'fair'. Like all network industries, such universal prices tended to favour certain groups. Business/urban users contributed much more to revenues and less to costs than residential/rural users (Vietor, 1989; Nguyen, 1988).

By the late 1970s, the distinction between communications and data processing was becoming blurred and once the possibilities of alternative long-distance networks arose and the simple handset came to be replaced by a complex terminal equipment market of computers, fax, email, and so on the case for dominant monopolies was undermined. Hence the 1980s saw the divestiture by AT&T of its 22 telephone companies which were reorganized into seven new regional Bell companies, the regulation of which required a review of the PUC's regulatory procedures (Temin, 1987). In Japan, the state-owned Nippon Telephone and Telegraph was privatized in 1985, albeit with 51 per cent of shares still in government hands and universal service retained. In Europe new telecommunications companies emerged as part of a threefold process: disentanglement of operations from being a section of a government department since the telegraph and telephone were no longer regarded as key sources of security concern; privatization of the new undertakings (Telecom Denmark, British Telecom, and so on); and deregulation of the markets. This was a complex process where ideological passions certainly affected the pace, while in other cases (Deutsch Bundespost) several parliamentary hurdles had to be jumped (Haid and Mueller, 1988). Such technology changes in telecommunications and airlines were mirrored in another sector, electricity supply, where, especially in the use of combined cycle gas turbines, technology changes facilitated the break-up of the generation market and was enhanced by the promotion of competition in the retail market.

Turning to the more general features of privatization, state budget deficits and similar economic factors were key elements, but these were common to all countries. It is clear that an important element of the privatization process was the disappearance, by the 1990s, of any major hangovers from the inter-war depression years 1919–39 in arm's length regulation of private natural monopolies. By the 1940s, that kind of solution was shunned but it became an acceptable alternative in the 1990s in Europe and Japan, manifest clearly in the privatization of natural gas and electricity transmission grids and of Japanese National Railways in 1987 and of British Rail (much less successfully) in the 1990s. The Japanese case is of interest because privatization and deregulation did not go hand in hand (Mizutani and Nakamura, 2004).

In all countries, the tension for state enterprise in the period 1945–90 between their public service obligations, on the one hand, and the requirement, on the other, to balance revenues and expenditures, is well documented. A central thread from that issue is that state enterprise financial losses were an important part of public sector deficits which plagued many countries in the Western world after the macroeconomic disturbances of the 1970s. The incidence of privatization can often be traced to specific financial crises in each country. Questions about security and about social and political unification were still important but a wider range of instruments were available and railways, telephone and airlines no longer seen as essential tools. Nonetheless the great technological changes of the last quarter of the twentieth century, the changing priorities in geopolitical strategies and the changing array of policy instruments have played an important element in the whole deregulation and privatization process. They are especially important in explaining the onset of privatization and deregulation in Europe, Japan and the USA in the last quarter of the century, rather than sooner or later.

It is interesting in this context to finally look at the incidence of privatization in Western Europe which suggests that national governments on the Continent were still, by 2006, much more strongly linked to their infrastructure industries by share ownership than the UK/USA (for detail see Köthenbürger et al. 2006; and Clifton et al., 2006). Thus telecommunications were still part owned in France, Germany, Ireland, Sweden,

Netherlands, Belgium, Italy. Railways were still fully state owned in most countries. State shares, sometimes golden, were still held in GDF, EDF, Air France, Endesa, Repsol and Iberia in Spain, ENEL and ENI in Italy, while state enterprise survived in Swedish power (Vattenfall), Spanish coal (Hunosa) and Portuguese airlines (TAP). There does, however, seem to be reduced usage of these sectors as instruments of strategic policy and we see a pattern of ownership and regulation more predictable from market forces. If one were to simply count the number of privatizations (a term which I use here to refer only to changes in ownership) then manufacturing, banking and commerce would dominate. Many of the firms were small. The UK was the exception to prove the rule with a larger proportion of enterprises in infrastructure industries like energy, telecommunications and transport where large firms dominate.

Manufacturing, commerce and banking are sectors without major scale economies and where competition might be expected to flourish. In the infrastructure industries, privatization has occurred mainly in the sub-sectors where the technology and economics would allow some element of competition to develop – electricity generation and retailing, airlines, airports, oil and gas exploration and refining. The grid networks, where natural monopoly conditions prevail, have seen the least privatizations with railways and postal services still generally in state ownership, governments retaining shares in natural gas and electricity grids, complete re-nationalization in Netherlands electricity and partial in British rail track. It is telecommunications where most privatization has occurred.

CONCLUSIONS

Strategic issues seem to account for much of the differences in the business – state interface in the nineteenth century with significant state interventions in continental European countries. For the UK government, it was overseas telecommunications that were important, joined in the early twentieth century by a dominant state-owned airline in Britain, and indeed in all other European countries. The struggle to develop national networks in electricity, telephones and railways together with the problems of capitalism in the 1930s, lead to increased regulation and state control. France and Britain witnessed widespread nationalizations in the late 1940s prompted by the regulatory problems, by the demands of reconstruction after the war and, in the case of France, by a determination to plan the development of strategic sectors against further military invasions. The outcome was that by the end of the 1940s, throughout Western Europe, airlines, railways, telecommunications, electricity, coal, oil distribution companies and gas supply were fully or partly state owned. By the end of the century however, changes in technology and in strategic considerations facilitated different market structures in airlines, electricity, gas and telecommunications, and with memories of the 1930s disappearing, the need to have state ownership of natural monopoly networks faded and gave way to arm's length regulation. Lingering strategic considerations seem to account for the fact that continental European governments still differ from the UK and the USA in holding on to shares (sometimes golden) in many sectors.

One final conclusion concerns performance studies (cf. Bogart, 2006; Baily, 1993; Wallsten, 2005). These authors do not seem to have recognized that in the nineteenth century, in several of the European countries where state ownership was present in railways and telecommunications, the state operated only trunk lines (Scandinavia, Italy, Belgium). Network spread is therefore not a test of the efficiency of state enterprises. More generally, given the strategic aims of many governments, it is inappropriate to compare public and private as if one was simply socialist inspired and the other driven by free enterprise and competition. In some cases, the true range of options should be different institutional settings for achieving specified security objectives.

NOTE

1. Some of these themes appear also in Millward (2008) but without the firm analytical framework which is pursued in this chapter.

REFERENCES

Amatori, F. (2000), 'Beyond state and market: Italy's futile search for a third way', in P.A. Toninelli (ed.), *The Rise and Fall of State Owned Enterprise in the Western World*, Cambridge: Cambridge University Press, pp. 128–56.

Baily, M. (1993), 'Competition, regulation and efficiency in service industries', Brookings Papers in Microeconomics, **2**, 71–159.

Barrett, S.D. (2006), 'Privatisation in Ireland', in M. Köthenbürger, H. W. Sinn and J. Whalley (eds), *Privatisation Experiences in the European Union*, CESifo Seminar Series, Cambridge, MA: MIT Press, pp. 265–88.

Barrett, S.D. (2007), 'Transforming air transport in Ireland', in J. Clifton, D. Fuentes and F. Comin (eds), *Transforming Public Enterprise in Europe and the Americas: Networks, Integration and Transnationalisation*, Basingstoke: Palgrave, pp. 117–30.

Bogart, D. (2006), 'Private ownership and the development of transport networks: cross country evidence for the diffusion of railroads in the 19th century', Conference on European Growth and Integration since the Mid 19th Century, CEPR/EU, Lund University.

Chick, M. (2002), 'Le tarif vert retrouvé: the marginal cost concept and the pricing of electricity in Britain and France 1945–73', Energy Journal, **23**(1), 97–116.

Chick, M. (2006), 'The marginalist approach and the making of fuel policy in France and Britain, 1945–72', *Economic History Review*, **59**(1), 143–67.

Clifton, J., F. Comin and D.D. Fuentes (2006), 'Privatising public enterprises in the European Union 1960–2002', *Journal of European Public Policy*, **13**(5), 736–56.

Crawcour, E.S. (1997), 'Industrialisation and technological change 1885–1920,' in K. Yamamura (ed.), *The Economic Emergence of Modern Japan*, Cambridge: Cambridge University Press, pp. 50–115.

Dienel, H.L. and P.J. Lyth (eds) (1998), *Flying the Flag: European Commercial Air Transport since 1945*, Basingstoke: Macmillan.

Dormois, J.P (1999), 'France: the idiosyncracies of *voluntarisme*', in J. Foreman-Peck and G. Federico (eds), *European Industrial Policy: The Twentieth Century Experience*, Oxford: Oxford University Press, pp. 58–97.

Dowd, T. and F. Dobbin (2001), 'Origins of the myth of neo-liberalism: regulation in the first century of US railroading', in L. Magnusson and J. Ottosson (eds), *The State, Regulation and the Economy: An Historical Perspective,* Cheltenham, LIH and Northampton, MA, USA: Edward Elgar, pp. 61–88.

Fernandez, A. (1999), 'Les Lumières de la Ville: L'Administration Municipale à l'Épreuve de l' Électrification', *Vingtième Siècle Revues d'Histoire*, **62**, 107–22.

Fremdling, R. and G. Knieps (1993), 'Competition, regulation and nationalisation: the Prussian railway system in the 19th century', *Scandinavian Economic History Review*, **41**(1), 129–54.

Hackett, J. and A. Hackett (1963), *Economic Planning in France*, Cambridge, MA: Harvard University Press.

Haid, A. and J. Mueller (1988), 'Telecommunications in the Federal Republic of Germany', in J. Foreman-Peck and J. Mueller (eds), *European Telecommunications Organisation*, Baden-Baden: Nomosverlagsgellschaft, pp. 155–80.

Hannah, L. (1979), *Electricity Before Nationalisation*, London: Macmillan.

Harada, K. (1993), 'Policy and railroads: transportation in the period of railroad priority (1892–1909)', in H. Yamamoto (ed.), *Technological Innovation and the Development of Transportation in Japan*, Tokyo: United Nations University Press, pp. 45–71.
Jacobson, C.D. (2000), *Ties that Bind: Economic and Political Dilemmas of Urban Utility Networks 1800–1990*, Pittsburgh, PA: University of Pittsburgh Press.
Keller, M. (1990), *Regulation and the New Economy: Public Policy and Economic Change in America 1900–1933*, Cambridge, MA: Harvard University Press.
Kosai, Y. (1995), 'Government–business relations and competitiveness: the Japanese case', in M. Hyung-Ki, M. Muramatsu, T.J. Pempel and K.Yamamura (eds), *The Japanese Civil Service and Economic Development: Catalysts of Change*, Oxford: Clarendon Press, pp. 188–212.
Köthenbürger, M., H.W. Sinn and J. Whalley (eds) (2006), *Privatisation Experiences in the European Union*, CESifo Seminar Series, Cambridge, MA: MIT Press.
Kuisel, R.F. (1981), *Capitalism and the State in Modern France*, Cambridge: Cambridge University Press.
McCraw, T.K. (1975), 'Regulation in American: a review article', *Business History Review*, **49**(2), 159–83.
Mierzejewski, A.C. (1999), *The Most Valuable Asset of the Reich: A History of the German National Railway Company*, Chapel Hill, NC: University of North Carolina Press.
Millward, R. (1997), 'The 1940s nationalisations in Britain: means of production or means to an end?', *Economic History Review*, **50**(2), 209–34.
Millward, R. (2005), *Private and Public Enterprise in Europe: Energy, Telecommunications and Transport c.1830–1990*, Cambridge: Cambridge University Press.
Millward, R. (2008), 'Business and the state', in G. Jones and J. Zeitlin (eds), *Oxford Handbook of Business History*, New York: Oxford University Press, pp. 529–57.
Mizutani, F. and K. Nakamura (2004), 'The Japanese experience with railway restructuring', in T. Ito and A.O. Krueger (eds), *Governance, Regulation and Privatisation in the Asia-Pacific Region*, Chicago, IL: University of Chicago Press, pp. 305–42.
Moch, J. (1953), 'Nationalisation in France', *Annals of Collective Economy*, **24**, 97–111.
Nguyen, G.D. (1988), 'Telecommunications in France', in J. Foreman-Peck and J. Mueller (eds), *European Telecommunications Organisation*, Baden-Baden: Nomosverlagsgellschaft, pp. 131–54.
O'Grada C. (1997), *A Rocky Road: The Irish Economy since the 1920s*, Manchester: Manchester University Press.
Schram, A. (1997), *Railways and the State in the 19th Century*, Cambridge: Cambridge University Press.
Schröter, H.G. (2007), 'When ugly ducklings grow up: cases from German utilities in energy, transport and telecommunications', in J. Clifton, D. Fuentes and F. Comin (eds), *Transforming Public Enterprise in Europe and the Americas: Networks, Integration and Transnationalisation*, Basingstoke: Palgrave, pp. 77–89
Singleton J. (1995), 'Debating the nationalisation of the cotton industry, 1918–50', in. R. Millward and J. Singleton (eds), *The Political Economy of Nationalisation in Britain 1920–50*, Cambridge: Cambridge University Press, pp. 212–33.
Supple, B. (1987), *The History of the Coal Industry: Vol. 4: 1913–46: The Political Economy of Decline*, Oxford: Clarendon Press.
Temin, P., with L. Galambos (1987), *The Fall of the Bell System*, New York: Cambridge University Press.
Toninelli, P.A. (2004), 'Between state and market. The parabola of Italian public enterprise in the 20th century', *Entreprise et Histoire*, **37**, 53–74.
Vietor, R.H.K. (1989), 'AT&T and the public good: regulation and competition in telecommunications, 1910–1987', in S. Bradley and J. Hausman (eds), *Future Competition in Telecommunications*, Boston, MA: Harvard Business School Press, pp. 27–103.
Wallsten, S. (2005), 'Returning to Victorian competition, ownership and regulation: an empirical study of European telecommunications at the turn of the century', *Journal of Economic History*, **65**(3), 693–722.

PART II

SECTOR STUDIES

7 Liberalization in the telecom sector

William H. Melody

INTRODUCTION

The liberalization process in telecommunication has evolved through several distinct phases of reform. It has proceeded at an uneven pace, not only across different regions and countries, but also in the implementation of specific policy reforms within countries. The main forces driving reform have varied among regions and countries, as have the particular methods of regulation adopted in attempting to implement liberalization policy objectives. But the fundamental policy goals of telecom liberalization have been remarkably similar, differing primarily in the priorities given to particular policy goals in the overall liberalization agenda. The general, although far from universal, positive results from early experience in the USA and the UK, and the strong advocacy from these countries and international agencies for liberalization in other countries, has stimulated the export of a variety of liberalization policies, institutional arrangements and regulatory practices to most other countries.

The basic principle underlying liberalization reforms has been opening access to telecom networks that had previously been monopolized to the detriment of participation of new players. The essential policy objectives have been to foster the extension of networks for greater coverage (universal access), to stimulate the introduction of new network technologies and services, and to introduce competitive efficiencies in the provision and pricing of services. But liberalization has been a far more complex and much slower process of institutional change than was initially anticipated. The policy issues at each stage of the reform process have been hotly contested among incumbents, new entrants, equipment manufacturers, large users and other interests. Economic theories, concepts and research studies have played a central role in the debates over liberalization policies and regulations, providing ammunition for contending interests.

Liberalization has involved considerable trial and error, especially as policies have been implemented by complex regulation under conditions of incomplete information and market uncertainty in a rapidly changing environment. Liberalization has been a learning-by-doing experience at each stage of the reform process, with countries learning both from their own good and bad experiences as well as those of other countries. After more than a generation of reform, the liberalization process is far from complete, but it has evolved to the point where policy-makers, industry players and researchers are beginning to consider the limits of liberalization and the conditions necessary to prevent a retreat from liberalization in some areas.

This chapter provides an overview of the foundations of telecom liberalization, the experience to date and the outstanding policy issues. This includes the introductory stage of the reform of spectrum management and the initial policy and regulation changes that provided the foundation for the Internet. Given the explosive growth of mobile communication and the Internet, and the many specific follow-on liberalization policy and regulatory issues that have arisen relating specifically to spectrum management and Internet development, separate chapters examine developments in these areas.

THE 'NATURAL MONOPOLY' HERITAGE

After the Bell patents expired in the USA in 1895, there was a 20-year period of vigorous competition in telephone network development. But political concerns about destructive competition, monopolization, refusals to interconnect, and a desire for a uniform public service led to the adoption of the *public utility model of a regulated private monopoly* – AT&T for the most populated parts of the country and smaller geographic monopolies for the rest (Gabel, 1969; Mueller, 1997). Many other countries had a degree of competition in the initial network development phase, but most countries soon adopted the monopoly model, generally placing it under the post office, creating post-telegraph-telephone (PTT) administrations.

The economic theory justifying the telecom monopoly policy was developed *after the fact,* a rationalization for the industry structure that had been advocated by AT&T and adopted as US government policy at both state and federal levels. Telecom was a 'natural monopoly' characterized by continuously declining unit costs of supply for a single network operator. These scale economies were technologically determined by the requirements for a coordinated single network and heavy front-end investments in telecom infrastructure. Thus economic efficiency would be greatest with a single supplier, and the end result of a competitive market would be a monopoly, as early experience in the industry seemed to have demonstrated.

Moreover telecom had quickly developed to become a public service, like the telegraph and post, which could best be provided by a monopoly, and in most countries by the government monopoly administering post and telegraph where, what has been later conceptualized as economies of scope, could be realized. This rationale provided not only the economic justification for geographical monopolies, but also for the formidable policy and regulatory barriers to entry that were established to preserve and in some cases extend the reach of the monopolies.[1]

Periodically an independent party would come up with a good idea for improving telecom service, most commonly an enhancement to the telephone, but these were summarily rejected as unlawful attachments. The natural monopoly policy guided telecom industry development for more than half a century until it was successfully challenged, first in the USA beginning in 1968, and then in other countries.

It is perhaps ironic that the monopoly policy was challenged first in the USA where universal access had been virtually achieved, AT&T was regarded as the best telephone operator in the world, and its Bell Labs was the scientific and technological leader. Prior to the era of reform, only a few other countries provided any semblance of universal access and most PTT provider countries offered inefficient service with potential customers on multi-year waiting lists. But in the USA the then unique institutional structure of an independent industry-specific regulator of a private monopoly made it more accessible to considerations of policy change, and AT&T's technologically up-to-date network attracted the interest of firms in neighbouring industries with ideas on how they could benefit from, or contribute to, this network if certain monopoly restrictions were liberalized.

TELECOM REFORM IN THE USA: 1968–84

Early Challenges at the Boundaries of the Monopoly

By the early 1950s the AT&T monopoly had grown to include not only the transmission and distribution network in the highly populated areas served by AT&T, but also the terminal attachments to the network (essentially telephones), the services provided over it, the equipment used in the network, most of which was manufactured by its Western Electric subsidiary, and the Bell Labs which did research and development for AT&T and the federal government. It included the telephone books and even their cover jackets. AT&T received about 85 per cent of industry revenues, but served only about 15 per cent of the land area. The rest of the country – the rural and less populated areas – was served by more than 10 000 local private, municipal and cooperative operators. Their interconnection 'privileges' were regulated by AT&T. From 1934, the Federal Communications Commission (FCC) regulated 'interstate and foreign' communication, which included part of AT&T's operations, but the majority was regulated by state public utility authorities, established between 1907 and 1920, with jurisdiction over local and intra-state communication.

With the advent of radio microwave technology for radar during the Second World War, video transmission for television broadcasting became an important new service after the war. AT&T decided that its voice telephone network monopoly extended to the new video network facilities and transmission services in the areas it chose to provide them, but it did not try to extend its monopoly geographically or to the provision of television broadcast services. Western Union, the telegraph monopoly then in decline, constructed microwave links between Washington, Boston and Pittsburgh. AT&T refused to interconnect and the FCC supported its decision on the basis of AT&T's claim of superior network and spectrum efficiency with a monopoly. The FCC noted that an AT&T objective was 'to discourage development of competition in the field' (FCC, 5 RR 639).

Railroads were permitted to provide communication networks essential to the safety and operation of rail transport, but their non-essential administrative traffic had to be purchased from AT&T. Interconnection with AT&T was by private contract. With new transmission equipment being developed using the new spectrum resource, the railroads and other 'right of way' industries (for example, pipelines, electricity, trucks) applied to the FCC for frequency licences. They wished to build their own private microwave networks so as to provide their respective industry communication requirements on a shared basis. In its *Above 890 MC* decision in 1959,[2] the FCC permitted this limited use for this limited class of industries. AT&T responded by introducing more favourable interconnection arrangements and dramatically reduced private (leased) line service prices for these special industries to reduce their economic incentive to construct private networks.[3]

Terminal Interconnection

Over the years there were many terminal devices developed outside AT&T, but they were considered 'foreign attachments' by AT&T, unacceptable as part of the AT&T

network, although some of the small local 'independent' operators either purchased them or allowed customers to attach them to the network. The regulators generally considered this to be a management decision, but occasionally they had to rule on formal complaints.

The idea of connecting terminals that required a direct electrical connection to the AT&T network was unthinkable, as AT&T claimed it would introduce an unidentifiable risk of technical harm to the entire network. However, the Hush-A-Phone attachment was a simple, cuplike device that snapped on to the telephone and confined the speaker's voice within its enclosure, reducing noise and providing privacy. In 1949 the FCC accepted AT&T's claim that it could affect the quality of service throughout the network, but was overruled by the court on appeal.[4] In 1954, the FCC supported AT&T's claim that there had been no demonstrated demand for attaching recording devices to phones, so interconnection should not be permitted.[5]

Terminal interconnection as a policy issue, rather than a matter involving only a specific terminal device, was addressed by the FCC in the Carterfone case. The Carterfone was an acoustic coupling device developed by Tom Carter that would connect phones from the private networks of railway, oil, gas, timber and mining companies in remote areas not served by AT&T, to the AT&T network. It was a cradle where two phones were set side by side, matching the speaker on one phone with the receiver on the other. When AT&T threatened to terminate service to users of the Carterfone, Carter brought a private anti-competition case to the court, which asked the FCC what its policy was. So the FCC had to come up with an interconnection policy.

The Carterfone decision, 27 June 1968,[6] established the principle that terminal interconnection should be permitted 'so long as the interconnection does not adversely affect the telephone company's operation'. Although it took some time to overcome the barriers AT&T threw up to slow down implementation of this decision and narrow its scope, it signified a major change in policy direction that provided for the participation of other players in the provision of customer premises equipment – the beginning of liberalization. Thereafter the policy debate focused on the justification, terms and conditions for liberalization in other areas that had been historically protected by AT&T practice and regulatory indulgence.

Network Competition

Radio microwave technology lowered the barriers to entry for constructing transmission networks. They could be constructed much faster than cable networks, required much lower initial investment, and proved economical at lower volumes of capacity and traffic. The major railroads and other right of way companies built private networks to enable better management of their services. Television transmission companies were established to deliver television signals in the areas not served by AT&T. A small new company, Microwave Inc (MCI), applied to the FCC for construction permits to establish point-to-point common carrier communication service between Chicago and St Louis using radio microwave to provide leased line services. After an extensive administrative hearing, in August 1969, the FCC approved the application, acknowledging that MCI would be competing with AT&T with its limited differentiated service, but with a probability of substantial public benefit.[7]

Following the MCI decision there was a groundswell of applications for permits to build similar systems all over the USA. It included an application by a new firm, the Data Transmission Company (DATRAN) to build a nationwide digital data network. The FCC was forced to establish a 'specialized common carrier' policy for treatment of all the applications.

Extending the basic principle of its Carterfone decision, the FCC concluded that applications would be approved unless there was evidence of potential harm to the public interest. Soon thereafter the new policy was applied to the licensing of another new spectrum-based technology, domestic satellite networks. These liberalization policy changes opened the leased line services market to potential competition, but the basic public service market which provided 85–90 per cent of AT&T revenue, was still protected. Nevertheless 10–15 per cent of the revenue of the then largest company in the world represented a significant market that was both growing and benefiting from opportunities associated with new technologies and service enhancements (Melody, 1972).

Services Liberalization: The FCC Computer Inquiries

The earliest signs of potential ICT convergence were recognized by the FCC in 1966 when it undertook a formal 'Inquiry into the Regulatory and Policy Problems presented by the Interdependence of Computer and Telecommunications Services and Facilities'. The FCC concluded in its 1972 decision that 'data processing cannot survive, much less develop further, except through reliance upon and use of communication facilities and services'.[8] It noted special issues of computer terminal interconnection (modems interfacing with the network), network technical standards (new digital data versus existing analogue voice), and formulating a liberalization policy relating to the supply of data and information services over AT&T's public network by service providers from the computer and other industries. But as it was virtually impossible to separate data traffic from voice traffic on the public network, it was impossible to implement a liberalization policy for public data services while retaining a monopoly policy for public voice services (Irwin, 1969).

The FCC continued to examine these issues throughout the 1970s in its Computer II Inquiry. Modems for data terminals came to be treated as other terminals with routine equipment certification. AT&T had begun the long process of gradually converting its network from analogue to digital standards. But the problem of separating AT&T's data from its voice services still could not be resolved satisfactorily despite attempts at technical, organizational and accounting separation.

A partial resolution of this problem came in an FCC decision in 1976 allowing the resale and shared use of common carrier facilities and services.[9] This liberalized the retail market for voice as well as data services. Related to this was a court decision interpreting the scope of the MCI licences that concluded (contrary to FCC intentions when it authorized the licences in 1969) there were no restrictions on MCI's rights to offer any communication service. These decisions meant that in effect all services were liberalized and could be provided over the facilities of any network. They created the foundation for the development of services markets untied to the provision of network facilities. The development of enhanced or value-added services, and later the Internet, now became possible.

But there remained the problem of AT&T's monopoly power in competing against the new providers of terminal devices and voice, data and combined services. AT&T's monopoly power could be expressed by offering competitors inferior technical quality, insufficient network capacity, high prices and unresponsive service. The FCC needed to be able to monitor these issues to ensure the new liberalized equipment and services markets, and early-stage computer/communication convergence, developed efficiently. In its 1980 Computer II[10] decision the FCC recognized that these and other issues required further examination. The Computer Inquiry was continued with a third inquiry which was finally completed in 1986,[11] after the break-up of AT&T, and helped prepare the network foundation for the development of the Internet.

The Break-up of AT&T

Liberalization did not arise from a policy decision to promote competition, but rather from a reassessment of the justification of specific monopoly restrictions that prevented other firms from participating in the industry. The FCC was a reluctant reformer in that it only began to confront the main issues when asked by the court what its policy was. It also had enormous difficulty trying to implement its new policies in the face of AT&T's formidable resistance.

After the Carterfone decision, it took eight years to establish a procedure for routine equipment certification, as existed in many other industries. MCI was licensed in 1969, but did not offer significant service until 1974. Its early revenue was primarily from court settlements over AT&T anti-competitive activities. It was jokingly known as a law firm with a telecom subdivision. The Data Transmission Company (DATRAN) was unable to negotiate an interconnection agreement with AT&T. Just before it was to receive its main tranche of investment funds for network construction, AT&T filed a tariff for a new service, Data under Voice. It had discovered that in its voice circuits there was unused frequency space below the voice signal that could be used for data. So it already had a nationwide data network. DATRAN lost its funding and closed. Soon after, AT&T withdrew its Data Under Voice tariff.

AT&T's systematic practices of delays, disruptions, misrepresentations and inferior quality interconnections, followed by price squeeze and predatory price reductions, put most potential competitors out of business before they could get started and slowed the development of competition to a snail's pace. FCC decisions requiring that one anti-competitive price or practice be stopped were immediately replaced with others, in some cases even more anti-competitive. Private anti-trust cases were begun by almost all of AT&T's competitors that could afford it. MCI won the first major case, and AT&T negotiated financial settlements in many others. It was clear that for AT&T the damages awarded to, or negotiated with, its victims were far less than the enormous benefits from destroyed competitors and restricted competition.

After monitoring industry developments throughout the 1970s, the Anti-trust Division of the US Justice Department took AT&T to court charging abuse of its monopoly power and seeking to break up the company as the only effective solution. After years of delay and an extensive hearing, AT&T negotiated a settlement rather than risk a court judgment. It would be broken up into seven regional Bell operating companies (RBOCs), with AT&T retaining its national and international networks and services, its manufacturing affiliate and the Bell Labs. When the break-up went into effect on 1 January 1984, it was anticipated that this would establish a liberalized industry structure that would allow competition to drive its future development (Brock, 1994).

THE ECONOMIC DEBATES: THEORY AND APPLICATION

Academic economists and their concepts, theories and applied research played a central role in the policy debates over liberalization. In the course of the reform period reviewed here, AT&T sponsored theoretical and applied research at Bell Labs, employed more than 100 academic economists, and more than 200 small consulting firms, many run by leading academic economists. Most economics departments at leading universities in the USA were represented (Owen and Braeutigam, 1978).

The FCC staff of a half-dozen economists, and later the Department of Justice (DOJ) Anti-trust Division with even fewer assigned to the AT&T case were supported by a handful of specialist consultants. The work by the AT&T economists was virtually all special applications or extensions of static neoclassical theory. The work for the FCC and DOJ was all within the dynamic paradigms of political economy as developed in institutional, evolutionary and Schumpeterian economics. As debates on these economic and policy issues continue today in many fora throughout the world and in the academic literature, the key elements are summarized briefly.

Policy Protection for the Natural Monopoly

The basic thesis for defending the AT&T monopoly rested on two interrelated propositions, presented as universally accepted and uncontestable facts. First, telecom networks are characterized by economies of scale that have no limit. This makes AT&T a natural monopoly. Any competition would have to be at a much smaller scale, be relatively inefficient and unable to survive. It would be a waste of resources to allow competitors to enter the industry.

Yet some entrepreneurs would see opportunities for profitable entry to particular parts of AT&T's market because AT&T did not price its services on a cost basis. The second proposition was that AT&T employed a uniform pricing policy for its services and subsidized the basic local and long-distance public services, including the telephone handset, in implementing the government's universal service policy. Thus, any proposals for new entry must be 'cream-skimming' the more profitable services and markets that generated the extra profits necessary to subsidize the high cost services and users. Should competitive entry be allowed, then AT&T would not be able to continue its social cross-subsidy that was essential to implement the universal service policy.

Beginning in 1966, FCC economists began to test these propositions in relation to new entry proposals and other evidence (Melody, 1971b, 1975). The evidence indicated that the Carterfone, answering machine, hush-a-phone and other terminals would provide benefits to users and to AT&T as increased use of its network would be stimulated.

AT&T produced a mammoth engineering-economic study demonstrating its major network economies of scale (Mandanis, 1976). FCC economists showed that proposed new entrants MCI and DATRAN had significantly different production functions derived from different applications of the technology, construction methods, proposed service offerings and organizational structures, resulting in totally different cost functions with much lower unit costs than AT&T even at much smaller volumes of output.

It was also demonstrated that economies of scale in telecom networks are not internalized to a single firm. They are external economies of the network achieved by interconnection and coordination among the participants providing it. Competition need not change that, and it could be the catalyst for generating improvements across the entire network.

AT&T's social cross-subsidy proposition was suspect as AT&T had declined to serve the rural, remote and high cost areas of the country. Independent studies of prices and costs for local exchange service and the basic residential handset showed that they were not being subsidized, but rather producing high profits being used to subsidize terminals and services for industrial users. Studies of the basic public inter-state long-distance service showed that it was subsidizing AT&T's leased line services, those that would be most competitive with the proposed services of new entrants. The conventional wisdom was wrong on every point. The subsidies ran in the opposite direction to that claimed by AT&T. There were no grounds for a policy decision to prevent entry.

Anti-competitive Cross-subsidy

However, there was reason to be concerned about predatory pricing and anti-competitive cross-subsidies from monopoly profits under the prevailing system of profit regulation. The FCC regulated AT&T's rate of return on investment in providing inter-state and foreign services. Thus, the greater the investment, the higher AT&T's allowed total profit. This provided an incentive to 'gold-plate' the network and to prefer capital-intensive technologies and biased capital/labour trade-offs in expansion decisions (Averch and Johnson, 1962).

Within this framework of overall profit regulation, AT&T could choose to subsidize services that were vulnerable to competition with its protected monopoly services without suffering any loss of its allowed profits. In fact, it had a powerful incentive to do this as a key element of a profit maximizing strategy. The first comprehensive study ever done of AT&T's rate of return by major service classification showed that AT&T's monopoly services realized excessive returns and its potentially competitive services realized very low or negative returns, while overall AT&T realized its allowed return (Melody, 1971a).

The Commission economists concluded that it would have to regulate the maximum rate of return for AT&T's basic public service monopoly, and monitor those services subject to potential competition to prevent anti-competitive cross-subsidies and predatory pricing. AT&T then turned to leading neoclassical economists to develop theories and costing approaches that would preserve its pricing freedom.

AT&T justified its very low prices to discourage competitive entry as reflecting marginal cost pricing in a network with significant economies of scale. The user industries receiving the benefits of AT&T's low prices also discovered marginal cost pricing. The FCC was presented with a veritable chorus of leading economists singing static, textbook, neoclassical market theory as justifying AT&T's discriminatory pricing favoring large industrial users.

The FCC economists rejected static neoclassical theory and marginal or long-run incremental cost (LRIC) as appropriate guidelines for regulating AT&T under dynamic market conditions where changes in technologies and services, and proposals for new entry were both evident and imminent. They cannot provide an objective standard for regulating AT&T prices as they are based entirely on subjective forecasts by AT&T that can neither be audited nor compared with actual costs. Moreover, AT&T presented no information to suggest it ever used or intended to use these concepts in making management decisions.[12] It was concluded that new entry could facilitate both consumer protection and universal access policy objectives.

LRIC and Ramsey Pricing

Despite the evident weaknesses that had been demonstrated, AT&T put forward long-run incremental cost as the appropriate regulatory standard by which its low prices should be judged. The remainder of its revenue requirement would be covered by higher prices for its protected monopoly services. Princeton economist Will Baumol came up with the theory of Ramsey pricing to justify this. Optimal pricing under conditions of rate of return regulation and economies of scale requires that prices for different services be set by mark-ups from marginal cost in inverse relation to their demand elasticities (Baumol and Bradford, 1970). Thus, low prices for competitive services and high prices for monopoly services are not only justified, but required to achieve 'second-best' efficiency.

Years later, long after the dust had settled on these issues, Baumol spelled out the full implications of his proposal in his article on Ramsey pricing in *The New Palgrave* (1987):

> That theorem states that if a monopolist who is constrained by a regulatory (or other) profit ceiling chooses to adopt the Ramsey price vector rather than some other set of prices that enable him to earn his allowed return, then under a fairly attractive set of assumptions the monopolist will be rewarded for his virtuous decision by being protected from entry by those prices. In other words, self-interest may impel a monopolist to adopt Ramsey prices because those prices are sustainable against entry, meaning that at those prices the monopolist will earn the profits that the constraint allows to him, but any rival firm that undertakes to enter the field will be predestined to lose money even if the incumbent undertakes no strategic (retaliatory) response. (p. 51)

With respect to its implementation of Ramsey pricing, AT&T was unable to provide the FCC with its market forecast data (too speculative) or its demand elasticity studies (too uncertain). Its LRIC studies were based entirely on unverifiable forecasts, reductions in the valuation of existing facilities justified by 'engineering judgement', and the omission of common costs shared with other services. AT&T never applied Ramsey pricing, but it has fostered a steady stream of neoclassical literature and justifications for monopoly pricing in the utility industries that continues to this day.

Theoretical Innovation in Defence of Monopoly

By 1972, in its Carterfone, MCI, Specialized Common Carrier and Computer Inquiry decisions, the FCC had prepared the ground for a new era of liberalization in telecom.

It had withstood the challenges from AT&T and its army of neoclassical economists. AT&T then attempted to reverse the tide of change by reasserting its case for monopoly in a major paper from its CEO that was distributed widely to politicians and others (de Butts, 1973). It claimed the FCC liberalization decisions were an attack on the 'common carrier principle' of a public interest monopoly dedicated to serving everyone on an equal basis. Competition would destroy this public service and raise costs and prices as network efficiency and social cross-subsidies were lost to cream-skimmers. The FCC decisions should be repealed.

AT&T also turned its neoclassical economists at the Bell Labs, Princeton and other leading universities loose on the problem to find some theoretical support to justify its monopoly. Throughout the 1970s and early 1980s a number of innovative theories were developed including:

1. For a policy reversal back to monopoly protected by regulation the 'sustainability of natural monopoly' theory showed that unrestricted entry in a multiproduct natural monopoly could set in motion the unraveling of economies of scale and scope and the loss of efficiencies (Baumol et al., 1977).
2. Then, as a justification for potential rather than real competition, 'contestable market theory' showed that real market competition was not necessary to achieve efficient resource allocation. The potential threat of competition was sufficient to force optimal pricing by the monopolist (Baumol et al., 1982).
3. For protecting monopoly prices and profit after entry has occurred, the 'efficient component pricing rule' (ECPR) was developed as a standard for determining interconnect prices that new competitors should pay incumbent monopolists (Baumol and Sidak, 1994). They should be based on the opportunity cost to the monopoly of losing business to the competitor, that is, cost plus the profit the monopoly would lose if the competitor took the business away. The efficient component pricing rule has been widely adopted by incumbent monopolists around the world in proposals to regulatory authorities.

For economists and their theories, the near 20-year struggle to liberalize telecom markets exposed some puzzling ironies. Leading neoclassical theorists actively participated in the support of the monopoly, in an attempt to reverse the competition policy, in the defence of AT&T's anti-competitive practices, and in developing special purpose theories to justify monopoly as competitively efficient. They never supported the liberalization of these markets before the FCC, the courts or in the economic literature – a strange role for a school of thought based on competitive markets as the optimal condition for efficient resource allocation.

It was political and institutional economists, traditionally associated with greater recognition of non-market and social policy objectives, that made the running for the introduction of a liberalization policy as promoting both efficiency and social policy goals. They provided no new neoclassical theories for optimal resource allocation, but rather provided applications of the economists' tools of analysis to dynamic issues and institutional change so as to develop policies and regulations appropriate for a new economic environment.

TESTING AND SPREADING LIBERALIZATION: 1984–98

The period following the break-up of AT&T was one of testing the new more competitive industry structure in the USA and the gradual spread of liberalization to other countries. Competition in the fixed network segment of the industry developed slowly with continuing battles between new regional monopoly incumbents (the RBOCs) and new operators before regulatory authorities and the courts. The major beneficiaries of liberalization were the neighbouring equipment, computing and electronics industries and the new telecom industry segments, mobile communication and network services.

Opportunities for Neighbouring Industries

The primary beneficial effect of the AT&T break-up was freeing up the RBOCs, as well as consumers, large and small, to buy their telecom equipment – ranging from basic telephones to corporate data networks and telecom operator cables, switches and radio equipment – from any supplier. Equipment manufacturers from the computer and electronics industries joined independent suppliers in the telecom industry in providing digital equipment for the gradual redesign and upgrading of the telecom network. Personal computers (PCs) could be connected to the network and data services introduced. Information and communication technology (ICT) convergence began to have a major impact on network development and the Internet was introduced.

When first licensed in the late 1980s, mobile phone service was seen as a specialized service with a limited market, not a competitor to fixed network voice service. In the USA and many other countries, two licences were awarded initially, one to the incumbent fixed line monopoly and one to a new operator. But as the technology improved (especially handsets) and the mobile market grew much faster than expected, additional licences were granted. Policy and regulation in mobile had to address old issues inherited from the monopoly era (especially interconnection), and new issues associated with the new technology (especially spectrum management policies).

New satellite networks enabled the extension of communication services to many previously unserved areas, provided capacity for the transmission of television signals throughout the world, and stimulated major price reductions for international communication as they competed with upgraded undersea cable networks. But they had a very limited impact on domestic network services competition.

Experience with Liberalization in Network Services

Throughout the 1980s, performance indicators showed positive results for increasing competition, reduced consumer prices and greater consumer choice in the inter-state long-distance voice communication market. AT&T's market share declined to about 65 per cent and MCI, Sprint and Worldcom grew to become significant players. Hundreds of voice service retailers and enhanced service suppliers came into the market. Competition was expanding significantly in the market for US-based international services.

But the FCC competition policy did not prompt a spread of significant competition to intrastate services, and virtually not at all to local services where the RBOCs vigorously protected their monopolies over access to customers. Only a few states were promoting competition. Many were resisting it. Under the Reagan administration, the US government's conception of both competition and regulation changed dramatically. Competition was no longer seen as a tool for achieving regulatory objectives. Rather the 'free market' was seen as a substitute for regulation. Policy shifted from liberalization from monopoly restrictions to liberalization from regulation. Most distinctions between competitive and monopoly markets lost significance. Through Reagan era appointments to commissions, utility regulation was considerably weakened at both the federal and state levels (Melody, 2002).

By the early 1990s, the positive trends towards increased competition in the telecom network sector began to falter and in some cases were reversed through mergers. Although the long distance operators (now an oligopoly), RBOCs (regional monopolies) and cable television companies (national firms with local monopolies) were making threatening noises about wanting to move into one another's markets, none did. The movement to competition in the telecom sector was gravitating to a 'tight oligopoly' structure and local monopolies remained (Trebing, 2004).

The Telecommunications Act of 1996 attempted to provide a basis for a major surge forward in competition, especially at the local level where the RBOC monopoly prevailed. It liberalized all telecom markets and specified detailed conditions to promote entry opportunities at several different possible levels within the monopoly local exchanges: as resellers of services; direct interconnection of networks; or access to any combination of unbundled network elements. The FCC was instructed to develop specific implementation standards, and the state commissions were to facilitate implementation of the new competition policy at the state level.

Liberalization Spreads to Other Countries

Liberalization in the European Union (EU) and most other countries required a much more radical set of policy changes than the USA. National policy commitments had to be made to undertake major institutional changes to facilitate liberalization. Most national telecom monopolies were government administrations. Some were jointly provided with postal service. The initial steps typically required that the telecom administration first be established as an independent commercial organization within the government, and then privatized, in whole or in part.

It was also necessary to establish an independent regulator to implement the new government liberalization and consumer protection policies in countries with no tradition for using independent regulatory agencies. In many countries, powerful government trade unions and ministers of finance who benefited from the established monopoly structure had to be convinced of the merits of this major institutional restructuring. Licensing competitors was the easier part of this package of institutional changes. This process generally took some time to implement.

Many countries had been monitoring US telecom market liberalization developments and began to move in the same direction. Canada cautiously followed some US liberalization policies a decade or so later. In 1984 the UK privatized its national telecom operator (British Telecom [BT] and established an independent telecom regulator (then OFTEL, now OFCOM). It experimented with a duopoly structure by giving Cable &

Wireless a domestic licence, but its subsidiary Mercury failed after a few years as a result of interconnection and other problems with BT under passive regulation.

In 1987 the European Commission (EC) issued a Green Paper announcing its intention to begin a process of telecom liberalization as an essential step for promoting European markets. As many EU countries were not enthusiastic supporters of the new policy objective, it was introduced by 'progressively lifting restrictions on Europe's telecom infrastructure and services' over 11 years until all remaining policy restrictions on public networks and voice telephony were removed (EC, 1998). However, implementation required that the policies be transposed into national legislation and enforced by new national regulators, which was done at varying speeds and degrees of enthusiasm. In 1997 an Independent Regulators Group (IRG) was established 'to share experiences and points of view' on the new regulatory issues. [13]

Throughout the 1990s mobile voice communication grew rapidly in all countries once it had been introduced. Global leadership in the field was taken by Europe as a direct result of an industrial policy decision by 13 European countries to adopt a common second generation (2G) technical standard for the first digital cellular mobile voice networks. The Global System for Mobile Communication (GSM) standard launched European equipment manufacturers, led by Nokia, to global leadership, as it was adopted throughout most of the world, but not in North America, and mobile penetration rates in Europe raced ahead of other countries (Gruber, 2005).

New Zealand, Australia, Hong Kong and a few other countries around the world began to make liberalization policy changes. By now the primary attraction of telecom reform policy was liberalizing the mobile market which offered significant growth opportunities (FCC, 'First Report and Order'). Liberalizing fixed networks and reforming incumbent monopolists had been shown to be a much tougher proposition.

LIBERALIZATION GOES GLOBAL: 1998–2010

In many respects 1998 marked a turning point in the telecom reform process taking place throughout the world. The US Telecommunications Act of 1996, which liberalized local markets (thus finally bringing full liberalization in the USA), seemed ready for implementation following an FCC ruling establishing the guidelines for implementation. The European Union (EU) policy on full liberalisation of telecom services in member states came into effect. The World Trade Organization (WTO) recorded an agreement on telecom services by which 69 countries representing 90 per cent of global telecom traffic committed themselves to specific programmes of liberalization over specified future periods. In addition, a large number of other countries took specific steps to liberalize their telecom markets, including a large number of developing countries.[14]

These changes can be seen as shifting the balance of telecom reform activity from the phase of policy development and basic institutional restructuring in preparation for liberalization in many countries, and a difficult period of experimentation in the USA, to a full-scale implementation of liberalizing policies.

The sector was being driven by explosive growth in mobile and the Internet. In the late 1990s telecom came to be seen by the investment community as the driving force for the future information economy. High-capacity fibre optic transmission links under the oceans and across North America and Europe added enormous amounts of capacity. Speciality firms were entering niche Internet markets daily and being bought up at high prices by larger players, including incumbent national operators. Mobile technology had improved to the point where the next generation equipment (3G) would be able to provide at least some Internet services. But shortly after the turn of the century the stock market bubble burst, setting a sombre environment for the implementation of additional liberalization reforms as most of the major players were saddled with unsustainably high levels of debt and severely constrained network expansion programmes.

Fixed Network Development: Liberalizing Access

The key to liberalization in most countries is providing reasonable access for enhanced Internet and mobile services. The expansion of network capacity to provide services requiring broadband connections is now seen as essential for growth in the future information economy. But the incumbent operators maintain monopolies over most fixed local network facilities and access to most customers, especially residential customers, for fixed voice and most Internet services. Both customers and service providers can be charged for access to one another. This monopoly power has been leveraged to enable the incumbents in most countries to be the dominant supplier of most major services. Wireless local networks have been licensed, and local fibre cable connections have been provided by some electric power companies, municipalities or private operators, but they have not made significant inroads to the incumbent's dominance of the local access market.

The market shares of incumbent operators in many countries have reversed a period of decline and begun to increase. In the USA, the seven RBOCs have been reduced to three. One, SBC, has purchased AT&T and adopted its name. The cable television industry is dominated by a single network operator, Comcast. The lack of effective competition has been recognized as a factor explaining the relatively low US ranking in global comparisons of broadband and enhanced mobile services development (Berkman Center, 2010). In Europe the ground is being tested for potential mergers among European incumbent national operators.

Unbundling the local network capacity of incumbent operators to allow direct access to customers by other service providers has been a difficult challenge for all regulators, although Europe and Asian countries have had greater success than North America. There is a concern, especially in the USA, that the local incumbents wish to use their monopoly power to establish a new local access pricing model that would include differential and discriminatory charges based on Internet traffic volumes and quality of service, for example, high- and low-quality Internet access. This would be a major change from the 'net neutrality' that now characterizes the Internet.

Anticipating a transition to fully liberalized and more competitive markets, the European Commission has introduced a new model for regulating the telecom industry intended to facilitate this transition. Rather than have national regulatory authorities (NRA) seek to liberalize telecom markets by focusing on the activities of incumbent operators, this model shifts the focus to national telecom services markets. It requires NRAs to measure market concentration in specific telecom services markets with the intention of directing regulatory attention to the steps needed to liberalize those markets where a firm has significant market power (SMP). The early study results documented SMP by incumbent operators in the vast majority of services markets, and by the largest mobile operators in key markets in many EU countries (EC, 2003, 2010).

In 2009 a new Body of European Regulators for Electronic Communication (BEREC) was established by the EU, consisting of the heads of the 27 NRAs from EU countries, with support from the EC. It is intended to be a stronger forum to exchange expertise and best practice and give opinions on the functioning of the telecoms market in the EU than its predecessors. In addition, a new European Telecom Market Authority has been established to support the Commission and national telecoms regulators in ensuring that market rules and consumer regulation 'are applied consistently, independently and without protectionism in all 27 EU Member States' (EU, 2009). The EU has decided to play a more forceful role in driving the telecom reform agenda in Europe.

The Mobile Revolution: Oligopoly Policy

In the years following the turn of the century, mobile market penetration has exceeded fixed lines in most countries. Mobile has begun to become a substitute for fixed line voice communication for users in developed countries, and it is the major source of new network development in developing countries.

As mobile markets grew, the liberalized licensing of additional operators depended on national and regional spectrum policies allocating essential radio frequencies to the operators. As all countries failed to anticipate the very rapid growth, new entry was restricted by a policy-created spectrum shortage, so most countries have ended up with three, four or, in a few cases, five operators. The liberalization policy adopted has been managed oligopoly, not liberalized entry. The industry has been enormously profitable everywhere.

By the late 1990s, third generation (3G) mobile network and terminal equipment had been developed that could provide data services, including lower capacity Internet connections. Although some of these services could be provided over upgraded 2G networks, 3G services required more capacity. In some countries the 2G licences were restricted to the supply of voice services, so an additional licence would be required for data services. Additional frequencies were needed to provide for the anticipated high growth.

Policy-makers realized that with 2G licences they had assigned extremely valuable frequencies as a free resource to mobile operators, and auctioning might be a better way for selecting among bidding applications and raising funds for the government. Generally, one more 3G licence was assigned than there were 2G operators, essentially a policy of one new entrant.

Unfortunately, the UK, France, Germany and the Netherlands did their auctions at the peak of the telecom stock market bubble. Although the governments made a fortune from the auction, they almost bankrupted their national incumbent operators and the other winners. The winners acquired such high debt mountains that planned investment in the roll-out of 3G networks had to be scaled back and delayed (Melody, 2001). Asian countries – particularly South Korea and Japan – took over leadership in mobile technical and new services development and rates of market penetration.

The industry, particularly in Europe, has exploited its monopoly over fees operators charge one another to terminate calls at the numbers of their customers, which get passed back to the originating customers. Regulators, competition authorities and the European Commission have been forced to intervene to require price reductions.

The concentrated oligopoly structure of mobile markets has prompted many regulators to seek further liberalization by requiring licensed operators to unbundle their networks to provide access to Mobile Virtual Network Operators (MVNO). This has increased retail market competition, but MVNO success depends very much on the terms of access provided by the mobile facility network operators.

For the future, mobile development is focused on expanding networks and terminals to provide ever-increasing broadband capacity to accommodate Internet applications and overcome its present significant bandwidth limitations in comparison with fixed networks. The extent of further liberalization will depend upon the allocation of additional spectrum to mobile, the licensing of new entrants and the policies for managing MVNO access to oligopoly operator networks.[15]

DEVELOPING COUNTRY EXPERIENCE AND CHALLENGES

Most developing countries did not begin the process of telecommunication reform and liberalization of their telecom sectors until the late 1990s when they were being pressed to make commitments associated with the WTO telecom services trade agreement. There seemed little reason. They would have to import equipment, capital and skilled people, and their incumbent national operators served only a small portion of the population and were too weak, inefficient and under-financed to be able to capitalize on the opportunities. Local business and professional users were not a powerful lobby.

The pressures for change in most countries came primarily from outside. Developed country governments and their donor agencies lobbied on behalf of their home industries, consultants and their way of doing telecom reform. The World Bank and the International Monetary Fund (IMF) were promoting the 'Washington Consensus' of open markets with loans to facilitate reforms in telecom and other sectors. The WTO was pressing liberalization in telecom as essential to future trade.

Typically the first step was licensing mobile operators as they did not seem to offer a threat to the incumbent's established fixed line business. This released pent-up demand as users on the waiting list for a fixed line connection became part of rapid mobile growth. When prepaid mobile was introduced there was an unanticipated overwhelming growth among the poor who could not qualify for subscriptions, but who could share, resell and receive calls, many in communities that had no fixed service. Mobile penetration quickly grew to exceed fixed line so that by 2010 in most developing countries mobile connections were at least three or four times the very limited fixed line connections. Prepaid has become the primary driver of voice and SMS mobile growth worldwide (Mahan and Melody, 2008, 2009).

Many developing countries have adopted auctions for future licensing, following the common developed country practices, mostly involving foreign operators or consortia. But as most developing countries do not have strong policy and regulatory institutions – that is, telecom regulators and competition authorities – many are having difficulty establishing and enforcing MVNO access, equal interconnection and reasonable termination prices.

With respect to fixed line operations, foreign investors have been interested in access to developing country markets, not promoting a liberalization agenda. Their concerns have been primarily in investment and control of privatized incumbent monopoly operators where they are viable businesses. The foreign investors, together with incumbents, generally have resisted the establishment of effective regulation, unbundling the network for access by Internet and other service providers, and the implementation of competition policies. As a result in most developing countries, fixed network development has proceeded at a slow pace, regulation has remained weak and liberalization very limited.

Although fixed line incumbents in most developing countries are considered to be the universal service operator, and in some countries they collect universal service financial contributions from the mobile operators to help fund fixed network expansion, it has been the mobile operators who are reaching out to the previously unserved. Prepaid cards, cheap handsets and micro-finance, as introduced by Grameen in Bangladesh, are enabling service to be provided in some of the poorest villages (Samarajiva and Zainudeen, 2008).

Where liberalized entry has been permitted, new fixed-wireless networks have extended coverage, and continuously improving WiMAX and Wi-Fi technologies are reaching further into remote areas. But success in extending networks to the bottom-of-the-pyramid will require cooperative interconnection with the dominant operators (fixed and mobile), as well as licensing and spectrum regulation that facilitates network development by non-established players, local cooperatives and public–private partnerships (Melody, 2010). This is an aspect of liberalization that has yet to be provided with supportive policies, but which is extremely important if Internet as well as voice access is to be extended to the bottom of the pyramid in developing countries.

CONCLUSION

Although policy changes are often seen as a direct response to changes in technologies and/or markets, the evolving path of telecom liberalization has been shaped primarily by changes in policies and regulations. They opened market opportunities and created an environment fostering innovation and the development of new technologies and services. These in turn have produced feedback effects – mostly unanticipated – that have required an ongoing reassessment and modification of policies and regulations.

Telecom liberalization has followed a long and somewhat torturous path to a point where it seems evident that the extent of liberalization is likely to continue to be contested on a number of fronts for the foreseeable future. The long-established defence for natural monopoly policy has been defeated in theory and practice. But it is also clear that complete freedom of entry cannot be provided and dominant operators with SMP are successfully resisting the full implementation of liberalization policies. The future effectiveness of liberalization policies will depend upon how well regulatory and competition authorities can enforce them in the face of opposition from the previously licensed operators.

Experience has shown that liberalization is not a substitute for regulation, but rather an important tool of regulation, and continuous testing of liberalization opportunities will be needed by proactive regulators in the face of new technology and service developments if a retreat from liberalization is to be avoided. This will require an improved performance from regulators than they have provided to date. As many national regulatory authorities do not have independent jurisdiction over new entry or spectrum management, the maintenance of liberalization policies will depend also on the ongoing commitment of communication ministries.

Fixed and mobile local infrastructure markets are likely to remain highly concentrated and subject to direct regulation. Consumer protection from the monopoly power of dominant operators will still be needed. Access regulation (unbundling, net neutrality) to accommodate the provision of an increasing number of Internet services and applications will be an ongoing issue. Many retail services markets should become sufficiently competitive to justify regulatory forbearance. But universal access to Internet services will become more important in the future information economy. Many countries are adopting subsidy programmes to achieve universal broadband access within the next few years.

NOTES

1. The natural monopoly theory was applied to other infrastructure sectors as well and became part of the conventional wisdom in the economics and law of public utilities in the USA and other countries. See, for example, Glaeser (1957).
2. 27 FCC 359 (1959); 29 FCC 825 (1960).
3. The Telpak tariff provided price reductions between 51 per cent and 85 per cent from the standard tariff. For an analysis, see Melody (1971a).
4. Hush-A-Phone corp., FCC Reports, Vol. 14, 282 (1949); and 238 FCC 2d 266 (1956).
5. Jordaphone Corp, FCC Reports, Vol 18, 644 (1954).
6. Carterfone Corp., 13 FCC 2d 420 (1968).
7. Microwave Communications Inc. (MCI), 18 FCC 2d 953 (1969)
8. Computer 1, 29 FCC 2d 870
9. FCC Docket No. 20097. Report and Order (1976).
10. FCC Computer II. Docket No. 20828, Final Decision (1980).
11. FCC Computer III, Final Decision (1986).
12. The FCC held a one-week conference to obtain agreement among all the economic experts on the appropriate theoretical conception of marginal cost to be implemented. The neoclassicists could not agree among themselves on the most basic parameters, such as long run, short run, the output unit or the size of the increment, let alone the production function. See Melody (2002).
13. See, www.irg.eu.
14. For a comprehensive review of developments, see ITU (1998).
15. The 2010 FCC annual report on the state of the wireless industry found that concentration had increased by 32 per cent since 2003. It could not conclude that the industry was 'effectively competitive' as it had in the past. See FCC (2010).

REFERENCES

Averch, H. and Johnson, L. (1962), 'The firm under regulatory constraint' *American Economic Review*, **52**, 1050–69.

Baumol, W.J. (1987), 'Ramsey pricing', *The New Palgrave: A Dictionary of Economics*, Vol. 4, London. Macmillan.

Baumol, W.J. and Bradford, D.F. (1970), 'Optimal departures from marginal cost pricing', *American Economic Review*, **60**, June, 265–83.

Baumol, W.J. and Sidak, J. G. (1994), *Toward Competition in Local Telephony*, Washington, DC: The MIT Press and American Enterprise Institute.

Baumol, W.J. Bailey, E.E. and Willig, R.D. (1977), 'Weak invisible hand theorems on the sustainability of prices in a multiproduct monopoly', *American Economic Review*, **67**, June, 350–65.
Baumol, W.J., Panzar, J.C. and Willig, D.G. (1982), *Contestable Markets and the Theory of Industrial Structure*, New York: Harcourt Brace Jovanovich.
Berkman Center (2010), *Next Generation Connectivity: A Review of Broadband Internet Transitions and Policy from Around the World*, Cambridge, CA: Berkman Center, Harvard University.
Brock, G. (1994), *Telecommunication Policy for the Information Age*, Cambridge, MA: Harvard University Press.
De Butts, J.D. (1973), 'An unusual obligation', speech before the Annual Convention of the National Association of Regulatory Utility Commissioners, Seattle, Washington.
European Commission (EC) (1998) Full Competition Directive, Directive 96/19/EC.
European Commission (EC) (2003), Commission guidelines on market analysis and the assessment of significant market power under the Community regulatory framework for electronic communications networks and services (2002/C 165/03).
European Commission (EC) (2010), 15th Progress Report on the Single European Electronic Communications Market, COM(2010)253 – 25 May.
European Union (EU) (2009), Regulation 2009 (EC) No 1211/2009 of the European Parliament and of Council, 25 November.
Federal Communications Commission (FCC), Charges and Regulations for Television Transmission Services and Facilities, Docket No. 8963, @ 62. 5 RR 639.
Allocation of Frequencies in the Bands Above 890 Mc., 27 FCC 359 (1959); 29 FCC 825 (1960).
Federal Communications Commission (FCC), Regulatory and Policy Problems Presented by the Interdependence of Computer and Communication Services and Facilities, Notice of Inquiry, Docket No. 16979, FCC 66-1004, November 9, 1966. Supplemental Notice of Inquiry, FCC 67-239, March 2, 1967. Report and Further Notice of Inquiry, FCC 69-468, May 1, 1969.
Federal Communications Commission (FCC), Use of the Carterfone Device, 13 FCC 2d 420 (1968).
Federal Communications Commission (FCC), Microwave Communications, Inc., 18 FCC 2d 953 (1969).
Federal Communications Commission (FCC), Computer 1 Final Decision 28 FCC 2d 267.
Federal Communications Commission (FCC), Regulatory Policies Concerning Resale and Shared Use of Common Carrier Services and Facilities. Docket No. 20097, Report and Order (1976).
Federal Communications Commission (FCC), First Report and Order re Implementation of the Local Competition Provisions in the Telecommunications Act of 1996, and Interconnection between Local Exchange Carriers and Commercial Mobile Radio Services Providers. 8 August, 1996.
Federal Communications Commission (FCC) (1949) Use of the Hush-a-Phone device, FCC Reports, Vol.14, 282.
Federal Communications Commission (FCC) (1956) 238 FCC 2d 266.
Federal Communications Commission (FCC) (1954) Jordaphone et al., 18 FCC 644.
Federal Communications Commission (FCC) (1980) Computer II Final Decision 77 FCC 2d 384
Federal Communications Commission (FCC) (1986) Computer III Report and Order 104 FCC 2d 958.
Federal Communications Commission (FCC) (2010), The State of the Wireless Industry, May.
Gabel, R. (1969), 'The early competitive era in telephone communication, 1893–1920', *Law and Contemporary Problems*, Durham, NC: Duke Law School.
Glaeser, M.G. (1957), *Public Utilities in American Capitalism*, New York: Macmillan.
Gruber, H. (2005), *The Economics of Mobile Telecommunication*, Cambridge: Cambridge University Press.
International Telecommunication Union (ITU) (1998), *General Trends in Telecommunication Reform 1998 – World (Vol. 1)*, Geneva: ITCL.
Irwin, M.R. (1969), 'Computers and communications: the economics of independence', *Law and Contemporary Problems*, Durham, NC: Duke Law School.
Mahan, A.K. and Melody, W.H. (eds) (2008), *Diversifying Participation in Network Development,* report on the World Dialogue on Regulation, Montevideo: LIRNE.NET.
Mahan, A. and Melody, W. (eds) (2009), 'Network development: wireless applications for the next billion new users', *info*, special issue.
Mandanis, G.P. (1976), 'An empirical analysis of economies of scale and specialization in communications', in H.M. Trebing (ed), *New Dimensions in Public Utility Pricing*, East Lansing, MI: MSU Public Utility Studies, pp. 324–88.
Melody, W.H. (1971a), 'Interservice subsidy: regulatory standards and applied economics', in H.M. Trebing (ed.), *Essays on Public Utility Pricing and Regulation*, East Lansing, MI: MSU Public Utilities Studies.
Melody, W.H. (1971b), 'Technological determinism and monopoly power in communications', Annual Meeting of the American Economic Association, December.
Melody, W.H. (1972), 'Interconnection: impact on competition, carriers and regulation', in S. Winkler (ed), *Computer Communications: Impacts and Implications*, New York: ICCC and IEEE.

Melody, W.H. (1975), 'Relations between public policy issues and economies of scale', IEEE Transactions on Systems, Man and Cybernetics, **1**, January, 14–22.
Melody, W.H. (2001), 'Spectrum auctions and efficient resource allocation: learning from the 3G experience in Europe', *info*, **3**(1), 5–10.
Melody, W.H. (2002), 'Designing regulation for 21st century markets', in E. Miller and W. Samuels (eds), *The Institutionalist Approach to Public Utility Regulation*, East Lansing, MI: Michigan State University Press.
Melody, W.H. (2010), 'Wi-Fi in developing countries: catalyst for network extension and telecom reform', in W. Lemstra, J.P.M. Groenewegen and V. Hayes (eds), *The Genesis of Wi-Fi and the Road toward Global Success*, Cambridge: Cambridge University Press.
Mueller, M.L. (1997), *Universal Service: Competition, Interconnection and Monopoly in the Making of the American Telephone System*, Washington, DC: AEI Press.
Owen, B. and Braeutigam, R. (1978), *The Regulation Game*, Cambridge, MA: Ballinger.
Samarajiva, R. and Zainudeen, A. (eds) (2008), *ICT Infrastructure in Emerging Asia: Policy and Regulatory Roadblocks*, New Delhi Sage and IDRC.
Trebing, H.M. (2004), 'Assessing deregulation: the clash between promise and reality', *Journal of Economic Issues*, **38**(1), 1–27.

8 Liberalization in radio spectrum management[1]

William H. Melody and Wolter Lemstra

INTRODUCTION

The radio spectrum is the universe of frequencies that can be used for a variety of types of electronic communication through the air, ranging from opening garage doors to the provision of emergency services and researching outer space. Specific frequencies are allocated by national and international spectrum authorities for particular uses, including the provision of telecommunication (telecom) services to the public. Prior to the recent era of telecom reform, the primary spectrum uses in telecom were for the local, national and international transmission of radio/television broadcast signals and the long-distance transmission of voice telephony by incumbent telecom operators using microwave technology.

In recent years, the explosive growth of mobile voice and data services and the Internet for a continuously increasing variety of services has created an ever-expanding demand for more frequencies to be allocated for the provision of these services. Thus, spectrum management, which used to be seen as a technical matter of no great consequence that could be best left to the engineers in an environment where the supply of available frequencies could comfortably accommodate new demands, has been transformed into a major economic and policy matter in an environment where new demands exceed supply and the historic policies and practices of spectrum management have been called into question.

The reform of spectrum management has been at the heart of the telecom reform process as it is directly associated with the increasing role of all forms of mobile communication. Like other dimensions of telecom reform, the liberalization of a previously highly restrictive spectrum management policy regime to permit the participation of new players in the sector and a more competitive market environment has been a central element in the reform process. The criteria for spectrum management decisions have been broadened to include market and other economic considerations, as well as technological parameters and public policy objectives. There have been reforms of both the management process and the criteria by which frequency allocations for particular uses and licence assignments to service providers and users are made. As the spectrum is comparable to a natural resource that must be managed for electronic communication to take place at all, liberalization does not replace spectrum management, but rather it has become a key element in the design and implementation of a new management framework that is intended to be more efficient and responsive to policy objectives for the expanding telecom/information and communication technology (ICT) sector. This chapter examines the new roles for liberalization in the spectrum management process, which is part of the larger process of liberalization in the telecom sector that is reviewed in Chapter 7 of this volume.

SPECTRUM CHARACTERISTICS AND GROWTH IN APPLICATIONS

Transmission over the air is governed by the technical communication properties of particular radio frequencies. As these basic spectrum parameters do not change, technical improvements must come from improvements in the signal processing capabilities in transmitting and receiving terminals, and in signalling systems that enable improved use of the spectrum resource. The term 'spectrum' refers to the range of frequencies of electromagnetic radiation, or radio waves, used for communication. Different frequencies have different wave lengths and therefore different propagation properties. They are identified by their frequency as measured in cycles per second, or Hertz.[2]

The standard measures are kilohertz (kHz, 1000 cycles per second), megahertz (MHz, 1 000 000) and gigahertz (GHz, 1 000 000 000). Lower frequencies have longer waves than higher frequencies. Lower frequency waves tend to pass through obstructions and bounce off surfaces, and so can be received inside buildings, beneath trees, or in rain, which is very important for some services like broadcast radio and mobile. Higher frequencies typically require a line of sight between the transmitter and receiver, which fits the requirements for communicating from geostationary satellites to rooftop antennae (Cave, 2002).

The technical quality and cost of providing services can vary widely depending upon the particular frequencies that are allocated and assigned, as can the cost of manufacturing equipment for use at different frequencies. It is the combination of the technical properties of signal transmission and reception at different frequencies, and the regulations and standards adopted in the spectrum management process that determine the capacity, quality and type of communication signals that can be transmitted, and therefore the services that can be provided, over different frequency ranges, or bands. (Aitkin, 1994).

Electronic communication through the air began with the award of a British patent to Marconi in 1897 for 'a practical wireless telegraphy system known as radio'. Signals sent over a particular radio frequency could be received and 'understood' by a terminal designed to receive signals on that frequency. The earliest uses of radio frequencies for communication were for ships at sea, local emergency services, broadcast radio, amateur radio and transmission links in fixed telecom networks where cable would be impossible or prohibitively expensive, for example, over water. During the Second World War the British military made advances in radar that led to the development of higher-capacity, inter-city, long-distance transmission links in fixed telecom networks using frequencies in the microwave band, and the transmission of television signals which required broadband capacity. Later, communication satellites, which were initially developed by the military, made both global television and telecom services to remote locations possible, (Douglas, 1987).

Technical improvements in the communication capabilities of terminals have provided the primary force for the explosive increase in communication capabilities over the air. The new terminal capabilities have increased the volume of communication possible on existing frequencies, made it possible to use previously little-used higher frequencies, and introduced new ways of technically managing signals across different frequencies in 'smart' terminals (ITU, 2004). The rapid growth in telecom services using the radio

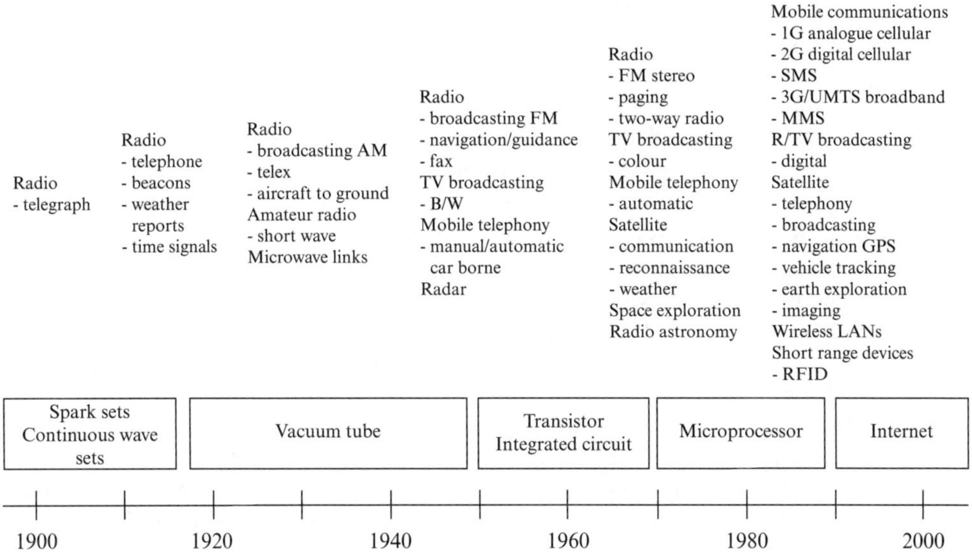

Source: Adapted from http://www.ceprc.ca/spec_e.html.

Figure 8.1 Growth in telecommunications applications and the use of the radio spectrum

spectrum over the past century is illustrated in Figure 8.1. Continuing growth at a rapid rate is expected for the foreseeable future.

THE EVOLUTION AND STRUCTURE OF SPECTRUM MANAGEMENT

The Establishment of Spectrum Regulation

The system of spectrum management employed today has arisen from the experience with radio communication in its earliest days. One of the first successful business models using radio-telegraphy, resulted from the entrepreneurial initiatives by Marconi. Following a period of manufacturing radio equipment, the Marconi Wireless Telegraph Company started to build and exploit coastal radio stations and the operation of radios on board ships. This initiative also led to the first 'market failure' when it appeared that Marconi had instructed his operators to refuse to relay messages received from competing operators, for example, from operators using Telefunken equipment.

In 1902 this became apparent to Heinrich von Hohenzollern, Prince of Prussia, when travelling from the USA back to Germany, as he could not send a courtesy message to President Roosevelt. This contributed to the first governmental meeting on the management of radio spectrum being convened in 1903 in Berlin, attended by representatives of nine countries including the USA. It was decided that: 'Coast stations should be bound to receive and transmit telegrams originating from or destined for ships at sea without distinction as to the system of radio used by the latter.' The subsequent meeting in

1906, bringing together representatives of 29 nations, can be considered the start of the international coordination of spectrum use with the formation of the First International Radiotelegraph Conference, modelled after the Convention of the International Telegraph Union of St Petersburg in 1875. Two wavelengths were allocated for public correspondence in the maritime services at 500 and 1000 kHz. Frequencies below 188 kHz were reserved for long-distance communication by coastal stations. The frequency range between 188 and 500 kHz was reserved for military and naval stations. Moreover, all details of stations, their frequency use, hours of operation, call signs and radio system being used were to be sent to the Bureau of the Union in Berne, Switzerland. The radio stations would require a government licence and operators had to have a certificate as to their competence (ITU, 1965).

The sinking of the *Titanic* on 14 April 1912 became the defining event for spectrum management. The *Titanic*'s SOS call was not received by a nearby ship that had shut down its radio for the evening, and in the immediate aftermath a flood of amateur radio messages jammed the airwaves and contributed to widespread confusion. A Special Committee of the US Congress and a special commission presided over by Lord Mersey in England investigated the disaster and established the rather perturbing facts (ITU, 1965).

Within four months of the tragedy, the USA enacted the Radio Act of 1912 which brought the spectrum under government control. The Act required that all operators be licensed, that stations adhere to specified frequency assignments, that distress calls take priority over all other calls, and that the secretary of commerce and labour be empowered to issue licences and make other regulations necessary 'to sort out the wireless chaos.' Similar legislation was adopted by other nations (Douglas, 1987).

Radio broadcasting began in the early 1920s and many entrepreneurs entered this new field. Although the frequencies were licensed, there were no strict rules on the transmitting power of the radio signal. Broadcasters realized that the greater the power, the stronger the signal and the greater the potential listening audience. But stronger signals overwhelmed weaker ones. As a result there was much signal interference and broadcasters were unsure where and how their signals were received. A good deal of 'wireless chaos' remained. In addition, the initially established private broadcasters were also concerned about the continuously increasing competition from new broadcasters, and many national governments were becoming concerned about the growing economic and political power of broadcasting to large public audiences. All of these concerns stimulated the establishment of a comprehensive regime of spectrum management in many countries, as well as government ownership or direct regulation of radio broadcasting. In the USA, the Radio Act of 1927 created the Federal Radio Commission (FRC) to manage the spectrum, a resource which by this time had come to be seen as a national asset. Most other countries established spectrum management at the national level, and participated in regional and international bodies to take up the task of coordinating international radio communication (Levin, 1971; Slotten, 2000).

Also in 1927, the Washington Radio Conference was held, with 80 countries being represented and 64 private broadcasting organisations participating on a non-voting basis. During the conference, charging and international accounting were being decided upon, and the Radio Consultative Committee was established, with one of its tasks to draw up the first Frequency Allocation Table; the useful spectrum was considered to be the range from 10 kHz to 60 MHz (ITU, 1965).

The Management Responsibilities

The major activities of spectrum management relate to the specification of the basic technical parameters permissible in using particular frequencies, and the coordination of usage rights and responsibilities across different frequencies and users. To illustrate, a mobile phone will not work unless the network operator has been assigned a licence to use a particular frequency band in a particular locality and both the user's phones (or the SIM cards in the phones) and the operator's network of transmitters and receivers are technically designed to communicate on that precise frequency band. Moreover, that operator needs some assurance that the same band of frequencies can not be used by others, as their signals could create interference, disrupt service and make business planning and service delivery unreliable. Mobile phones and network equipment will not be manufactured, and research and development funds will not be invested unless there is some assurance that appropriate frequencies will be allocated for mobile services, and there is likely to be a demand for this equipment from operators and users. Potential mobile operators will need to obtain spectrum licences for the appropriate frequencies in order to supply services. Thus, the development of both new technologies and new services depends heavily upon spectrum management decisions.

The core of spectrum management is a two-step process by which government authorities decide, (1) how frequency bands will be allocated among different types of use, for example, mobile phone networks, broadcast television, aeronautical and ship navigation, taxis, emergency services and defence communication; and then (2) how the specific frequencies within each allocated band of spectrum will be assigned to different service providers or users, for example, competitive mobile operators, police and fire departments or amateur radio enthusiasts (Melody, 1980).

The most important technical parameters considered in spectrum management are: the specific frequency band; the geographical location; the power that can be used, that is, the strength of the signal; and the time when the spectrum will be used for sending signals. Radio frequency licences typically specify the standards permissible for each. Generally, the wider the frequency band, the greater will be the communication capacity and the lower the frequency the lower the cost of the equipment. The larger the geographical area, the larger will be the market that can be served on the assigned frequencies. The greater the power, the farther the transmitted signal can be sent; and for broadcasting, receiving-only terminals can be cheaper. When different signals are sent on the same frequency, the one with the greatest power will have the strongest signal and experience less interference. If communication needs of different users are at different times of the day, month or year, the same frequencies can be shared. With lower power and larger distances the same frequencies can be reused geographically, as in cellular communications. Spectrum management is largely about balancing and coordinating these spectrum parameters as part of frequency allocation and assignment processes that attempt to accommodate ever increasing demands for spectrum (Arnbak, 1997).

Spectrum management, new technological development, and the demands of existing and new services are the three interdependent elements that underpin the development of spectrum-based technologies and services. Spectrum management can facilitate or frustrate development. It can be proactive and promote particular technical and market developments, or reactive and try to follow them, or even obstructive and hinder development. Historically, one can find evidence of each approach, although in most circumstances spectrum managers tend to react to specific technical and market developments and proposals, with varying degrees of efficiency and effectiveness (Levin, 1971; Melody, 1980).

The International Structure of Spectrum Management

Spectrum management takes place at three levels of government – global, regional and national. The International Telecommunications Union (ITU) facilitates the international coordination of the basic allocation of frequency bands for generic types of uses. Its main activities are inter-country frequency notification, coordination to minimize potentially harmful interference, and facilitating discussion and agreement about frequency allocations for new services made possible by new technologies. It promulgates radio regulations which have treaty status among the organizations' member states.

The ITU maintains an international frequency table that today allocates spectrum from 9 kHz to over 275 GHz for a range of radio services. These allocations typically do not specify a particular technology or application but rather provide generic classifications to the services that may share specific allocated ranges of spectrum (for example, mobile allocations could be used for any mobile service – paging, public, private, and so on) With the growth of global markets for some mobile services, the ITU has begun to pay more attention to greater global coordination of frequency allocations for some services, most notably the recent spectrum identified for 3G and 4G mobile services, which are seen to be global in nature.

The ITU divides the world into three regions for the purposes of managing the global spectrum. Each region establishes its own set of frequency allocations within the broad parameters recommended for global communication. Region 1 comprises Europe, Africa, the Middle East west of the Persian Gulf and including Iraq, the former Soviet Union and Mongolia. Region 2 comprises the Americas and some of the eastern Pacific Islands. Region 3 comprises most of non-former Soviet Union Asia east of and including Iran and most of Oceania.

In addition, within these three regions there are sub-regional organizations that manage spectrum issues common to particular regions. This structure of spectrum management attempts to provide the necessary international harmonization, while preserving as much flexibility as possible for spectrum management at the national level, which is where the final frequency allocations to services, and all frequency assignments are made, primarily by issuing licences.

Spectrum Allocation and Assignment at the National Level

Spectrum management at the national level is undertaken by government authorities, usually in the ministry responsible for communication policy or the National Regulatory Authority (NRA). These agencies implement national policies, make or recommend changes in allocations, issue licences, monitor developments, enforce licence conditions and resolve conflicts. The debates about the efficiency of the spectrum management process and alternative mechanisms that might improve the process have focused primarily on spectrum management at the national level (Gruber, 2001).

Although the allocations of frequency bands for generic types of use are influenced by international recommendations and guidelines from the ITU, and further recommendations and agreements negotiated at the regional level, the final allocations to specific services are made by national authorities based on national conditions. These national frequency plans determine how much spectrum will be made available for each of the many different types of services and how much frequency bandwidth will be included in its assignments.

After frequency bands have been allocated, the national spectrum authorities assign particular frequencies to specific network operators or users generally by granting licences. The licences provide certain guarantees in terms of exclusive use, protection from harmful interference and other protections. They also carry specific responsibilities with respect to signal strength, not creating harmful interference for others, and other considerations. The number of licences awarded for particular communication uses at any point in time depends on many factors, including:

- the amount of spectrum allocated for each type of use;
- the bandwidth (that is, number of frequencies) allocated for each licence;
- the number of licences previously assigned;
- the amount of spectrum reserved for future assignment; and
- the current demand for new licences in the allocated frequency bands.

Allocation decisions by spectrum managers are based on a weighing of both public and private needs for frequencies to provide the variety of desired communication services. Frequency assignment decisions are based on selection processes designed to license those particular operators or users best qualified to provide the designated services. Thus, public needs and market influences bear heavily on both the allocation and the assignment processes. In recent years, with the rapid growth in mobile, wireless and satellite communication creating increased demand for spectrum for commercial services, the influence of market forces on the spectrum management process has increased significantly.

The spectrum management activities of frequency allocation and assignment at the national level are associated with the more static aspects of allocating a fixed set of frequencies among different communication services, and then assigning this fixed supply of frequencies to specific applicants for licences. National spectrum managers have a fairly wide range of flexibility and discretion in allocating frequencies for different services, and they consider the market needs of commercial service providers and users in different industries, as well as a range of public needs, in making their allocations. This requires a balancing of technical, economic and public interest considerations.

For each type of communication, some frequency bands will be more desirable than others on technical grounds. Although these will vary among services, there are certain frequency bands (usually the lower ones) that are most attractive for several types of service, so not all types of communication will be able to use the optimal frequency band. When economic considerations are brought into play (that is, the cost of developing and producing equipment for the various types of services, and of providing those services), further compromises from optimal frequency selection must be made. When national and public interest considerations such as defence and emergency services are introduced, the equation changes again. All this must then be balanced against the spectrum that has already been allocated by past decisions, the uncertain forecasts of how much spectrum is needed for the new allocations, and the fact that overall there is greater demand for spectrum than there is spectrum available. It is a scarce resource. Thus, spectrum allocation activity is an ongoing balancing act in an ever-changing environment.

As spectrum managers cannot have perfect knowledge, there are always circumstances where too much or too little spectrum is allocated for particular types of communication. Therefore, one will almost always find unused frequencies in some bands, and shortages in others. In most cases, it is not possible simply to transfer frequencies from one type of use to another, say from television or national defence to mobile phones. Once frequencies are allocated and major investments are made in equipment, networks and services, it generally becomes extremely complex and costly to consider reallocation.

Frequency assignments today must be done under constraints of greater scarcity than in the past and, despite continuing technological improvements, the anticipation of increasing scarcity in the future. For many types of services the demand for frequency licences exceeds supply because spectrum managers underestimated the growth of a particular class of service and too few frequencies were allocated to meet the demand. Spectrum allocation then not only allocates frequencies for different types of services, it allocates frequency scarcity as well. The increasing concern about frequency scarcity has focused attention on whether both the allocation and assignment processes could be improved.

SPECTRUM MANAGEMENT PERFORMANCE

The spectrum has always been regarded as a scarce resource because its supply is limited, and its use requires the application of technical skill and specially designed equipment under conditions of coordination and control. The effectiveness of the coordination and control has a major influence not only on the efficiency of utilization of the spectrum directly, but also indirectly on the incentives to improve spectrum-based technologies. However, the early days of spectrum management were characterized by conditions where, for the most part, spectrum supply and capacity satisfied or exceeded demand. Both allocations of frequency bands to services and specific assignments of frequencies to operators and users were made on a first-come, first-served basis, and new demands were fitted in where there was still unused spectrum or where accommodations could be easily made with a few technical adjustments to licence conditions. The processes of decision-making were administrative with no sense of urgency, and the criteria for decisions were almost entirely technical and based on spectrum engineering. But as the demand for spectrum increased faster than the supply, scarcity became an ever more important issue, fitting in new demands became more complex, time-consuming and costly, and the economic scarcity value of the spectrum escalated. Critical attention began to be directed to the performance of the spectrum management process (Melody, 2008).

Allocation Processes and Standards

As spectrum allocation processes are intended to take place sequentially at the global, regional and then national levels, it takes considerable time for proposals for new frequency band allocations for existing and new services to work their way through the system. As more and more countries joined the international spectrum management organizations and the problems of managing increasing demands for frequencies became more complex and in some cases contentious, the spectrum allocation process became far too slow for the more rapid changes in the technologies and industries that require spectrum. Industry lobby groups became more active at the national level and country positions at the international level increasingly reflected domestic industry interests, not just technical accommodation. The traditional engineering issues were broadened to include economic, political and social concerns as well. But the established system for making allocation decisions was not well adapted to this more dynamic environment.

The weaknesses of the established system for spectrum allocation can be summarized as follows:

- It takes too long to make decisions. This slows down both technical development and economic investment, and increases risks and costs for those parties waiting for allocation decisions.
- It is primarily an ad hoc, reactive decision-making system that simply responds to requests and recommendations, and is therefore poorly equipped to examine spectrum allocation in a holistic manner or plan future allocations in a systematic way.
- Decision-making processes are extremely opaque. It is unclear how most decisions actually get made, who participated in making them, and what factors and evidence were considered. Accountability is minimal.
- The standards and criteria for decision-making are rarely specified. It can best be characterized as politically negotiated engineering accommodation, increasingly reflecting the lobbying power of various industries.
- The future (or even present) economic value of alternative uses of different spectrum bands is seldom estimated or incorporated into allocation decisions.
- The monitoring of spectrum use in frequency bands already allocated, in order to assess how efficiently they are being used and the costs and benefits for reallocation, is rarely undertaken.

Overall it can be concluded that the established system of spectrum allocation has failed to respond effectively to the new dynamic environment. The decision-making structure is uniquely 'path dependent' as past allocations heavily constrain the possibilities for future development. Only minimal reallocation of spectrum resources is possible. The costs to many economies and societies of this ineffective management have become substantial and are increasing.

One illustration is the allocations of spectrum bands to third generation (3G) mobile services (capable of providing voice and data services) in many European countries. (Melody, 2001). At the World Radio Conference (WRC) of 1992, an initial allocation was made for development of the then undefined future 3G services (IMT-2000). Throughout the 1990s the demand for 2G mobile services exploded, so most countries were forced to restrict the number of 2G spectrum licences they could award well below the demand for market entry to provide mobile services. This spectrum management barrier to entry fostered the establishment of concentrated oligopoly mobile markets across Europe. By 1998 it was clear that mobile growth still had a long way to run and that planned 3G services would require much more spectrum than 2G services in order to accommodate the new information services and a rapid growth rate. Establishing additional frequency bands for 3G was placed on the agenda for the planned WRC in 2000.

Yet the UK, Germany, the Netherlands and other countries chose not to examine ways of expanding the spectrum allocation to 3G, or to wait until the WRC 2000 meeting at which possibilities for expanding the 3G allocation would be taken up.[3] They knew that in relation to the anticipated demand for licences and for frequency bandwidth required by 3G services, a limited amount of spectrum had been allocated, but chose to go ahead with assigning 3G spectrum in 1998 under circumstances of restricted frequency allocations.

Knowing the enormous economic value of 3G spectrum and the great demand for it, and that only a few licences could be assigned, most national spectrum managers were content to plod along the traditional path for making new spectrum allocations with no attention to the artificial tight oligopoly industry structure they were creating. This stimulated intense interest in how this restricted supply of 3G spectrum would be assigned to a few privileged licensees. Combined with the prevailing view that 3G licences represented an entry ticket to the future of telecommunications, fuelled by the Internet boom and the large amounts of money made available by the financial sector, the first application of spectrum auctions led to prices well in excess of the economic value of the licences being granted.[4] Would it be possible to solve the problem of an insufficient and inefficient allocation of spectrum for 3G services by a more efficient method of assigning frequency licences?

Assignment Processes and Methods

As with allocations, frequency licences historically have been assigned on a first-come, first-served basis to qualified applicants. When the demand for frequencies for a particular class of use exceeded the supply that had been allocated for that type of service, and decisions had been taken not to allocate more spectrum to meet that demand, the limited number of licence assignments has been determined on a competitive basis. The primary criteria used for the selection of licensees have been based upon the bidder's technical and financial qualifications, the quality of technical proposals for using the frequencies, and plans for developing the specific spectrum-based services.

These selection processes have sometimes been called 'beauty contests', especially where the proposals have been very similar and the basis for selection of winning over losing proposals is very subjective and unclear (OECD, 2005). In some countries it is also vulnerable to political interference and is vulnerable to corruption. Moreover as key segments of the spectrum have become scarce, issues of congestion and interference have become more acute, the number and complexity of licence applications have increased, and the time period for assessing and making assignments has lengthened, in some cases causing delays in investment and new services development.

It has also been recognized that licensing valuable spectrum at no significant direct cost to the licensees other than the costs associated with the preparation and submission of proposals fails to recognize the economic value of the spectrum and provides an artificial stimulation to the demand for this valuable 'free' resource. An administrative process that assigns valuable spectrum without charging a 'price' to licensees provides incentives to promote the wasteful use of the spectrum resources and to encourage the uneconomic 'banking' of frequency licences. The incentive to conserve spectrum in the design and use of equipment will be very weak and biased towards spectrum-using options if the spectrum resource has virtually no cost associated with it (Melody, 1980).

When spectrum authorities have monitored the actual usage of spectrum, they have discovered that considerable licensed spectrum was not being used, and when a 'use it or lose it' rule was put in place, significant numbers of licences were returned. When administrative fees were imposed on licensees to cover the costs to government of its spectrum management activities, more licences were returned. It has become clear that putting a price on the spectrum resource helps stimulate greater efficiency in spectrum assignments and use. This has stimulated interest in considering options for charging higher prices more in line with the economic value of the spectrum. But the economic value of frequencies in any particular spectrum band depends upon the relative efficiency of using that particular band as compared to other bands for the particular service, as well as the number of frequencies allocated for that service. In economic terms, it is how much scarcity has been created or alleviated by prior spectrum allocation decisions. When too few frequencies are allocated, then the economic scarcity value will be driven up. When more are allocated, the scarcity value will decline. If there are more frequencies available than demanded, the value will be minimal. However, spectrum allocation decisions do not reflect estimates of the economic value of allocating more or fewer frequencies to different classes of service; and there have been no guidelines or objective criteria established for valuing the different frequency bands if they were to be allocated or reallocated to different services. However it is possible to raise or lower frequency licence fees to reflect the scarcity that has been created by the spectrum allocation decisions.

PROPOSALS TO IMPROVE SPECTRUM EFFICIENCY AND MANAGEMENT

Proposals to improve spectrum efficiency fall into several different categories. Some are technically based and see new technologies as providing major improvements if spectrum management will make the necessary changes in its decision criteria to accommodate them. Others involve the more effective implementation of economic criteria into the spectrum management process, and in some cases the establishment of spectrum trading in private markets as a substitute for certain spectrum management decisions. Still others involve changing the organizational structure, processes and methods of allocation and assignment to improve efficiency and effectiveness.

Implications of ICT Convergence

An important factor influencing the thinking behind most proposed spectrum reforms is a recognition that as the pervasive network technologies of digitalization and Internet protocol continue to permeate communication networks of all kinds (both fixed and mobile), full services convergence will soon be possible. Across a wide variety of commercial communication and information services in particular, any service, or combination of services, will soon be able to be delivered over networks constructed of many different technologies. Quadruple-play offerings – fixed and mobile, Internet and television services – are already being introduced in some countries.

Therefore distinctions between technologies and services in the allocation and assignment of frequencies will no longer be necessary or appropriate in many cases. New flexibilities are made possible for both allocation and assignment decisions. A new spectrum management problem is implementing these flexibilities in an environment where specific frequency rights and protections have been given in existing licences, which are mostly technology or service specific, and which will require modifications that most licensees will not necessarily find in their interest. For example, most 2G mobile licences are restricted to the provision of voice services when the 2G frequencies could be used to provide at least some data, that is, 3G services, and reduce the scarcity. But this would reduce the value of the 3G licences that have been assigned. Resolving these inherited inefficiency problems is a major challenge for spectrum management.

Another consequence of technological convergence is that content regulation cannot anymore be made specific by linking it to broadcasting infrastructure in the traditional sense. Moreover, convergence of markets will lead to convergence of regulation of the wired and the wireless environments.

Technology-based Improvements

Other technological developments relate primarily to the technical capabilities of transmitters and receivers. This includes for instance the transition from time division multiple access (TDMA) in 2G systems to code division multiple access (CDMA), a form of spread spectrum technology, in 3G systems, as well as the transition to an ALL-IP (Internet protocol) based infrastructure as foreseen in 4G mobile systems, also denoted as Long-Term Evolution (LTE) technology.

Spread spectrum technologies code and send the radio signal over a range of frequencies thereby enabling more efficient use of frequencies. Under the present system the permissible range of frequencies must be licensed to the particular service provider or user. However, it has been successfully applied in a special case where designated frequencies have been specified as 'unlicensed', WiFi – based on the IEEE 802.11 series of standards – now being used in an increasing variety of locations and circumstances. (Lemstra, 2010).[5] This unexpected success of a licence-exempt regime has stimulated the debate whether spectrum allocations should be based on exclusive usage or property rights, or can be better allocated in the form of collective usage rights in a 'commons' (Bauer, 2002; Benkler, 2002; Noam, 1995).

Ultra wide-band technology allows transmission at high data rates sending the information across a range of frequencies at a low power level that is below the 'noise floor' for creating harmful interference. But this requires permission to use frequencies that have been licensed to others and changes the rights and responsibilities of licences that have already been given (spectrum easements).

'Smart' antennae can send and receive pinpointed transmissions that reduce potential interference. Examples include switched beam antennae that can adjust sensitivity of transmission to a sector where the user is located, and adaptive array antennae that can dynamically track the user and transmit the signal to an actual position.

Software-defined radio (SDR) offers the potential for computer-programmed transmission and reception. Though still at a nascent stage of development, SDR has the potential to make adaptive spectrum management possible through dynamic channel assignments. A type of SDR that has attracted considerable interest is 'cognitive radio', which can track the user's 'frequency environment' over time and space with respect to other radio frequencies in its environment, and frequencies momentarily available for communication. Initial applications are foreseen in so-called television white spaces. These white spaces refer to guard bands between television channels, and television channels that have been allocated but may not be in use in certain geographical areas. The efficiency of the radio spectrum can be increased significantly if these white spaces can be used by cognitive radio devices, (Doyle 2009; Anker, 2010). In 2008 the FCC made an initial ruling to use these white spaces for improving broadband access in rural areas on a licence-exempt basis. In September 2010 the FCC took a next step towards implementation, based on the use of a database for coordinating the spectrum use.

New 'mesh' networks operate on the principle that each terminal located in the network can simultaneously connect and communicate with all terminals within its range. Thus they all can be used to pass along messages in the transmission path from sender to receiver. This can provide a cost-effective and robust way of expanding network access and communication capacity within networks where all users have sophisticated mesh network terminals. The theoretical possibilities of large mesh networks, together with potential widespread applications of the other technological improvements, have prompted some observers to suggest that portions of the spectrum may be most efficiently used if treated as a spectrum 'commons', freely accessible to anyone with a compatible terminal. But this assumes that congestion, interference, capacity limitations and efficient frequency use can all be automated effectively in these networks (Lansford, 2004). Although this seems highly unlikely given the explosive growth in demand for spectrum-based communication, experimentation with the possibilities may yield further opportunities for improving spectrum management efficiency.

These new technological developments introduce improved technical parameters for communication that differ from those upon which the existing spectrum management system has been established and existing licences awarded. They introduce increasing capabilities for flexibility in the selection and use of frequencies by transmitting and receiving terminals that make the existing technology and service-specific allocations and assignments unnecessarily restrictive and inefficient rather than orderly and efficient. They raise issues of reassessing the existing structure of technical standards, allocation criteria and assignment methods, and moving to a more flexible system for allocation and licensing, and of fashioning a transition path from the established system of spectrum management to a new system, a tall order that has never been previously attempted.

Regional Harmonization

In Europe, as an example, cooperation in the field of telecommunications regulations became formalized in 1959 through the establishment of the European Conference of Postal and Telecommunications Administrations (CEPT). Meanwhile CEPT membership has expanded to include the Central and Eastern European countries; with 48 members it covers almost the entire geographical area of Europe.

Within CEPT the European Radiocommunications Committee (ERC) addresses all radio communication related matters, with the prime objective to develop European harmonization measures for the allocation and the use of radio frequencies. The ERC pursues consensus among administrations for the development of decisions and recommendations; the implementation decision of these policy measures remains at the national level.

In the context of promoting a common market in the European Union, the European Commission considers the management of the radio spectrum as a means to that end. In 1998 it published the 'Green Paper on Radio Spectrum Policy' aimed at fostering the development of the internal market for and competition in radiocommunications equipment and services, in particular pan-European systems and services for which radio spectrum availability should be secured. Moreover, coherence between radio spectrum and standardization policies should be ensured. The latter reflects the success of the harmonization effort associated with the global system for mobile communication (GSM) project, being initiated in 1982 and having led to a coordinated launch of digital mobile communication services in seven countries by 13 operators in 1992.

As an example of an important step towards harmonization, the European Commission published the R&TTE Directive in 1999. This Directive obviated the need for national approval regulations. All equipment that is placed on the market in the European Union must comply with a set of essential requirements, covering the protection of health and safety, electromagnetic emission and immunity of the equipment and effective use of the radio spectrum so as to avoid harmful interference. Equipment manufactured in accordance with a harmonized standard may be placed on the market based on the manufacturer's own declaration of conformity (self-declaration).

Another intervention of the European Commission in radio spectrum management came with the introduction of the new regulatory framework. This framework was aimed at further liberalization, harmonization and simplification of the regulations in the telecommunications sector. The Framework Directive (2002/21/EC) states that the allocation and assignment of radio frequencies by national regulatory authorities are to be based on objective, transparent, non-discriminatory and proportionate criteria. The related Authorization Directive (2002/20/EC) specifies the circumstances under which the granting of an individual licence is being allowed and limits the conditions that may be attached to the rights of use for radio frequencies. In November 2006 a Decision of the European Commission was published on the harmonization of the radio spectrum for use by short-range devices.[6]

Digital Dividend

As a consequence of the transition from analogue to digital encoding of television broadcasting signals, less frequency bandwidth is required to transfer the same amount of information, hence, previously assigned spectrum is becoming available for reallocation and assignment. In Europe, this so-called Digital Dividend is earmarked for regional harmonization and in support of pan-European services. Unfortunately, the harmonization effort is made complex by national differences in timing of the transition.[7]

INTRODUCING ECONOMIC CRITERIA

Economists have criticized the traditional method of allocating spectrum and assigning frequency licences under the first-come, first-served system at a nominal cost to the licensee because there is no incentive to conserve the use of spectrum (Herzel, 1951), (Coase, 1959; Melody, 1980).[8] Essentially it is a valuable free resource for those who obtain licences. It artificially stimulates demand to use frequencies inefficiently and hoard them for the future. These inefficient practices may have been acceptable when there was ample supply of spectrum relative to demand, but as demand has grown the cost of this inefficiency has escalated dramatically. This has stimulated attention to possibilities for applying economic criteria to encourage the more efficient use of the spectrum (Forge and Blackman, 2006).

Administration Fees

Beginning in the 1970s, spectrum managers in some countries began placing more stringent requirements on licensees, imposing a 'use it or lose it' condition within a specified time period. Administration fees were charged to reflect the costs of spectrum management, thereby forcing recognition on users that frequency licences carry a 'noticeable' ongoing cost. This improved the situation significantly, but these fees were still far less than the economic value of the assigned frequencies. Economic analysis suggests that the charges for frequency licences should bear some relation to their economic value if a high degree of spectrum efficiency is to be achieved, especially as the greatest demand for licences is now coming from the private sector where the licences are a valuable asset that benefits the companies and their stockholders. As the spectrum is a public resource, the economic value of the spectrum should be captured for the public, and the licence fee is the simplest and most direct way to claim that value. The unsettled debate is about how the economic value should be determined.

Spectrum Markets

This has opened a discussion of the extent to which spectrum markets might be introduced into the spectrum management process to improve the efficiency of the assignments and utilization of spectrum. (Cave et al., 2007). But this is not an easy task as spectrum markets inherently are highly imperfect and tend to reflect monopoly/oligopoly rather than competitive market conditions. Thus, there are severe constraints on possibilities for introducing the economic criteria necessary for making spectrum markets efficient, as judged by competitive market standards. This severely limits the extent to which markets can be used in spectrum management.

Despite the growth in private sector demand for frequencies, the great majority of spectrum is still used by public and non-market sectors ranging from the military, emergency and security agencies to public broadcasting. The demands by these non-market actors and their capabilities for bidding for spectrum are not determined by market conditions, but for the most part by government budget allocations (Melody, 1980). Further, if a market system of spectrum management is imposed only on private sector users, this raises issues of establishing justifiable criteria for the very different treatment between public and private sector uses and users, and within the very large public non-market sector. Thus markets cannot be expected to improve efficiency for the great majority of spectrum use and other new approaches will be needed for them.

For private sector spectrum uses and users where market forces are being considered, economic efficiency requires that the essential conditions for effective competition be approximated. These include minimal barriers to entry, that is, a sufficient number of competitors bidding for licences to ensure effective competition in the markets where the frequencies will be used, as well as opportunities for free entry or exit at any time. These are difficult and often impossible conditions to meet when allocating and assigning frequencies.

The most significant factor determining entry conditions is the amount of spectrum that will be allocated for particular uses, for example, 4G mobile. Although the equipment manufacturers and operators in an industry that uses the spectrum will lobby heavily for a large allocation, there is at present no way of subjecting the spectrum allocation decisions to market determinations. The most important precondition for implementing efficient market criteria cannot be met. The amount of spectrum allocated for particular uses, such as 4G mobile, will be determined not by changing supply and demand conditions in dynamic markets, but rather by the traditional administrative criteria. This casts serious doubt on how markets in frequency assignments could improve efficiency if the allocated spectrum to be assigned is not determined by the same market conditions.

If spectrum managers could introduce new criteria into the allocation process whereby they allocate an appropriate amount of spectrum to ensure that competitive market efficiency will be achieved in their frequency assignment process, the application of market criteria in the assignment process might be expected to improve spectrum efficiency. But such criteria have yet to be developed. Nevertheless, recognizing the inherited inefficiencies in the traditional system of spectrum management, the application of market criteria has been introduced into the frequency assignment process in the USA and in some countries in Europe, New Zealand and Hong Kong for a limited portion of the spectrum, primarily on assignments for mobile and wireless services to private sector commercial operators. Suboptimal economic arrangements still may be a significant improvement over an inefficient administrative system that pays even less attention to economic efficiency in managing the spectrum resource (Melody, 1980).

Assigning Frequencies by Price Auctions

A price auction for assigning frequency licences eliminates the weaknesses of administrative assignment characterized by subjective judgements of a spectrum manager, by applying a single clear criterion for selection. The parties that place the highest economic values on particular frequency licences at the time of auction will obtain a licence. But this use of a single market criterion for assignment also requires judgements on a number of factors in designing the market parameters for the auction. These include the amount of spectrum to be allocated for the auction, the bandwidth that will be allocated for each licence, the specific geographical coverage for each licence, the duration of the licence, as well as technical specifications such as power levels and interference standards. Other important considerations for bidders include such matters as the flexibility for using previously allocated frequencies for new services and possibilities for future allocations of additional frequencies for the same services in later auctions (Cave et al., 2007).

To be effective in promoting spectrum efficiency, the market conditions for price auctions must be carefully established to meet the efficiency objective. In addition the particular method of auctioning employed must be selected to facilitate both spectrum efficiency and the effective rollout of the new communication services for which the licences are being granted. Thus, the markets for auctioning frequency licences are not free or natural markets in any sense. They are markets designed in every respect by spectrum policies and regulations. Unfortunately the importance of the many normative factors necessary to establish the appropriate market conditions for an efficient auction have not been recognized in most of the early experimentation with price auctions (Akalu, 2007).

In recent years, in some countries spectrum authorities have adopted price auctions as the standard for determining certain licence assignments under conditions of significant scarcity. This was initially undertaken by the US Federal Communications Commission (FCC) in the early 1990s, and most notably for 3G mobile licences in some European countries in the late 1990s and elsewhere. The early experience with price auctions has produced inconclusive results. The 3G auctions were based upon an insufficient allocation of spectrum for 3G services, creating a condition of policy-determined scarcity. Price auctions under these conditions brought enormous sums of money to some national governments, but placed the winning bidders in financial difficulty so that investments in 3G network and market development were delayed by several years (Grimm et al., 2003; Klemperer, 2003; Van Damme, 2001)

With only a few high priced 3G licences in each national market, new entry is virtually closed and government policy-makers have been forced to consider modifying other policies, including exemptions from certain competition policies and regulatory responsibilities (for example, by allowing sharing of infrastructures), in order to assist the 3G licensees having difficulty meeting their network rollout obligations. By changing the frequency assignment method to price auctions, spectrum authorities eliminated the inefficiencies of the non-price administrative assignment methods, but introduced inefficiencies associated with the monopoly power of spectrum managers, restricted competition, delayed market development and ultimately higher prices for 3G services.

These price auctions for the 3G licences took place within a structure that was designed to be far away from the efficient competitive market model, approximating concentrated oligopoly. National governments, acting as monopoly owners of the spectrum resource, restricted the amount of spectrum allocated for 3G services so there were only a few licences, and then set an auction process designed to maximize the payment of monopoly rents. These payments were far greater than would be required by the efficiency standard of competitive markets. An unfortunate side effect of these 3G price auctions is that many national governments now see spectrum management primarily as a superb method for raising funds for the government rather than facilitating the development of new networks and services. What began as an attempt to achieve the competitive market efficiency standard for pricing frequencies, ended as a simple taxation device for maximizing the monopoly rent to government from government-created spectrum scarcity (Melody, 2001).

Fortunately not all governments chose to use simple maximum price auctions for assigning 3G licences as a vehicle for maximizing revenue to the government. Some, for example, Hong Kong, adopted a method combining price auction, royalty payments as services were rolled out, and minimum rollout criteria in selecting licensees. This brought relatively lower initial payments, but the linking of royalty payments to market development provided a better alignment of incentives between licence payments and the rollout of 3G networks and services. Other countries, for example, Sweden, continued to have applicants compete on the traditional beauty contest criteria of capabilities, technical plans and network rollout commitments.

Nevertheless the negative effects of the 3G price auctions in Europe had a significant impact, not only in delaying the rollout of 3G networks and services, but also on the growth of the 3G equipment market where manufacturers were forced to delay their development and production plans. The lessons learned from the early experiences with 3G spectrum auctions for assigning frequency licences have been very costly. They do not provide sufficient evidence to justify abandoning price auctions. But they do clearly demonstrate the importance of attending much more carefully to the institutional arrangements that establish the structure and conditions for auctioning frequencies, including all aspects of the related allocation process and the network and services development purposes for assigning the frequencies.

Trading Frequency Licences

Under traditional spectrum management regulation, holders of frequency licences may not sell them. The frequencies must be used or returned. If a firm that holds licences is purchased by another, the transfer of licences normally must be approved by spectrum authorities, and generally are approved. A few countries are exploring the possibilities for allowing market trading in spectrum licences, for example, Australia, El Salvador, Guatemala, New Zealand, and the UK. This proposal also arises from the theory of competitive markets where market exchanges can facilitate increased efficiency as resources are sold for higher value uses. Herzel (1951). This has been proposed as a change in spectrum policy that would help firms that had won 3G licences at price auctions but were not in a financial position to build out new 3G services networks (Analysys et al., 2004).

Where frequency trading has been permitted, for example, in the UK, there has been surprisingly little trading taking place, raising questions about how these markets are likely to function (Akalu, 2010). Clearly there needs to be more attention to the specific market conditions under which trading does, or would take place. If they are such that trading values are likely to be determined by standards of competitive market efficiency, the trading can be expected to improve the efficiency of spectrum utilization. If the trading values are governed by artificial scarcity, speculation, gains from holding rather than using scarce resources, or anti-competitive objectives, then trading will not yield the efficiency results desired and direct regulation may be needed (Cave, 2010). The experience to date suggests that the consequences of trading in spectrum markets will be determined by the institutional arrangements governing allocation and assignment practices, and the design and regulation of the trading markets.

CONCLUSIONS

In light of the increasing realization of the weaknesses of the traditional system of spectrum management, in recent years steps have been taken to improve spectrum management decision-making processes at all levels by liberalizing participation in spectrum management processes and introducing economic criteria into spectrum management decision-making. Experience has shown that if economic efficiency standards and market mechanisms are to be adopted effectively in future spectrum management, then significant institutional reforms must be implemented at all levels – global, regional and national.

Changes are under way to speed up decision-making processes and introduce greater flexibility for both spectrum managers and licensees to take advantage of the full range of opportunities associated with ongoing technological improvements. Greater attention is being paid to forecasting future spectrum needs and more comprehensive spectrum planning is being undertaken in some countries. Possibilities for introducing greater flexibility into spectrum allocation and assignment structures are being considered and implemented.

However, there remains a serious question whether spectrum management institutions will be able to adapt in a timely and effective manner to the rapidly changing technologies and services markets. Price auctions in particular could turn out to be a diversion from the real challenges of spectrum management reform unless they are carefully designed to meet efficiency criteria and integrated into the larger set of reforms for spectrum allocation, frequency assignment and trading. Both the traditional methods of administrative assignment and the newer methods of price auctions and spectrum trading have particular strengths and weaknesses that need more detailed examination than they have received to date. The challenge for the future is designing assignment methods that can capture the strengths and avoid the weaknesses of each. This will become increasingly important as new opportunities for classifying frequencies as 'unlicensed' become possible, as the US FCC has proposed to do with the 'white space' frequencies made available with the conversion of television broadcast signals from analogue to digital.[9]

Experience has demonstrated that spectrum managers should never undertake licence assignments without first examining the spectrum allocation possibilities, as allocation decisions are the most important factor that structures the markets in which the frequency assignments will take place. Improved allocations provide the essential foundation for improved assignments. Assignment methods cannot compensate for poor allocation decisions. In the more liberalized and flexible system of spectrum management of the future the tasks of management will be more complex than in the past, require more rapid decision-making and be more important for sector development.

NOTES

1. Thanks to the radio spectrum management group at the Section Economics of Infrastructures, Delft University of Technology, and especially Rajen Akalu, Peter Anker (Senior policy adviser Ministry of Economic Affairs), André Arts (Economic adviser Radiocommunications Agency), Helmut Leonhard (Scientific advisor Radiocommunications Agency) and Vic Hayes for discussions on many of the issues reviewed in this chapter.

2. In 1873 James Clerk Maxwell posited the concept of electro-magnetic (E-M) waves and in 1888 Heinrich Herz demonstrated the creation and detection of E-M waves.
3. The allocation of the 900 MHz band to land mobile radio was made by the WRC in 1979. The GSM project started in 1982 and the decision to go digital was formalized in 1986 (Manninen, 2002). The initial allocation for GSM was 2 x 25 MHz (paired up and down channel) in the 900 MHz band, later an extension with 2 x 10 MHz in the 900 MHz band was made and 2 x 72.5 MHz was allocated for GSM1800 (DCS1800) in the 1800 MHz band. The negotiations in the WRC and ERC resulted in an allocation in Europe of 2 x 60MHz + 35 MHz (unpaired) in the 2.1 GHz band for UMTS. To be extended with 190 MHz in the 2.6 GHz band in a proposal to WRC2000. Note that taking into use of radio spectrum following an allocation agreement in WRC typically takes several years.
4. The combined proceeds of the auctions in the UK, Germany, the Netherlands and Italy, which took place at the peak of the telecom/Internet bubble in 2000, were approx. US$ 90 billion. The UK auction raised US$600 per inhabitant.
5. The global success of Wi-Fi can be traced back to the 1985 decision by the FCC to allow radio communications in the 900 MHz, 2.4 and 5 GHz bands designated for Industrial, Scientific and Medical applications provided spread spectrum technology would be used. The NCR Corporation initiated the standardization effort within IEEE.
6. Increasing flexibility is addressed in, for instance, 'Radio Spectrum Policy Group Opinion on Wireless Access Platforms for Electronic Communications Services (WAPECS), A more flexible spectrum management approach', RSPG05-102, November 2005.
7. See for instance 'Report for the European Commission, exploiting the digital dividend – a European approach'. Analysys Mason. 14 August 2009.
8. For more recent literature on the topic see for instance Hazlett (1990) and Moss and Fein (2003).
9. New York Times 2010.

REFERENCES

Aitkin, H. (1994), 'Allocating the spectrum: the origins of radio regulation', *Technology and Culture*, **35**(4), 686–716.

Akalu, R. (2007), 'Opportunity costs and spectrum reform', *info*, **9**(6), 3–9.

Akalu, R. (2010), 'Why have there been so few spectrum trades in the UK: lessons for Europe', *info*, **12**(1), 10–17.

Analysys, DotEcon and Hogan & Hartson (2004), 'Study on conditions and options in introducing secondary trading of radio spectrum in the European Community', report for the European Commission, May.

Anker, P. (2010), 'Cognitive radio, the market and the regulator', paper prepared for the IEEE DySPAN 2010 symposium, Singapore, 6–9 April.

Arnbak, J. (1997), 'Managing the radio spectrum in the environment,' in W.H. Melody (ed.), *Telecom Reform Principles, Policies and Regulatory Practices*, Lyngby, Denmark: Technical University of Denmark, pp. 131–9.

Bauer, J. (2002), 'Spectrum management: private property rights or commons', in R. Mansell, R. Samarajiva and A. Mahan (eds), *Networking Knowledge for Information Societies: Institutions and Intervention*, Delft: DUP Science Delft University Press, pp. 118–26.

Benkler, Y. (2002), 'Some economics of wireless communications', *Harvard Journal of Law and Techology*, **16**(1), 25–83.

Cave, M. (2002), *Review of Radio Spectrum Management*, London: HM Treasury and DTI.

Cave, M. (2010), 'Anti-competitive behaviour in spectrum markets', *Telecommunications Policy*, **34**(June–July), 251–61.

Cave, M., C. Doyle and W. Webb (2007), *Essentials of Modern Spectrum Management*, Cambridge: Cambridge University Press.

Coase, R. (1959), 'The Federal Communications Commission', *Journal on Law and Economics*, **2**, 1–40.

Douglas, S. (1987), *Inventing American Broadcasting*, Baltimore, MO: Johns Hopkins University Press.

Doyle, L. (2009), *Essential of Cognitive Radio*, Cambridge: Cambridge University Press.

Forge, S. and C. Blackman (2006), 'Spectrum for the next radio revolution: the economic and technical case for collective use', *info*, **8**(2), 6–17.

Grimm, V., F. Riedel and E. Wolfstetter (2003), 'The third generation (UMTS) spectrum auction in Germany', in G. Illing, and U. Klüh (eds), *Spectrum Auctions and Competition in Telecommunications*, Cambridge, MA: MIT Press.

Gruber, H. (2001), 'Spectrum limits and competition in mobile markets: the role of licence fees', *Telecommunications Policy*, **25**(1–2), 59–70.
Hazlett, T. (1990), 'The rationality of U.S. regulation of the broadcast spectrum', *Journal of Law and Economics*, **33**(1), 133–75.
Herzel, L. (1951), 'Public interest and the market in color television regulation', *Chicago Law Review*, **18**, 802–18.
International Telecommunications Union (ITU) (1965), *From Semaphore to Satellite*, Geneva: International Telecommunication Union (accessed 10 May 2010).
International Telecommunications Union (ITU) (2004), 'Background paper: advanced wireless technologies and spectrum management', Document RSM/08 February, available online at www.itu.int/osg/spu/ni/spectrum/RSM-AWT.pdf.
Klemperer, P. (2003), 'How (not) to run auctions: the European 3G telecom auctions', in G. Illing and U. Klüh (eds), *Spectrum Auctions and Competition in Telecommunications*, Cambridge, MA: MIT Press.
Lansford, J. (2004), UWB coexistence and cognitive radio, *Ultra Wideband Systems, 2004*, Joint with Conference on Ultrawideband Systems and Technologies, Joint UWBST & IWUWBS, pp. 35–9.
Lemstra, W., V. Hayes and J. Groenewegen (eds) (2010), *The Innovation Journey of Wi-Fi, the Road to Global Success*, Cambridge: Cambridge University Press.
Levin, H. (1971), *The Invisible Resource*, Baltimore, MD: Johns Hopkins Universities Press.
Manninen, A.T. (2002), 'Eleboration of NMT and GSM standards: from idea to market', dissertation, University of Jyvåskylå, Jyvåskylå, Finland.
Melody, W. (1980), 'Radio spectrum allocation: role of the market', *American Economic Review*, **70**(2), 393–7.
Melody, W. (2001), 'Spectrum auctions and efficient resource allocation: the 3G experience in Europe', *info*, **3**(1), 5–10.
Melody, W. (2008), 'Spectrum management for information societies', in J. Ure (ed.), *Telecommunications Development in Asia*, Hong Kong: Hong Kong University Press.
Moss, D. and M. Fein (2003), 'Radio regulation revisited: Coase, the FCC, and the public interest', *The Journal of Policy History*, **15**(4), 389–416.
New York Times (2010), 'F.C.C. likely to open airwaves to wireless', 13 September.
Noam, E. (1995), 'Taking the next step beyond spectrum auctions: open spectrum access', available online at: www.vii.org/papers/citinoa9.htm (accessed 22 March 2010).
Organisation for Economic Co-operation and Development (OECD) (2005), 'Secondary markets for spectrum: Policy issues', available online at: www.oecd.org/dataoecd/59/2/34758854.pdf (accessed 10 May 2010).
Slotten, H. (2000), *Radio and Television Regulation. Broadcast Technology in the United States, 1920–1969*, Baltimore, MD: Johns Hopkins University Press.
Van Damme, E. (2001), *The Dutch UMTS Auction in Retrospect*, The Hague: CPB.

9 Liberalization and the Internet

Milton L. Mueller and Wolter Lemstra

The Internet has brought about a new era in communications infrastructure. Since around 1994 the Internet protocols have provided the dominant networking platform for an expanding digital economy and digital society. The Internet is the basis for an open economy that links a growing array of digital devices to providers of information, software and services. It is no longer just a computer network, but also a way of linking computers with television, gaming systems and handheld mobile devices. Also, the Internet has led to new forms of communication (increasingly computer mediated), new modes of transacting (electronically mediated), new perspectives on property rights, new modes of organizing (networked) and new modes of social engagement (virtual networks).

INTERNET: DEFINITION AND DISTINCTIVE FEATURES

The Internet is not a hardware standard or a physical infrastructure. The Internet is a virtual infrastructure based on a set of software instructions (known as 'protocols') for transmitting data over networks.[1] The Internet protocols – TCP/IP – can be used on wireless and wired networks, on copper and fibre networks, which can be used as the underlying communication platform for almost any kind of higher-level software application, such as the World Wide Web, word processing, streaming video, voice communication or games. It is a myth that the Internet was developed to provide the military with a communications capability that could survive a nuclear war. In reality, the Internet protocols were designed to facilitate the movement of all kinds of data across independently managed networks and different physical media (Abbate, 2000). Indeed, to speak of 'the' Internet is actually a misnomer, though one that is probably impossible to eradicate. Internetworking is a *process* that links many networks; it is not a single thing.

Like many contemporary ICT standards (for example, computer operating systems, the major application suites, mobile phone handset specifications), TCP/IP is transnational if not completely global in scope. Although the Internet enabled bottom-up development of local networks, services, content and applications, it is based on a common set of open, global standards. Moreover, because it is based on software, the Internet creates a virtual space that operates independently of physical geography. From the standpoint of the Internet protocol, it does not matter whether the server with the IP address 98.128.123.155 is next door or 10000 miles away; it is just another host to connect to. The address space is based on autonomous systems – that is, on the organization with management responsibility for routing – not on geography. Of course, the virtual space created by the Internet is dependent on physical facilities that must be located somewhere, but unless special measures are taken to limit and confine communications within specific boundaries, the Internet will operate on a borderless basis.

INFRASTRUCTURE COMPONENTS FOR INTERNET SERVICES

In this section we distinguish between four elements of the infrastructure for Internet services, namely, the organizations and processes for defining Internet standards, the institutions that regulate and coordinate the allocation and assignment of the virtual resources created by the Internet standards, Internet service providers (ISPs) who supply access and connectivity, and providers of content, applications and services. For the sake of brevity our discussion groups the first two (standardization and resource allocation/assignment) together in the same section on support for the Internet's logical infrastructure.

Institutional Support for the Internet's Logical Infrastructure

The core Internet standards are developed and maintained by the Internet Engineering Task Force (IETF). The IETF started in the early 1980s as a small and informal group of computer science researchers funded by the US government. It has since evolved into a more elaborate standards development and maintenance organization. Along the way it pioneered a new mode of organization, relying heavily on email lists, in which participation is open and based more on the expertise and willingness of an individual to contribute than on any formal role as representative of a government or organization. The standards have become identified by the initial Request For Comments (RFC) and a serial number. In this model, publication of the standards documents is open and free (Vincent and Camp, 2004).

Two appointed groups within the IETF, the Internet Architecture Board (IAB) and the Internet Engineering Steering Group (IESG), take responsibility for overseeing certain aspects of the standards development process. Standards that emerge from the working groups with a 'rough' or declared consensus are notified to the entire IETF. They must then be approved by the IESG. To the extent that IETF has any legal identity, it is derived from the Internet Society (ISOC), which funds the RFC Editor, reviews and approves the selection of IAB members (which in turn reviews and approves IESG members), and generally provides something close to a central organizational focal point for many of those active in IETF.[2]

The basic Internet standards establish several critical virtual resources, such as internet protocol (IP) addresses, autonomous system numbers (which are used in routing), domain names and the domain name system root, and protocol port numbers. In order to maintain global compatibility, the use of these resources must be coordinated. These coordination functions are performed by a transnational regime that consists of the Internet Corporation for Assigned Names and Numbers (ICANN) and five regional Internet address registries (RIRs). ICANN acts under the political oversight of the US government (Mueller, 2002).

ICANN was incorporated in 1998 in response to a request from the US Department of Commerce.[3] Regional Internet address registries were created in the 1990s to regionalize address allocation and assignment as the Internet grew.[4] The European region registry, RIPE-NCC, was created in 1991 and an Asia-Pacific region registry was created in 1995. In 1997, parallel to the creation of ICANN, the address administration functions performed by several US government contractors were privatized and placed in the hands of a new nonprofit entity known as the American Registry for Internet Numbers (ARIN).

Two new RIRs, one in Latin America and another in Africa, were created this decade. Like ICANN, all five of the RIRs rely on a private sector-based, contractual model of governance. Organized as nonprofit membership organizations, their members are ISPs, hosting companies, and large organizations with private networks (corporations, universities, and so on).

Control over the allocation and assignment of critical Internet resources gives these organizations some leverage over economic and even social policy. For example, ICANN's ability to approve any new top-level domain names gives it control over entry into the domain name registry market and leverage over the registries policies and practices. Likewise, a regional address registry's authority over the registration and assignment of address resources gives it some ability to ration or conserve addresses and attach conditions to their usage. Just as radio spectrum managers at the national level can attach conditions to and regulate to some degree the entities to which they grant spectrum, so the granting of IP addresses and top-level domain names can assume a regulatory function.

Internet protocol address assignments are made in a hierarchical fashion. At the top of the hierarchy is ICANN, which distributes large blocks of 16 777 216 addresses to one of five RIRs. The RIRs then accept applications from organizations with networks that need addresses within their territory. Some larger blocks may be assigned directly to end-user organizations, but most will go to Internet service providers who will then reassign them to their customers.[5]

Domain names, another virtual resource created by the Internet standards, are the user-friendly labels that we attach to computers, web servers, mail servers and the like. The domain name system (DNS) is a protocol for hierarchical assignment of these names in a way that maintains global uniqueness while allowing local administrators to choose their own names at lower levels of the hierarchy.

The top of the DNS hierarchy consists of a single authoritative list, known as the 'root zone file', that tells any computer in the world which top-level domains exist and at which IP address it can find the domain names registered under them. Global interconnection of websites, email and all other services using domain names are dependent upon the maintenance of a unique entry for their top-level domain in the root of the DNS so that they can remain globally interconnected.

In addition to ICANN, there are 13 operators of root servers around the world. Put in its simplest terms, root servers are the computers that propagate the authoritative contents of the root zone file to all users of the DNS in the world. The root servers handle billions of queries every day generated by Internet users of the DNS. Ten of the authoritative root servers are located in the USA; the other three are in London, Sweden and Japan. The initial set of root server operators is simply a holdover from the early days of the Internet; there is no formal, institutionalized process for selecting or replacing them. And there is only a loose and relatively informal agreement among the root server operators that they will distribute only the root zone file generated by ICANN.

Internet Service Providers

As the Internet evolved in the realm of public and private research institutes, its users typically gained access to the Internet through terminals and PCs connected to a local area network, which through a server would provide access to a nearby Internet exchange (IX). Around 1990 the Internet started to open up to access by the wider public. In the first stage of the public Internet's evolution, access was provided through the telephone network using dial-up modems. Data rates started at 9.6 kbits/s; as modem technologies improved the speed eventually reached a plateau at 56 kbits/s. At the other end of the connection, modem banks were installed; these were connected to an access server which would be linked to a nearby node of the Internet using a dedicated leased telecommunication circuit. Organizations providing these Internet access facilities were called internet service providers (ISPs).[6]

To manage access, the ISP provides authentication and authorization services as well as accounting and billing services. Note that the dial-up call charges fall to the local telephone network operator. As the Internet assumes that computers are permanently connected the ISP typically acts as a proxy agent providing a store-and-forward function for, for example, sending and receiving email and hosting web pages. Through the creation of a 'virtual presence' the user may access the ISP from different locations (Huston, 1999).

As telecommunication technology evolved, higher speed, digital access lines were introduced under the name of DSL (digital subscriber line) or ADSL (asymmetrical-DSL).[7] These provided speeds from 384 kbit/s into the Mbit/s range (actual speeds depend on the attenuation of the signal on the copper cable, that is, the distance between the customer premises and the telephone exchange). The DSL signal is carried above the frequency band used for analogue telephony and provides for a permanent connection between the customer premise equipment and the network access server of the ISP, hence the 'always-on' notion of the Internet. The data rate on cable television networks has also been increased regularly, through new versions of the standards for high-speed data transmission, such as data over cable service interface specification (DOCSIS).

According to the Organization for Economic Co-operation and Developmant (OECD) dial-up access subscribers accounted for 91 per cent of all fixed line Internet subscribers in 2000. By the end of 2005 that number had dropped to 40 per cent, and by the end of 2007 it had declined to 10 per cent (OECD, 2009).

In this set-up the serving area of an ISP is in principle unlimited. However, to limit access costs to the users, typically an ISP will establish a point-of presence in each (major) local serving area.[8] In serving a larger area the ISP will become a regional ISP using multiple network access servers, and acquire or lease their own long-distance transmission facilities to connect their local access points.

In connecting to the Internet, each computer terminal needs an IP address and for a website a domain name may be required. Internet service providers will typically serve as intermediaries for their customers to obtain these unique identifiers from a regional address registry, see the previous section.

In the telephone network, connections are based on circuit switching, that is, for the duration of the call a dedicated two-way connection of 64 kbit/s is established between origin and destination. The Internet in contrast uses packet switching, which is much more suitable for the varying length of messages being transmitted and the differences in upload and download data rates (email, file transfer, video downloads). Each message to be transferred is broken down in fixed length packets (IP datagram) to which the originating and destination IP address is added. The packets are then transferred from the host server to the nearest Internet exchange or network to be routed on a link by link basis towards its destination on a best effort basis. The forwarding of the packets is based on routing tables that keep a current image of the topology of the network, being provided with information on the available links towards the desired destination. The actual speed of the transfer depends on the capacity of the underlying transmission network and the traffic load. Between Internet exchanges, use is made of dedicated high-capacity transmission links, often optical fibre with data rates into the Gbit/s range.

As noted earlier, the TCP/IP protocol is software and thus decouples the higher level application, for instance email or Skype, from the properties of the underlying communication media, such as copper, fibre, radio waves. Hence, in principle the Internet renders obsolete communication networks optimized for a single application, such as the telephone network or the cable television distribution network.[9]

As the Internet is a collection of independent networks, ISPs need to cooperate and negotiate to interconnect with each other; ISPs need each other to provide universal connectivity. Interconnection arrangements among ISPs depend on their relative size and thus the value obtained from exchanging traffic. If ISPs are of unequal size, the smaller one pays its larger interconnection partner to carry its packets (a transit agreement). If they are of roughly equal size or obtain about the same amount of value from interconnecting, they enter into what is known as a 'peering' agreement. Peering is based on the principle of 'sender keep all'; that is, an assumption that the amount of originating and terminating traffic is approximately equal and does not warrant a costly accounting arrangement.

Peering and transit interconnection agreements are bilateral. Another method of interconnection involves an IX and is multilateral. Many ISPs converge on a single facility to exchange traffic. Internet exchanges are usually neutral nonprofit entities. The Amsterdam Internet Exchange (AMS-IX) Association is a typical example. In 1997, about 20 internet service providers and telecommunication carriers formed the AMS-IX Association. In 2009 there are over 300 members of the Association, the largest number of exchange connected parties worldwide (AMSIX, 2009).

Content and Application Providers

Content and services are the consumption goods that complement the demand for Internet bandwidth. One of the revolutionary aspects of the Internet was its enhanced interactivity. Its design was based on the end to end principle, which meant that the transmission network was architected to be a simple mover of bits, allowing the applications and services to be defined by the computers at the end points (Saltzer et al., 1984). To develop and offer new services one does not need permission or even the participation of the network intermediary. There is no need for a gatekeeper. These design principles have fostered a tremendous amount of innovation in this segment of the market.

In addition to a movement of familiar services and products to the online environment, such as music and movie distribution, the Internet has also fostered the phenomenon of user-generated content and what is known as peer production (Benkler, 2006; Wunsch-Vincent and Vickery, 2007). User-generated content refers to online content that is produced by users of websites as opposed to traditional, formally organized media producers such as broadcasters and production companies. Blogs, social networking sites, wikis and various forms of peer to peer collaborative networks are examples.

As a platform of electronic commerce, the Internet can impact both informational content and services, and provide a new transactional channel for trade in physical goods. Internet commerce includes services and applications (email, photograph storage, maps, blogs, online applications, and so on), computer and video games, streaming video and DVDs, digitized music, digitized books, online newspapers, advertising, online banking and financial services, online retailing and online sales of software.

A growing share of the Internet-based market for content and applications is offered for free, and money is made on advertising. Google, Yahoo and similar portals offer users a growing array of online tools and services, ranging from maps and directions to simple email, all at no charge. User-generated video content sites such as YouTube now host hundreds of millions of short videos posted for free. These user-generated content sites, like social networking sites such as Facebook, make money from exposing these users to advertisements.

LIBERALIZATION AND THE INTERNET

There is a close relationship between the liberalization of physical telecommunications infrastructure and the subsequent rise and growth of a new Internet infrastructure. By privatizing, unbundling, opening to competition and deregulating telecommunications services across the world, regulators were – mostly inadvertently – paving the way for the emergence of an open, competitive, mostly unregulated Internet.

Historical Process

The research and development activity that eventually led to the commercial Internet originated in US government research contracts with universities and commercial firms. This was not, however, the highly classified research one associates with weapons. It was an outgrowth of what some have referred to as the *government-industrial-university complex.* In the case of the Internet, military agencies such as the Defense Advance Research Projects Administration (DARPA), looking for new ideas that would enhance American technical prowess, gave long-term funding for data communications development to computer scientists and engineers starting in the 1960s (National Research Council: Committee on Innovation in Computing and Communications, 1999). Elite universities such as the Massachusetts Institute of Technology (MIT) and the University of California Los Angeles (UCLA) were favoured in this programme, as were the private research firms clustered around them, such as Stanford Research Institute (SRI) and Bolt, Beranek, and Newman (BBN) Probably the most consequential result of government support was that the resulting standards were nonproprietary. The Internet standards became a relatively neutral arena around which the multiplicity of firms involved in data communications and networks could coordinate.

The Internet and Telecommunications Liberalization

One of the most important relationships between telecommunications infrastructure and Internet was forged through the US Federal Communications Commission's Computer Inquiries.[10] It was already evident in the mid-1960s that computers and telecommunications were converging. The FCC Computer Inquiries (there were three separate proceedings from 1966 to 1985) were first initiated to separate data processing services from basic telephone conveyance service in order to create a freer market not distorted or dominated by the huge AT&T monopoly (Brock, 1994). In order to foster competition and growth in the nascent computer services industry, the FCC created two distinct classes of service. One, known as basic telecommunications, continued to be regulated as a common carrier; the other, which became known as *enhanced services* or *data processing*, was left unregulated. The FCC defined as *telecommunications* a situation when the content of the message is transmitted over the network with no change in the content or form of the message. Data processing, on the other hand, was declared to involve 'the use of the computer for operations which include, inter alia, the functions of storing, retrieving, sorting, merging and calculating data, according to programmed instructions'.[11]

The FCC concluded that data processing services were 'essentially competitive.' '[D]ata processing services of all kinds are becoming available in larger volume and . . . there are no natural or economic barriers to free entry into the market for these services.' Consequently there was 'no public interest requirement for regulation by government of such activities.' The Commission also tried to carefully segregate the offering of competitive data-processing services by the monopolistic, regulated, territorial telephone companies.

This regulatory distinction between telecommunication (basic service) and data processing (enhanced or value-added services) diffused throughout the developed world in the 1980s, alongside the broader process of telecommunications liberalization. The concept of 'value-added services' as a more competitive, less regulated domain that depended in some way on underlying 'basic' conveyance service, typically more regulated and less competitive, became widely accepted. The revolutionary transformation this implied crept into the telecommunications regime almost by stealth; well into the early 1990s value-added services constituted a tiny portion of the overall telecommunication services market. Thus, trade negotiators and regulators seeking to protect their national telecommunication companies from foreign and domestic competition would eagerly offer market access in value-added information services as a concession to the free trade demands of other countries, even as they carefully sheltered basic voice services. Once the Internet was opened up to the public and the web browser made it accessible to a mass market after 1993, the Internet spread through the inviting channel. As a software-based protocol that moved packetized information among computers, the Internet was clearly an information service under the internationally agreed definitions of 'value-added' services.

The involvement of the European Commission in the development of the telecom sector, emerging in the 1970s, follows naturally from the objective to create a 'common market', to increase economies of scale to meet worldwide competition.[12] Moreover, next to being an important 'high tech' sector, telecommunications could be an important means of achieving 'economic unity' in Europe.

So far, telecommunications services had remained firmly in the hands of the national operators, but this would change with the 1987 landmark document 'Green Paper on the development of the common market for telecommunications services and equipment' (EC, 1987). The first and politically acceptable step in the process of liberalization was aimed at introducing competition in terminal equipment and at the services level, while the infrastructure could remain under monopoly control. A series of directives followed that would provide the legislative framework for the implementation of this first phase of liberalization on the national level (Cawley, 2001):

1988: Competition in the markets for telecommunication terminal equipment.
1990: Competition in the markets for telecommunication value-added services.
1990: Establishment of the internal market for telecommunications services through the implementation of open network provisioning.
1995: CATV Services liberalization, allowing the provision of Internet and voice services.
1996: Mobile telecom liberalization.

But, '[t]he Commission recognised that the gains in innovation, productivity improvements and price re-structuring would only come about through competitive entry in infrastructure, be it at a local level by up-grading cable networks or building new ones, or more immediately through alternative backbone investments' (Cawley, 2001, p.4). Following the publication of the Bangemann Report addressing the implications of the 'information society' in early 1994, the European Council by the end of 1994 officially recognized the principle that telecommunications infrastructures should be liberalized and it set 1 January 1998 as the date 'by which all remaining restrictions on services competition would be lifted' (Cawley, 2001, P.4).

By 1998 all formal legal barriers to enter the telecom services market had been removed and many new players did emerge. Regulation was put in place to assure a 'level playing field' for new entrants, that is, an *ex-ante* regime to prevent the use of the 'significant market power' that incumbents had acquired during the monopoly period. This involved the creation of national regulatory agencies to monitor and enforce the implementation of the new regulations.

The principle of 'unbundling' was introduced to open up network access to new entrants, to allow business models to develop based on a 'mix and match' of own and leased infrastructure elements. Cable television networks were considered as one of the few if not the only alternative infrastructure that would provide for 'infrastructure based' competition. In many countries this implied the divestment of the cable operations owned by the incumbent telecom operators.

Terminal equipment liberalization, which occurred in parallel, also contributed to the environment for the Internet to develop. By eliminating the telephone monopoly's vertical integration – which extended all through the switching office, the local line and the customer premises equipment – terminal equipment liberalization allowed new companies, businesses and private networks to attach whatever they wanted to the end of the telephone line – including mainframe computers and (later) the home personal computer.

Because of the Internet and other forms of telecom liberalization, trade in ICT-related services among OECD countries increased from around US$70 billion in 1996 to more than US$235 billion in 2006 – a 13 per cent annual rate of growth. Over the same period, the share of ICT services in total OECD services trade increased from 3.4 per cent to 6 per cent (OECD, 2008).

Privatization of Internet Functions

As the economic potential of Internet services started to become evident in the early 1990s, the Internet was gradually opened up to the general public. Prior to 1995 the Internet had been restricted to US military users, US government agencies and National Science Foundation (NSF)-supported universities, subject to an 'acceptable use policy' that more or less prevented e-commerce from taking place. From 1991 to 1995 the NSF's support for a single transport backbone for the Internet was phased out. The backbone function was privatized in a way that was designed to usher in a market structure of multiple, competing Internet service providers, connected at designated Internet exchange points. The acceptable use policy was eliminated in 1995. In conjunction with these changes, the first commercial Internet service providers were created. Many evolved out of the regional Internet providers that the NSF had supported prior to its withdrawal; in many other parts of the world, the first ISPs were computer bulletin board operators who converted their operations to an Internet service.

From 1996 on, the US federal government also took steps to privatize the central coordinating functions for Internet names and addresses. As noted earlier, the university-based coordinator of DNS and IP addresses, Jon Postel, was replaced by the ICANN, and the address allocation and assignment function was privatized into the ARIN.

Competition in ISP Markets

From the above one may conclude that the early Internet service provider market had relatively low barriers to entry, as use was made of the existing telephone network and investments in a server platform were relatively modest. Moreover, Internet services were being considered value-added information services and therefore restrictions applied with respect to telecommunications regulation. From 1995 onward the Internet could be navigated much more easily through the introduction of the web browser and the hypertext mark-up language (html). Demand for internet access started to grow exponentially, leading to a plethora of ISPs emerging. Some ISPs targeted the residential market, others aimed at professional and business users.

Using as a rough indicator for the number of ISPs, the assignment of so-called autonomous system numbers (AS) to networks that form collectively the Internet (denoting ISPs, military or government networks, or firms with independent networks) in late 1997, 2899 systems were visible in the routing tables, in late 2007 there were 26 606 autonomous systems on a global basis. The USA accounted for 43 per cent of the world total, which includes ISPs offering services globally. In the USA the number of ISPs is approximately twice the number of telecom operators.

As incumbent telephone operators recognize the new means of communication they start to provide ISP services, often obtaining a quick 'footprint' by acquiring smaller players, while retaining the established brand name(s).

As part of the liberalization of the telecommunication sector, cable television network operators were also being allowed to provide access to the Internet. By making their network two-way and by using a cable modem, access was being provided to the Internet. Initially, the coaxial cable networks of the cable operators allowed for higher data rates than those possible on the telephone wire. In countries with a well-developed cable network infrastructure, such as the USA, this resulted in the cable networks taking the early lead in providing broadband Internet access. The telephone operators followed with the introduction of DSL. The result was and still is a 'tit-for-tat' competition in data rates between the telephone network operators and cable network operators.[13]

As a next step of Telecom Reform the incumbent telephone operators were required to open up the local loop to allow other operators to provide access services using the existing telephone wires (local loop unbundling). This provided the opportunity for ISPs to install DSL equipment at the local exchange (co-location) to compete with the incumbent operator, further intensifying the competition for Internet access. Following the crash of the Internet/telecom bubble in 2000 many new entrants had to file for bankruptcy protection, leading to foreclosures as well as a consolidation of the ISP market around the incumbent operators of cable, telecommunication and wireless infrastructures and a few larger ISPs.

Infrastructure-based competition is continuing as cable operators introduce DOCSIS v3 with data rates up to 200 Mbit/s. This requires the public switched telephone network (PSTN) operators to respond with Fibre-to-the-Curb (FttC) combined with very high bit rate DSL (VDSL) or Fibre-to-the-Home (FttH).

Interestingly, in 2005 the US Supreme Court upheld an FCC decision that Internet access over cable modems is an 'information service' under the Telecommunications Act. This forced the US Federal Communication Commission to eliminate the requirement for DSL Internet sharing, leaving the resale of DSL to the discretion of the incumbent local operators.[14]

Over the period 2000 to 2005 the number of DSL connections in the OECD area increased from fewer than 6 million to 98 million, and cable modem Internet subscribers from 7.6 million to 49 million. The share of DSL increased from 43 per cent in 2000 to more than 63 per cent by the end of 2005, surpassing cable in 2001 (OECD, 2007). By June 2008 the OECD area accounted for 151.3 million DSL connections, 72.5 cable connections, 23.0 million fibre and 4.4 million other types of broadband connections (OECD, 2009).

With the introduction of packet access on mobile networks, mobile network operators have also assumed the role of an ISP. Initially Internet access was provided using general packet radio service (GPRS), with data rates up to 171 kbit/s. As part of the next generation mobile services much higher data rates are provided through (HSPA) high speed packet access in Europe and other GSM-dominated operators, and the evolution-data optimized (EVDO) standard in the USA. High speed date packet access advocates claim to provide data rates up to 7.2 Mbit/s on the downlink and up to 5.7 Mbit/s on the uplink (Beming, 2007; GSMA, 2007). In July 2009, 132.1 million HSPA connections were being served (GSMA, 2009). The transition to the next generation of wireless communication infrastructure denoted as LTE (long-term evolution) will be All-IP.

Internet as Agent of Information Society Liberalization

The emergence of the Internet has been, and continues to be, an agent of change and catalyst of liberalization in other sectors of the information economy. One reason is its non-territoriality, which fosters more open trade in information products and services. Another reason is that it fosters the separation of content and conduit. In other words, its architecture has helped to de-couple the market for network bandwidth (conduit) from the market for services that are delivered through networks (content). It does this by providing an open, nonproprietary platform for compatible communications. Thus as the Internet diffuses it is closely followed by widespread innovation in content generation, information services, networked devices, and network-enabled applications.

Google is the most convenient poster child for this kind of innovation. It developed a powerful and useful search engine, leveraged that to develop a lucrative interactive advertising business, and then moved to ancillary functions, ranging from free maps and directions to free email. It has also proposed a new mobile handset standard, Android – in an attempt to ensure that the mobile Internet will be as open as the fixed Internet – as well as new browser software.

The Internet has brought with it new kinds of retail storefronts (Amazon, eBay); an explosion of user-generated content and social network sites; disintermediation and new forms of intermediary services in banking, finance and travel. The market for voice telephone services is radically changed by the fact that on the Internet, telephone service is just another application that anyone can install on their computer (for example, Skype). Likewise, Internet access opens up the possibility of globalized audio-visual services, allowing websites to deliver video programming to customers to either complement or compete with regulated broadcasting services.

For the scholars and policy-makers with a prior involvement in the liberalization of telecommunications, the Internet might be seen as the hyper-liberalization of communications infrastructure. The Internet represents everything that the telephone system was not. Where the telephone system was vertically integrated and monopolistic, the Internet was decentralized, is non-integrated and is competitive. Where the telephone system was national and territorial in scope, the Internet was global. In telecommunication policy, regulators and legislators were preoccupied for three decades with the problem of interconnection between dominant incumbents and new entrants who would be unable to survive without regulated access to the incumbent's facilities and customers. Interconnection of Internet service providers, on the other hand, took care of itself; private operators negotiated contractual agreements to achieve universal connectivity without government involvement or supervision. Of course, the Internet still relies on physical infrastructure, which is often provided by the existing telephone, mobile and cable television networks.

RECENT DEVELOPMENTS AND TRENDS

In this section we briefly recount some of the key policy and governance issues that will define the shape of the Internet infrastructure going forward.

Broadband Policy and Network Neutrality

Although we have characterized the Internet as a software or virtual infrastructure, it remains dependent on physical telecommunications infrastructure. The evident success of the Internet as a converged platform for digitized products has stimulated growing demand for more bandwidth. The increased reliance on the internet for video distribution especially puts pressure on the ISPs to improve their speed and performance. As we move toward a high-speed, 'broadband' Internet some of the traditional regulatory problems of telecommunications infrastructure are beginning to impinge on the Internet.

The debate over network neutrality (NN) is created by the way in which the construction of broadband networks in a competitive environment might undermine the Internet's separation of content and conduit (Hahn et al., 2007; Wu, 2003; Yoo, 2006). In other words, some fear that as competing cable and telecommunication companies invest in new high-speed networks, they will start to exert vertical leverage over the content, applications and services delivered over the Internet. Departures from network neutrality are defined as the exercise of *vertical tying* between the supply of bandwidth (in which the incumbent is presumed to have market power, if not a monopoly) and the supply of content, applications and terminal equipment, all of which require bandwidth to be used. Tying refers to the behavior of selling one product (the tying product), conditional upon the purchase of another product (the tied product) (Tirole, 2005).

Not all blocking of access to Internet resources is done to achieve competitive advantage. From a global perspective, governmentally mandated Internet blocking equals or exceeds the threat to open access to Internet resources from economically motivated discrimination by carriers. The case for NN is deeply informed by concepts of freedom of expression. True NN advocates would object to efforts by private carriers to regulate the content that can be accessed by their subscribers regardless of whether the interference gives them a competitive advantage, or a competitive disadvantage. The point is simply that the basic infrastructure of Internet communication should enable end users to decide what they receive and what they send and that it is improper for a public, network intermediary that aggregates large amounts of heterogeneous traffic to insert itself into that role.

The WSIS Debate over US Control of the ICANN

From 2002 to 2005, a United Nations (UN) Summit (the World Summit on the Information Society, or WSIS) became the flashpoint for a coordinated attack on ICANN (Klein, 2004; Kleinwachter, 2004). ICANN was singled out because of its unique status as a private-sector global governance organization that relegated national states to a secondary role, and because of its unilateral creation and oversight by the US government. Underlying this debate was a clash between two distinct models of global governance (Mueller et al., 2007). On the one hand, there is the traditional, nation-state-based approach in which governments solve international coordination and policy problems through multilateral agreements; on the other, there is the transnational, contractual approach rooted in the private sector and civil society. The WSIS process provoked a wide-ranging debate and discussion around global internet governance, but did not push the US into yielding its control of ICANN nor did it bring ICANN into an intergovernmental regime. It did, however, create a new international institution, the UN Internet Governance Forum, to continue discussions of Internet governance on a non-binding basis, in an environment where the private sector, civil society and governments are given roughly equal status. In general, the WSIS process legitimated the multi-stakeholder governance principle although the problem of US unilateral control is still contentious.

Reactions against Internet Liberalism

Just as WSIS represented a reaction against the institutional innovations underlying Internet governance, so there have been many other reactions against the freedom and openness of the Internet in a variety of policy domains.

Security issues

Security has been the watchword of many of these reactions. The diffusion of powerful computing devices globally interconnected through an open network has not only facilitated innovation in services and content, but has also facilitated innovation in crime and vandalism (Van Eeten and Bauer, 2008; Zittrain, 2008). The largest and most notable problem is email spam. Once merely a nuisance, spam emails have become the distribution channel for an underground economy of organized crime. Fraud, identity theft and software exploits that allow criminals to enslave large numbers of computers and put them into their own service as 'botnets' are now commonplace. Quite apart from the openly criminal element, the growing importance of the Internet has made it into a critical infrastructure – a classification that often brings with it more intensive and confining security and economic policies.

Attempts to make the Internet more secure take many different forms. Some are simply market-based products and services: anti-virus software, the engagement of security service firms and consultants, or the purchase of more powerful and sophisticated firewalls and intrusion detection systems. Other remedies are based on new technical standards that try to engineer more secure, encryption-based methods into the basic internet protocols. Still others are based on transnational cooperation to rapidly identify and take down illegal sites (Moore, 2008).

Greater efforts to regulate content

Having essentially lost control of content regulation to the advent of the Internet in its early years, national governments have since invested in the acquisition and deployment of technologies that allow them to block access to content deemed illegal or objectionable. An increasing number of governments are mandating the blocking of specific Web URLs as a form of content regulation (Deibert et al., 2008). This applies not just to authoritarian governments such as China or Saudi Arabia, but also to Australia and many European governments interested in preventing access to prohibited forms of political expression such as pro-Nazi materials. Child pornography has provided a strong pretext for intensifying censorship of Internet materials. Some have questioned, however, whether Internet blocking offers politicians an easy substitute for actually locating and prosecuting the producers and consumers of these images of child abuse (Deibert and Rohozinski, 2009; Stol et al., 2009). With its new policy process to add new top-level domains to the DNS, ICANN also stepped into the field of content regulation. Its new top level domain (TLD) policy[15] would censor TLD applications that governments or members of the public object to on the basis of 'morality and public order.' Because it is impossible for a short TLD string to threaten morality and public order by itself, this can only mean that it would really be concerns about the content of websites under a top level domain that would motivate the objections.

Deep packet inspection

Concerns about security, content regulation and copyright protection often come together in an attempt to impose more regulation on Internet service providers. Sometimes legal tools are proposed, such as greater intermediary liability for ISPs. At other times technological mechanisms are applied. One tool that is being promoted for greater regulation of ISPs is deep packet inspection (DPI). With its origins in intrusion detection systems to detect viruses and malware, this is a technology that opens up the data packets passing through the network to detect their contents. Although its use and effectiveness is still a matter of contestation, copyright owners have proposed to use it to attack peer-to-peer file sharing of movies and music, would-be censors have proposed to use it to attack child pornography and other forms of illegal content. It has also been proposed or used by ISPs themselves to ration bandwidth. At best, such bandwidth management could optimize bandwidth utilization by differentiating between applications and services that require real-time, high-speed packet delivery and those that can move more slowly. Critics motivated by net neutrality concerns, however, fear that this capability might be used to discriminate among Internet services in an anti-competitive manner.

Resource issues

Critical Internet resources continue to be an area of policy concern and change. Whereas during the last decade domain names were the focus of most political contention, in 2008 IP address resources became a policy issue. The key causal factor bringing IP address policy to the forefront was the depletion of the IPv4 address space. The 1981 standard that defined Internet protocol created a fixed address field of 32 bits, which creates a mathematical possibility of about 4 billion unique addresses. Because of continuing Internet growth, IPv4 addresses have become scarce and valuable resources. Put simply, we are running out of them. Most projections (as of 2009) state that we have only about a two-year supply left.

In principle, a new Internet standard, IPv6, solves the problem of address scarcity because it has a very large address space (2^{128} addresses). But the new Internet standard is not compatible with the old one. Thus, anyone who implements IPv6 and throws away their IPv4 capabilities is going to be isolated, cut off from the vast majority of communication partners, websites and services associated with the old Internet. One must think of the transition from IPv4 to IPv6 as the (possibly temporary) coexistence of two distinct 'Internets.'

One policy response to the problem of address scarcity has been proposals to allow organizations holding IPv4 addresses to sell address blocks to other organizations willing to buy them (Mueller, 2008). Internet protocol address transfer markets, as they are called, have been proposed as a pragmatic way to extend the life of the IPv4 address space. One important benefit of such a policy is to provide incentives for existing holders of addresses to release unused address resources. Another possible benefit is the way it might rationalize and make more transparent an underground economy in address resources. Transfer markets also increase the autonomy of Internet users by providing an alternative to the centralized administrative processes that currently control address allocations (and for that reason is opposed by many within the RIRs).

Another key area of change in the Internet resource space is the attempt to institute multilingual domain names, often known as internationalized domain names (IDNs). The early, global Internet was an English-language Internet based on the Roman alphabet. Domain names were confined to a restricted set of ASCII characters in order to simplify addressing. It is now possible to have domain names in Chinese, Japanese, Arabic and other scripts. While making domain names more usable for the world's non-English reading users, the implementation of multilingual names poses a number of important policy issues. The domain name trademark disputes of the past are extended to a more complex, multi-script environment. More importantly, since new ASCII top-level domains have failed to dent the dominance of .com, it is likely that IDNs offer the best opportunity for new entrants to gain a strong foothold in the market for domain names. Thus, competition policy issues are raised as ICANN must consider whether to give incumbent TLD registries an equivalent of their existing domain in other scripts, or whether giving them to newcomers will lead to user confusion. Geopolitics also enters the equation, as many national governments are demanding that ICANN give them country code TLDs in multiple scripts. The political and sovereignty-based demand for, say, the word "India" in the multiple official languages of that country overlaps with competition policy. If the incumbent country code operator is the first to offer the public domain names in their native script, its dominance of the market would be locked in and entrenched, possibly forever.

CONCLUSION

In the near future, the fate of the Internet as an infrastructural platform will play out in several arenas. At the technical level, the migration to a new Internet protocol with a larger address space, IPv6, is projected by many. But it is still unclear how long that migration will take and how disruptive the change will be. There are also important efforts to institute stronger security protections in core Internet routing and domain name protocols. These changes raise governance issues as well as technical issues, such as who will control the encryption keys or who will stand at the root of a public key infrastructure. It will be interesting to see whether markets are allowed to develop for IP address resources. At a more political level, the world is still digesting the implications of institutional innovations such as ICANN and the Internet Governance Forum. On 30 September 2009, the US government took a small step toward globalizing ICANN's position by allowing its Joint Project Agreement with ICANN to expire and putting the Governmental Advisory Committee chair in a position to supervise recurring reviews of ICANN's performance. In 2010, the United Nations must decide whether to renew the mandate of the Internet Governance Forum for another five years. At the regulatory level, we see growing pressures to make Internet service providers responsible for mitigating the ills of the Internet. With net neutrality debates, the physical bandwidth of ISPs and its management becomes a regulatory and policy battleground as well.

The Internet was 'born' liberalized, as it were, and now becomes more institutionalized and regulated. As economy and society become more dependent on the Internet the economic stakes become higher, the political battles more entrenched. As this happens, the policies and practices applied to the Internet tend to become more formalized and restrictive. Nevertheless, there are still strong policy norms that link Internet freedom to its capacity for wealth-creation, innovation and change. Even when we grumble about its consequences, there is a widespread recognition of the tremendous social value of an open Internet, transnational interconnection and user-generated content and innovation.

NOTES

1. Internet protocol (IP) works by dividing messages up into 'packets' and attaching sender and receiver addresses to those packets so that they can be routed to their destination. A closely related protocol, transmission control protocol or TCP, governs error control and the rate at which packets are sent.
2. Standards for the World Wide Web – which are not the same thing as the Internet, but rather a specific application that generally relies on the Internet protocols for transport – are developed by the World Wide Web Consortium (W3C). However, the W3C is relevant to Internet governance because so much of Internet traffic is now based on the Web (http) protocol.
3. US Commerce Department, National Telecommunications and Information Administration. Management of Internet Names and Addresses: Statement of Policy. Docket Number: 980212036-8146-02. June 5, 1998.
4. See RFC 1174 (1990) and RFC 1466 (1993) for the earliest documents providing the rationale for creating Regional Internet Registries.
5. An 'allocation' is an address block given to an Internet service provider for intermediate use in selling Internet service to other users. An address 'assignment' is a block given to end users for their own use; for example, corporations or universities with private networks.
6. For an extensive discussion of the function of the ISP see, for instance, Huston (1999) and Knight (2000).
7. Asymmetrical refers to the download data rate being higher than the upload data rate.
8. In Europe this would imply within each local tariff zone, in the USA within each WATS area as local calls are subject to flat fee.
9. The current Internet is still limited in the capability to broadcast television information in real time, that is, in full replacement of terrestrial and satellite services.
10. Regulatory and Policy Problems Presented by the Interdependence of Computer and Communication Services, *Notice of Inquiry*, 7 FCC2d 11, 8 Rad. Reg.2d (P & F) 1567 (1966); *Tentative Decision*, 28 FCC2d 291, 18 Rad. Reg.2d (P & F) 1713 (1970); *Final Decision*, 28 FCC2d 267, 21 Rad. Reg.2d (P & F) 1561 (1971).
11. That distinction proved to be slippery, because as the network technology became more digitized the function of basic conveyance involved signaling and switching functions that relied on data processing.
12. The European Union, formed in 1992, has evolved from the 'European Community on Coal and Steel', established in 1951, to become the 'European Economic Community' in 1957. The European Monetary System was established in 1979 and the decision to create a single European Market was made in 1985, to be effective in 1992.
13. For example in the Netherlands between incumbent KPN and cable operators Ziggo (a merger of Casema and Essent Kabelcom) and UPC.
14. National Cable & Telecommunications Ass'n v. Brand X Internet Services, 545 US 967 (2005).
15. ICANN, Draft Applicant Guidebook, v.2 (February 18, 2009), available at: http://www.icann.org/en/topics/new-gtlds/draft-rfp-clean-18feb09-en.pdf (accessed 15 May 2009).

REFERENCES

Abbate, J. (2000), *Inventing the Internet*, Cambridge, MA: MIT Press.
AMSIX (2009), 'AMS-IX organisation', http://www.ams-ix.net/about/, accessed 3 March 2009.
Bangemann Group (1994), 'Europe and the global information society: recommendations to the European Council', Brussels: European Commission.
Beming, L. (2007), 'The HSPA eco system', http://www.gsmworld.com, GSMA, accessed 13 August 2008.
Benkler, Y. (2006), *The Wealth of Networks: How Social Production Transforms Markets and Freedom*, New Haven, CT: Yale University Press.
Brock, G.W. (1994), *Telecommunication Policy for the Information Age*, Cambridge, MA: Harvard University Press.
Cawley, R.A. (2001), 'The European Union and world telecommunications markets', in G. Madden and S. Savage (eds), *International Handbook of Telecommunications Economics*, Cheltenham, UK and Northampton, MA, USA: Edward Elgar.
Deibert, R. and Rohozinski, R. (2009), *Risking Security: Policies and Paradoxes of Cyberspace Security*, New York: Online Paper Archive.
Deibert, R., Palfrey, J.G., Rohozinski, R. and Zittrain, J. (eds) (2008), *Access Denied: The Practice and Policy of Global Internet Filtering*, Cambridge, MA: MIT Press.
European Commission (EC) (1987), 'COM(87)2 290: Green paper on the development of the common market for telecommunications services and equipment', COM(87)2 290, Brussels, European Commission.
Group Special Mobile Association (GSMA) (2007), 'GSM Statistics', http://www.gsmworld.com, accessed 4 February 2008.
Group Special Mobile Association (GSMA) (2009), 'GSM World', GSM Association, accessed 3 March 2009.
Hahn, R.W., Litan, R.E. and Singer, H.J. (2007), *The Economics of Wireless Net Neutrality*, Washington: AEI-Brookings Joint Center.
Huston, G. (1999), *ISP Survival Guide – Strategies for Running a Competitive ISP*, New York: John Wiley & Sons.
Klein, H. (2004), 'Understanding WSIS: an institutional analysis of the UN world summit on the information society', *Information, Technology and International Development*, **3-4**, 3–14.
Kleinwachter, W. (2004), 'Beyond ICANN vs ITU? How WSIS tries to enter the new territory of internet governance', *Gazette*, **66**(3-4), 233–51.
Knight, C.M. (2000), *ISP Marketing Survival Guide – Proven Strategies and Secrets for Outmaneuvering the Competition*, New York: John Wiley & Sons.
Moore, T. (2008), 'Cooperative attack and defense in distributed networks', Cambridge University Computer Laboratory Report UCAM-CL-TR-718, June, available at: http://www.cl.cam.ac.uk/techreports/UCAM-CL-TR-718.html.
Mueller, M.L. (2002), *Ruling the Root: Internet Governance and the Taming of Cyberspace*, Cambridge, MA: The MIT Press.
Mueller, M.L. (2008), *Scarcity in IP Addresses: IPv4 Address Transfer Markets and the Regional Internet Address Registries (NO. IGP08-002)*, Syracuse, NY: Internet Governance Project.
Mueller, M.L., Mathiason, J. and Klein, H. (2007), 'The internet and global governance: principles and norms for a new regime', *Global Governance*, **13**(2), 237–54.
National Research Council: Committee on Innovation in Computing and Communications (1999), *Finding a Revolution: Government Support for Computing Research*, Washington, DC: National Academy Press.
Organisation for Economic Co-operation and Development (OECD) (2007), *Communications Outlook 2007*, Paris: OECD.
Organisation for Economic Co-Operation and Development (2008), *Information Technology Outlook*, Paris: Organisation for Economic Co-operation and development.
Organisation for Economic Co-operation and Development (2009), *OECD Communications Outlook 2009*, Paris: Organisation for Economic Co-operation and Development.
Saltzer, R.H., Reed, D.P. and Clark, D.D. (1984), 'End-to-end arguments in system design', *ACM Transactions on Computer Systems*, **2**(4), 277–88.
Stol, W.P., Kaspersen, H.K.W., Kerstens, J., Leukfedt, E.R., and Lodder, A.R. (2009), 'Governmental filtering of websites: the Dutch case', *Computer Law and Security Review*, **25**, 251–62.
Tirole, J. (2005), 'The analysis of tying cases: a primer', *Competition Policy International*, **1**(1), available at: http://ideinfi/doc/by/tirole/tying-cases2.pdf.
Van Eeten, M. and Bauer, J. M. (2008), *Economics of Malware: Security Decisions, Incentives and Externalities (No. STI Working Paper 2008/1)*, Paris: Organisation for Economic Co-operation and Development.
Vincent, C. and Camp, J. (2004), 'Looking to the internet for models of governance', *Ethics and Information Technology*, **6**(3), 161–73.

Wu, T. (2003), 'Network neutrality, broadband discrimination', *Journal of Telecommunications and High Technology Law*, **2**, 141.
Wunsch-Vincent, S. and Vickery, G. (2007), *Participative Web: User Created Content (No. DSTI/ICCP/IE(2006)7/FINAL)*, Paris: Organisation for Economic Co-operation and Development.
Yoo, C. (2006), 'Network neutrality and the economics of congestion', *Georgetown Law Journal*, **94**, 1847.
Zittrain, J. (2008), *The Future of the Internet: And How to Stop It*, New Haven, CT, and London: Yale University Press.

10 The liberalization of electricity markets[1]

Jean-Michel Glachant and Yannick Perez

INTRODUCTION

The competitive reform of electricity industries has recently experienced a wide expansion worldwide, with over 200 new instances of sectoral deregulation between 1990 and 2008 (Newbery, 2000; Glachant and Finon, 2003; World Bank, 2006; Sioshansi and Pfaffenberger, 2006; Glachant and Lévêque, 2009). Nonetheless, subsequent to the California electricity crisis (2000 – 2001), there has been an increasing dissatisfaction with regards to the limitations, and in some cases the failures, of these new ways of framing electricity industries (Kessides, 2004; Joskow, 2006b). We are witnessing a slowdown or, in some cases, a blocking of the reforms, as if the progression of competition policy in electricity industries had a cyclical component.

Previously, the unique characteristics of electricity industries appeared to set them apart from most other industries, deemed 'competitive'. These electricity industries notably feature: significant economies of scale or scope (extending to natural monopolies); far-reaching externalities (positive or negative) in production or consumption; and extensive vertical and horizontal integration (either under a single corporate umbrella or in the form of long-term ad hoc contracts (Brousseau and Glachant, 2002, 2008)). Within this very specific framework, the successful introduction of competitive mechanisms, substituting for administered regulation or internal corporate management hierarchies, along with the creation of open markets either upstream or downstream of the formerly integrated networks, created both disruptions and innovations in equal measure (Joskow and Schmalensee, 1983; Baumol and Sidak, 1994).

The purpose of this chapter is to present, discuss and offer the road ahead in the liberalization of the electricity sector worldwide. The chapter is structured in the following way: the next section gives the empirical evaluation of reforms and will stress the huge diversity of these processes. The third section explains this diversity as the combination of the modular nature of the electricity industry in different institutional environments. The role of existing abilities to change both property rights, industrial structures and the rules of the game shape the worldwide reforms. The fourth section provides a research and policy-making agenda in the field. The final section concludes.

EMPIRICAL DESCRIPTION OF ELECTRICITY LIBERALIZATION

The main characteristics of electricity reform across the world is its huge diversity. The list of actions needed to make a reform is given in Box 10.1. In fact the electrical industry has proven unable to present a robust competitive market design that has universal acceptance or, for that matter, that is capable of instantaneously and simultaneously

BOX 10.1 THE TEN TEXTBOOK REGULATORY ELEMENTS WHICH MAKE A SUCCESSFUL COMPETITIVE REFORM OF THE ELECTRIC INDUSTRY

1. *The privatization* of the electric industry. It is understood at the same time as a means to introduce competition, give strong incentives for industrial and financial performances and work as a barrier to the governmental economic and political pressures.
2. *The vertical separation* of the competing segments (production and supply) from the segments having to remain under public regulation (transmission grids and distribution networks). This separation can be carried out structurally or legally and must guarantee that cross-subsidization between the two segments will be impossible and that the access to the grids will not be discriminatory.
3. *The horizontal dismantling* of the competing segments must be operated to create a sufficient number of participants in the various markets (wholesale and retail markets). These industrial surgeries are necessary to guarantee *ex ante* reasonably competitive mechanisms.
4. *The creation of an independent management of the transmission grid.* Its functions include: balancing of electricity in real time and the guarantee of the quality of the good (frequency voltage and stability). This separation withdraws the temptation of discriminatory practices in the operational management of network.
5. *The creation of electricity markets in real time* coupled with competing mechanisms of allowance of the physical constraints of electric transport are also necessary. They allow that the risks of production and consumption are managed effectively.
6. *Rules of the game are needed for setting:* (a) third party access on a non-discriminatory basis; (b) allowance of the transmission capacities in a market base; (c) signals of localization and connection for the new entrants.
7. *The identification and the separation of the costs of electric supply* must be made. Thus each consumer has to know in its consumption: (a) the cost of the network, (b) the cost of the supply of energy and, finally, (c) the cost of the services associated with the electric supply.
8. *The creation of an independent regulator* is a crucial point of reforms: regulator must be quite informed (costs, quality of service, compliance with the rules of competition) and must hold the sufficient authority and means to effectively implement the monitoring of the operators on both network and competitive activities.
9. *Mechanisms of transitions* and *ex post* amendments of the initial scheme of the reforms are always necessary.
10. *Coherence of decisions over time.* The absence of decision-making in the first nine topics is largely prejudicial. This has two consequences: the electric reforms are evolutionary process of improvement, and correction of errors and omissions. Finally, the authorities must support diaries of reform in the long run.

Source: Adapted from Joskow (2006c) and Littlechild (2006).

Table 10.1 Milestones of reforming in continental Europe

1996	EU-15	European Council of Energy Ministers and Parliament reached agreement on a market liberalization Directive
February 1997	EU-15	This 'Directive concerning common rules for the internal market in electricity' (Directive 96/92/EC)
2001	EU-15	Approval of the 'Directive of the European Parliament and the Council on the promotion of electricity from renewable energy sources in the internal electricity market (RES-E Directive)' (European Parliament and Council, 2001 – Directive 2001/77/EC)
2003	EU-25	Approval of the 'Directive concerning common rules for the internal market in electricity' (officially Directive 2003/54; usually named "the Second Directive")
2004	EU15+10	Extension of the EU to 25 member countries, new CE member countries to open their market with minimum of 30 per cent
2004	EU 25	Electricity Directive 2003/54 due to be transposed by member states All non domestic customers made eligible in the EU in July 2004 An EU Regulation on cross-border electricity trade came into effect (Regulation 1228/2003) in July 2004
2007	EU 25	Due to Electricity Directive 2003/54, 100 per cent market opening in all EU-25 countries in July 2007
2009	EU 25	Third package amending the electricity industry and introducing the creation of ACER

coping with all the new problems arising in the course of reform. But according to discussion between Joskow (2006c) and Littlechild (2006), it is now possible to define a 'Textbook model for electricity reforms' as the combination of ten complementary elements described in Box 10.1.

The empirics of the reforms across the world however show a puzzling diversity and the distance taken by reformers from the textbook model (Glachant and Lévèque 2009; Green et al., 2008; Sioshansi and Pfaffenberger, 2006; Kessides, 2004; Glachant and Finon, 2003). Here we want to stress seven points of difference in the conduct and outputs of electricity reforms:

1. In electricity market reforms, we faced major differences in the timing of the process itself, from full market opening in one night in Germany to more gradual reforms over a decade in the rest of Europe (see Table 10.1 for the milestones of EU reforms).
2. The methods used are also different, that is, from radical surgery in generation in England and Wales to no major change in France; from mandatory reforms in continental Europe to voluntary approaches in the US states. In the USA, the impossibility of passing a federal electricity reform bill has allowed many states (about half!) to escape these changes and refuse to comply with the federal energy regulator's ambitious plans.
3. The level of unbundling of monopoly activities from competitive ones goes from simple legal unbundling to more deeply securing property unbundling (Joskow, 2005).

4. The depth of market based tools in the replacement of command and control mechanisms (*unit commitment and optimal dispatch*) to run the network activities reflect also major differences from real time markets to administrative mechanisms (Baldick and Niu, 2005; Saguan et al., 2009).
5. The size of the privatization processes, from the selling of all generation part of industry to an absence of change (Kessides, 2004).
6. In the requirement to use market-based auctions in day ahead market for energy, from voluntary markets (like Texas, California, Italy, Poland) to multilateral markets (like the EU Nordic countries in the Nordpool, the Pennsylvania, Maryland and New Jersey (PJM), the Trilateral market coupling (France – Belgium – Netherlands) to pure mandatory ones (former UK Pool and the actual Spanish market).
7. The regulatory task can take diverse forms, from no specific regulatory body at the time of creating the reform (Glachant et al., 2008) to some very independent national regulator with large powers (Green et al., 2008) or to Federal organization with limited state intervention (Prosser, 2005).

Making a precise mapping of all these worldwide differences is a very difficult task for two reasons: first, over 150 countries have performed some sort of electricity reform across the world. Secondly, these reforms are characterized by processes of reforms to correct the mistakes made in the first place (Hogan, 2002).

The Reform Process in Europe: A Continuous Process of Adaptation

Before 1990, almost every electricity supply industry in continental Europe was vertically integrated with a captive franchise market, either state-owned (the majority of cases) or under price-regulated mixed private/public ownership (as in Belgium, Germany and Switzerland). Regulated area monopolies prevailed in all countries. Until the end of the 1990s, the standard model was an effectively vertically integrated franchise monopoly under either public ownership or cost-of-service regulation. Although electricity networks were typically synchronized over wide areas, interconnections of areas under different transmission system operators (TSOs) were frequently guided by security, rather than by economic considerations. Real electricity liberalization in Europe started with Britain's restructuring and privatization in the 1990s, demonstrating that vertical unbundling and the creation of wholesale electricity markets was actually feasible (Newbery, 2005).

The EU national diversity is first and foremost a predictable result of the nature of compromises among its member states, formalized by the European Community in the First Directive of 1996. According to the insightful commentary of L. Hancher (1997), this First Directive allowed nearly everything (including 'single buyer'), except an integrated internal market. The Second Directive (2003) and its regulations managed to reduce the scope of this diversity, but not to eliminate it. The effect of the last 2009 Directive will probably go in the same direction with the future creation of the Agency for the Cooperation of Energy Regulators (ACER). It is worth noting that the last Directive is a clear output of the DG Competition inquiry in the Energy Markets between 2005 and 2007. In this new dynamic coming from the EU competition authority, the *ex post* action takes the lead on the energy market creation in continental Europe. The milestones of reforming the electricity sector in continental Europe are given in Table 10.1. In the following section, we explain why this diversity exists and where it originates.

INTRODUCING COMPETITION 'WHERE FEASIBLE'

The basic idea behind all economic analyses in favour of the competitive reforms is that the creation of markets within electricity industries presupposes *ex ante* acts of 'industrial surgery'. Prior to creating these markets or seeing them appear spontaneously, it is necessary to end the traditional vertical and horizontal integration of the incumbent monopolies. Thus, those industries need to be redefined in elements that will permanently be monopolistic and must be separated from those with competitive potential. This cannot be accomplished overnight. Rather it requires incremental experimentation with new procedures for the unbundling of activities that have been integrated for decades. Thus, there is a transition period during which the new markets are weak and the incumbent monopolies remain quite strong. Consequently, a governance structure reflecting the competitive reform throughout this transition period is useful, even indispensable.

The duration of this transition period depends upon many conditions, including the characteristics specific to each network industry. As early as 1985, Williamson foresaw that aviation[2] and roadways would be easier to reform along sustainable competitive lines than railways or electricity. In fact, with regard to electricity reforms, which began in Great Britain in 1990, the architecture of competitive market design proved to be an unstable hodgepodge of market and non-market mechanisms (Newbery, 2000). In keeping with the principle of separating monopolistic activities from those that are potentially competitive,[3] the industry divided into several distinct operational and transactional groups of activities that, in accordance with Baldwin and Clark (2000) and Baldwin (2008), we call 'modules'. However, the entire chain of all new unbundled modules often required a more comprehensive and far-reaching governance structure than that provided by the initial road map of the electricity reforms.[4]

Thus, the building of competitive markets combines three main dimensions: (1) the overall separation of potentially competitive activities from inherently monopolistic electricity activities (*unbundling*); (2) the segregation of all the operations and transactions of the industry into modules organized around various mechanisms for internal coordination (*modularity*); and (3) the implementation of the various modules in the chain to carry the competitive transactions (*sequentiality*). The following subsections explain the theoretical foundation of electricity market reforms and provide an empirical illustration for them.

Unbundling between Competitive and Regulated Activities

As emphasized by Littlechild (2006), the first British regulator and the inventor of the notion of the price cap as applied to telecommunications, the principle of 'competition where possible' is central to the reform of electricity industries and their transformation into competitive markets. This type of separation is expected to free competitive forces on one side of the new boundary and to concentrate the regulatory activity in the electricity monopolies on the other side. However, the principle of unbundling assumes that there exists a 'natural' demarcation, clear and robust – by nature almost technical or at least technico-economic – between these two domains: the market for services on one side, and infrastructure monopolies on the other side. Sometimes this is true for roads and highways,[5] and it remains the case, though a little less unambiguously, in aviation[6] and telecommunications.[7] However, we generally consider that competitors to incumbent telecommunication operators have no difficulty duplicating their infrastructures and creating their own private grid, at least beyond the 'copper' local loop. The question is quite tricky in the case of electricity, because the service rendered is not storable and because there are no queues. Furthermore, the entire supply-demand equilibrium is a global phenomenon, common to the entire industry and extending beyond the ownership boundaries of dozens of different generators or sellers of electricity.[8]

In practice, it is not difficult to see why this issue of global equilibrium in electricity must be ensured by a third party with decision authority over all immediate and very short-term time horizons.[9] Thus, the transmission grid must directly administer very short-term imbalances between the consumption and generation of power (*balancing management*) and between the flows of current and line capacity (*congestion management*). We observe here that activities specific to the network monopoly are very strongly enmeshed with, and weakly separable from, all activities that are characteristic of the competitive activities.

The precise allocation of tasks and decision-making rights between competitive and monopolistic modules, as well as the detailed design of the interface mechanisms connecting these two types of module are central and decisive questions about the real nature of the competitive reforms.

Modularity and Boundaries of Market Activities

Boundaries between monopolistic activities and potentially competitive activities, like the boundaries between the firms themselves, between their respective tasks, and between their real or potential transactions and the corresponding markets, are thus not given once and for all. Quite the opposite, these boundaries are primarily defined over the course of the long process of creating the reform. They are the result of segregating the industry into new operational modules. The competitive reform is thus a 'modularization' of the electricity industry.

According to Baldwin (2008), modularity is a particular design structure, in which parameters and tasks are interdependent within modules and independent across them. This technical definition of modularity is well suited to the new modularity of electricity industries. It nicely complements the work of Williamson (1996) on 'technological separability', which distinguishes between the technological constraints within non-separable clusters of tasks and a strong institutional constraint on the design of interfaces connecting task clusters that are technologically separable.[10] To Baldwin and Clark (2000, p. 89): '*The ideal of perfect modularity is full "plug and play" flexibility all along the value chain of the industry*'. They then add, '*but in a complex design, there are often many levels of visible and hidden information* (ibid.)'. Perfect modularity is thus not universal.[11] In the competitive reforms of electricity industries, the ideal of 'perfect modularity', the hermetic separation of task clusters having different natures, is far from universally implemented.

In practice, some boundaries between modules remain porous. Some modules retain interdependence among each other in their operational functioning, even if, of course, the interdependencies are stronger and more pervasive within the modules than between them.

Imperfect Modularity in Electricity Industries

Perfect modularity would define 'independent task blocks', build 'clean impermeable interfaces', and separate 'hidden and visible information'. Three invaluable characteristics would result for the process of performing these tasks. First, perfect modularity would increase the potential for managing complex chains of operations. Second, perfect modularity would allow the various modules of a complex chain to operate in parallel with a certain degree of autonomy. Third, and finally, perfect modularity would make it easier to react to uncertainty, provided uncertainty was confined to a single module. We here recognize the motivation for separating the professions and tasks in the initial implementations of competitive reforms. However, we must acknowledge that market building often fails to reach that degree of perfect modularity. Conversely, actual modularity analysis of the competitive reforms of electricity industries frequently reveals nothing else than a flawed chain of imperfect modules and faulty interfaces. Porous borders and non-exclusive interfaces have been inserted between the monopolistic and competitive module clusters, as well as between the specific modules. At the same time, incomplete rules of operation have been imposed within the various modules. It follows that all of this modularity remains flawed, notably with numerous operational 'leaks' across modules. Thus, many direct dependencies persist in the operational functioning of a number of these modules, which are seen as externalities or incompleteness.

The first aspect of the issue of weak modularity is the diversity of module organizations for monopolistic transmission activities. The pivotal architecture of electrical networks is the transmission grid, since this transports the energy generated by power plants over long distances and on a huge scale.[12] Comparing the typical organization of transmission in the competitive reforms of the USA (the independent system operator, or ISO) with its European analogue (the transmission system operator, or TSO) immediately reveals the diversity of the 'Transmission' modules put into place,[13] as Table 10.2 summarizes.

In England, the transmitter is a private firm that is listed on the stock exchange, owns its own transmission facilities and plans and finances investments in the grid. It manages congestions with the physical method known as 'redispatching'. However, it does not transmit a direct price signal to the users of the grid who are liable to be at the source of this congestion. The cost of congestion is socialized across all grid users during periods of congestion.[14] A direct consequence of this method for managing congestion is the existence of a real border, both physical and price based, that completely surrounds the zone administered by the transmitter. Furthermore, the transmitter charges the costs of transmission[15] in fees that are socialized[16] across a regional grid, with a dozen or more 'postage stamp prices' for generators and a similar number of other 'postage stamp prices' for consumers. Finally, the cost of new connections to the grid are also largely socialized, since the hook-up fee does not account for the cost of adapting the network upstream from the point of connection.[17]

In contrast, in the USA, most typically in the Pennsylvania, New Jersey, Maryland

Table 10.2 An example of sub-modularity within the module 'monopoly transmission network'

English transport system operator (TSO)	American independent system operator (ISO)
Owns the assets and is a "for profit" company	Doesn't own the assets and is a "not for profit" entity
Plans & builds new lines	Doesn't plan or build new lines
Manages internal congestion with physical redispatching	Manages internal congestion with nodal pricing
Manages connections with other TSOs as boundary	Manages connections with other ISOs as new nodes
Prices access with regional 'postage stamp'	Prices access by calculating prices at each node
Charges new generator connection with shallow costs	Charges new generator connection with deep costs

Source: Rious et al. (2008).

(PJM) control zone, the transmitter is comprised of a club of electricity professionals. Thus, it functions as a cooperative, making no profits and distributing no dividends. This club does not own the transmission grid facilities, which remain the property of the incumbent operators. It is, however, their only operator. It is the system operator (SO), and is distinct from the proprietor of the network, the transmission owner (TO). From the point of view of ownership of the network equipment, this system operator is designed to be independent of the incumbents, making it an 'independent SO', or ISO. This ISO neither plans nor finances investments on its grid. The users, generators and distributors, take the initiative of requesting modifications or extensions to the transmission grid, and then pay for them fully. This ISO manages congestion with an economic method known as 'nodal pricing', transmitting a direct and individualized price signal to each grid user liable to have an impact on congestion (by creating, exacerbating or easing it).

The cost of congestion is thus only borne by those who directly contribute to it, and only for as long as they do so, being calculated in very short time frames that are recomputed every 10 minutes. Each of the thousands of nodes in the grid is handled independently, with a vast technical and economic programme of costing congestion for each entry and exit node on the transmission system. That is why this pricing is called 'nodal'. A direct consequence of this method for managing congestion is that no real border exists, either physical or price based, around the zone administered by the transmitter. Its zone is nothing else than a collection of computation nodes. To the extent that adjacent transmitters practice the same nodal method of pricing and collaborate in its application, there are no real borders between neighbouring transmission zones. This ISO does not charge users the other costs associated with transmission (notably the cost of infrastructures). These costs are recovered through fees that are socialized across a local grid and administered by state Public Utilities Commissions, or PUCs. Finally, the costs of new connections to the transmission grid are not socialized. The hook-up fee imposes all the costs created by this connection, in terms of upstream development, on the new user (called 'Deep Cost' pricing).[18] In conclusion, Rious et al. (2008) apply a systematic modular approach to the management of energy flows in electricity networks.[19]

Imperfect Modularity in Electricity Regulation

A second aspect of the 'weak' nature of the organization of the modules in the competitive reforms of the electrical industry is found in the allocation of responsibilities and decision-making power within the regulatory functions (Green et al., 2008; World Bank, 2006; Castalia Strategic Advisor, 2005; EBRD, 2004).

Nearly every conceivable variation on the definition and allocation of regulatory functions has already been tried somewhere: regulatory functions are shared between federal and local regulators (the USA, Belgium); between the federal executive power, the association of local regulators, and representatives from local governments in a formula called 'Comitology' (European Union); between stakeholders who administer a mandatory pool and a strong regulator (England and Wales); between transmitters who own a voluntary exchange, stakeholders, and ministers from local governments (the Nord Pool of the four Scandinavian countries); between stakeholders[20] administering the ISO and a strong local regulator (Texas); between a weak or semi-weak regulator and the minister of energy (Spain and France); between a weak regulator and the transmitter (Sweden); self-regulation[21] by a national committee of stakeholders overseen by the competition watchdog and the courts (Germany from 1998 to 2005).

This veritable patchwork of formulas has been characterized as 'regulatory modularization' by Midtun (2005). It is noteworthy that not one of these structures has proven able to provide adequate *ex ante* guarantees to simultaneously manage the classical risks of 'regulatory capture'[22] and governmental opportunism (Holburn and Spiller, 2002), while effectively countering the exercise of market power by the dominant operators (Smeers, 2004). Here one theoretical assumption from the microeconomics of incomplete contracts rings true: the public third party ('the judge') does not know what needs to be understood or done, even when the feuding parties to the contract provide all the databases relevant to the matter. Private order and private justice get things done, for better or worse, where both public order and public justice have been displaced by arbitrary decisions, lobbyists' manipulations and, finally, poor administration and legal uncertainty. The cognitive framework for contemporary government regulation is given by an explosion of private activity in the production, warehousing and transmission of informational signals. However, these are not easily evaluated by third party administrations, who eschew permanently creating new codes and new languages embedded in communities of professionals or users, or communities of practitioners (Glachant et al., 2008).

In order to be more efficient, regulators and regulation can open forums on 'production in the public sphere' in which assorted private actors can continue competing in the creation of formal frameworks for future cooperation. In this competition, private actors will have to externalize knowledge to have an influence on publicly produced legitimized standards and mechanisms. This is because legitimate new operational codes will be actionable before all the general institutions of society (agencies, courts, administrative bodies and parliament). Therefore, they exercise a long-term influence on private conflicts over definitions and coding, appropriation and valuation, as well as access and usage. Public regulation based on forums, deriving from 'green papers or white papers', from public hearings or soft-regulation tools (such as statements of intent or interpretation) thus constitute a broad mechanism for revealing information and knowledge. Here, communities, interests, and lobbies come to meet or clash, compete or collide

– and, in so doing, enrich or derail the creation of public legitimacy (Eberlein, 2005). These 'open' regulatory processes reduce informational asymmetry and the embedding of knowledge, thus favouring information revelation and knowledge transfer (Libecap, 2002; Kleindorfer, 2004). This third, and final, transformation of the modern regulatory economy thus contributes to the creation of a new regulatory mix combining a wealth of information and a dearth of comprehension into an imperfect, but viable, regulatory process. This is reminiscent of J.M. Clark's (1940) 'workable competition', with which he reconciled the obvious imperfections of market failure with the administrative capacity expected of the Welfare State nearly 60 years ago.

The Timing and Order of Modularity Decisions

Newbery (2000) stresses the fact that the sequencing of the reform is critical. The sequencing of a reform is the way by which the reform agenda is decomposed into different steps to be undertaken at successive periods. Sequencing structures the behaviour of stakeholders by creating new interests and new rights over the various modules of activity and over the transactions that come into play between these modules. One of the most important consequences of this sequential modularity is that certain models of network industry reform, while working well under some circumstances and in some areas, are not easily transferable elsewhere. Newbery (2002) emphasizes the importance of a robust reform strategy, which must include all of the following: the privatization process, the type of unbundling between monopolistic and competitive activities, the initial market design, the powers and functions of the sectoral regulator, and so on. Any creation of new institutions occurs in a world that is already 'saturated' with other institutions. Consequently, the compatibility and complementarity between the new institution and other pre-existing institutions are fundamental objective characteristics that define the new institution. This is why, in theory as much as in fact, the *ex ante* choice of a good competitive reform strategy for entire blocks of industry is more difficult than some optimists had prematurely announced. In fact what is called 'reform' is a set of interactions between the 'new' and the 'given' institutions.

In conclusion, we want to stress the fact that understanding all the dimensions and impacts of weak modularity in the electricity industry remains analytically problematic. This challenge is rendered more complex by the issue of choosing an appropriate sequence for key decisions.

THE AGENDA AHEAD IN ELECTRICITY REFORMS

In this section the objective is to highlight the main remaining issues in the electricity industry around the world, namely, the question of investment in electricity markets, the treatment of long term contracts and the support of Renewable Energy Sources for Electricity (RES-E) in energy markets.

The Question of Investment

During the design of electricity market reforms, the issue of investment in generating capacity generally received insufficient attention. The model was a vertically and horizontally de-integrated industry facilitating entry and allowing effective competition on each market from wholesales to retail sales, with regulation discouraging vertical arrangements and long-term contracts. The canonical business model in generation is the merchant plant, a standalone producer which sells all this production on short-term markets and without long-term contracts at fixed prices and develops its new capacities under project financing by non-recourse debt.

Regulation disallows vertical integration and long-term contracts between producers and suppliers, as well as between suppliers and consumers, and tends to incite historic producers to divest in generation. In the first post-reform period in the UK, the regulator imposed constraints on vertical integration in generation (no more than 15 per cent in own generation assets within their own area). Europe is not different. In 2005 and 2007, the Directorate General of Competition of European Commission underscores that incumbents' vertical integration generation-supply and the historic suppliers' long-term sales contracts create a risk of foreclosure (European Commission, 2005, 2007). In some of the liberalized US markets there are regulatory restrictions on long-term purchase contracts to provide an incentive for the suppliers to minimize their purchase cost of wholesale electricity in relation to the changing conditions of the market (Joskow, 2002; 2006a).

This insufficient attention was highlighted by the crises on electricity markets, which were partly due to inadequate capacity as well as the focus of generators' investment decision on gas generation technologies which could create an excessive specialization of the technology mix. Then, after these crises, theoretical and practical considerations on generation investment largely focused on incentives for developing peak generating capacity and ensuring a reserve margin to guarantee reliability, that is, short-term security of supply. An abundant literature developed on this issue, in particular on the different ways of capacity payment (see, for instance, Oren, 2003; Cramton and Stoft, 2006; Joskow 2006b; De Vries, 2007).

But little attention was paid to the conditions for other investments in baseload and semi-baseload equipment, since untimely development and non-optimal technology mix distorted in favour of low capital-intensive technologies do not present the same risks as inadequacy of total capacity and its impact on system reliability. The consequence will be a non-timely development of capacities and a non-optimal orientation of the overall technology mix in the different liberalized electricity markets, as new equipment will be added to the fleet of the competitors to follow the demand growth and to replace old equipment at the end of its working life. As shown by Green (2006), if the mix of capacity is wrong and characterized by a lack of baseload equipment, marginal price will be unduly high during a large part of the year comparative to a situation with an optimal mix and, finally, it will be at the expense of the social surplus, the loss of the consumers being higher than the supplement of net profit of the producers. For a deeper discussion of these issues, see Chao et al. (2008) and Finon and Perez (2008).

The Issue of Long-term Supply Contracts (LTC)

Long-term supply contracts remain a pervasive feature of most European energy markets despite the progress of liberalization (DG COMP, 2007). To get out of the monopoly era, the current refining and harmonization of European market designs may be pointless if incumbents continue to use these contracts as market control devices (Glachant and Lévêque, 2006). Indeed, these contracts frequently have anti-competitive foreclosure effects when competition is imperfect (Fumagalli and Motta, 2006) and these effects are likely to be worse in a market-building context. However, there is growing acceptance that their positive impact on investment makes them desirable as long as spot market competition remains unsatisfactory (Chao et al., 2008).

The institutional structure of the European Union does not give to the European Commission the power to *ex ante* alter property rights in the different member states and thus to promote an aggressive policy of horizontal de-integration which would probably deliver better and faster results (Green and Newbery, 1997; Newbery et al., 2003). In addition, the lack of an EU-wide energy regulator with effective power to monitor and regulate market developments *ex ante*, especially cross-border trade issues, is particularly detrimental to the integration and the well-functioning of European energy markets (Glachant and Lévêque, 2006).

Conducting market building through antitrust does not go without uncertainties about the suitability of the policy tool itself and about the ability of the judges to enforce it. Antitrust policy remains constrained by the legal process, and especially judicial review. This also raises questions both about the discretion which a process of market building requires to be conducted efficiently and about the time and cost of dispute resolution through court trial (Newbery, 2005). In addition, antitrust is usually enforced in sectors where competition is more mature and a limited knowledge of competition dynamics may result in significant error costs (Smeers, 2005; Hautecloque and Glachant 2009).

The Support of Renewable Energy Sources for Electricity

Since the 1990s, multiple objectives have driven governments to develop renewable energy policies: preserving a set of collective goods; climate stability; local environment and energy security.[23] Government support is necessary for Renewable Energy Sources in Electricity (RES-E) because, although desirable from a social welfare perspective, whose private costs are not competitive in power generation systems dominated by large electricity generation plants. Three reasons account for the bias against RES-E in the electricity market: (1) environmental costs are not adequately internalized for conventional electricity generation technologies; (2) the absence of scale effects on costs, due to the small size of the plants[24] and (3) the random nature of their intermittent production of some major sources of RES (wind power, mini-hydraulic), which creates negative externalities.

Following failed attempts using systems of voluntary purchases of green electricity by consumers, as well as direct investment subsidies, demand-side strategic deployment policies have emerged as the preferred instrument in most countries. By imposing obligations to purchase RES electricity or to meet a RES-E quota on a clearly specified type of agents, these policies are designed to allow demand-side forces to determine the allocation of RES-E production subsidies to pre-commercial and commercial RES-E technologies. There are three instruments with such common characteristics: feed-in tariffs (FIT), bidding instruments (BI) for the assignment of long-term purchase contracts and exchangeable quotas (EQ).

Feed-in tariffs oblige electricity distributors (or the incumbent suppliers in a spatial

Table 10.3 Comparison of performances of the three instruments

	Feed-in tariffs		Bidding		Exchangeable quotas	
Criteria	*Ex ante*	*Ex post*	*Ex ante*	*Ex post*	*Ex ante*	*Ex post*
Investment safeguards	++	–	–	++	+	+/–
Ex ante adaptability of the instrument	+		0		0	
Incentives intensity and cost control	+/0		+		++	
Technological progress	++		0 (Best available per technology band)		– (The most mature technique)	
Conformity with electricity market regime	0		0/+		++	

area) to purchase electricity from any new RES-E plant in their service area and pay a minimum guaranteed tariff per kWh, which is fixed over a long period of time. Bidding instruments select by auction RES-E projects and oblige local electricity distributors (or the incumbent supplier in the market regime) to buy electricity from the successful plants by a long-term contract on the basis of bid price in the reference design (or the marginal price in some countries). Exchangeable quotas Q introduce future obligatory targets for electricity suppliers to buy either green electricity directly from the RES-E producers, or green certificates issued to RES-E producers, targets being defined in terms of a percentage of their electricity deliveries. A complementary trait is a compensation mechanism for the opportunity cost for obligated purchasers, which is different in the monopoly regime and in the market regime.

Following Finon and Perez (2007), the outcome of these three support mechanisms can be differentiated in terms of investment safeguard, adaptability, positive impact on technological progress and conformity with market-based instrument. Authors show that the absence of a first best solution raises the important question of creating a new smart energy policy to stimulate RES-E technologies in the long run, as Table 10.3 summarizes.

These challenges remain on the top of the agenda for both academics and policy-makers around the world willing to achieve the creation of electricity markets. Finding solutions to investments, long-terms contracts and RES-E support will make the energy markets sustainable in the long run.

CONCLUSION

In this conclusion we want to stress three main ideas; first, the international diversity of electricity market reforms is a logical result of the combination of industrial, technological and institutional constraints shaping the policy-making process of electricity market creation. We doubt that this diversity will be drastically reduced in the future, even if the definition of the textbook model of electricity reforms helps better understand the remaining problems in this industry. The second idea is the absence of a perfect architecture of electricity markets as Wilson (2002) has demonstrated. Then the diversity of reforms brings a very interesting field of experiments and paves the way to international comparisons to adapt the existing electricity reforms. Finally, a long list of questions still need to be investigated more deeply, and continuous and interdisciplinary work between lawyers, economists and engineers is needed to foster a deeper understanding of electricity market reforms.

NOTES

1. The Authors want to thank, for helpful comments, Adrien de Hautecloque, Vincent Rious and Marcelo Saguan.
2. Aviation reorganized itself independently and durably – until the appearance of low cost airlines. – on the 'Hub & Spokes' model with large airports and 'private' interconnections between the flights of a single company or a pool of affiliated companies.
3. The design of the competitive electricity market remained heterogeneous and unstable, made up of many distinct industrial actors and transactional linkages, variously disassembled and reassembled and, sometimes, though not always, associated with competitive mechanisms or true markets.
4. We here recall the operational difficulties encountered by California's electricity markets between the summer of 2000 and the spring of 2001. We also think of the comprehensive redesign of the English system in 2002, leading to the closing of the Electricity Pool of England and Wales, which was mandatory for all generators and resellers as of the beginning of the reform on 1 April 1990.
5. Both are infrastructures that can easily be differentiated from taxis, buses and trucks.
6. Flight corridors and airport runways are clearly distinct from the airplanes chartered by airlines. However, it is also necessary to prescribe how these air routes and runways are to be allotted to the various users when the sum of all possible usage slots is less than the airlines' demand, in particular in the case of new entrants. Are the airlines' large hubs private infrastructures with strictly controlled access, or are they private domains built on essential infrastructures freely accessible to all?
7. The issue is similar to that in the previous note.
8. It is as if all airlines operating in the same control space were obligated to continually equate the number of seats on all their airplanes to the exact number of passengers having boarded them!
9. From 'real time' to one or three hours before real time
10. In other words, transactions arose in specific locations because designers created technologically separable interfaces that made transactions cost-effective at those points.
11. The plug and play image is helpful, compatibility between modules is not naturally given or cannot be seen as given without a huge work of harmonization, definition of standards and enforcement mechanisms to stabilize the design.
12. This component also underlies the spectacular 'blackouts' that have shaken up this industry on several occasions since the beginning of the twenty-first century (USA and Canada, Italy, Denmark, Germany and France, and so on)
13. Littlechild (2006) brings two aspects to the debate: he shows that, in Australia, merchant transmission companies have been allowed to compete with incumbent transmission monopolies for the building of new lines; while in Argentina transmission line expansion decisions have to be proposed, approved and paid for by market participants and not by the regulator or the regulated transmission company.
14. Via a half-hourly 'postage stamp' price.
15. Especially the costs of infrastructures.
16. The fees are paid on average by all the users.
17. This method of pricing connections is called 'shallow cost'.
18. For a discussion of the economic consequences of the various methods of recovering connection costs in the electrical industry, see Hiroux (2004).
19. They show that, for each of these three modules, three fundamentally different management modes are possible, yielding 27 options for organizing the basic functions of the TSO. The paper also proposes a definition of a first-best TSO as the optimal combination of the three most efficient ways of performing the three tasks.
20. Another example is provided by Littlechild (2006) and the settlement of disputes organized in Florida:

instead of a traditional litigated process, settlements are often reached between utilities and the Public Council and/or users, and are typically approved by regulator.

21. The basis of self-regulation is reciprocity: Individuals recognize the benefits they will derive from behaving in accordance with others' expectations. Such reciprocity may be reflected in individual agreements but, as standards of behaviour, will spread to other members of a group as property rights when the benefits of doing so exceed the costs of defining those rights. The expectation is that such cost savings will be significant where the group is small enough for informal control – generally requiring continuing face-to-face interaction – but also where power is broadly dispersed.
22. Thus, proponents of the theory of capture demonstrate how repeated exchanges between the regulatory agency and the firms can culminate in collusion between them.
23. Ambitious targets have been set at both national and European Union (EU) levels. If the voluntary objectives adopted by EU Directive 2001/77/EC on the promotion of electricity produced from renewable energy sources are met, the share of 'new' renewables in the electricity production are projected to increase from 1 per cent to 8 per cent of the total electricity production on average in the EU by 2010 (European Commission, 2001).
24. This character could be discussed for certain RES-E technologies, as the biomass electricity which could benefit from the scale economies in the supply of the biomass up to a certain size.

REFERENCES

Baldick, R. and Niu, H. (2005), 'Lessons learned: the Texas experience,' in J. Griffin and S. Puller (eds), *Electricity Deregulation: Where to From Here?*, Chicago, IL: University of Chicago Press, pp. 182–224.

Baldwin, C., (2008), 'Where do transactions come from? Modularity, transactions, and the boundaries of firms', *Industrial and Corporate Change*, **17** (1), 155–95.

Baldwin, C. and Clark, K. (2000), *Design Rules: The Power of Modularity*, Cambridge, MA: Massachusetts Institute of Technology Press.

Baumol, J.W. and Sidak, J.G. (1994), *Toward Competition in Local Telephony*, Cambridge, MA: Massachusetts Institute of Technology Press.

Brousseau, E and Glachant, J.-M. (eds) (2002), *The Economics of Contracts: Theories and Applications*, Cambridge: Cambridge University Press.

Brousseau, E. and Glachant, J.-M. (eds) (2008), *European School for New Institutional Economics: A Reader*, Cambridge: Cambridge University Press.

Castalia Strategic Advisor (2005), *Explanatory Notes on Key Topics in the Regulation of Water and Sanitation Services*, World Bank eds, Washington, DC: World Bank.

Chao H.-P., Oren, S., and Wilson, R. (2008), 'Reevaluation of vertical integration and unbundling in restructured electricity markets', in F.P. Sioshansi (ed), *Competitive Electricity Markets: Design, Implementation, and Performance*, London: Elsevier, pp. 27–65.

Clark, J.M. (1940), 'Toward a concept of workable competition', *The American Economic Review*, **30** (2), Pt 1, 241–56.

Crampton, P. and Stoft, S. (2006), *The Convergence of Market Designs for Adequate Generating Capacity with Special Attention to the CAISO's Resource Adequacy Problem*, White Paper for the Electricity Oversight Board, Berkeley, California

De Vries, L. (2007), 'Generation adequacy: helping the market do its job', *Utilities Policy*, **15** (1), 20–35.

DG COMP (2007), Competition report on energy sector inquiry (SEC(2006)1724), 10 January.

Eberlein, B. (2005), 'Regulation by cooperation: the "third way" in making rules for the internal market', in P.D. Cameron (ed.), *Legal Aspects of EU Energy Regulation*, Oxford: Oxford University Press, pp. 59–88.

European Bank for Reconstruction and Development (EBRD) (2004), *Transition Report 2004*: *Infrastructure*, London: EBRD.

European Commission (2001), Directive 2001/77/EC of the European Parliament and of the Council of 27 September 2001on the promotion of electricity produced from renewable energy sources in the internal electricity market, *Official Journal of the European Communities*.

European Commission (2005), Cases COMP/B-1/39.172 (electricity) and 39.173 (gas).

European Commission (2007), DG Competition report on energy sector inquiry (SEC(2006)1724), 10 January.

Finon, D. and Perez, Y. (2007), 'Transactional efficiency and public promotion of environmental technologies: the case of renewable energies in the electric industry', *Ecological Economics*, **62**, 77–92.

Finon, D. and Perez, Y. (2008), 'Investment risk allocation in restructured electricity markets: the need of vertical arrangements', LARSEN working paper, Paris.

Fumagalli, E. and Motta, M. (2006), 'Exclusive dealing and entry, when buyers compete', *American Economic Review*, **96**, 785–95.
Glachant, J.-M. and Finon, D (eds) (2003), *Competition in European Electricity Markets: A Cross Country Comparison*, Cheltenham, UK and Northampton, MA, USA: Edward Elgar.
Glachant J.-M. and Lévêque, F. (2006), 'Electricity internal market in the European Union: what to do next?', EPRG working papers 0605, University of Cambridge.
Glachant, J.-M. and Lévêque, F. (2009), *European Union Electricity Internal Market: Towards an Achievement?*, Cheltenham, UK and Northampton, MA, USA: Edward Elgar.
Glachant, J.-M., Dubois, U. and Perez, Y. (2008), 'Deregulating with no regulator: is Germany electricity transmission regime institutionally correct?', *Energy Policy*, **36** (5), 1600–10.
Green, R. (2006), 'Investment and generation capacity', in F. Lévêque (ed.), *Competitive Electricity Markets and Sustainability*, Cheltenham, UK and Northampton, MA, USA: Edward Elgar, pp. 21–53.
Green, R. and Newbery, D.M. (1997), 'Competition in the electricity industry in England and Wales', *Oxford Review of Economic Policy*, **13** (1), 27–46.
Green, R., Lorenzoni, A., Perez, Y. and Pollitt, M. (2008), 'Benchmarking in the European Union', working paper, University of Cambridge.
Hancher, L. (1997), 'Slow and not so sure: Europe's liberalisation', *The Electricity Journal*, **10** (9), 92–101.
Hautecloque, A. de and Glachant, J.-M. (2009), 'Long term energy contracts in European competition policy: fuzzy not crazy', EUI working paper, RSCAS 2009/6, Florence.
Hiroux, C. (2004), 'Shallow cost and deep cost, an assessment', working paper GRJM, Université Paris-Sud 11.
Hogan W.W. (2002), 'Electricity market restructuring: reforms of reforms', *Journal of Regulatory Economics*, **21** (1), 103–32.
Holburn, G.L.F. and Spiller, P.T. (2002), 'Institutional or structural: lessons from international electricity sector reforms', in E. Brousseau, and J.-M. Glachant (eds), *The Economics of Contracts: Theories and Applications*, Cambridge: Cambridge University Press, pp. 463–502.
Joskow, P. (2002), 'Electricity sector restructuring and competition: a transaction cost perspective,' in E. Brousseau and J.-M. Glachant (eds), *The Economics of Contracts: Theories and Applications*, Cambridge University Press.
Joskow, P. (2005) 'Regulation and deregulation after 25 years: lessons for research', *Review of Industrial Organization*, **26**, 169–93.
Joskow, P. (2006a), 'Patterns of transmission investment', in F. Lévêque (ed.), *Competitive Electricity Markets and Sustainability*, Cheltenham, UK and Northampton, MA, USA: Edward Elgar, pp. 131–86.
Joskow, P. (2006b), 'Market electricity markets and investment in new generating capacity', working paper CEEPR-MIT 06-009, Massachusetts Institute of Technology, April.
Joskow, P. (2006c) 'Electricity sector liberalization: lessons learned from cross-country studies', in F.P. Sioshansi and W. Pfaffenberger (eds), *Electricity Market Reform: An International Perspective*, Amsterdam: Elsevier.
Joskow, P. and Schmalensee, R. (1983), *Markets for Power*, Cambridge, MA, USA: Massachusetts Institute of Technology Press.
Kessides, I.N. (2004), *Reforming Infrastructure: Privatization, Regulation, and Competition*, a World Bank Policy Research Report, Washington, DC: World Bank and Oxford University Press.
Kleindorfer, P. (2004), 'Economic regulation under distributed ownership: the case of electric power transmission', working paper, Wharton School of the University of Pennsylvania, January.
Libecap, G. (2002), 'A transaction-cost approach to the analysis of property rights', in E. Brousseau, and J.-M. Glachant, (eds), *The Economics of Contracts: Theories and Applications*, Cambridge: Cambridge University press, pp. 140–56.
Littlechild, S. (2006), 'Foreword', in F.P. Sioshansi and W. Pfaffenberger (eds), *Electricity Markets around the World*, Amsterdam: Elsevier.
Midtun, A. (2005), 'Deregulation: design, learning and legitimacy', in J. Gronewegen and R. Kunneke (eds), *Reform of Infrastructure Industries*, Cheltenham, Uk and Northampton, MA, USA: Edward Elgar.
Newbery, D. (2000), *Privatization, Restructuring, and Regulation of Network Utilities*, Cambridge, MA: Massachusetts Institute of Technology Press.
Newbery, D. (2002), 'Issues and options for restructuring electricity supply industries', DAE working paper WP 0210, University of Cambridge.
Newbery, D.M. (2005) 'Electricity liberalisation in Britain: the quest for a satisfactory wholesale market design', *Energy Journal*, special issue, 43–70.
Newbery, D.M. van Damme, E. and von der Fehr, N.H. (2003), 'Benelux market integration: market power concerns', report of the Market Surveillance Committee, Dutch Competition Authorities.
Oren, S. (2003), 'Ensuring generation adequacy in market electricity markets', working paper UCEI, EPE 007, University of California Energy Institute.

Prosser, T. (2005) 'Regulatory contracts and stakeholder regulation', *Annals of Public and Cooperative Economics*, **76** (1), 35–57.
Rious, V., Glachant, J.-M. Perez, Y. and Desante, P (2008), 'The diversity of design of TSOs', working paper, Cambridge University, and working paper, MIT.
Saguan, M., Glachant, J.-M. and Perez, Y. (2009), 'Le marché du temps réel comme le plus petit multiple commun des architectures de marché électrique', *Revue d'Economie Industrielle*, no. 127, 69–88.
Sioshansi, F.P. and Pfaffenberger, W. (eds) (2006), *Electricity Markets around the World*, Amsterdam: Elsevier.
Smeers, Y. (2004), 'TSO, electricity markets and market power', *European Regulatory TSO Benchmarking*, Den Haag, Mai.
Smeers, Y. (2005) 'How well can one measure market power in restructured electricity systems?', communication to SESSA Conference, published in J.-M. Glachant and F. Lévêque (eds) (2009), *Electricity Reform in Europe. Towards a Single Energy Market*, Cheltenham, UK and Northampton, MA, USA: Edward Edgar, pp. 207–68.
Williamson, O.E. (1985), *The Economic Institutions of Capitalism*, New York: Free Press.
Williamson, O.E. (1996), *The Mechanisms of Governance*, Oxford: Oxford University Press.
Wilson, R. (2002), 'Architecture of power markets', *Econometrica*, **70** (4), 1299–340.
World Bank (2006), *Handbook for Evaluating Infrastructure Regulatory Systems*, World Bank edition, Washington, DC: World Bank.

11 Trends in gas

Maarten J. Arentsen

INTRODUCTION

'In the last decade, natural gas has grown to become a truly global energy source.'

This quote from the 2006 gas market review of the International Energy Agency (IEA) clearly indicates the significance of natural gas in today's global energy supply. In all parts of the world natural gas has become a much wanted energy resource. It has the highest hydrogen to carbon ratio of all fossils and the carbon dioxide emissions from its conversion are lower than oil and coal (IPCC, 2007). It can easily be transported by pipelines, tankers or trucks and there is little need for additional treatment before application and consumption (Hawdon, 2003). No wonder the demand for natural gas is increasing in a world facing climate change and lacking cost-effective, non-fossil, alternative energy resources. Increasing demand for natural gas combined with greater flexibility in long-distance transport will allow the gas market to become global (Correljé and van der Linde, 2006; Stern, 2006).

Regulatory reform preceded the worldwide increase in gas demand; it was undertaken on all continents, but it took different routes. In many parts of the world the reforms have strengthened liberalization, deregulation and privatization, especially in the wholesale and retail segments of the market. The different liberalization models all reflect the idea of strengthening competition and market based transactions, the introduction of commercial incentives, and free consumer choice of gas supplier. Countries and continents made their own regulatory choices, quite often with reference to the US gas market, which is generally recognized as the world's most competitive gas market and is therefore considered the reference model for reforms in Australia, Europe and South America. Reforms also took another, state-oriented, direction, especially in the gas-producing regions of world. In countries like Russia, Qatar and Iran, the state strengthened its position in gas production and started using the natural resource as a strategic instrument in international relations. Regulatory reform therefore not only took a turn towards favouring the market, but towards favouring the state too.

A third and final trend in natural gas markets, which is especially visible in Europe, is the upcoming diversification of gases in gas pipeline systems. Climate change and such political ambitions as sustainable development are clear drivers of this search for competitive alternatives to natural gas. Gasification of biomass and coal as well as the production of hydrogen are currently being explored as future alternatives to natural gas. Biogas, synthetic gas and hydrogen are being explored and experimented with in the expectation that they will become a full alternative for natural gas sometime in the future. In combination with liberalization, this diversification of gases opened minds to more flexibility in the technology and operation of the gas pipeline system.

This chapter discusses the three above-mentioned trends in the world's leading gas markets: regulatory reform, globalization of competition, and climate change. The trends are visible in all parts of the global gas market, but they differ in terms of intensity, scale and scope. They have allowed some scholars to suggest the emergence of a new energy paradigm (Helm, 2007). A multidimensional rather than a one-dimensional paradigm, reflecting the emerging complexities of today's globalizing gas market. The chapter discusses the trends and draws conclusions about their future impact on the regulation of natural gas.

The chapter is structured as follows. The next section discusses trends in regulatory reform in the world's leading gas markets. As indicated, the reforms took two opposite directions in gas-producing and gas-consuming regions of the world. The third section discusses the causes and implication of the emerging globalization of natural gas competition. Economic growth and climate change are important causes underlying the increasing gas demand in the world, which is clearly transforming competition into a contest between buyers rather than suppliers of gas. The production of green gases (biogas and synthetic gas) and hydrogen is leading to a further diversification of gas types and gas properties, which is especially challenging to the technology and management of the gas pipeline infrastructure. The causes and implications of the diversification trend are discussed in the fourth section. The final section draws some conclusions in terms of prospects.

TRANSFORMATION OF THE GAS VALUE CHAIN: LIBERALIZATION AND POLITICIZATION IN NATURAL GAS

During the last two decades, natural gas supply has undergone major regulatory changes everywhere in the world. The reforms took two different directions: liberalization on the one hand, strengthening the market in gas supply, and politicization on the other, thus strengthening the state's role in gas supply. This section discusses both directions of reform.

Liberalization

Until the late 1980s the industrial organization of the gas value chain was rather straightforward. For reasons of scale and scope the dominant model was the vertically integrated industry model, covering production, transmission and supply. From the mid-1980s onwards, this dominant industrial model based on vertically integrated companies and regulated monopoly supply and long-term contracts started changing in the direction of regulated competition. Although the reforms followed different routes in the world's gas markets, the core idea underlying the reform is everywhere similar: establishing competitive demand and supply relations by providing third parties with access to the gas pipeline infrastructure in combination with the unbundling of gas production/supply/trade and the transmission and distribution of natural gas. Implementation of this core idea of liberalization required the unbundling of vertically integrated gas companies everywhere in the world. These vertically integrated companies offered wholesale, retail and transport of the natural gas in one single package. In the USA, for instance, high-pressure gas pipelines (interstate and intrastate)[1] were initially owned by affiliates of the large gas and oil companies, which produced and supplied the natural gas. In the USA, gas pipeline infrastructure comprising about 300 000 miles, the Henry hub in Louisiana is a well-known gas hub interconnecting nine interstate and four intrastate pipelines. The hub facilitates gas futures trading on the New York Mercantile Exchange (NYMEX). In 2005 the USA had 76 gas trading hubs (Murry and Zhu, 2008, p. 249). The North American market uses natural gas as an energy resource for power production, as feedstock and as energy resource for residential heating and hot water supply.

Before the mid-1990s, the gas trade in the USA was actually 'regulated delivery' and not a transactional activity at all (Leitzinger and Collette, 2002). At that time the Federal Gas Market Regulator, FERC, decided on almost all aspects of transactions offered as a package deal of volume, transport and services. From the early 1990s onwards, FERC facilitated the introduction of competition in the US gas market and third-party access by a few traders. Currently the US gas market is among the most competitive in the world, allowing the full trade of natural gas and with no entry barriers to the pipeline system. Despite its huge size, today's US gas market can be considered an integrated, competitive market (IEA, 2008; Murry and Zhu 2008).

This cannot be said for Europe's continental gas market. Before liberalization the European gas supply was dominated by vertically integrated companies offering volume, transmission and services as a package deal. There was strong resistance to any change in this centralized national model. Except for the UK, all other European countries therefore only started reforming the gas market after the implementation of directive 98/30/EC in 2000. At that time the UK was already far ahead in gas market reform, which had its origin in the 1986 privatization of British Gas (Thomas, 2003). It took the European Union (EU) member states nine years to develop a minimal set of reforming rules, which they laid down in the gas directive of 1998 (Hanscher, 2003). In the years following the implementation of the directive, the member states all made the minimum changes required by the gas directive, with the result that by the 'mid-2000 the larger part of the EU region shared some common focal points in gas market regulation as required by the Directive' (Arentsen, 2004, p. 75; Haase, 2008). The reform of the European gas market is still continuing, while encountering persistent problems such as market concentration, lack of transparency and adequate unbundling[2] of transmission/distribution and supply, all of which still await final settlement. As the explanatory note to the latest energy market directive says:

> The internal electricity and gas market is suffering from a lack of liquidity and transparency hindering the efficient allocation of resources, limiting risk hedging possibilities and blocking new entrants. Trust in the market, its liquidity and the number of market participants need to increase, by increasing the information that is available to the market. (EC, 2007)

Consequently, today's EU gas market is not yet liberalized and integrated as it is in the USA (Egging and Gabriel, 2005). The lack of integration is signaled by a diversity of regulatory models and significant price differences between the major gas markets in Europe (Eikeland, 2007; Haase, 2008; Spanjer, 2008).

Deregulation also transformed Latin American gas markets, a reform initiated by general economic reforms in the early 1990s. 'Widespread structural reforms and efforts towards macroeconomic stabilization helped South American countries recover sustained levels of economic growth in the 1990s, after a "lost decade", as the 1980s have been called' (IEA, 2003, p. 13). Huge public debts blocked gas infrastructure investments on the continent and reform of public debts was needed. The reform wave started in Argentina early in the 1990s with the unbundling and privatization of national gas companies (Honoré, 2004). The reform opened up all links in the gas chain to private companies and made Argentina today's most liberalized gas market on the Latin American continent. Other Latin gas markets liberalized too, but here the reforms were not as radical as in Argentina. Argentina, Chile and Trinidad and Tobago have now privatized the exploration and production of natural gas. Other Latin American countries with indigenous gas resources have mixed public–private structures. All Latin American countries are more open to deregulation in transmission and distribution and privately owned gas pipelines. The pipeline connections between the countries are basically owned and run by consortia of companies. In South America it is an ambition to connect the North and South of the continent through the Amazon forest. Apart from this very ambitious connection, Brazil is investing in new pipeline connections between the Rio de Janeiro region (the major load center) and the Northern Brasilia region. The country is also considering new cross-border pipeline connections with Paraguay and Uruguay. Unbundling is less clearly organized in Latin America. Some countries do allow vertically integrated companies to own and run gas pipelines, as in Chile, for instance. Argentina and Bolivia, on the other hand, explicitly prohibit vertically integrated companies, while Brazil and Venezuela require functional unbundling of companies. Chile is also an exception in energy market regulation. It is the only Latin American country to have integrated energy policy and energy market regulation in a single governmental organization. All other Latin American countries have separate institutions for policy and regulation, as in the USA and Europe, for instance.

The Australian gas market reform is rather similar to Europe. Australia started reforming in the mid-1990s but the continent's wholesale as well as the retail market is still not genuinely competitive. Part of the problem is the continued high degree of concentration of the regional markets of the Australian continent. There is no single Australian gas market but, as on the European continent, only regional markets: a Western, Northern and the integrated Victoria, New South Wales, Southern Australia, Queensland and Tasmania market. All markets are highly concentrated both upstream and downstream and lack liquidity due to the relatively limited domestic consumer market. The domestic segment of the market is well covered by the gas grid, but the Australian population is small and concentrated in a few regions on the huge continent. So the physical and geographical conditions add to the degree of concentration in the Australian market. Most gas was and still is traded under bilateral, long-term contracts and the regional markets are basically supplied by one or two suppliers. Pipelines are basically point-to-point connections, between the Northern gas fields and the Western load centres where the population is concentrated. Moreover, Australia's access regime still involves a lot of uncertainties for both the pipeline owners and shippers. The conditions are predominantly negotiable and quite often it is unclear whether a pipeline is eligible for third party access (TPA) or not. Like Europe, Australia too still faces major problems and barriers in gas market reform.

Indonesia is the major gas producer in the Pacific region, exporting liquefied natural gas (LNG) to Japan, Korea and natural gas by pipeline to Singapore. Indonesian gas production is dominated by private foreign firms in an upstream segment deregulated in 2001. The motivation underlying the reforms was to attract foreign investments to the country's oil and gas sector. With the help of these investments Indonesia tried to transform its export-oriented gas industry into a domestic gas supply. For too long the domestic market has been neglected, but is reappearing as an attractive supply area now the country is facing global competition in LNG. Developing the domestic market is quite challenging, though. The country lacks a well-developed gas system and has no funding for the investment needed to develop one.

Politicization

Politicization covers a second reform trend in natural gas: strengthening the state's role in the national gas industry. This trend, which is especially visible in significant parts of the gas producing regions, is leading to what is called 'resource nationalism' (Stern, 2006, p. 12). It reflects the re-establishment of the state's authority in natural resource policy. Russia under President Putin provides a clear example of this statism in natural gas production. But the model is also emerging in other production regions of Africa and the Middle East. On the European continent, the Norwegian state has strengthened its position by merging the two largest energy companies and taking a majority position in the new company. In this way Norway is maximizing the social benefits of its oil and gas reserves.

In Russia, as in the Middle East, the state dominates the country's gas sector and this domination is effectuated by Gazprom, the country's largest gas company, active in all segments of the gas value chain and with a dominant stake in the exploitation of the huge Russian gas fields. Gazprom dominates the whole Russian gas chain, but the number of independent companies having access to the Gazprom pipelines has increased from 6 in 1998 to 33 in 2004 (IEA, 2006, p. 32). Gazprom also dominates the downstream segment of the Russian gas market, but here too the openness to independent companies is growing. According to the IEA gas market analysis of 2006, one of the reasons for opening up the market to third parties is to involve independent producers more actively in developing the domestic Russian consumer market. Gazprom is facing huge challenges in exploitation of the current gas fields and keeping up with the exploitation of new fields and the maintenance and extension of the extended Russian pipeline infrastructure. The pipeline infrastructure is highly inefficient, being among the major leaking gas infrastructures in the world.[3] However, despite all goodwill and current initiatives for third-party access, while it is supported by nationalistic Russian gas politics, Gazprom continues to be the dominant player in Russian gas and therefore one of the leading gas companies in the world. Under Putin, the Russian state has recalibrated investment participation by foreign companies, who are no longer allowed to hold majority positions in gas projects. Existing contracts have been 'renegotiated' to the benefit of state dominance.

The statism in gas-producing regions is in contrast to the deregulation trend in most liberal economies of the world. Both trends are taking place simultaneously, but in different parts of the world. State control in gas production is strengthened, whereas in gas consumption state domination has faded away in favour of the market. The question is how these diverging trends in different parts of the world should be perceived? It is not so much the difference in organization as well as the diversification of the geopolitical relations that frightens analysts.

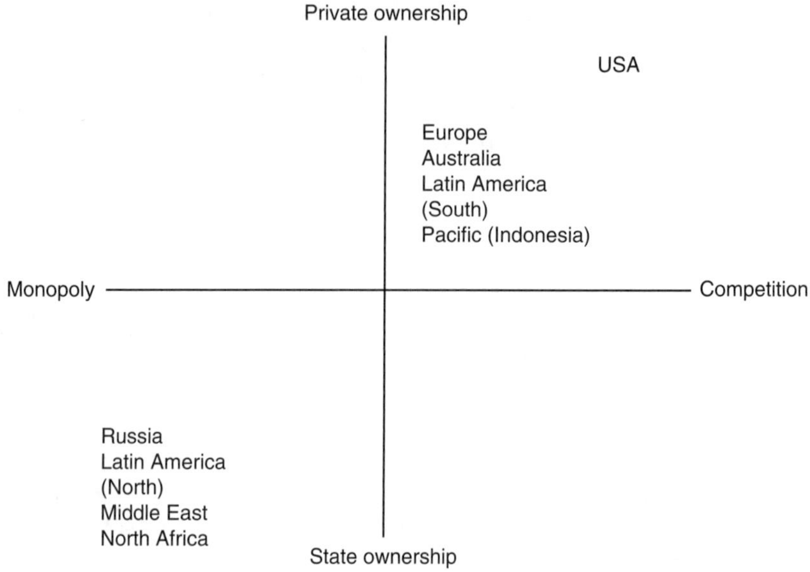

Figure 11.1 The world's dominant gas market models

> Geopolitical scenarios, such as the Clingendael Institute's 'Regions and Empires', and Shell International's 'Low Trust Globalisation', have produced comprehensive storylines that are strongly negative for oil and gas trade. Correlje and Van der Linde have observed that under 'Regions and Empires' there is likely to be '. . .a slowly emerging [gas] supply gap, as a result of lagging investments as a consequence of ideological and religious contrasts, particularly with regard to the North African suppliers, the potential supplies in the Persian gulf and the Caspian Sea region'. (Stern, 2006, p. 13)

In the short term the consumer markets can still rely on the long-term gas contracts, but the world of natural gas is destabilizing and experts expect greater uncertainty in the medium and longer term. In combination with intensified global competition for natural gas, this is not a tempting prospect. Figure 11.1 visualizes the relative positions in gas market organization.

Figure 11.1 roughly indicates a demarcation along the lines of the world's gas reserves with the USA occupying an exempt position. The world's future gas reserves are found in the regions advocating the state-owned, regulated-monopoly model. Private participation is allowed but not in the 'driving seat'. The continents at the other end of the spectrum all adopted variations of the regulated-competition model. These continents also have gas reserves and gas production activities carried on by private or hybrid types of companies. The USA is an outsider. The continent's gas market is carried on by private companies and is considered to be the ideal type of competitive gas market. Supply and demand are driving the market and gas-to-gas competition instead of oil indexation is what determines gas prices. Voluminous liquidity and technology act as core facilitators of the gas trade. This is what the regulatory reforms in Europe, Latin America and Australia have not yet achieved. The reforms on these continents show how difficult it is to establish real competition in the gas market. Liquidity and competition between pipelines is what explains the successful introduction of competition in the USA gas market and what is still missing in the other consumer markets of the world. Only when the gas flow is diversified and voluminous enough to provide a real consumer choice can gas-to-gas competition develop. Except for the USA, this has not yet been realized on the other continents. Europe is preparing for it technically, but it will take time to establish the additional technical infrastructure. Even then, the question is whether the physical connections in Europe will be mature enough to have an integrated competitive market like the one in the USA. In Europe quite often politics turns out to be a tougher barrier to integration than technology or the economy. Australia and Latin America are facing similar infrastructural problems, but here the modest size of the consumer market is an additional source of uncertainty and risk for investors. Some Latin American countries also have shown political risks lately by changing the conditions for private investors and private participation in investments.

But if the current investments in the physical and technological conditions for gas trade, like LNG trains and regasification plants, gas storage and gas pipeline connections pay off, the prospects for competition outside the USA might improve. Technology is clearly needed everywhere in the world to take the next steps towards competition in wholesale and retail markets. The crucial role of technology is perhaps the major lesson of the first wave of gas market reform in the world. It was politics, political rhetoric and political ambition that initiated the first wave of gas market reform, but it is technology that has carried the reforms. The results thus far show that gas market competition is only possible with alternative modes of transportation and application of the natural gas, and only now is this diversification and flexibilization developing. This technology-driven diversification and flexibilization might bring gas market reforms into a second phase. The next steps in gas market reform can be taken only if the availability of natural gas diversifies economically and technically. The current gas market dynamic is a clear sign that this might happen, but it is still too early to conclude that it will.

EMERGING GLOBAL COMPETITION

Trend watchers, scenario developers, international energy organizations, journalists and scientists all agree about a further boost in the demand for natural gas. This generally recognized increase in demand will turn the gas market into a global one in the coming decades. More buyers will compete with each other and LNG technology will move the gas more flexibly around the globe, turning demand for natural gas into a global market. Competition in the demand market will be tough. Two of the major gas markets in the world, the USA and Europe, are facing serious cuts in indigenous gas production, forcing them to contract new gas imports. At the same time emerging economies in Asia, especially China and India, want natural gas as one of the energy resources to facilitate further economic growth and development. Liquefied natural gas technology is the facilitator of the expected globalization of the natural gas market (Correljé and van der Linde, 2006; Linde et al., 2006; Stern, 2006; IEA, 2008).

The World Energy Outlook scenarios forecast an increase in global demand for natural gas by an average of 2 per cent per year till 2030 up to 4.7 tcm (trillion cubic meter) global consumption in 2030 (IEA, 2007). Several factors support this increasing demand for gas. Natural gas is one of the facilitators of the increase in world economic growth. Especially economic growth in emerging economies in Asia (China, India) and Africa are expected to boost global gas consumption. The impressive annual growth rates of the Chinese and Indian economies require constant energy flows and both countries are putting massive efforts into gaining access to energy resources such as coal, oil and natural gas. The IEA expects increases in natural gas demand of 4.8 and 6.5 per cent for China and India, respectively, between 2005 and 2030 (IEA, 2008). Natural gas demand in the Organisation for Economic Co-operation and Development (OECD) countries is expected to contribute to three-quarters of the world's gas demand increase until 2030, whereas the OECD member countries are only able to contribute 3 per cent to the expected annual growth in gas production through 2030. Besides economic growth, increased production of natural gas is considered as a second driver in the growth of world's gas demand.

Everywhere in the world natural gas is expected to grow in power production, including countries where natural gas is only a niche in power production thus far. Brazil shows the role of natural gas in diversifying the fuel base of a country's power production. Gas-based power production is a well-proven, efficient technology with the lowest CO_2 emission of all fossils used for power production. Moreover, gas-based power plants are flexible and easy to manage, which makes them favourable everywhere in the world. Besides power production, (chemical) industry is a second growth market for natural gas. Here the prospect of ever-increasing oil prices is a powerful incentive to adapt to natural gas as an addition to the oil base of the industry. The gas industry is anticipating these developments with the gas-to-liquid mode of natural gas transportation. A final segment of the consumer market that is expected to grow in the short term is the transport and mobility sector. Natural gas is increasingly coming to be considered as an alternative fuel for oil-based petrol and diesel. Rising oil prices in combination with emission reduction are strong drivers here. Concentrated natural gas, as well as LPG,[4] are both cheaper and cleaner than petrol or diesel. In several European and Latin American countries gas-based car and truck fuels have already been available for a long time, with consumption on the increase. The natural gas infrastructure for transportation can also be fed by green gases or biogases, which is an additional reason for the growing popularity of gas as a fuel. In a country like Sweden, for instance, which lacks indigenous gas reserves and where the power system is dominated by nuclear and hydro, biogas is rising as a fuel in both public and private transportation. In a small country like the Netherlands, gifted with indigenous gas, LPG is well developed as fuel but natural gas is not. Only recently is the prospect of natural gas as a fuel for private and public transportation starting to change in the context of climate change policies (see also below).

Increase in demand needs to be met by more production. In contrast to gas demand, gas production is rather concentrated in certain regions (see Figure 11.2), with the largest proven gas reserves in the Euro-Asian region, especially in the former Soviet Union, and in the Middle East (BP, 2008). However, everywhere in the world the hunt for natural gas has intensified and this has resulted in newly found reserves and a world overall reserve-production ratio in 2007 of 60.3 years, meaning that at current production and consumption rates the world has enough gas for about 60 years.

The production of natural gas has increased and is still increasing. According to the World Energy Council (WEC, 2007) natural gas production strongly benefited from

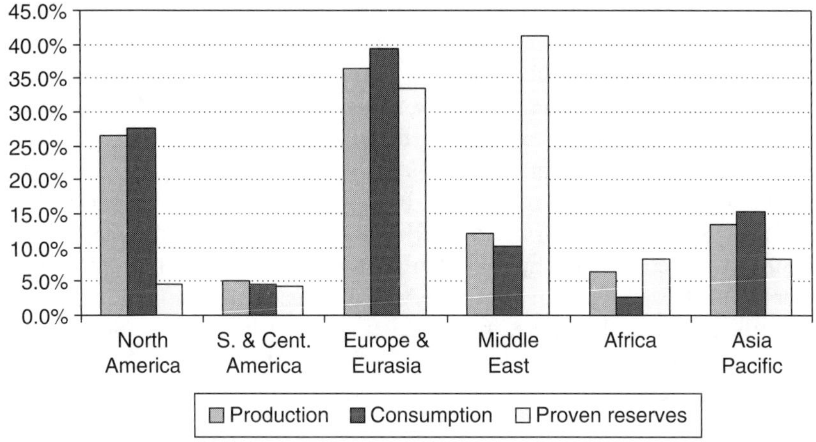

Source: BP (2008).

Figure 11.2 Regional share in global gas production, consumption and proven reserves 2007 in BCM

technological innovations in oil production, but natural gas liquefaction, long-distance pipelines and deep offshore pipe-laying in particular have improved the gas supply. More specifically, three-dimensional technologies in geological research significantly improved the search for and validation of the world's gas reserves. This has improved the reliability of the information on the proven gas reserves everywhere in the world. In 2006 these reserves were estimated at 177 trillion cubic metres (tcm). The specific search for natural gas[5] also led to unexpected discoveries of gas fields in, for instance, Bolivia and Egypt. For a long time the search for natural gas was only considered as a 'by-product' of the oil industry as, for instance, Venezuela shows; a country that uses the wet gas predominantly to increase the efficiency of the country's oil production. Intensified exploration in combination with improved technology are expected to trace new gas fields in the deeper regions of the earth (which have hardly been explored) and in the Arctic basins, which are assumed to contain huge volumes of hydrocarbons, especially natural gas. The USA, Canada, Russia and others have already expressed claims on Arctic natural resources and paradoxically, climate change is accommodating access to the resources in the region.

Thanks to innovative technology it has already become possible to produce gas in areas that are difficult to access. For instance, Norway and Mexico have started producing natural gas in the deep sea. Until recently offshore production was only possible in waters up to 300 meters depth, but Norway is now producing gas in waters up to 1100 meters deep (Orman Lange gas field), thanks to almost completely automatic drilling and production technologies.[6] Everywhere in the world increasing demand has intensified upstream activities (exploration and production) (Linde et al., 2006; Stern, 2006).

The LNG boom the world is currently preparing for has made competition basically a game between buyers of natural gas.[7] The three LNG markets in the world – Europe, North America and the Asian-Pacific region (CIEP, 2008) – are competing with each other for LNG. For the USA, LNG is the only real alternative, given the prospect of declining gas reserves in Canada and Mexico. Analysts forecast a further growth and diversification of the LNG market due to substantial cost reductions in all phases of the LNG chain, increasing import dependence in Europe and the USA, slow pace of liberalized access to pipelines in Europe and the need to diversify gas supplies for overall security of the gas supply, (Stern, 2006, p. 19).

The Pacific region is in a similar position. In this region the LNG supply is already well developed in the absence of a connecting pipeline system. Europe, the USA and Asia are preparing an increase in LNG supply and all three have to compete for the same group of suppliers. The USA is preparing for a new oversea LNG supply from the Middle East and Latin America. The USA is constructing new LNG trains and is an attractive market due to the relatively high (spot) market prices. In Europe and the Pacific the long-term contracts, indexed to the oil price, are considered disadvantageous to LNG by the experts. Liquefied natural gas is like oil, flexible enough to respond to price differences.

Europe is facing an increasing import dependency (up to 85 per cent in 2030) in natural gas and is investing in new pipelines to the Russian and North African gas fields (Stern, 2006; Kjärstad and Johnson, 2007; CIEP, 2008; Finon and Locatelli, 2008; Winrow, 2009). Projects currently under construction are North stream and South stream, connecting the North and the South of Europe to Russian gas fields, the North European gas pipeline crossing the Baltic, and the (planned) Nabucco pipeline bringing gas from the Middle East and the Caspian region to the European market. Access to and trading in gas is also facilitated by gas hubs and gas storage. Gas hubs are connection points between different gas pipelines and LNG terminals, which facilitate the physical exchange and economic trading of natural gas. The best known gas hub in Europe is the Zeebrugge hub in Belgium, with connections to the UK, Norway, the Netherlands, Germany, Russian Federation, Luxembourg, France and every LNG-producing region in the world. Another gas hub on the European continent is the Central European Gas Hub in Austria, which is still under development. Apart from gas hubs (virtual) gas trading is undertaken on European exchanges. For the western European market, trade concentrates in Amsterdam, while Nordpool facilitates trade in the Nordic gas markets. Storage capacity is also expected to grow in Europe (Höffler and Kübler, 2007).

The investment requirements are tremendous and are predicted to rise to US$3.9 trillion by 2030 (IEA, 2007, p. 78). Liquefied natural gas trains are under construction or planned in many production locations in the world. Thus far, Russian gas was exported to Europe, but Russia is considering the construction of more LNG trains and is also planning pipeline connections between the East Siberian gas field and China (IAE, 2006, p. 104).[8] China and Russia are considering pipeline connections between Tomsk and Urumuqi in China with a direct connection to the existing east–west pipeline of China. Another pipeline between Chita and Harbin in the North of China is under consideration. It is assumed that LNG will respond to the rising gas demand in China's coastal areas, for which reason China has planned nine re-gasification sites along the coast. The country's hunger for natural gas is motivated by economic growth and to improve air quality, especially in China's crowded coastal areas.

The import dependency of two major gas markets in the world is on the increase and will intensify competition. Europe is still well served with natural gas, but the old gas reserves need to be combined with new imports. Algeria and Russia are the two major

European gas suppliers. There are new pipeline connections planned, but Russia in particular is focusing on LNG too, enabling the country to serve markets other than Europe. China could develop into an alternative market for Russia's EU supply. Demand is high and China also wants to import more LNG. Russian gas delivery to the USA is not an option, but the EU might suffer from intensified competition from the USA. The US market is considered an attractive short-term LNG market because of the relatively high prices.

Increasing competition in the demand for natural gas has intensified the energy diplomacy of the major geopolitical players. China has concluded bilateral gas contracts with Iran, Venezuela and Sudan, and Europe is trying to respond to the Russian Ukraine crisis to secure the Russian gas supplies. 'The conflict has greatly impacted EU–Russian relations' (CIEP, 2008, p. 19), even though the EU–Russian energy relationship is already rather complex. Since the fall of the Berlin Wall in 1989, Russia and the EU have been focusing on securing the gas corridors between the two regions. Within the corridor the alliances have been diversified and this makes it problematic and challenging for both the EU and Russia to develop sustained relations. The Russian Ukraine crises are manifestations of this search process. To Russia's annoyance, former Council for Mutual Economic Assistance (COMECON) states in the Baltic and the Caucasus are developing a more pro-Western orientation, which is being used by the EU to develop gas corridors outside Russia (CIEP, 2008). On the other hand, Russia is diversifying its entrance points to the European market (North and South stream pipelines), in this way decreasing the dependence on the Ukraine which transits 60 per cent of the Russian gas flow to Europe and strengthening its monopoly position as European gas supplier (Winrow, 2009). At the same time, Russia is strengthening its position in the European gas chain by taking over Western companies. Back home, Russia has amended investment policies by no longer allowing foreign companies majority shares in upstream exploration and exploitation activities. Russia is clearly strengthening its position throughout the entire gas value chain.

The developments show the dynamics of the market, with both consumers and producers recalibrating their position, faced with the prospect of globalization of the gas trade. Not only trade is becoming more complex, so is energy geopolitics. Scholars suggest at least two competing models: the markets and institutions model, and the regions and empire model to cope with these complexities (Correljé and van der Linde, 2006). Both models draw different perspectives and predict different prospects for trade relations. In the markets and institutions model, the gas trade continues as an institutionalized but global effort. In the regions and empire perspective the world is divided in different ideological and religious blocks, contesting access and control over energy resources (ibid., p. 535). Both perspectives require specific approaches in energy diplomacy to secure a country's energy interest. For instance, the EU continues to work in the markets and institutions model in its energy relations with Russia, whereas Russia actually perceives its geopolitical position from the regions and empire perspective (Finon and Locatelli, 2008).

The attempt to strengthen the position of gas-producing countries in the global market by the Gas Exporting Countries Forum[9] has not worked out to date. The forum brought together the three countries holding the largest world gas reserves – Russia, Qatar and Iran – but their interests differ too much to allow them genuinely to manage the world gas market (Hallouche, 2006; CIEP, 2008). The Forum's biggest threat 'could come from an agreement between the pre-eminent LNG exporters to the Atlantic Basin (Qatar, Algeria, Nigeria and Egypt) which, by acting together in a tight LNG market, could exert significant market power over importers' (Stern, 2006, p. 17). At the same time, gas-consuming countries are diversifying gas supplies, *inter alia* by investing in LNG terminals. The short-term supply position of the EU is still relatively comfortable. The Chinese market as an alternative for the Russian gas supply to Europe still lacks infrastructure, while the number of current, planned and considered pipeline connections between Russia and the EU is substantial (CIEP, 2008). For the US market, the LNG supply will become more important due to the ever-increasing length of the pipelines between the Canadian gas fields and the US market.

The increasing import dependency of the two world's leading gas markets, the US and Europe, and economic growth ambitions in the emerging world's economies clearly puts pressure on the demand for natural gas. It will intensify competition between buyers, who will meet each other in the very few significant production regions in the world. Liquefied natural gas technology will facilitate the globalization of the world's gas trade and will intensify competition. In the short term, countries and continents are preparing for the medium term when all will become genuinely dependent on imports. In such a world of global competition and the prospect of increasing shortage, alternative gases might become a tempting perspective. Driven by shortage and climate change the search for and experimentation with alternatives has already started in some parts of the world.

CLIMATE CHANGE AND GREEN GASES

In November 2007 the Intergovernmental Panel on Climate Change (IPCC) published its fourth assessment of the extent, causes, and consequences of climate change. The report concluded that climate change is now 'unequivocal', warming up the earth by 0.56 to 0.92 °C in the last century, an effect 'very likely' due to anthropogenic causes. The emission of carbon dioxide (CO_2) resulting from the conversion of fossil fuels is considered a major cause of climate change. Reduction of fossil fuel consumption for heat and power production and in transportation, therefore, is one of the core challenges of climate protection. At the same time it is a most controversial theme because of the world's dependence on gas, oil and coal. As previously indicated, natural gas has the lowest CO_2 emission of all fossils and is also for that reason a most desired energy resource in all parts of the world (IPCC, 2007). Natural gas-fired power generation has grown since the 1980s in the western world. Gas-fired plants are superior to other fossil fuel-fired plants in terms of investment costs, fuel efficiency, flexibility of operation and environmental impact. Likewise, gas turbine technology and conversion efficiency have improved significantly over the last few decades (Hofman and Marquart, 2001). Natural gas is forecast to be the fastest growing fossil fuel in the coming decades, attaining levels of up to 161 exajoules in 2025. The industrial share of natural gas consumption is expected to increase to 23 per cent, with 31 per cent in power production in 2030 (IPCC, 2007, p. 266).

The climate change challenge is framing a new perspective on energy resources in general and natural gas in particular. Until now fossil fuels have satisfied the world's energy needs, but the perspective is moving. Countries and regions are recognizing a need to change the energy mix in favour of low-carbon energy resources like renewables, but this recognition is still fragmented and is far from generally shared throughout the world. The EU, for instance, agreed a reduction of greenhouse gas (GHG) emissions of 20 per cent and an increase of renewables in the EU energy mix of 20 per cent by 2020, but other continents like the USA and Canada are more hesitant to really express any commitments. Europe has put itself in a leading position in the world's climate change dialogue vis-à-vis these hesitant industrialized continents.

In the efforts to develop alternative energy resources, fossil fuels will continue to dominate in the coming decades. New technologies are considered to facilitate the further continuation of fossil fuels. One of the serious options is the conversion of fossil fuels in combination with CO_2 capture and storage (CCS). This so-called clean fossil route is currently on scientific and political agendas in many parts of the industrialized world. Maturing CCS technology implies a continued use of coal, natural gas and oil with minimal or no CO_2 emissions. By means of thermochemical reactions, carbon and CO_2 are separated, with the carbon being used and the CO_2 stored in underground caverns. Carbon dioxide capture and storage is generally recognized as transition technology and will draw on current practices in the oil, gas and chemical industry. It is operational at full scale at several locations in the world. The technology is proven, but has two disadvantages: it increases the costs of energy and it reduces the conversion ratio of the fossil fuels, due to the energy needed for capture and storage. The captured CO_2 can be stored in exhausted oil and gas fields everywhere in the world, where it should remain for many years. In particular in the USA and in some parts of Europe, CCS is being applied at full scale and the expectation is that the number of locations will increase in the coming years.

Clean fossil fuel has also renewed the hydrogen alternative in the energy debate. After the publication of Rafkin's book on the hydrogen economy, this energy carrier started to lose prominence. But proponents of hydrogen stress its benefits, such as its high energy density, its energy storage capacity and its flexibility of application, stationary and mobile. However, hydrogen is an energy carrier and not an energy resource. Producing hydrogen requires both energy and energy resources, since hydrogen does not exist in the natural state, but only as water and fossil fuel. Natural gas is still by far the most commonly used fossil resource for hydrogen production. The carbon is separated from the hydrogen by steam-based technology. In combination with CCS technology, the production of hydrogen can become climate neutral, since hydrogen has no hazardous emissions when utilized and consumed. Producing the steam requires energy and therefore the conversion of fossils into hydrogen is still not very efficient from an energetic point of view. The electrolysis of water by means of renewables or nuclear energy has been suggested as alternative to production with fossil fuels. However, electrolysis technology is not yet competitive with the steam reforming of natural gas. The general expectation is that hydrogen will only be able to penetrate as an energy carrier on a large scale by reducing the costs of renewable energy production. To date, hydrogen is predominantly being experimented in transport and mobility.

Biogas is another alternative to be initiated by the climate change problem. The production and usage of biogas is developing only in some regions of the world, in particular Europe. Of all biomass and waste-based energy production in the EU in 2002, about 5 per cent originated from biogas. In 2006 biogas was produced at some 4000 sites in the

EU, landfill sites, and urban and industrial sewage treatment plants (Fagernäs et al., 2006). In 2004 the total European gas production was estimated at 179 petajoules, representing a 9 per cent increase over 2003. On the European continent the UK and Germany are the leaders in biogas production and consumption. Most biogas is used for electricity production in both countries. The potential of biogas production in Europe is estimated to be high, particularly in countries like Sweden where biogas is used for transport and heating (Lantz et al., 2007). Biogas is compensating for Sweden's lack of natural gas and is seen as a serious option to diversify the country's energy mix. More countries are considering biogas for the same reason, in combination with securing the supply position. As indicated above, Europe's import dependence will increase, forcing countries to look for alternatives. Biogas is considered to be one of them. Even in the Netherlands, which still has substantial indigenous production, biogas is booming, a dynamic driven by climate change rather than security of supply. Climate change clearly added arguments to intensify the search for alternatives for natural gas, not only in the Netherlands, but in other European countries too.

PROSPECTS

This chapter has discussed three trends in natural gas and shown how they appear in different mixtures and combinations in different regions of the world. In one way or another, the world's gas markets have reformed and have exposed us to the prospect of declining gas reserves owned by a few countries in the world. The owner countries are well aware of their position and have intensified state control to benefit their resources, faced with high demand, to the maximum extent possible. The world's largest consumer markets, the USA and Europe, have reformed in favour of the market and have put the state in a referee position as market regulator. Private companies are expected to solve the security of supply issue. The supply issue, in combination with the climate change challenge, will increase demand competition but it is still uncertain whether private companies will be able to deal with the issues on their own. Huge investments are needed in both the production and transmission of natural gas. Investors will not invest in a risky and uncertain environment. Guaranteeing a good investment climate in a liberalized gas market, therefore, also requires smart government strategies and instruments. Such strategies might also affect the reform process itself, for instance by refining the market-based regulatory regime. This is already becoming visible in Europe in the derogations on third-party access to gas storage and LNG trains (Haase, 2008; Spanjer, 2008). These refinements show that the regulatory reform is a continuing process and how difficult it really is to accomplish the institutional reform of gas markets. Especially in the absence of competing gas infrastructures (pipelines) it turns out to be extremely difficult to introduce real gas-to-gas competition. At the same time more investments in pipeline infrastructures require securities for the investor and this will require more state intervention. In a liberalized market, therefore, the public–private configuration as well as the regulatory model continue to be subject to change.

Geopolitics and climate change are two more drivers of the refinement of regulatory models and public–private collaboration in gas. Drawing on leading experts, the chapter has shown how the international power relations in the world are changing, with possibly a severe impact on the international gas trade. The world of regions and empires is less predictable than the world of markets and institutions. Developing and continuing gas trade in this changing international environment requires smart diplomacy. For that reason, too, state and private companies are developing new relations and searching for new collaborations in order to confront the geopolitical challenge. From a physical point of view, the world's gas reserves might appear quite comfortable; current knowledge predicts gas supplies for the next 60 years. From a political perspective the prospects are less clear, and rather mixed for the different regions in the world. The gas-producing countries, especially Russia, Qatar and Iran, are in a rather comfortable position. Liquefied natural gas technology allows them to consider alternatives for current supply contracts. The state dominates gas production in these countries, with private companies participating as minority shareholders. The large consumption markets, the USA and Europe, on the other hand, are facing increasing import dependence in the medium and longer term. They will face increased competition from emerging consumption regions in the world, especially in Asia. Experts, therefore, expect a renewal of the public–private collaboration when contracting for new gas. Smart energy diplomacy is required, which will also refine the current liberalized regulatory gas market models.

A third driver of the regulatory refinement is climate change. Climate change brings new and thus far unknown ambitions to the gas market. The short climate change message is that gas consumption rates need to decrease and gas conversion rates need to increase. Despite natural gas's climate friendliness relative to other fossils, the search for and the experimentation with alternative gases has clearly begun, with Europe as prime mover. Europe has agreed rather ambitions climate goals and this, again, will refine the current public private relations in natural gas. The state sets the ambition and provides the facilities and support necessary to allow private companies to develop and commercialize alternatives, such as hydrogen and synthetic gas in combination with CO_2 capture and storage, and biogases from biomass.

In combination the three trends analysed here mean that the regulatory reforms in the world's gas markets are a continuing process. But the refinement and direction of the reform will be different for production and consumption. Figure 11.3 visualizes and summarizes the argument.

Figure 11.3 is a two-dimensional model describing the type of gas market, on the one hand, and the actor dominance in the regulatory model, on the other. The market dimension lists two extremes: the production and the consumption market. The two extremes of the actor dimension are the state on the one hand and the private company on the other. Both dimensions result in four different regulatory configurations: two for each type of market. Models II and IV represent the ideal typical model of the liberalized gas consumption market and the politicized gas production market. Models I and III are hybrids, representing mixtures of type of market, type of regulation and type of actor configuration. Our expectation is that these hybrid models will become dominant in both production and consumption regions in the leading gas markets in the world. Geopolitical dynamics, changing power relations, security of supply and climate change will urge for much more flexibility and adaptability in regulatory models. The future challenges no longer allow fixed regulatory models. All parameters of the world's gas market will continue to be on the move in the coming decades and will require regulatory flexibility both in the direction of the state and in the direction of the market. In

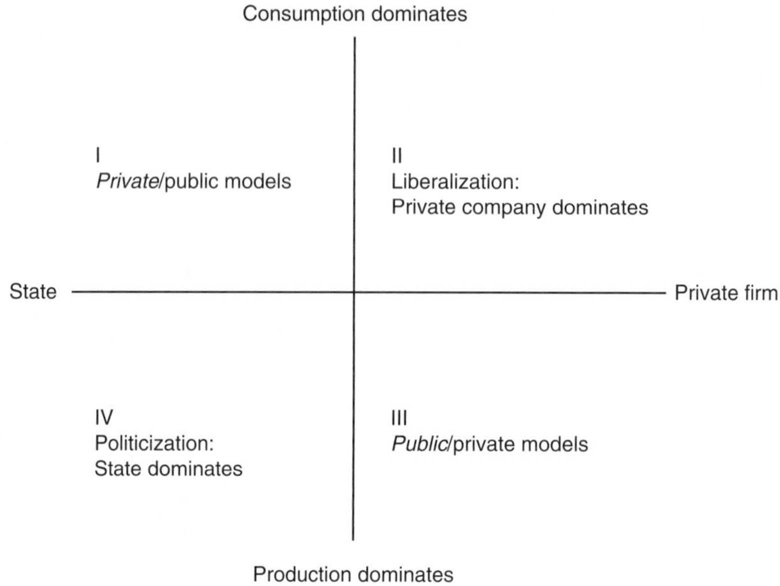

Figure 11.3 Regulatory gas market models

short, the new dynamic of the global gas market requires smart concepts, strategies and collaboration to cope with the huge challenges incorporated in the three trends discussed in this chapter.

NOTES

1. In North America the Alaska pipeline between the North and the South of Canada is among the world's longest pipelines (6000 km), the Rocky Mountain pipeline connecting the Mid West and the Eastern states of the USA.
2. The latest EU directive on natural gas requires ownership unbundling of system operators (see EU Commission, 2007).
3. The leakage in the Russian gas transmission and distribution system is held accountable for 15 per cent of the country's greenhouse gas emission (IEA, 2006)
4. LPG stands for liquefied petroleum gas which is a derivative of oil and gas production.
5. For many years gas discovery was basically a by-product of exploration of oil fields. Only lately has exploration focused specifically on natural gas fields.
6. Innovative technology also improved the production of existing gas fields. One of the approaches is called Smart Field and improves production by using smart sensor and IC technologies. These technologies enable more detailed and specific monitoring of the dynamics in the gas field during production and this is very effective in extracting more gas out of gas fields (Van Velsen, 2008, p. 31).
7. 'Natural gas, especially liquefied natural gas (LNG), consumption and trade are booming. The US is becoming an important importer, the EU is liberalizing and expanding its regasification capacity and China and India are emerging as new LNG buyers. The market is slowly moving from the traditional long-term rigid contract model to a relatively more liquid state.' This quote indicating the boosting of the globalized gas market was taken from Hallouche (2006). See also CIEP, 2008, p. 28.
8. According to Stern the Russian ambition 'to divert European supplies to North America or Asia and – in the most optimistic of all possible scenarios – will not have such capability for a decade' (Stern, 2006 p. 5).
9. 'The Gas Exporting Countries Forum was set up in Tehran in 2001 to generate 'tangible cooperation among gas producing and exporting countries'. According to a document proposed by the experts' meeting

(3–4 March 2001), and endorsed as a Mandate for the Forum by the Ministerial Conference in Tehran (19–20 May 2001), the aims of the forum are (a) to foster the concept of mutuality of interests by favoring dialogue between producers, between producers and consumers and between governments and energy-related industries, (b) to provide a platform to promote study and exchange of views and (c) to promote a stable and transparent energy market' (Hallouche, 2006 p. 12).

REFERENCES

Arentsen, M.J. (2004), 'Politics and regulation of gas in Europe', in D. Finon and A. Midttun (eds), *Reshaping European Gas and Electricity Industries*, London: Elsevier, pp. 103–33.
British Petroleum (BP) (2008), *Statistical Review of World Energy 2008*, London: British Petroleum.
Clingendael International Energy Programme (CIEP) (2008), *The Gas Supply Outlook for Europe. The Role of Pipeline Gas and LNG*, The Hague: Clingendael.
Correljé, A. and C. van der Linde (2006), 'Energy supply security and geopolitics: a European perspective', *Energy Policy*, **34**(5), 532–43.
Egging, R. and S. Gabriel (2005), 'Examining market power in the European natural gas market', *Energy Policy*, **34**, 2762–78.
Eikeland, P. (2007), 'Downstream natural gas in Europe – high hopes dashed for upstream oil and gas companies', *Energy Policy*, **35**, 227–37.
European Union (EU) Commission (2007), Proposal for a directive of the European Parliament and of the Council amending Directive 2003/55/EC concerning common rules of the internal market in natural gas, European Commission, Com (2007) 529 final.
Fagernäs, L., A. Johansson, and C. Wilen (2006), *Bioenergy in Europe. Opportunities and Barriers*, Helsinki: Technical Research Center Finland.
Finon, D. and C. Locatelli (2008), 'Russian and European gas interdependence: could contractual trade channel geopolitcs?', *Energy Policy*, **36**, 423–42.
Haase, N. (2008), *European Gas Market Liberalisation: Are Regulatory Regimes Moving towards Convergence?*, Oxford: Oxford Institute for Energy Studies.
Hallouche, H. (2006), *The Gas Exporting Countries Forum*, Oxford: Oxford Institute for Energy Studies.
Hanscher, L. (2003), 'Harmonisation of European gas markets: the EU Gas Directive', in M. Arentsen and R. Künneke (eds), *National Reforms in European Gas*, London: Elsevier, pp. 31–45.
Hawdon, D. (2003), 'Efficiency, performance and regulation of the international gas industry – a bootstrap DEA approach', *Energy Policy*, **31**(11), 1167–78.
Helm, D. (2007), 'Introduction: the return of energy policy', in D. Helm (ed.), *The New Energy Paradigm*, Oxford: Oxford University Press, pp. 9–36.
Höffler, F. and M. Kübler (2007), 'Demand for storage of natural gas in northwestern Europe: trends 2005–30', *Energy Policy*, **35**, 5306–19.
Hofman, P. and E. Marquart (2001), *Electricity in Flux; Sociotechnical Change in the Dutch Electricity System 1970–2000*, Bilthoven, Dutch National Research Programme on Global Air Pollution and Climate Change.
Honoré, A. (2004), *Argentina: 2004 Gas Crisis*, Oxford: Oxford Institute for Energy Studies.
International Energy Agency (IEA) (2003), *South American Gas. Daring to Tap the Boundary*, Paris: OECD/IEA.
International Energy Agency (IEA) (2006), *Optimising Russian Natural Gas. Reform and Climate Policy*, Paris: OECD/IEA.
International Energy Agency (IEA) (2007), *World Energy Outlook 2006*, Paris: OECD/IEA.
International Energy Agency (IEA) (2008), *World Energy Outlook 2007*, Paris: OECD/IEA.
International Panel on Climate Change (IPCC) (2007), *Fourth Assessment Report*, Cambridge: IPCC.
Kjärstad, J. and F. Johnson (2007), 'Prospects of the European gas market', *Energy Policy*, **35**, 869–88.
Lantz, M., M. Svensson, L. Bjornsson and P. Borjesson (2007), 'The prospects for an expansion of biogas systems in Sweden – incentives, barriers and potentials', *Energy Policy*, **35**, 1830–43.
Leitzinger, J. and M. Collette (2002), 'A retrospective look at wholesale gas: industry reconstructuring', *Journal of Industrial Economics*, **21**(1), 79–101.
Linde, C. v. d., A. Correljé, J. de Jong and C. Tönjes (2006) *The Paradigm Change in International Natural Gas Markets and the Impact on Regulation*, The Hague: Clingendeal/IGU.
Murry, D. and Z. Zhu (2008), 'Asymmetric price responses, market integration and market power: a study of the U.S. natural gas market', *Energy Economics*, **30**, 748–65.
Spanjer, A.R. (2008), 'Structural and regulatory reform of the European natural gas market. Does the current approach secure the public service obligations?', PhD thesis, University of Leiden.

Stern, J. (2006), *The New Security Environment for European Gas: Worsening Geopolitics and Increasing Global Competition for LNG*, Oxford Oxford Institute for Energy Studies.
Thomas, S. (2003), 'Gas as commodity: the UK gas market: from nationalism to the embrace of the free market', in M. Arentsen and R. Künneke (eds), *National Reforms in European Gas*, London: Elsevier, pp. 181–213.
Velsen, T. van (2008), 'Meer in de pijplijn', *De Ingenieur*, **120**(2), 27–33.
Winrow, G. (2009), *Problems and Prospects for the 'Fourth Corridor': The Positions and Role of Turkey in Gas Transit in Europe*, Oxford: Oxford Institute for Energy Studies.
World Energy Council (WEC) (2007), *Deciding the Future: Energy Policy Scenarios to 2050*, London: World Energy Council.

12 The oil industry: a dynamic patchwork of approaches?

Aad Correljé and Lucia van Geuns

INTRODUCTION

In the context of this book, the *liberalization of infrastructures* means: 'to allow for more competition and involvement of the private sector, thus redefining the role of government'. Applying this notion to the oil, or petroleum, industry, yields a rather opaque perspective. There is no clear and consistent pattern observable throughout the industry as a whole. It is huge and complex, with a large variety in country-specific public–private relationships. Unlike other 'infrastructures', it is not organized on a national basis. Oil is found in some countries. Sometimes it is also refined into products there, but more often the crude oil is moved abroad for processing. Refined products may be marketed locally, but also exported to markets all over the world. These activities are organized according to the strategies of the companies involved and the policies and traditions of the countries where they take place. The operators are the numerous larger and smaller public and private oil companies, or public–private joint ventures. Together, these firms operate the loosely connected international network of flows, processes, storage and other activities that constitute the structure, or the value chain, of the industry. Transactions may be organized in a variety of ways, covering anything between the spot market and vertical integration. Depending on the countries involved, the public–private interaction is organized in different ways, while the pattern does not seem to show any unilateral movement into a more or less liberal 'direction'. Rather, historically, the dominant feature seems to be the cyclical nature of the supply–demand–price relationship, moderated by specific political, economic and technical factors.

In this chapter we try to capture the meaning of liberalization of infrastructures in the context of the oil industry. The next section presents the value chain, explaining the main technical, economic and strategic characteristics of the segments. The interdependence between these segments has specific consequences for the role of markets and public intervention in the sector. The third section provides an overview of the forms of state intervention in the industry. The fourth section examines how the industry has been organized in terms of public–private relationships, historically. It sketches the evolution of the industry, dealing with its organization in terms of the control over investments, price formation and production and marketing strategies.

By highlighting the consequences of these specific forms of interaction for (groups) of countries involved and for the functioning of the oil industry as a whole, we shed light on the dynamics behind the liberalization of infrastructures in the oil industry. Given the particular organization and functioning of the oil market in the periods distinguished, specific local outcomes emerged in terms of the distribution of the rents, the security of supply and demand, and so on. We show how the public and private actors involved sought to adjust the prevailing system to their advantage, by allowing more or less competition, by influencing the involvement of the private sector and by redefining the role of governments. Obviously, they were constrained in their actions by the cyclically evolving context of the international energy market, and by local and (geo)political and economic circumstances.

THE PETROLEUM VALUE CHAIN AND THE STRUGGLE FOR RENTS

The value chain is a useful device to represent the oil industry. Commonly, three main segments are distinguished: first, the upstream exploration and production (E&P) activities; secondly, the midstream transport and refining activities; and, thirdly, the downstream distribution, storage and retail activities. The value chain highlights the several interdependent flows between these segments. A physical flow links the exploration for oil reserves in the ground, their production, the transport and storage of crude oil, the refining process and the storage, distribution and sales of finished products in down-stream markets. Parallel to the physical flow, there is the return monetary flow, associated with the transactions that take place between the segments. In each segment more or less value is added and finally the products are sold to consumers, at a price reflecting the value attached to their specific end use (Ellis Jones, 1988; Frankel, 1976, p. 11–41). Because the activities often take place in different countries, the value chain is international by nature (Odell, 1997).

A classic problem in the oil industry is that it requires a certain degree of coordination between the evolution of demand of petroleum products and the use of capacity in the several segments, to avoid bottlenecks and shortages or unused excess capacities. Crucial to maintaining this balance are upstream and downstream investments, adjusting the segments to supply and demand. Investments depend on the expectations of the firms regarding the value added and rents created, and the market risk to which they are exposed. As is generally acknowledged, in the oil industry there are serious impediments to the functioning of an ideal market in which supply and demand interact in a balanced manner, swiftly reacting upon the information embedded in crude oil and product prices (see Bindeman, 1999a).

The origins of these reservations are various but generally they are labelled as market failures or market imperfections. As for the impact of *market failures*, it is argued that the oil market malfunctions because of the interaction of huge sunk investments involved, the lack of information, long lead times of investments, economies of scale, a weak price elasticity of demand, the geological, technical, economic and political risk, the small number of producers and the possibilities for opportunistic behaviour (Ellis Jones, 1988, pp. 115–19).

A crucial aspect of the petroleum market is that the demand for oil products is a *derived* demand. Oil products deliver energy, enabling end-users to secure specific services, for example, transportation, heating and lighting. As such, there is no objective demand for petroleum, but for the most appropriate form of end-use energy, which may be a petroleum product, given end-use characteristics and the market context. Over the shorter term, generally, there are no readily available alternatives, as users will have invested in their appliances and installations. When, however, they have to decide upon new investments there is a possibility to switch to other energy inputs. As a consequence, the short-term price of elasticity of demand for petroleum products is fairly low. The amounts of energy consumed are generally dependent on levels of income and economic activities (Ellis Jones, 1988, pp. 61–4).

Adjustment of the production of crude oil and fuels to shifts in demand does not happen easily either. The investments in exploration and production assets and in the transport and storage systems are sunk and the capital costs are fixed. Crude producers and refiners keep on going, as long as their revenues are sufficient to cover the relatively modest variable cost. So, despite oversupply and low prices in the market, firms continue producing while not recovering their full costs. Yet, the industry is also slow in committing investments in new capacity when demand surges (see Eden et al., 1992; Frankel, 1976).

Moreover, the production of the several products, by refining an amount of crude oil, is a typical multi-product process, which yields several fuels (fuel oil, kerosene, gasoil, gasoline, and so on) in more or less fixed proportions. Yet, demand for those fuels does not evolve in such fixed proportions. So, there is always a shortage or an excess production of these fuels, with an impact on the prices they command in the market. The industry has to cover its costs and earn a profit on the revenues of selling all these products at their going price (Frankel, 1976; Ellis Jones, 1988, Bacon et al., 1990).

The structure of the production processes in the value chain implies that there are a number of interfaces, where volumes of crude oil (either produced or as reserves in the ground), semi-finished and end-use products change hands between the firms active in the several segments of the industry. In principle, these interfaces could take the shape of 'markets'. Yet, as stated above, the market is not trusted.

This implies that forms of explicit coordination have always been sought by the industry to protect their investments and to generate 'appropriate' margins to survive the business cycles. Historically, a variety of contractual and ownership structures have been used to coordinate these exchanges, ranging from vertical and horizontal integration, to long-term contracts and spot markets. Firms and states have sought to integrate, forward and backward, into those segments of the value chain where high rents are generated, or withdrew when rents were too low (Frankel, 1976; Penrose, 1969). Moreover, horizontal cartels have been established between firms to ban competition and coordinate investments (Bindemann, 1999a). Prominent examples of private 'market coordination' are Rockefeller's Standard Oil in the USA at the end of the nineteenth century and the Red Line Agreement and, the Seven Sisters' cartel of the international oil companies (IOCs) between 1928 and 1959.

Also among governments, there is a strong distrust that reliance on the market will yield a maximum of welfare to their economies. So, many governments from producing and consuming countries have intervened in the oil industry either at home or abroad, via regulation and state ownership or through particular (tax) arrangements with national private companies. Indeed, there is a strong notion that the exploitation of petroleum resources and/or the provision of energy and oil products belong to the 'national interest' of states, as a driver and precondition for economic activities and social and political stability. Hence, as a means to protect the national interest and security, governments all over the world have intervened in the operation and organization of the upstream and downstream industry (Baker Institute, 2007; Bohi and Toman, 1996; Claes, 2001, Mommer, 2002; Stevens, 2005, 2008a, 2008b). The Organization of Petroleum Exporting Countries (OPEC), founded in 1960, is the most well-known example of collective state intervention from the producer's side in the market, while the International Energy Agency (IEA) was established as a response by the consumer countries in 1974.

At the same time, however, such coordinative mechanisms and cartels – either public or private – have been interpreted as *market imperfections*, as producers' and firms' attempts to abuse their market power to collect high monopoly rents, by curbing industry output or by fixing prices. This interpretation has brought about other forms of state intervention, in response to these 'monopolistic' practices. These interventions range from competition policy and the regulation of private firms' monopolistic behaviour, to industry nationalization and the establishment state-owned enterprises.

Obviously, the interpretation given by different governments at different times has been influenced and inspired by their ideological perspective and by their interests regarding the distribution of the rents. Indeed, a crucial element in the evolution of the oil value chain is the struggle over the rents between producer countries, consumer countries and national and international oil companies, governments and consumers. This rivalry is a consequence of the important distributional effects that emerge from the different forms of organization and coordination in the value chain and the resulting performance of the firms in the segments and the host countries (Stevens, 2008a, 2008b).

STATE INTERVENTION IN THE OIL MARKET

This section provides an overview of five basic types of state intervention in the supply and the demand side of the oil market. The first, most general form of intervention involves the establishment of norms and standards in respect of accounting, safety, environmental protection, land-use and spatial planning, health impacts, emergency stocks, fuel composition and quality, and so on. Part of these norms is enforced via the firms' permits and concessions for undertaking specific activities in the industry; others are generically applicable to all parties and substances involved in specific activities. Such rules apply, in varying ways, to all segments of the value chain. In general, it can be stated that there are large differences in stringency of these norms and standards between the categories of developed, developing and less developed countries, but also within these categories. This variation in stringency does not only apply to the norms as such, but also to the local enforcement of such norms (see O'Rourke and Connolly, 2003).

The second form of intervention involves the establishment of taxes and levies on specific products and activities, or their subsidization. Such instruments may serve a number of purposes, such as: (1) the redistribution of rents between the several types of consumers; (2) the stimulation or discouragement of specific behaviour and activities in the industry and among consumers; and (3) generating income to the state. Important examples are the differences in taxation and levies on the export and import of specific fuels, either as a way of replenishing the state budget, or as a means of protection of the local oil industry. More recent are carbon dioxide (CO_2) emission taxes (see Bacon, 2004; Dunn, 1993; Energy Charter, 2008; Gupta and Mahler, 1995).

The third form of intervention implies the outright regulation of activities of the industry (see Bindeman, 1999b; Mommer, 2002). Examples are:

- the granting of (partial) monopoly rights to firms undertaking specific activities, like exploration, crude production, refining, distribution and retail trading
- quantative import and export controls for crude oil and specific fuels
- the establishment of production, supply and other quota to specific firms
- the regulation of prices of inputs, such as crude oil, pipeline and other transport tariffs, and outputs, via pre-tax product prices, at the wholesale or retail level
- the regulation of profits, returns on investments and other financial elements of the firms
- investment controls
- local content rules.

The fourth approach involves public ownership in the oil industry, either directly controlled by a Department or Ministry, or at arm's length, via ownership of the shares of firms. In the former case, generally, the aim is to actively influence the industry and/or control the local market, reducing the power of other (foreign) firms. In the latter case, generally, the objective is revenue generation. Other arguments for state ownership are the acquisition of technology and access to up and downstream markets. Public firms may also establish joint ventures with national or foreign private firms (Baker Institute, 2007; Grayson, 1981).

The fifth approach involves competition policy, under which the state seeks to reduce the market power of firms, consortia and cartels. This may happen, either through the traditional remedies of competition policy, like a forced fragmentation of the dominant firms or competitive bidding for retail and other concessions, or via the establishment of a countervailing power; often a state-owned firm.

Table 12.1 summarizes the most common forms of state intervention in the different segments of the value chain. Norms and standards for safety, environmental protection and spatial planning have been neglected, as these apply to all activities. Norms may be used strategically, in protecting markets or forcing firms to engage in specific forms of behaviour and rent sharing with governments.

Typically, in upstream exploration and production, state intervention by host countries aims to balance, on the one hand, the need to attract (foreign) firms and capital to engage in these activities, while on the other securing that the rents generated are appropriated to an acceptable degree. Essentially, this balance has been established via concessionary schemes, taxation, royalty and profit-sharing regimes and the regulation of foreign investments and by establishing state-owned oil companies (NOCs) (Baker Institute, 2007; Bindemann, 1999b; Dam, 1976; Energy Charter, 2008; Johnston, 2008; Mommer, 2002; Parra, 2005).

Moreover, depletion policies have been used to control the supply of oil to the market (Baker Institute, 2007; Claes, 2001; Lovejoy and Homan, 1967). Increasingly, the decommissioning of installations is becoming an issue for regulation and state intervention – particularly offshore (Osmundsen and Tveterås, 2003). State intervention in exploration and production (E&P) is not limited to producer governments. Many consumer governments have sought to support national public and private firms in exploring and

Table 12.1 Common forms of state intervention in the oil industry

Activities in the value chain	Main forms of intervention
Upstream	
Exploration and drilling test wells	– Permits and concessions for access to exploration acreage – Exploration levies and fees – Subsidization and support for national firms, exploring abroad
Production of crude oil	– Taxation and royalties – Regulation of production capacity – Regulation of well depletion and crude production quota – Local content requirements – Profit sharing – Rules on foreign investments – State ownership and joint ventures – Subsidization and support for national firms, producing crude abroad – Decommissioning of installations
Midstream	
Transport and storage of crude oil	– Export quota and taxes – Mandatory use of national shipping
Crude refining and product manufacturing	– Subsidization – Investment controls and local content rules – Regulation of crude purchases (by origin) – Regulation of input prices – Wholesale price regulation – Output quota – State ownership and joint ventures
Transport, wholesale and storage of oil products	– Subsidization – Regulation of mandatory fuel emergency stocks – Regulation of cost and tariffs – Product specifications – Import and export controls – State ownership and joint ventures
Downstream	
Local storage, trading and distribution of products to retail outlets	– Subsidization – Regulation of mandatory emergency stocks – Regulation of investments, cost and tariffs – Import and export controls – State ownership and joint ventures
Retail trade in products	– Regulation of locations and market shares – Regulation of mandatory purchasing quota – Subsidization and taxation – Regulation of end-user prices and tariffs – Competitive bidding and licensing of new outlets – State ownership and joint ventures

producing crude abroad (Baker Institute 2007; Grayson, 1981; Pirog, 2007). As illustrated below, over time, this balance has been achieved in different manners; each with its particular pattern of coordination and outcomes for the different parties involved, depending on the evolution of the oil market and the geopolitical context.

In the midstream segment, state intervention – if any – has sought to achieve two different objectives. Either the aim was to curb the dominant market power of the large international oil companies, forcing them to supply their products at 'acceptable' prices, while appropriating a share in the rents generated via end-use levies, or to support the construction of 'national' refineries, to reduce the dependence on foreign fuel supplies and to stimulate a national industry for strategic and economic reasons (Baker Institute, 2007; Grayson, 1981; Molle and Wever, 1984).

In consuming countries, the import of crude oil instead of ready-made products may provide a balance-of-payments advantage. Moreover, national refineries and distribution systems are often seen as a strategic asset in times of international conflict. Sometimes, after being abandoned by the IOCs for lack of profitability, they were taken over by the state involved. In oil producing countries, the construction of export refineries had to enhance the export value, as oil products command a higher value in the market than crude (Al-Moneef, 1998). Hence, the construction and operation of national refineries and petrochemical complexes was often supported via state ownership and joint ventures with (local) private firms under preferential conditions, subsidization, the regulation of crude purchases and wholesale prices, guarantees for market share, and so on (see Bacon et al.,1990; Correljé, 1994).

Downstream, consumer countries' governments try to strike a balance between securing the provision of petroleum products, achieving acceptable end-use prices for the several types of consumers and generating their own revenues. Combinations of subsidization and taxation are used to redistribute rents between the several types of consumers. In high-income countries it is common to tax 'luxury' fuels with a low price elasticity of demand, like gasoline and automotive diesel for passengers transport, as a source of income to the state. Often economically essential fuels, like gasoil for freight transport, navigation, agricultural use and power generation, are not taxed at all. Kerosene is provided at subsidized tariffs to consumers, as a fuel for the poor. Fuel oil may be provided at a rebate to power generation and infant industries. (see Bacon, 2004; Gupta and Mahler, 1995; IMF, 2006). In many oil producing countries fuels are made available to consumers at extremely low prices. End-use price regulation is also applied to achieve macroeconomic objectives, like the reduction of the rate of inflation and the provision of stability to consumers by shielding them from sharply rising crude oil and product price movements (Baker Institute, 2007, pp. 14, 15; Contín et al., 2009). In addition to fiscal measures and end-use price regulation, quotas and concessionary and licensing arrangements are used to support the development of retail networks, either in less attractive peripheral regions, or to bring about competition in regions where the retail branch was dominated by the large international oil companies (Correljé, 1994).

It is evident that the particular combinations of these different forms of state intervention in the oil industry have varied at different times and by country. Nevertheless, as illustrated in the historical overview below, a sequence of periods can be distinguished in the organization of the oil industry, in which specific forms of state intervention – or the absence thereof – were characteristic. New circumstances regarding access to resources, supply and demand arose with specific consequences for (the relationships between) the segments of the oil value chain, the role and strategies of the firms in the industry and the interventions through which national governments have sought to secure their position in the value chain.

THE ORGANIZATION AND REGULATION OF THE OIL INDUSTRY

Crude oil is currently produced all over the world. Large-scale exploitation started in the mid-nineteenth century in the USA. Subsequently, oil was discovered in Russia, Romania, Indonesia and Mexico. In the early twentieth century, the oil reserves in the Middle East and Latin America came into the picture. In the second half of the century, new oil provinces were discovered and brought into production in the North Sea, Alaska and Africa. The geographical expansion of the petroleum resource base is an ongoing process, facilitated by the continuous development of exploration and production technology (Pinder, 2001).

The consumption of petroleum products also experienced a remarkable evolution. Starting with kerosene as a fuel for illumination and cooking; oil derivatives became a main source of energy almost everywhere in the world. Fuel oil became a main fuel for navigation, industrial heating purposes and power generation, until this role was partly taken over by natural gas. Gasoline and diesel became dominant fuels in transport. Gas oil is used as a multifarious fuel in many mobile and static services, such as heating, power generation, industrial and agricultural uses. Kerosene became the main fuel in air transport, with petroleum gases substituting for kerosene in domestic uses. Remaining fractions are processed in the petrochemical industry. Similar to the developments upstream, this also happened in a dispersed geographical pattern. The USA converted to oil early. In Europe this happened in the 1950s, later followed by the Soviet empire and Latin America, Asia and Africa. Patterns of use showed large variations, often depending on the stage and structure of economic development, modes of transport and power generation and the availability of alternatives, such as low-cost coal, natural gas, nuclear energy and hydropower (Darmstadter et al., 1977; Lucas, 1985; Schurr et al., 1960; Schipper et al., 1992).

Pre-1927: Unfettered Competition and Rivalry

Not long after the discovery of oil in Titusville, Pennsylvania, in 1859, the industry began to experience the typical boom and bust cycles. From 1870 onwards, John D. Rockfeller managed to bring a large part of the US refining industry under control of his Standard Oil, by his dominance of the railroads as the essential transport facility. Interestingly, during this period the price for kerosene fell by about 50 per cent, to the enthusiasm of the consumers. By 1890, however, the American Congress passed the Sherman Act, which forbade every contract, scheme, deal or conspiracy to restrain trade. Standard Oil attracted the attention of the antitrust authorities and in 1911 it had to be broken up in to 34 companies, including Exxon, Mobil and Chevron.

Standard Oil was not the only large US oil company, however. It had to compete with Gulf Oil (established in 1890) and Texaco (1901) inside the USA and in Asia and Europe. Royal Dutch Shell (1907) was producing oil in the Dutch East Indies and also selling in Asia and Europe. Firms, like Nobel and Rothschild produced considerable volumes of oil in the Baku region in Russia to supply Europe. In the first decade of the twentieth century, the Anglo-Persian Oil Company (later British Petroleum – 1909) and the Turkish Petroleum Company (1910) started E&P activities in the Middle East, in Persia

(Iran) and the Turkish Empire. The relations between these companies and their 'host countries', were relatively easy; most countries were still under colonial rule. Countries without colonial rule, like Argentina, Mexico and Romania, put more pressure upon the oil companies (see Philip, 1984).

In Europe, tension developed between countries with access to oil, like the UK and the Netherlands, and Germany, France, Spain and Italy without oil-rich colonies. The strategic role of oil was underlined by the conversion of the British Royal Navy to fuel oil and the establishment of the Anglo-Persian Oil Company to secure oil supply (Sampson, 1985). In the USA, motor fuel taxation began in 1919 and spread to all states within ten years, to finance roads and highways to accommodate growing car traffic. Also, European countries started to raise levies on petroleum fuels.

1928–59: The Seven Sisters and Private Coordination

By 1928, the seven largest internationally operating oil companies came to sign the 'As Is' agreement to coordinate the industry. Parties were Standard Oil of New Jersey (Exxon), Gulf, Texaco, Socal and Mobil from the USA, and British Petroleum and Shell from the UK and the Netherlands. The French firm CFP played a minor role. They agreed to respect each other's market shares. Their main market was in transport fuels, initially, but rising wages in the coal industry were gradually shifting the comparative advantage to fuel oil in industry and heating oil in domestic households. The IOCs set posted prices for crude oil based on the Mexican Gulf basing point system. While abstaining from price competition, they set end-use prices and coordinated the evolution of the industry, via investments in crude production, transport facilities and refineries. The Seven Sisters did coordinate the entire oil value chain in the non-communist world (Blair, 1978). At the time, the USA was the main producer of crude, with a share of about 60 per cent of the total production. Yet, the USA was essentially self-sufficient, as was the Soviet Union (10 per cent of total). A main exporting country was Venezuela (15 per cent). The discovery of substantial oil reserves in Iraq, Iran, Saudi Arabia and Kuwait turned these countries into important suppliers with a share in total supply of about 4 per cent each. Generally, the concession agreements involved a 10 per cent royalty share for the colonial powers in the host countries.

Some countries, notably Mexico as a producer, and France, Spain and Italy, tried to achieve independence from the IOCs by establishing state-owned oil companies (Correljé, 1994; Grayson, 1981). After the Second World War, the USA established its long-term oil relationship with Saudi Arabia by forming Aramco (Arabian American Oil Company), a conglomerate built upon private US companies. By the end of the war, the Seven Sisters' market coordination began to be questioned in the USA and Europe (see Adelman, 1972, p. 136; Blair, 1978, pp. 38–43; Hartshorn 1962; Jacoby, 1974, pp. 37–40; Odell, 1986, p. 118; Painter, 1984; Yergin, 1991, p. 424).

1959–73: The Independents Break Down the Oligopoly

After 1959 the dominant position of the Seven Sisters came gradually under pressure because of the arrival of a number of new firms. These were smaller US firms, such as Marathon, Continental, Amereda, Occidental, independent from the majors, plus some

European NOCs, such as ENI/AGIP (Italy), Deminex (Germany) and Hispanoil. Three developments infused competition in the fast growing post-war oil market. First, the US firms had managed to get concessions for the production of crude oil in new, recently decolonized, countries, such as Libya and Algeria, competing with the arrangements offered by the majors. Secondly, in 1959, the USA decided to establish import quotas for crude oil, to protect its indigenous oil industry and to avoiding dependency on imported crude. The US independents, thus, were blocked from delivering the crude to their home market and had to find alternative outlets. Thirdly, Italy, Germany and Spain had ambitions to become less dependent on the IOCs and managed to arrange production concessions too, via their own E&P firms.

As a consequence, the number of oil-producing countries increased. Excess oil production, as compared to the evolution of demand, caused a gradual erosion of the prices for crude and products. Substantial rebates were given on the prices posted by the majors, which eventually only functioned as a fiscal reference price, determining the royalties and income taxes payable to the host countries (Adelman, 1972). The expanding world economy, with a fast growth of automobility and low fuel prices stimulated the consumption of oil. Coal was displaced by fuel and heating oil. Yet, low prices kept the large oil companies from investing in new exploration and production ventures (Clark, 1990; Odell, 1986).

The decline in real crude prices and the falling value of the US dollar also brought about a reduction of the purchasing power of the host countries. This induced a growing discontent. By 1960, main oil producers, such as Venezuela, Iran, Iraq, Saudi Arabia and Kuwait established the Organization of Petroleum-Exporting Countries (OPEC). Later on Indonesia (1962), Libya (1962), Qatar (1961), the United Arab Emirates (1967), Algeria (1969) and Nigeria (1971) joined. In the early 1970s, the first serious conflict took place. Libya forced the US independents, Occidental, Marathon and Continental, to enhance their royalties (see Adelman, 1972, 2002; Hammer, 1988; Odell, 1986; Parra, 2005; Yergin, 1991).

1973–83: The Oil Crises Break Down Vertical Integration

By the end of 1973, the conflict between Israel and its Arab neighbours ignited the first oil crisis. Following a cutback in production and driven by political unrest, prices skyrocketed from a level of less than US$2 per barrel to US$12. After the crisis, OPEC appeared on the market as the new coordinating mechanism, posting the real crude oil prices.

In addition to this change in the pricing regime, the market structure underwent a radical change. Host countries began to nationalize the production activities of the IOCs, replacing them with their own NOCs. This effectively shifted the coordination of the market to the OPEC countries. The traditional vertical integration of the majors was broken. Crude production was in the hands of the NOCs, while refining and marketing remained with the majors and independents. Initially, consumer countries tried to conclude state-to-state contracts with producers, hoping for a favourable treament. Yet, gradually a spot market began to develop in which increasing volumes of crude were sold.

On the consumers' side, the crisis brought about the shock of being dependent upon an OPEC cartel, controlling the prices and the availability of oil. Generally, it was believed that the price shock was caused by the looming depletion of the reserves. Only few recognized that the actual cause was the lack of investments in new exploration and in production facilities (see Odell, 1986). Consumer countries responded in a number of ways. A first response was the establishment of the International Energy Agency, to develop a mechanism to redistibute oil in case of further supply disturbances. Secondly, countries tried to isolate their economies from rising fuel prices by introducing end-use price regulation. Thirdly, policies were set up to save energy, which inspired the development of energy efficiency norms and standards. Fourthly, the IOCs invested in substitute sources of energy, like nuclear energy, natural gas, coal and hydro power, and in sun, wind and biomass. Fifth, large investments were made in E&P ventures in new, relatively unexplored, non-OPEC areas, such as Alaska, Canada, Venezuela, Mexico and offshore in the North Sea, the South Chinese Sea and the Barents Sea. In part these investments were made by the majors, to replace their assets lost to the OPEC countries. Yet, some OECD countries set up state-owned companies to produce indigenous oil, such as Statoil (Norway), BNOC (UK), Veba (Western Germany). These investments created large amounts of reserves, to be produced at higher costs than in the OPEC area, but justified by the high oil prices at the time (see Venn, 2002, pp. 113–43; Clark, 1990, pp. 323–9).

In 1979/80, a second price shock hit the market, following the Iranian revolution against the reign of the Shah and the Iran–Iraq War. The oil price went up from US$12.70 to $40 at the height of the crisis. These events induced fear among the consumers, as demand had been growing again during 1978 and further growth was expected. Although oil supply did meet consumption levels, additional panic purchases to replenish stocks and speculation put pressure on the market. The IEA was not effective, as it was geared towards physical disruptions and not against price shocks. Moreover, Saudi Arabia refused to increase its output to replace the Iranian production (Adelman, 2002, p. 175). Producers in OPEC began to sell increasing volumes of crude on the emerging spot market, instead of via lower-priced, long-term contracts. As a consequence, consumers saw themselves forced to buy on the spot market, thus putting more pressure on the price. This motivated OPEC to put a premium upon its posted prices (see Venn, 2002, pp. 24–7). In 1981, Iraq tried to gain control over the border river Shatt-al-Arab, taking advantage of the weakened Iranian army. The resulting war lasted for seven years and seriously hampered the oil production in both countries. It was also feared that a blockade of the Strait of Hormuz might shut in the oil produced by Kuwait (Claes, 2001, pp. 101–7; Clark, 1990, pp. 323–9).

Overt rivalry had grown among consumers and producers. Consumers faced an increasingly complex oil market. The Organization of Petroleum-Exporting Countries was blamed for driving up the prices. The variety in producers' pricing strategies contributed to uncertainty. Consumers were competing among themselves, trying to secure highly politicized state-to-state deals. What was completely lacking was coordination between the governments of consumer and producer countries and their oil companies, paralysing both OPEC and the IEA (see Adelman, 1995; Claes, 2001; Hartshorn, 1993; Odell, 1986; Venn, 2002; Yergin, 1991).

1983–2001: Over-supply and Under-investment

By the end of the 1970s, gradually, the tide began to turn. Newly discovered oil from Alaska, Canada, Mexico, the North Sea, the UK, Russia and China was reaching the market. In parallel, demand continued falling, as a consequence of the substitution of other sources of energy and because of the economic recession and the process of deindustrialization in the OECD world. During the early 1980s, spot markets had replaced long-term contracts for crude oil and oil products. The Organization of Petroleum-Exporting Countries could no longer dictate posted prices. Prices were determined by supply and demand on the exchanges of New York, Rotterdam and Singapore. As a consequence, the oil market was becoming increasingly liquid and sensitive to perceptions, sentiments and real world events in its price formation (Adelman, 2002, p. 176).

The emergence of a surplus in crude oil production and refinery output drove the prices down. In response, OPEC began to reduce its supply of oil to the market. The decline in the demand for oil continued, however, and the share of OPEC in total oil supply fell significantly. From March 1983 onwards, OPEC introduced crude production quotas for its members states, to influence the oil price via the control over supply. Saudi Arabia got a key role as the OPEC swing supplier, maintaining the balance between demand, the growing volume of oil supplied by non OPEC producers and the rigid production of the other OPEC members. This proved an increasingly unrewarding task, because the kingdom had to reduce its production time and again, while the price kept falling. Eventually, in 1985, Saudi Arabia refused to continue this any longer, causing a free fall of the oil price. From then onwards, coordination of the oil market was a function of the degree of control of OPEC, the 'clumsy cartel', over its production (Adelman, 1982).

There is an important difference between the behaviour of OPEC and non-OPEC producers. In essence, firms in the latter countries keep on producing oil from existing wells, as long as the variable production costs, including taxes, are covered by their revenues. If the medium-term supply-demand outlook suggests an acceptable oil price, they invest in new E&P ventures. The Organization of Petroleum-Exporting Countries, by managing its output, attempted to maintain the oil price at a level of about 22 and 25 dollars per barrel. It had great difficulty in making the right judgements and was plagued by incidents and disturbances in the world economy and the oil market. Some members did not adhere to the quota and secretly sold more oil, driving down the price. The Organization of Petroleum-Exporting Countries had great difficulty in meeting the diverging interests of its members. Small and densely populated countries with moderate reserves, like Algeria, Libya and Iran, aspired high prices to harvest shorter-term gains. In contrast, producers with large reserves, like Saudi Arabia, Kuwait and the United Arab Emirates, sought to maintain the position of oil over the longer term via a stable, moderate price, avoiding substitution by alternative sources of energy (see Adelman, 1990; Claes, 2001; Noreng, 2002; Hartshorn, 1993; Parra, 2005; Van der Linde, 1991, 2000; Verleger, 1990; Yergin, 1991).

Moreover, by the mid-1980s, the perception that the world was running out of oil and energy had withered away. High prices and supportive government policies made available a wealth of oil and natural gas, while energy efficiency and economic restructuring had reduced the energy intensity of the OECD countries. Outside OPEC, fundamental changes were taking place. Responding to the nationalizations, the shift in the rents towards OPEC and the falling oil price, the IOCs were forced to reduce their costs. This brought about a wave of mergers, starting with the US companies. In 1984, Gulf was taken over by Chevron, which merged in 2001 with Texaco, absorbing Unocal in 2005. Another US major, Mobil Oil, merged in 1999 with Exxon. In 1998 BP took over AMOCO. In Europe, in 1998, Total (France) and PetroFina (Belgium) became TotalFina, which merged in 1999 with Elf (France) into TotalFinaElf, later Total. Over the 1980s and 1990s, state oil companies in OECD countries were privatized, partly in reaction to the changed ideas about energy and oil scarcity, and partly as a consequence of the general tendency of liberalization and the need to replenish the state budget. Examples of (partial) privatizations are BNOC, BP, Repsol from Spain, ENI from Italy, Statoil from Norway, and Neste from Finland. In some production countries, notably Russia after the fall of the communist regime, the industry was opened up to foreign involvement. Most countries relaxed their resources and depletion management upstream; downstream competition policy ruled and international schemes for crisis management were dismantled. Wholesale and retail price regulation was abandoned.

The market had taken over. End-use levies in consumer countries were increased, however. And the preservation of vulnerable natural areas, spatial planning, quality and environmental standards were becoming increasingly important in determing the oil industry's evolution.

The New Millennium: An Uncoordinated Oil Market in Turmoil

Driven by economic growth in the USA, Asia and the European Union (EU) oil demand began to rise again, over the 1990s, causing an upward pressure on prices. In November 1997, though, OPEC missed the decline in demand following the Asian crisis and expanded its output with about 10 per cent. Moreover, Iraq had started to produce again, and the winter in the Northern hemisphere was mild. The result was a dramatic decline in the oil price below US$10 by early 1998. OPEC engaged in negotiations to achieve a reduction in output. Yet, 'the market' did not expect that OPEC would succeed. The Organization of Petroleum-Exporting Countries members did cheat and the impact of the UN sanctions on the contribution of Iraq was uncertain. The amount of oil in storage remained large during 1998 and, eventually, OPEC could stop the falling price only after three rounds of output reductions between March 1998 and March 1999. Main non-OPEC producers, such as Norway, Egypt, Russia, Mexico and Colombia adhered by not increasing their output and, eventually, the price went up again.

By the end of 1999, OPEC was urged to expand its production again. The Organization of Petroleum-Exporting Countries thought it too early because of the large stocks in place. The depletion of stocks during the cold 1999 winter, in addition to the decline in Iraqi output and the fear of a millennium crisis put upward pressure on the oil price. By March 2000, both OPEC and non-OPEC producers agreed upon a higher level of production, bringing about a small price fall. Yet, in July and September additional output was necessary to keep the price at bay. In September 2000, US President Clinton decided to sell 30 millions of barrels from its strategic petroleum reserve (SPR) (Horsnell, 2000).

During the 2000/01 winter, output expansion had little effect as the demand forecast was high. Moreover, in most OPEC countries, with the exception of Saudi Arabia, production capacity was fully employed already. Also the stocks had dwindled, while no replenishment had taken place in expectation of lower prices.

Yet, the oil price started falling because of weak demand driven by an economic slump in Asia and the USA, by OPEC overproduction and by the deployment of the strategic reserves. In turn, OPEC reduced its output in January, March and July 2001.

But the situation in Iraq remained uncertain. The attack on New York's Twin Towers in September 2001 caused a further decline in economic growth. In November, OPEC announced output reductions, also because of the rapid growth of the Russian export of oil. Eventually, other non-OPEC producers also followed suit and reduced their output (Kohl, 2002).

The political perspective had changed. Since the Twin Towers and the intervention in Iraq in 2003, the relationship between the USA and Saudi Arabia and between OPEC and the oil-consuming countries had cooled. Over the post-1985 period, OPEC had aimed at an oil price that was acceptable to both the producers and the consumers. Yet, it did revert to its position in 1973/74 and the early 1980s, taking advantage of the market circumstances, such as the US hurricanes, labour conflicts and political instabilities, which caused volatility and high prices.

From early 2002 onwards, oil prices began to increase again. This caused confusion and many conflicting explanations, including an output reduction of OPEC and non-OPEC producers, in combination with tensions in the Middle East over the anticipated intervention in Iraq, political turbulence in Venezuela and Nigeria, the cold winter in the USA and declining trading stocks. Yet, after the intervention in Iraq, in March 2003, it appeared that the oilfields had not been affected and the price fell for a while. So, early in 2004, OPEC decided to lower its output to avoid the price collapsing. This, however, was not necessary, because over the following period the oil price doubled to a level of US$70 by 2006, and again to about US$150, by July 2008.

The third oil crisis was the consequence of a number of mutually supportive factors. A fundamental aspect was that the large surplus crude production capacity, used by OPEC to manage supply and demand, had evaporated over the 1990s. The unanticipated level of economic development in China and India, but also in the USA and the EU, created a steady growth in demand. This, however, was not met by sufficient investments in crude production and refining capacity, either by OPEC or by non-OPEC producers. Hence, price behaviour in the oil market became extremely volatile. Single events or rumours had immediate effects on prices, which shot up to unprecedented levels until the summer of 2008. Moreover, the instable political situations in many regions fostered a so-called 'fear premium' and speculation, reflecting the threat of supply disruptions. It was expected that demand growth would outstrip the industry's supply capacity quite soon (Jesse and Van der Linde, 2008). Capacity problems also emerged in the refining segment, operating at high levels of throughput. Moreover, the shift in demand towards light products, such as gasoline, gas oil and kerosene, in combination with more stringent sulphur standards, required substantial capacity adjustments. Yet, after more than 30 years of excess capacity and low refining margins, oil companies were reticent in making those investments (Pieterse and Correljé, 2008).

LOOKING AHEAD

Predictions of demand outstripping the supply capacity and pushing up the oil price to levels as yet unseen did not materialize. The financial/economic crisis did reduce the demand for energy and oil products and moderated the future projections, causing a rapid decline in the West Texas Intermediate (WTI) oil price from US$145 in July 2008 to about $33 by the end of the year. Over 2009, the price recovered to around $60. The reduction in demand may have saved the world from a third oil crisis; it has not solved the underlying coordination problems in the oil market. The combination of the 'struggle for the rents' and the inherent uncertainty about the evolution of demand and supply yields a daunting perspective as regards the future role of the oil industry and the role of oil in energy supply.

The struggle for rents, within the context of fairly unpredictable cycles in the supply–demand relationships, has important implications for the position and the policy objectives of the countries involved and the strategies of firms. The 'free' market of the 1990s and the perception of an oversupplied market led the IOCs to rationalize their operations and made them risk averse in terms of investments in E&P. They were constrained in engaging in new ventures, as they chose to go for short-term shareholder value, under pressure from the financial markets. And, of course, they were uncertain about the need for new capacity (see Skinner, 2006; Stevens, 2005, 2008b).

The IOCs have only limited access to areas with significant undeveloped oil reserves, as these are controlled by local NOCs. Their once indispensable role in providing technology and know-how has also been reduced, because these assets are made available by specialized subcontractors and general subcontractors, staffed with former IOC employees who were laid off, offering their services to any paying customer.

In those countries where the IOCs managed to get a foothold, such as Russia, Venezuela, Mexico, and some provinces in Africa and the Caspian Sea area, the investment climate is hardening, as governments try to get the most out of their contractual relationship. The IOCs also experience a growing competition from smaller private and state-owned companies from consumer and producer countries. On the one hand, privatized NOCs from OECD consumer countries, like RepsolYPF (Spain), ENI (Italy) and Total (France), operate as 'national champions' in supplying their home and other markets, while an increasingly important role is also played by the NOCs from China and India, sometimes in joint ventures with producing countries, with the explicit task of securing access to untapped oil reserves. For these firms, commercial considerations seem less important than their strategic role in securing energy for their growing economies. Yet, also the NOCs from producing countries, such as Statoil and Petronas, have expansion strategies abroad. The recent bidding procedure for the promising oilfields in Iraq is a case in point. The IOCs had to compete with Russian, Chinese and OECD firms, but they all were offered only low-yielding service contracts.

So, oil-producing countries have strengthened their grip on the exploitation of their resources. Many OPEC states do not allow foreign firms or capital in their oil industry. In 2005, national oil companies controlled about 77 per cent of the total proven oil reserves, to which the IOCs have no equity access. Russian private companies controlled another 6 per cent. Exploration and production is reserved to these firms. In a number of countries, such as Russia, Venezuela and Kazakhstan, governments have managed to shift the balance from the IOCs to their NOCs.

Yet, the NOCs and Russia are reticent to invest for several reasons. Often NOCs have only limited resources, as their governments are fully dependent on the oil revenues to maintain their sovereign power vis-à-vis their own population or neighbouring countries. They have to spend much of their income on subsidizing the standard of life of their population, on military expenditure and other politically important oulays. Moreover, the producers fear a return of a situation of excess supply capacity, in which a divided OPEC will not be able to stabilize the market (Stevens, 2008b).

In parallel, in many OECD countries oil exploration and production is increasingly affected by requirements and rules related with environmental protection and resource management, including CO_2 emission reduction policy and local licensing procedures. So far, the security of supply issue has not led to the (re)establishment of public oil companies in the OECD region. However, given the exclusion of the IOCs from promising, low-cost oil provinces, combined with the huge uncertainties as regards the evolution of the energy market and the future availability of oil, this may be only a matter of time. Paradoxically, the solution of bringing about a quick transition towards a sustainable energy system only aggravates the situation. If it succeeds, investments in petroleum E&P will prove worthless over time.

In the introduction to this chapter, we argued that the interaction of logistic and technical characteristics in the value chain, with the evolution of the supply and demand side of the market and (geo)political developments are crucial in understanding the role of governments in the several segments of the industry. In the sections above we have shown that, over the subsequent periods, changes in these aspects have had specific consequences for the interaction for (groups) of countries involved and for the role of the state in the oil market. This interaction cannot be described as a simple cycle, however. During each phase new technical, geographical, economical or ideological drivers emerged, bringing about a long-term pattern of development that looks like a helix. The cyclical pattern repeats itself in a different shape, time and again.

This also means that the notion of the liberalization of infrastructures in the oil industry has a different content and meaning in the different phases (see Table 12.1). The liberalized, uncoordinated, market malfunctions, inviting new forms and ways of intervention by new actors, who for some time managed to control it. But during the 'controlled' period, the seeds are sown for the ensuing disturbances, involving new geographical patterns, new technologies or new economical paradigms, giving way to a new 'liberal' period. What happens is that, while there is an inherent need for coordination, in each phase market developments break down the prevailing patterns of coordination. Standard Oil privately coordinated an unfettered boom-and-bust market. Yet, the Standard Oil Trust was curbed by US 'antitrust' law, giving rise to a competitive oil market. The coordination of the Seven Sisters cartel was severely criticized, but it was broken by the actual competition of the US newcomers and NOCs of consuming governments. This created turmoil, at the expense of the producing countries uniting in OPEC. The OPEC monopoly was broken by IOCs and OECD NOCs entering into new oil provinces with new technologies, giving way to the 'free market' of the 1990s. The consequences are here. New forms of coordination will evolve in response to the market failures and the uncertain circumstances of today. Yet, this new phase has only just begun.

REFERENCES

Adelman, M.A. (1972), *The World Petroleum Market*, Baltimore, MD: Johns Hopkins University Press.

Adelman, M.A. (1982), 'OPEC as a cartel', in J.M. Griffin and D.J. Teece (eds), *OPEC Behavior and World Oil Prices*, Boston, MA: Allen and Unwin.

Adelman, M.A. (1990), 'The 1990 oil shock is like the others', *The Energy Journal*, **11**(4), 1–13.
Adelman, M.A. (1995), *The Genie Out of the Bottle: World Oil since 1970*, Cambridge, MA and London: MIT Press.
Adelman, M.A. (2002), 'World oil production & prices 1947–2000', *The Quarterly Review of Economics and Finance*, **42**, 169–91.
Al-Moneef, Majed A. (1998), 'Vertical integration strategies of the international oil companies', *The Developing Economies*, **36** (June), 203–22.
Bacon, R. (2004), 'Taxation of energy', *Encyclopedia of Energy*, vol. 6, Dordrecht: Elsevier.
Bacon, R., M. Chadwick, J. Dargay, D. Long and R. Mabro (eds) (1990), *Demand, Prices and the Refining Industry: A Case Study of the European Oil Products Market*, Oxford: Oxford University Press.
Baker Institute (2007), *The Changing Role of National Oil Companies in International Energy Markets*, The James Baker II for Public Policy of Rice University, Number 35, April.
Bindemann, K. (1999a), 'Vertical integration in the oil industry: a review of the literature', *The Journal of Energy Literature*, **5**(1), 3–26.
Bindemann, K. (1999b), *Production-Sharing Agreements: An Economic Analysis*, WPM 25, October, Oxford: Oxford Institute for Energy Studies.
Blair, J.M. (1978), *The Control of Oil*, New York: Vintage Books.
Bohi, D.R. and M.A. Toman (1996), *The Economics of Energy Security*, Boston, MA, Dordrecht and London: Kluwer Academic.
Claes, D.H. (2001), *The Politics of Oil-Producer Cooperation*, Boulder, Co: Westview Press.
Clark, J.G. (1990), *The Political Economy of World Energy: A Twentieth Century Perspective*, Hemel Hempstead: Wheatsheaf Harvester.
Contín, I., A. Correljé and B. Palacios (2009), 'Competition, regulation, and pricing behaviour in the Spanish retail gasoline market', *Energy Policy*, **37**, 219–28.
Correljé, A.F. (1994), *The Spanish Oil Industry: Structural Change and Modernization*, Amsterdam: Thesis Publishers.
Dam, K.W. (1976), *Oil Resources: Who Gets What How?*, Chicago, IL and London: University of Chicago Press.
Darmstadter, J., P.D. Teitelbaum and J.G. Polach (1971), *Energy in the World Economy, A Statistical Review of Trends in Output, Trade and Consumption since 1925*, Baltimore, MD: Resources for the Future by the Johns Hopkins Press.
Darmstadter, J., J. Dunkerley and J. Alterman (1977), *How Industrial Societies Use Energy*, Baltimore, MD and London: Johns Hopkins University Press.
Dunn, James A. Jr (1993), 'The politics of motor fuel taxes and infrastructure funds in France and the United States', *Policy Studies Journal*, **21**(2), 27–84.
Eden, R., M. Posner, R. Bending, E. Crouch, J. Stanislaw (1992), *Energy Economics: Growth, Resources and Policies*, 2nd edn, Cambridge: Cambridge University Press.
Ellis Jones, P. (1988), *Oil: A Practical Guide to the Economics of World Petroleum*, Cambridge: Woodhead-Faulkner.
Energy Charter (2008), *Taxation along the Oil and Gas Supply Chain: International Pricing Mechanisms for Oil and Gas*, Brussels: Energy Charter.
Frankel, P. (1976), *The Essentials of Petroleum: A Key to Oil Economics*, 2nd edn, London: Frank Cass.
Grayson, L.E. (1981), *National Oil Companies*, Chichester: John Wiley and Sons.
Gupta, S. and W. Mahler (1995), 'Taxation of petroleum products: theory and empirical evidence', *Energy Economics*, **17**(2), 101–16.
Hammer, A. (1988), *Hammer, Witness to History*, London: Coronet.
Hartshorn, J.E. (1962), *Oil Companies and Governments: An Account of the International Oil Industry in Its Political Environment*, London: Faber and Faber.
Hartshorn, J.E. (1993), *Oil Trade, Politics and Prospects*, Cambridge: Cambridge University Press.
Horsnell, P. (2000), 'The strategic petroleum blunder?', *OIES Monthly Comment*, October, Oxford Institute for Energy Studios, Oxford.
International Monetary Fund (IMF) (2006), *The Magnitude and Distribution of Fuel Subsidies: Evidence from Bolivia, Ghana, Jordan, Mali, and Sri Lanka*, IMF Working Papers 06/247, International Monetary Fund.
Jacoby, N.H. (1974), *Multinational Oil, A Study in Industrial Dynamics*, New York: Macmillan.
Jesse, J. and J.G. Van der Linde (2008), *Oil Turbulence in the Next Decade*, The Hague: Clingendael Institute.
Johnston, D. (2008), 'National oil companies and international oil companies in the Middle East: under the shadow of government and the resource nationalism cycle', *The Journal of World Energy Law & Business*, **1**(1), 31–54.
Kohl, W.L. (2002), 'OPEC behaviour, 1998–2001', *The Quarterly Review of Economics and Finance*, **42**, 209–33.
Lovejoy, W.F. and P.T. Homan (1967), *Economic Aspects of Oil Conservation Regulation*, Baltimore, MD: Johns Hopkins Press, for the Resources for the Future.

Lucas, N.J.D. (1985), *Western European Energy Policies: A Comparative Study of the Influence of Institutional Structure on Technical Change*, Oxford: Clarendon Press.
Molle, W. and E. Wever (1984), 'Oil refineries and petrochemical industries in Europe', *Geojournal*, **9.4**, 421–30.
Mommer, B. (2002), *Global Oil and the Nation State*, Oxford: Oxford University Press.
Noreng, O. (2002), *Crude Power, Politics and the Oil Market*, London and New York: I.B. Tauris.
O'Rourke, D., S. Connolly (2003), 'Just oil? The distribution of environmental and social impacts of oil production and consumption', *Annual Review of Environment and Resources*, **28**, 587–617.
Odell, P.R. (1986), *Oil and World Power*, Harmondsworth: Penguin Books.
Odell, P.R. (1997), 'The global oil industry: the location of production – Middle East domination or regionalization', *Regional Studies*, **31**(3), 311–22.
Osmundsen, P. and R. Tveterås (2003), 'Decommissioning of petroleum installations – major policy issues', *Energy Policy*, **31**, 1579–88.
Painter, D.S. (1984), 'Oil and the Marshall Plan', *Business History Review*, **58**(3), 359–83.
Parra, F. (2005), *Oil Politics: A Modern History of Petroleum*, London and New York: I.B. Tauris.
Penrose, E. (1969), *The Large International Firm in Developing Countries: The International Petroleum Industry*, London: Allen and Unwin.
Philip, G. (1984), *Oil and Politics in Latin America: Nationalist Movements and State Companies*, Cambridge: Cambridge University Press.
Pinder, D. (2001), 'Offshore oil and gas: global resource knowledge and technological change', *Ocean & Coastal Management*, **44**, 579–600.
Pirog, R. (2007), *The Role of National Oil Companies in the International Oil Market*, US Congress, 21 August, KL 34137.
Pieterse, C.W.and A.F. Correljé (2008), *Crude Oil Demand, Refinery Capacity and the Product Market: Refining as a Bottleneck in the Petroleum Industry*, The Hague: Clingendael International Energy Programme.
Sampson, A. (1985), *The Seven Sisters, the Greatest Companies and the World they Made*, London: Coronet.
Schipper, L., S. Meyers, R.P. Howarth and R. Steiner (1992), *Energy Efficiency and Human Activity: Past Trends, Future Prospects*, Cambridge: Cambridge University Press.
Schurr, S.H., B.C. Netschert, V.F. Eliasberg, J. Lerner and H.H. Landsberg (1960), *Energy in the American Economy 1850–1975*, Baltimore, MD: Johns Hopkins Press for Resources for the Future.
Skinner, R. (2006), *Strategies for Greater Energy Security and Resource Security*, Background Notes, June, Oxford: Oxford Institute for Energy Studies.
Stevens, P. (2005), 'Oil markets', *Oxford Review of Economic Policy*, **21**(1), 19–42.
Stevens, P. (2008a), 'National oil companies and international oil companies in the Middle East: under the shadow of government and the resource nationalism cycle', *The Journal of World Energy Law & Business*, **1**(1), 5–30.
Stevens, P. (2008b), *The Coming Oil Supply Crunch*, a Chatham House Report, London: Royal Institute of International Affairs.
Van der Linde, J.G. (1991), *Dynamic International Oil Markets*, Amsterdam: University of Amsterdam.
Van der Linde, J.G. (2000), *The State and the International Oil Market, Competition and the Changing Ownership of Crude Oil Assets*, Boston, MA, Dordrecht and London: Kluwer Academic.
Venn, F. (2002), *The Oil Crisis*, London: Longman.
Verleger, P.K. (1990), 'Understanding the 1990 oil crisis', *The Energy Journal*, **11**(4), 15–33.
Yergin, D. (1991), *The Prize*, New York: Simon and Schuster.

13 Reforming railways

Marc Laperrouza

INTRODUCTION

While railroads initially started as private undertakings in many countries, most were nationalized in the late nineteenth century. Railway companies thus remained integrated public monopolies for most of the twentieth century and it is only towards its end that the industry's landscape started to change. Yet, technical, economic and domestic political factors have and still severely limit the potential for a radical restructuring of the sector.

Problems of interconnection and interoperability, the idea that the sector operates as a natural monopoly or the defensive attitude of governments towards 'their' rail infrastructure have largely hindered the scope, the scale and the speed of reforms. In addition, there has been little change in basic operational and cost characteristics. The sector remains characterized by heavy capital investment and an important part of the value chain remains a natural monopoly. These substantial investments – terminals and in rolling stock usually considered as sunk costs since they are not easily tradable[1] – not only limited the entry of new actors in the sector but also the possibilities and modalities of restructuring. At the same time, pressures for reforms originated from many sides: deregulation of competing sectors (airlines and trucking), profound transformation of the state's role and broad ideological trends favouring liberalization all participated in one of the major upheavals the sector is undergoing in its history. The restructuring of the sector has been accompanied and sometimes driven by additional factors. There have been some advances in technology and performance, including the size and speed of trains, the scale of operations as well as the information technology to manage the railway system. Increasing budgetary constraints, the railway's loss of market share against other transport modes, heightened environment concerns coupled with the worldwide trend towards liberalization of network industries ensure that the reform of the sector remains on the top of the agenda of many governments.[2] Not surprisingly, reform experiences across the world are varied both in terms of the reform models chosen and in the market structure that resulted. In the USA there has been a shift from a heavily regulated to a deregulated environment. In the European Union the accent has not been placed on reducing regulation but rather on increasing harmonization across member states and on introducing more competition, a trend that can be witnessed in the rest of the world as well.

This chapter is structured as follows. We first look at the specificities of the railway sector before turning to reform experiences in the world with a particular emphasis on Japan, the USA and a couple of European countries. A number of reform models are developed before discussing the actual performances of the reforms undertaken. We conclude with some of the key future issues faced by the sector.

SPECIFICITIES OF THE SECTOR

Before detailing the reforms that took place over the past two decades and in order to capture the changes they imply, it is important to list some of the sector's key characteristics. Knieps (2005) argues that railways are *technical systems* that can be divided into tracks and stations, traffic control systems and train services (transportation of goods and passengers). This technical definition of a railway can be supplemented by a number of features. First, railways are characterized by a multi-output structure (that is, freight and passenger, long-distance and commuter rail, and so on) while exhibiting a high degree of interdependence (see Table 13.1).[3]

Secondly, like many other network industries, railways were long believed to be natural monopolies. High maintenance and important sunk costs would make the duplication of infrastructure prohibitive, almost naturally leading to vertically integrated monopolies. On the one hand, the integration of the infrastructure and operations makes it easier to plan the future investment and to program the operations. Moreover, it is possible to have a simple compatible tariff policy over the whole network and eliminate or reduce to a minimum any contract with other firms. As we will see, vertical unbundling causes coordination problems – vertical unbundling works best when there is excess infrastructure capacity (something not found in many European countries).[4] On the other hand, such a monolithic structure leads to a lack of competition and, as a result, it is difficult to provide incentives to improve efficiency and productivity in the industry. In addition and as with many government-owned or government-run industries, objectives of profitability are often made difficult by the imposition of social objectives.

As pointed out by Nash (2008), the rationale behind vertical separation is to allow

Table 13.1 Multi-product nature of railways

Types of transport	Substitutability/complementarity	
Freight transport		
– Bulk (coal, raw materials, oil, chemicals, wood and derivatives, steel and metal profiles, cars and heavy machines)	– Inter-modal substitutability (pipelines, waterways, road) – Possible inter-modal complementarity – Intra-modal competition	
– Parcel/postal (small volumes, fast deliveries, food stock)	– Inter-modal substitutability (waterways, road)	
– Dedicated (food; foodstock)	– Intra-modal substitutability	Network effects
Passenger transport		
Types of transport	Substitutability/complementarity	
– Long-distance • Inter-city • High-speed – Commuter rail • Urban • Suburban/regional	Intra-modal substitutability and inter-modal substitutability (depending on elasticity to price and time) Possible complementarity with other modes	

Source: Di Pietrantonio and Pelkmans (2004).

competition by ensuring that infrastructure managers have no incentive to favour one train operator over another.[5] Vertical separation creates a level playing field for competition-in-the-market, although problems concerning the determination of access rights and charges remain. With appropriate anti-trust policy, competition-for-the-market is possible for vertically integrated structures. However, under this kind of market structure, and in order to avoid the abuse of monopoly power, railways have mostly been operated as public firms or, in the cases of markets with private companies, have been heavily regulated. In practice, this reduces the incentives to improve productive efficiency and excludes the possibility of bankruptcy, except of course in the case where portions of the sector are in the hands of private firms (Nash and Rivera-Trujillo, 2004).[6] Vertical separation of operations (as mandated in Europe) also creates potential conflicts in the areas of investment, access and timetabling as well as real-time operations (Nash and Matthews, 2009). As we will see later, the organization of railroads has been fundamentally transformed in many countries with the unbundling of operations.

Thirdly, rail services are subject to economies of density (Seabright and Ivaldi, 2003). Railway systems are characterized by asset indivisibilities. Capital units (tracks, stations) can only be expanded in discrete, indivisible increments, while demand fluctuates marginally in much smaller units. Consequently, increases (decreases) in supply can exceed increases (decreases) in demand, resulting in excess or shortage of capacity on specific routes (Campos and Cantos, 2000). This can easily result in imbalances (either excess or shortage) in capacity, since supply is much less responsive than demand. Indivisibilities in capital investment and scale of operations result in decreasing average costs.[7] The existence of indivisibilities in inputs and outputs makes the expansion of networks and services for the railway industry efficient only if performed in absolute discrete order,[8] a further reason explaining why railroad companies have historically often been integrated firms.[9]

In addition to these economic characteristics, the railway sector faces a number of social constraints (Table 13.2). For instance, in Europe, approximately 90 per cent of domestic passenger transport is currently provided within the context of public service arrangements (CER, 2005). This is largely explained by some specific dimensions of railway transport, including redistribution concerns such as inter-regional equity (for example, financing low-demand local and regional infrastructure), cross-subsidization, access, alternatives (for example, 'free' roads), safety and national security issues, as well as congestion. Public service obligations imposed on railway companies coupled with a high attention to the safety aspect have favoured public ownership or at least strong public supervision of the sector. For instance, a number of countries still award public service transport contracts without a competitive tendering procedure to the national (monopolistic) rail operator on the basis of reimbursement for public service obligations (for example, in Finland, France, Greece, Hungary, Ireland, Slovenia and Spain).

REFORM MODELS

The appropriate role for structural reform in the rail industry depends upon the particular form of competition and regulation being pursued. According to the OECD

Table 13.2 Features of railway systems

Features	Typology	Effects
Natural monopoly (essential facility)	High sunk costs in infrastructure construction	– Non-duplication of infrastructure – Non-competition among Infrastructures (except in North America) – If vertically controlled, risk of barrier to entry
	High maintenance costs Renewal costs	– Costs have long-run perspective – Renewal in discrete units – Possible complication of the system
Network externalities and density effects	Positive and negative network externalities	– Value increases with increase of routes length – Doubling infrastructure substantially more than doubles capacity – Positive externalities in the value of carrying extra units of transport – Negative externalities in non-marginal adjustments of supply and demand (non-marginal adaptability of capacity)
	Asset indivisibility	– Excess capacity or bottlenecks
Public good	Public Service Obligations (PSOs)	– Widespread benefit / non-marginal cost allocation – Non-appropriability and non-excludability (of consumption)
	Safety aspects	– Essential requirement. Acceptance of risks. Heavy technological burden for risk reduction
Economies of scale and scope	High level of fixed costs to total costs (roughly 90%) of infrastructure. Average cost curves flat on high volumes	– Fewness of market players – Monopolistic behaviour – Difficult entry into market – Inflexibility
Asymmetric information	Presence of high share of common costs Difficulty to assess 'true costs' Knowledge accumulation in the sector	– Difficult cost apportionment – Possible regulatory capture

Source: Di Pietrantonio and Pelkmans (2004).

Competition Committee (2006) the three different modes of competition in the rail sector are:

1. Competition in-the-market between vertically integrated rail companies. This form of competition requires the existence of at least two separate rail infrastructures capable of providing substitute rail services (for example, two different rail routes

between a given city-pair). This is the predominant form of competition in rail freight services in North America.
2. Competition in-the-market between train operating companies with regulated access to track infrastructures (which may or may not be owned by one of the companies providing train services). This is the predominant form of competition in freight services in Europe.
3. Competition for-the-market between rail companies (either for integrated track-plus-train services or just for train services alone, operating under a regime of regulated access to the track infrastructure). This is the predominant form of competition for regional passenger services in many EU countries.[10]

Direct competition among railroads is rare because of the capital intensity of rail lines. Competition can nevertheless arise in several ways: intramodal (with concerns about stability and sustainability of competition between firms with long-lived sunk assets and substantial overhead costs), intermodal (truck and so on), competition or through contestability (competition *in* the market versus competition *for* the market).

Promoting competition within the rail sector through mandated access to the track and associated infrastructure raises many new and important regulatory issues. In a regime of regulated infrastructure access, vertical integration is a key issue. It will often be difficult to control anti-competitive behaviour by the integrated firm. In practice many countries have accompanied mandated access with various forms of separation of the infrastructure from train operations, but very few have completely cut the link.

Finally, two features make price competition in the railway sector different from other industries: (1) short-run cost elasticities between transport modes are rather low – price competition against airlines will not allow railway companies to reap quick profits; and (2) economies of density – when holding the route system or kilometres of rail line constant, unit costs of rail services decline as output increases (Ivaldi, 2005).[11]

REFORMS ACROSS THE WORLD

The technical complexity of railway systems associated with the political sensitivity of the sector and real economic challenges has meant that rail reforms have often lagged behind other network industries. While not immune to the wave of liberalization that swept across the world in the early 1980s – the Staggers Act was passed in 1980 in the USA – railroads seem to have been resisting the liberal orthodoxy. The fact that railways employ many workers, that there is a strong history of unionization and that there is a very strong 'rail-industry culture' is probably not unrelated to such resistance.[12] Paradoxically, the sector has long played an important and avant-garde role in the general thinking about network industries. Since the second half of the nineteenth century, railways have been at the core of major themes in economics including competition, pricing for multi-product enterprise or regulation. For instance, in the USA the importance of the industry – at the time railroads dwarfed most institutions in American society (McCraw, 1984) – and the issues it raised led to the creation of the first regulatory commission.[13]

Over the past two decades the railway sector has undergone significant reforms in virtually all Organisation for Economic Co-operation and Development (OECD) countries.[14] The reforms recently conducted contrast notably with the relative apathy which characterized the sector for most of the twentieth century. The push for reforms was largely driven by inefficiency and poor performance within the rail sector and long-term loss of market share to other transport modes. For instance, between 1970 and 2000 the percentage of freight carried by rail in Europe dropped from 21 per cent to 8 per cent while the share of passenger traffic fell from 10.2 per cent to 6.3 per cent Two additional drivers behind the reforms efforts were the reduction of the extent of government involvement in railroad management and the increase or introduction of competition in railroad services via non-discriminatory access to the railroad infrastructure.

The current cycle of reforms is resulting in a radical transformation of the entire sector, including a complete redefinition of the institutional framework governing the industry. In terms of ownership, and with the notable exception of the USA, railways were usually government owned. In the early 1980s, a number of governments transferred their ownership to the private sector either by sale or by the granting of concessions for ten years and longer. In terms of operations, for most of the railway's history, track and rolling-stock ownership were under the same roof. In addition, railroads tended to remain within national borders – railroads were horizontally separated in that they served different territories but vertically integrated in that they provided all the key functions needed to serve their territories (Gomez-Ibanez, 2004). Nowadays, vertical unbundling is becoming de rigueur in Europe with the hope that market forces will be strong enough to generate efficiency gains in the provision of services. Finally, the era of self-regulation is ending. Regulatory bodies are created to come to grips with the issues created by the unbundling of operation and introduction of competition in several segments.

REFORM PRACTICE

While reform and regulation often go hand in hand, there is a broad variety of approaches to reforming and regulating the sector.[15] Differences in reform and regulation are driven by the vision and the expectations of railways' role in the economy, in the perceived role of the state in relation to railways and in rail management strategy. The reforms carried out also vary along the type of separation of infrastructure from operations and the level of privatization (see Figure 13.1).[16] For example, in countries like the USA, Japan or New Zealand, companies remain vertically integrated. In Europe, the general trend is towards separation between the infrastructure management and the operations of passengers and freight (Gomez-Ibanez, 2004). This, in turn, impacts the type of regulation, degree of access or models for infrastructure charging. Some countries like the USA have favoured a regulatory approach (until the 1970s), whereas most EU countries and Australia favoured a state-ownership approach (the UK swings between both models) a pattern also found in most Asian countries.[17] Unlike other network industries, the railway sector does not have a United Nations (UN) agency (like the International Telecommunication Union [ITU] for telecommunications or the Universal Postal Union (UPU) for the postal sector) which establishes global standards or offers a meeting point for governments.[18] On the contrary, railways has very much been a sector where protec-

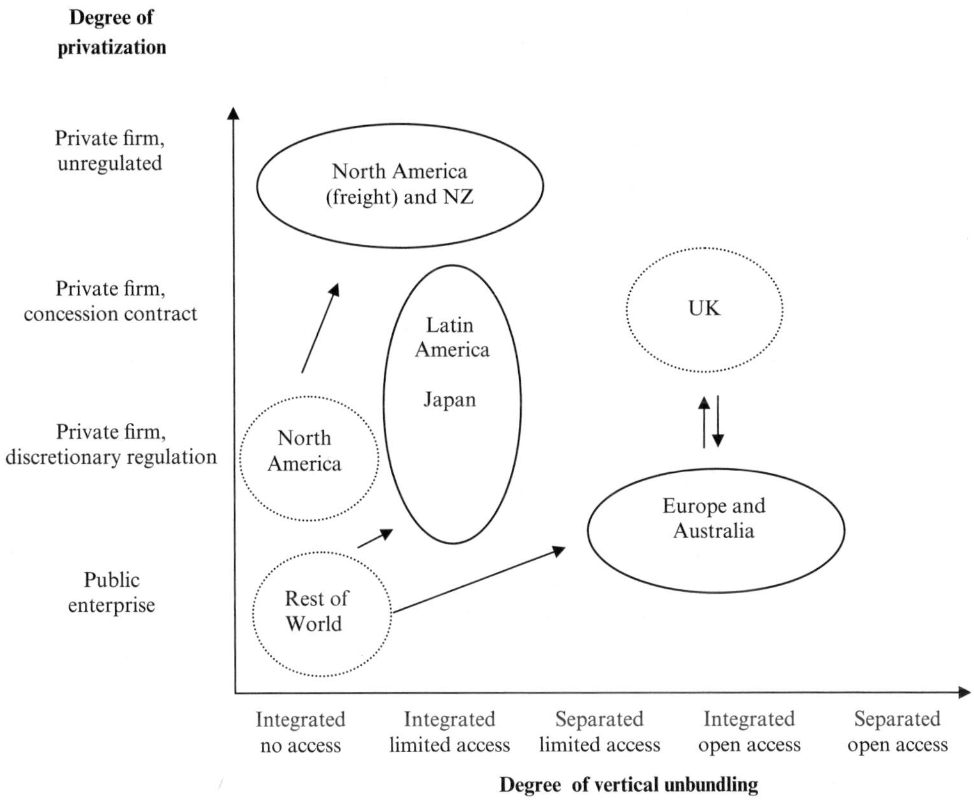

Source: Adapted from Gomez-Ibanez (2004).

Figure 13.1 Overview of reform options in the railway sector

tion and domestic concerns ruled and it is only in the last decade, thanks in part to the Directives passed by the European Union, that national barriers have started falling.

So, countries have experimented with three reform models. The first relies primarily on privatization and is found mainly in North America and New Zealand (until 2004). The second relies primarily on unbundling and is most closely associated with Europe and Australia. The third combines privatization with selective track access as practiced in Latin America and Japan (Gomez-Ibanez, 2004). These models are detailed hereafter through the examples of four countries.

USA

In the USA, the network is owned by private operators (Spychalski and Swan, 2004). There, the driving force for reform has been a combination of mounting evidence of the failure of regulation, compelling arguments that markets could work in the transportation sector and sinking financial conditions of the US railroads. Alexandersson and Hultén (2008, p. 19) noted that 'the American railway market is predominantly a freight market and the principal goal of the deregulation was to allow the private railway operators more freedom of action'. It culminated with the Staggers Rail Act (1980)[19] that largely deregulated the industry[20] and caused a dramatic improvement in efficiency.[21] The Act allowed companies to compete with each other and set transportation prices freely (Hillmola and Szekely, 2006). The restructuring of freight allowed for dramatic improvements in efficiency, reducing workforce and track mileage by 50 per cent (Ivaldi, 2005).[22]

Europe

The restructuring of the US freight sector contrasts markedly with the European case (Ivaldi, 2005).[23] European railway networks serve both passenger and freight markets, and both these markets have been deregulated. Like in other countries the aim of reforms in the European Union has been the improvement of efficiency and competitiveness of the railway system (Nash, 2008).[24] European rail policy has concentrated on the introduction of competition into the rail transport market via separation of infrastructure from operations (at least in an accounting sense), by the progressive opening up of entry to the market for new operators and by rules regarding the allocation of slots and the pricing of infrastructure use, administered by an independent regulator. This is all in marked contrast to the USA, where concentration has been on parallel competition between vertically integrated railroads (Nash and Rivera-Trujillo, 2004). Deregulation in Europe has meant both changes in the regulatory structure and a gradual transformation (leading to complete privatization in the case of the UK) of the former vertically integrated state monopolies. Reforms in Europe were initiated by Directive 91/440 ('first package') with the objective to achieve market opening of rail transport and create a single European railway market. They are designed around three axes: unbundling the infrastructure from the activity of railway transport,[25] opening the access to the network to competition and achieving a regionalization of passengers regional railway transport. The European Commission adopted in January 2002 a new package ('second package') of measures to revitalize the railways by rapidly building an integrated European railway area. In March 2004 the Commission adopted a third railway package containing measures to further the liberalization of the market.[26]

The European Commission published a White Paper in 2001 (European Commission, 2001) outlining its ambition to revitalize the sector through the introduction of competition – freight was to be fully liberalized by 2007 and passenger service by 2012 – and the reform of institutions.[27] The idea was to increase both the capacity and the safety of the sector. That said, some countries preceded the Commission's effort. In 1988 Sweden, the first country to reform its railway market, demanded complete unbundling of infrastructure from operations and the empowerment of regional governments for planning and funding of regional services. The UK followed in 1993 with the passing of the Railways Act which led to the franchising of all passenger operations to private operators. Germany followed with the creation of Deutsche Bahn, a government-held holding company grouping freight, passenger and infrastructure services.[28]

The European railway reform is implemented at this time among the member states of the European Union with some delays depending on the speed (and the political

Table 13.3 Models of regulatory bodies

Model	Description	Countries
Special regulatory body	Independent authority with decision-making powers specialized in regulatory matters	Austria, Belgium, Germany, Denmark, Greece, France, Italy, Luxembourg, Netherlands, UK
Regulation body within a railway authority	Regulatory duties primarily addressing licences, safety and other railway-specific administrative tasks	Bulgaria, Switzerland, Czech Republic, Hungary, Latvia, Norway, Poland, Portugal, Sweden, Slovakia
Regulatory body within a ministry	Weakest form of regulatory independence	Estonia, Spain, Ireland, Lithuania, Romania, Slovenia

Source: Adapted from IBM (2011).

will) of the implementation of EU regulations and laws into national legal systems. While the European Commission has taken the driving seat in reforming the sector, Member States battle hard to retain control over their national networks. In many European countries there are important delays in transposing EU regulation into domestic laws, and even more delays in implementing the new legal frameworks.[29] When looking at the EU in general, member states differ in terms of how they have interpreted requirements set by EC legislation. As noted by Nash (2007), the extent of actual reform does not match the formal transposition of European Directives. In fact, the majority of the countries only implemented the EU's minimum requirements. The Commission has repeatedly notified member states for failing to carry out Directives.

In terms of regulatory bodies, some of the EU's new member states, in comparison with many 'original' member states, had already set up better organized regulatory bodies. In fact, almost all countries that have made relatively good progress in terms of opening up their rail transport services markets have special regulatory authorities. In practice, most of the regulatory bodies have not yet been required to make decisions on discrimination cases. A growing number of countries have regulatory bodies with specially trained staff dealing exclusively with regulatory matters. In addition, the regulatory bodies were provided with far-reaching powers to enable them to enforce their decisions and had already accumulated considerable experience with regulatory cases (Table 13.3).

As a result, there are large variations in the extent to which railway markets are open to potential entrants. Whereas the freight market has been open to competition since January 2007 the passenger market remains to be opened. For the time being, international trains circulating in Europe are operated under joint ventures between national companies (for example, the defunct Cisalpino which was a 50-50 joint venture between Switzerland's SBB and Italy's FS). With the partial opening of the passenger market in 2010, a number of operators (both incumbent and newcomers) have applied for slots outside of their home markets.[30]

Sweden

The Swedish deregulation of railway services started as a process of reforms without any clear intentions to introduce competition or increasing the involvement of private actors, neither national nor international (Alexandersson and Hultén, 2008). The driving forces of the reform were the incumbent's (SJ) recurrent problems, coupled with political objectives to save railway lines from closing down, improve sector efficiency, increase railway travel and transfer freight transportation from the roads to the railways. Competition has gradually been introduced and spread in the Swedish railway markets. In freight, the first new entrants appeared in the early 1990s – generally, these were minor freight operators working as subcontractors to SJ. In 1993, the state-owned ore company LKAB became the first company to get its own operating licence on the state's railways, thereby being able to take control of its own transportation needs. The country has introduced competitive tendering for regional passenger transport since 1989 and non-profitable inter-regional passenger transport since 1992 (Alexandersson and Hultén, 2008). Deregulation in Sweden can be said to have been relatively successful. On the one hand, productivity levels have risen and rail's intermodal market share has increased. Since vertical separation in 1988, there has been a substantial surge in public spending on infrastructure investments and renewal. The transportation volumes (in terms of passenger kilometres) have increased by more than 40 per cent between 1990 and 2003. On the other hand, competition remains marginal (Hillmola and Szekely, 2006). While BK Tåg became the first regional passenger train operator in 1990 it was not until 1995 that another company entered this segment of the market. For inter-regional services, despite being tendered since 1992, the breakthrough for competing operators did not happen until the year 2000. Whereas competitive tenders led to subsidy reductions by around 20 per cent, operators have put very low bids in recent tenders to secure the contract in the competition.

United Kingdom

The British deregulation model combines two privatization measures mentioned above – privatization of monopoly rights and contracting out of public services (Alexandersson and Hultén, 2008). In fact, the UK approach is rather unique since it has pursued both vertical separation and privatization – a path that few countries have followed in the world. Vertical separation was decided on grounds that a separate infrastructure authority was needed to take the *sunk cost* element out of rail provision so as to facilitate franchising (Jupe and Crompton, 2006). The country's early experience with the comprehensive competitive franchising was very positive with strong growth in passenger numbers and reduced costs (lower subsidy payments). However, since 2001 there have been significant increases in costs for railway undertakings, in part explained through use of contracts with reduced incentives for cost containment and also linked to the problems regarding the organization of the infrastructure manager.

In summary, the deregulation of the railway sector in the EU member states seems to have been driven by different types of economic, institutional and legal concerns. (Table 13.4) Alexandersson and Hultén (2008) find that in the UK the pursuit of a liberalized

Table 13.4 European reform eras

	Previous era (until 1990)	Transition era (1990–2005)	New era (2010)
Organizational structure	Vertically integrated	Voluntary unbundling	Mandatory unbundling[1]
Regulatory policy and legislation	National	National with supra-national transport policy and directives	National with EU Directives (railway packages)
Drivers	Public service	Public service, productivity and financial sustainability	Public service, productivity, financial sustainability, environmental concerns
Market structure	Monopoly	Monopolistic (infrastructure) and market (services) segments	Monopolistic (infrastructure) and market (services) segments
Market opening	Closed with limited international traffic	Ad hoc opening of domestic markets	Freight open Mandated opening of passenger market
Ownership	Public ownership[2]	Mostly state-owned	State-owned (infrastructure) Some private rail companies
Regulatory arrangements	None (Ministry)	None (Ministry)	Independent railway authority
Scale	Regional and national	National to international	Increasingly international

Notes:
1. Under the mandatory unbundling, European countries have opted for different options, for example total vertical integration, competitive access, vertical separation.
2. Governments generally became system owners during the first half of the twentieth century.

market agenda has dominated. In Sweden, key concerns have been the need to find new ways to finance investments in the railway sector, and to increase efficiency through competition. In countries like France and Portugal, the initial most important factor seems to have been the necessity to act in accordance with EU legislation.

Japan

Japan presents an interesting case of railway restructuring. The process of privatizing JNR began in 1987 and consisted in partitioning the monopoly into six regional passenger companies and one nationwide freight company (Mizutani and Nakamura, 2004). The Japanese approach to reform has the following features: horizontal separation, functional separation, vertical integration, lump-sum subsidies for low density areas, creation of a regulator and a yardstick competition scheme (Mizutani, 2005; Mizutani and Nakamura, 2004).

Table 13.5 Restructuring railways in four countries

Restructuring	USA	Sweden	UK	Japan
	1980	1988	1993	1987
Degree of unbundling	Vertical integration	Vertical separation	Vertical separation	Horizontal separation
Ownership infrastructure	Privately owned	Publicly owned	Publicly owned	State enterprise with commercial statutes
Ownership operators	Privately owned	Privately owned	Joint stock companies privately owned	Partially privatized
Degree of competition	Freight services	Regional passenger and freight services	All markets	Regional markets
Public service obligation	None	Competitive tendering	Competitive tendering	None legally but negotiated with local communities

Source: Adapted from Mizutani and Nakamura (2004) and Nash (2006).

REFORMS AND PERFORMANCE

As noted above, reform models vary greatly across the world. It is therefore not surprising that the outcome of the reforms have differed widely as well (Ivaldi, 2005). According to one study on liberalization (IBM, 2007) the modal share of rail only increased in a few European countries between 1995 and 2005. Despite signs of market opening, strong entry barriers persist.

A number of empirical studies have looked at performance under de-regulation, most of them attempting to establish a link between reforms and efficiency in railways.[31] Among others, these studies looked at the impact on firms' financial condition (that is, revenue versus cost control), productivity and cost-efficiency. Studies were also conducted to analyse the impact on customers and shippers – substantial real decline of rail prices was found *on average.* Consolidation of the sector (that is, through horizontal or vertical mergers) raised concerns about foreclosure or control over bottlenecks.[32]

Despite an abundant literature on railway reforms, there seems to be relatively little consensus about the impact of reforms on efficiency. Friebel et al. (2003) find that *full separation* of infrastructure from operations is not a necessary condition for improving railroad efficiency. More recently, using a production frontier model Friebel et al. (2008) find that efficiency increases when reforms such as third-party network access, introduction of an independent regulator, and vertical separation are implemented.[33] In addition, reforms have positively impacted railroad technical efficiency. Higher reform intensity does not necessarily increase technical efficiency; rather the improvement of efficiency depends upon the sequencing of reforms. In fact, railroads are sensitive to changes in the regulatory framework and one-size-fits-all may not be a fruitful way to enhance efficiency.

Efficiency aside, some of the major themes that emerged from the deregulation experiences are: (1) concern for maximum price regulation of a monopolist and guidelines for maximum rate regulation, (2) restrictions on rail operations and pricing in connection with mergers and (3) the promotion of competition by enabling access to other railroads over a shared track.[34] For example, European policy defined that rail infrastructure charges should be related to costs actually incurred, place incentives for a more efficient use of infrastructure, avoid discrimination of users and allow public authorities to verify compliance with these objectives.[35] So, while the jury is still out on the impact of deregulation, it is safe to say that most railway undertakings have become more efficient and improved financially.

KEY ISSUES FOR FUTURE REFORMS

Further reforms of the sector will have to tackle a number of key points, notably infrastructure charging, third-party access and forms of market competition. Infrastructure charging has two core issues. First, determining the organization responsible for setting the charges. This can either be the responsibility of the infrastructure manager (for example, in the case of Germany, Lithuania, Poland and Austria) or an independent charging body. Secondly, and more importantly, sustainable access charging mechanisms still need to be developed. In the EU charges are set 'at cost directly incurred as a result of operating the train service' although, from a legislative point of view,[36] a number of different charges are allowed: costs reflecting scarcity, charges to cover environmental costs and mark-ups.

Pricing

Railways (like some other utilities) became special cases in economics, as standard theory about competition, costs and prices could not apply. There was a need for some differentiated pricing of railway outputs but the problem is how to identify and prevent exploitation of monopoly power. With economies of scale, competition in the market is not feasible. If marginal cost pricing (MCP) was to be followed, the firm would incur a financial deficit and go bankrupt. As a result, a subsidy for private firms is needed so that they can survive or one has to operate railway firms as a public enterprise (government budget). As noted by Waters (2007) railway cost analysis has led to refining the distinction between economies of scale (behaviour of costs if the network size increases as output *expands*) and density (behaviour of costs as output expands over a *given network*), as well as firm-specific effects and industry-wide characteristics (economies of scale and density).[37] But despite efforts to bring some clarity to the cost structures of the sector (for example, some countries have introduced accounting separation), ambiguity in measuring costs still makes it difficult to compare prices and costs.

In practice, a variety of charging mechanisms have emerged: (1) marginal social costs[38] (applied in the Netherlands and Portugal), (2) marginal cost pricing with mark-ups (applied in Austria, Bulgaria, Denmark, Finland, France, Sweden, Switzerland and the UK), (3) full cost recovery after receipt of grants[39] (applied in Belgium, Germany and Italy) and (4) full cost recovery model (applied in Estonia, Hungary, Latvia, Poland, Romania and Slovenia).

That said, rail access pricing is still a relatively new approach to capacity management, following vertical separation of infrastructure and operations and/or increased scope for different operators to gain access to rail infrastructure. For the time being there is only limited empirical evidence regarding relative performance of alternative access pricing regimes available. For instance, in the UK high fixed charges are combined with relative low variable charges. This led to overuse of rail infrastructure, and despite increasing traffic levels only very limited rail infrastructure investment made over the relevant period. In Germany two-parts tariffs charging contribute to ensure cost recovery but there are problems with ensuring fair and non-discriminatory access for non-DB operators (OECD, 2005).

Third-party Entry and Competition

As noted earlier, two principal forms of competition have emerged in the railway sector. First, *in-the-market competition* (on-track competition) where several railway companies compete by providing services on the same network during the same time period; the model is based on open access arrangements for third-party undertakings (rail freight and long-distance passenger services). Second, *for-the-market competition* (for-the-tracks competition) where railway undertakings are in competition prior to start of operations through alternative offers for a contract to provide services.[40]

Encouraging entry by independent train operators (after vertical unbundling) has proved harder than expected (Gomez-Ibanez, 2004). In the EU market opening and introduction of competition in rail market only occurred in last few years. A typical period required from market opening until an entrant gains access to the network and starts services is two years.[41] For the passenger market there is already the possibility of associations of operators wishing to offer international services between the countries in which they are established (for example, Thalys, a joint undertaking between France's SNCF, Germany's DB and Belgium's SNCB which offers a liaison between Paris, Brussels, Amsterdam and Köln). In practice, competition on the passenger side mainly takes place through the introduction of competitive tendering.[42]

Knieps (2005, pp. 21–2) argues that 'efficient competition in Europe is conditional upon the existence of non-discriminatory access to rail infrastructure for all active and potential service providers'. In short, infrastructure companies and train operating companies must be completely separated in order to enable 'real' competition. Weidman and Nash (2008) argue that the separation of railway operations and infrastructure is not the aim of European transport policy but a consequence of the open access principle. However, the reorganization of the relationship between the players goes much further than simply separating network infrastructure from operations; it is an entirely new way to organize and manage railway systems, akin to what is found in road transports or aviation (that is, separation of the system in the three levels of network maintenance, network operation and train operation). Weidman and Nash (2008) find that such an organization would not better support the non-discrimination objectives, but would create additional boundaries.

Another ongoing challenge is to try to balance the benefits of commercial freedom for railroads with desires for some public oversight. Safety issues as well as a more general concern for the provision of public obligation and the public's perception that railways are a service of general interest tend to restrict the margin of manoeuvre. Finally, the potential for discrimination remains important. Such opportunities can be found in investment planning, network access conditions, path allocation or technical standards, (Weidmann and Nash, 2008).

Institutional Settings to Enable Further Reforms

Railway reform also remains plagued by institutional constraints (Yvrande-Billon and Ménard, 2005). After more than two decades of reforms in the railway sector, it has become clear that one of the key challenges lays in the coordination between the various stakeholders. The trend towards vertical disintegration raises central challenges to the economic viability but also to the incentive to innovate.

In spite of the European Commission's attempts to create a single railway market through a number of Directives the approaches of European member states to reforming their railway markets differ importantly and there are limited signs towards convergence – a major difference lies in their approach to vertical separation. Whereas the movement of reforms was usually initiated by having better performing railway systems, the technical complexity of the railway sector has often been at odds with the desire to introduce more competition. In many countries the railway sector remained characterized by little investment, heavy subsidies and a high level of indebtedness.

Interoperability

One of the central elements facilitating access to a networked system – and thus making competition in the market possible in the first place – is technical interoperability. In the European railway context interoperability has been mandated by two EU Directives for high-speed lines and conventional lines.[43] These Directives formally define interoperability as 'the ability of the trans-European rail system to allow the safe and uninterrupted movement of trains which accomplish the specified levels of performance. This ability rests on all the regulatory, technical and operational conditions which must be met in order to satisfy the essential requirements' (High Speed Directive 96/48/EC). A further step towards interoperability was achieved through the passing of the second railway package in 2004 – a Directive for the harmonization of safety requirements and certifications that were different in all member states, and a regulation for the creation of a European Railway Agency (ERA) for Safety and Interoperability.[44] The need for such an agency arose from the magnitude, the scope and the nature of the problems that the European railway system faces in a liberalized business environment. While significant progress has been achieved towards greater interoperability within Europe – a precondition for cross-border competition – much work remains to be done until a train can drive from Stockholm to Naples in one go.

CONCLUSION

For most of its history, the railroad sector was conceived as a national system. As a result, railroads developed in isolation, leading to interoperability and interconnection problems. In the not-too-distant future, it should not be a surprise to have an airline company operating a high-speed passenger train across Europe or competitive freight operators offering to drive a train from Sweden to Slovenia without countless changes and delays. These fundamental changes of the railway landscape will have largely been made possible by both technological and regulatory changes. They will also mark another phase in the ongoing reform process of the sector.

The variety of reform models found across the world, as well as the variety of performance levels associated with them, make it extremely hard to single out an optimal approach to reforming a country's railway sector. As pointed out by Nash and Rivera-Trujillo (2004) the best structure in any one country is likely to depend on factors such as the importance of international traffic, the scope for open access entry and the willingness of government to subsidize infrastructure charges. Despite the wide range of experience, the appropriate role of vertical separation in the overall reform of the rail industry is not yet clear. Governance and subsidy mechanisms need further examination. The necessary presence of the government in the sector (either via subsidies or direct ownership) calls for introducing competition via competitive tendering or franchising rather than through open access.[45] In other words, future sector restructuring should be aware of the challenges created by vertical separation, and governments should carefully think about their relative level of disengagement from the sector.

NOTES

1. In certain markets, notably the UK, one sees the development of leasing rolling-stock.
2. Reforms in the sector often have much wider implications since railways tend to have strong links with broader economic development.
3. Economies of scope are said to exist vertically and horizontally in the railway sector.
4. Gomez-Ibanez (2004) suggests that the key to resolving these coordination issues is to establish charges and other terms of access that provide appropriate incentives for both the infrastructure manager and train operators.
5. The major problem associated with vertical separation lies in high transaction costs. Growitsch and Wetzel (2009) find slight efficiency advantages for integrated companies on average and observe economies of scope for a majority of observations.
6. Another problem with vertical separation is the hold-up problem (Ivaldi, 2005). The potential for extra competition offered by vertical unbundling is limited by the intense competition from alternative modes, locations and products (Gomez-Ibanez, 2004).
7. In other words, railroads have high fixed-costs for their infrastructure and constant or falling marginal cost of supplying transportation service.
8. It is important to note that track infrastructure and passenger operations are cost substitutes (higher track costs will lead to lower operation costs by permitting faster services) but track infrastructure and freight operations are cost complements (higher track costs lead to higher freight operation costs due to higher maintenance costs).
9. The intermediate input 'network' is produced by the inputs labour and land; additional labour and rolling stock are then used in the production of the final outputs (passenger kilometres and freight-ton kilometres).
10. Such competitive tendering may change but not eliminate the need for regulation.
11. The size of the firm has little to do with economies of density: a small firm only serving one or a few routes with high traffic density might well have lower average costs than a large firm, running a network including lines with low density traffic.
12. The challenges are particularly large in Europe with Brussels' aim of creating a single integrated European market and member states keen to retain decision-making and regulatory powers.
13. The Interstate Commerce Commission was established in 1887 to prevent railroads from exercising 'undue prejudice' in pricing.

14. This is not to say that reforms did not take place in non-OECD countries. For instance, China has been collaborating with the World Bank and other agencies to reform its railway sector since the early 1990s.
15. Regulation has usually been imposed on rail industry but rail companies benefited from the stability. At times, they welcomed and even sought regulation.
16. Nash (2006) identifies two dominant models for restructuring: complete separate organizations for infrastructure management and railway operations within the public sector (the Swedish model) or maintaining a single, vertically integrated publicly operator (German model).
17. As the railway sector is not covered by any specific World Trade Organization (WTO) agreement, it has not seen a worldwide trend towards liberalization as experienced in telecommunications.
18. The International Union of Railways (UIC), whose role is to promote rail transport, comprises integrated railway companies, infrastructure managers and railway or combined transport operators and is the organization that comes the closest to a UN agency.
19. Like in Europe the Staggers Act was preceded by another measure – the Railroad Revitalization and Regulatory Reform Act of 1976 – which removed many regulations over price-setting.
20. Key factors in the overall process were the Rock Island (1975) and Milwaukee Road (1977) bankruptcies and the Carter Administration's reform agenda (including reforms in the trucking and airline sector). In 2009 there were attempts to remove antitrust exemptions enjoyed by freight railroads and reform the Surface Transportation Board that oversees rail economic regulation in the USA – the antitrust bill S. 146 would eliminate a longstanding railroad exemption to antitrust laws and would bring railroad mergers and rate changes under the purview of the Justice Department.
21. One important question is why rates fell after deregulation. On one hand it is argued that the decline is due to changes in traffic characteristics (length of haul, percentage of bulk commodities). On the other hand, it is argued that the decline reflects cost savings and increased shipment size (Dennis, 2001).
22. Cost studies turned up little evidence of economies of scale (which had been a rationale for regulation).
23. The European deregulation context 15–25 years later is also radically different from the US deregulation.
24. Nash identifies four additional objectives: (1) improve rail's market share in international traffic, (2) clearly differentiate the role of government, train operator and infrastructure manager, (3) put inter modal competition on a level playing field and (4) introduce direct 'on-the-track' competition.
25. Three different models for separation have emerged in Europe: (1) complete separation, (2) holding company and (3) separation of key powers.
26. The second and third packages respectively came into force in 2007 and 2009. A fourth package is in the works; it contains a revision of the interoperability and safety directives
27. The Commission should adopt a Communication on the Future on Transport in June 2009.
28. With 280 operators, Germany is the most open market in Europe but Deutsche Bahn retains more than 90 per cent of train kilometres (Nash, 2007).
29. Countries are routinely sanctioned by the Commission for failing to transpose or implement new laws pertaining to railways.
30. Cabotage will be allowed. Trenitalia has applied for slots to operate a high-speed train in France.
31. An increasing number of *econometric studies* are looking at railway efficiency and performance (Growitsch and Wetzel, 2009; Jupe and Crompton, 2006; Smith, 2005; Yu, 2008).
32. This was mostly the case in the USA with concerns in late 1990s about monopoly powers (mergers, downsizing and growth in demand).
33. While reforms seem to impact positively technical efficiency, higher reform intensity does not necessarily increase technical efficiency which depends more on the sequencing of reforms. This last finding is similar to what Wallsten (2002) finds in the telecommunication sector.
34. There are two approaches to shared track access: (1) a regulatory approach requiring granting access to one's railroad tracks by some other carrier for an access fee (stimulate) or (2) separate the ownership of rail track (and way) from rail operations. Both raise various transition issues, including property rights and compensation.
35. Directives 2001/12/EC and 2001/14/EC have set a general framework for setting rail infrastructure charges but enormous difficulties have arisen in the calculation of marginal costs and in the allocation of full costs of the infrastructure use to the different cost drivers.
36. See Directive 2001/14/EC.
37. Cost analysis is used for other transport modes and helped rethinking the need for regulation.
38. It represents additional cost imposed on society as a whole from a marginal addition to train kilometres.
39. In this model the infrastructure manager is considered a commercial organization and therefore required to recover the costs incurred. Costs not funded by the state are shared out among the infrastructure users.
40. Driessen (2006) finds that competition for the tracks tends to improve productive efficiency while competition on the tracks could diminish productive efficiency. In addition more managerial independence from the government may lower productive efficiency. It is important to note that the intensity of competition

in railways is one of the factors that make vertical unbundling less attractive in comparison with other network utilities.

41. In Germany, 12 per cent of the volume of public service contracts in effect in 2005 were awarded following a public tendering procedure; improvements in labour productivity.
42. Only the UK has introduced competitive tendering (franchising) for all passenger services (since 1996–97).
43. Directive 96/48/EC of 23 July 1996 (OJ L235 of 17 September 1996) and Directive 2001/16/EC of 19 March 2001 (OJ L 110 of 20 April 2001).The Interoperability Directive 96/48/EC Article 4(1) required that the trans-European high-speed rail system, subsystems and the interoperability constituents including interfaces meet the essential requirements set out in general terms in Annex III to the Directive. The essential requirements were: safety, reliability and availability, health, environmental protection and technical compatibility. The Directive allowed that the essential requirements may be applied to the whole trans-European high-speed rail system or be specific to each subsystem and its interoperability constituents.
44. Directive 2004/50/EC of 29 April 2004 (OJ L164 of 30 April 2004) modified Directives 96/48/EC and 2001/16/EC: it updated the provisions of these two directives in line with the directive on safety and the role of ERA and stated the principle of interoperability for the whole railway system, to be implemented progressively as of 2008.
45. Waters (2007) identifies two paths for railway public policy: strict marginal cost pricing (but with need for ongoing subsidy) or a requirement for the rail industry to be self-financing (demands full cost recovery).

REFERENCES

Alexandersson, G. and Hultén, S. (2008), 'The Swedish railway deregulation path', *Review of Network Economics*, **7** (1), 18–36.

Campos, J. and Cantos, P. (2000), 'Rail transport regulation', in G. de Rus and A. Estache (eds), *Privatization and Regulation of Transport Infrastructure: Guidelines for Policymakers and Regulators*, Washington, DC: World Bank Institute Development Studies, pp. 171–234.

Community of European Railway and Infrastructure Companies (CER) (2005), *Public Service Rail Transport in the EU: An Overview*, Brussels: Community of European Railway and Infrastructure Companies.

Dennis, S.M. (2001), 'Changes in railroad rates since the Staggers Act', *Transportation Research Part E: Logistics and Transportation Review*, **37** (1), 55–69.

Di Pietrantonio, L. and Pelkmans, J. (2004), 'The economics of EU railway reform', *Competition and Regulation in Network Industries*, **5** (3–4), 295–346.

Driessen, G. (2006), 'Competition design and efficiency in railways', graduate thesis, University of Groningen.

European Commission (2001), *White Paper Submitted by the Commission on 12 September 2001: 'European Transport Policy for 2010: time to decide'*, COM (2001) 370 final, Brussels; European Commission.

Friebel, G., Ivaldi, M. and Vibes, C. (2003), *Railway (de)Regulation: A European Efficiency Comparison*, IDEI report no 3 on passenger rail transport, University of Toulouse.

Friebel, G., Ivaldi, M. and Vibes, C. (2008), 'Railway (de)regulation: a European efficiency comparison', *Economica*, **75**, 1–15.

Gomez-Ibanez, J.A. (2004), 'Railroad reform: an overview of the options', paper presented at the Proceedings of the Railway Reform Conference, Madrid, 8 September.

Growitsch, C. and Wetzel, H. (2009), 'Testing for economies of scope in European railways: an efficiency analysis', *Journal of Transport Economics and Policy*, **43**, 1–24.

Hillmola, O.-P. and Szekely, B. (2006), *Deregulation of Railroads and Future Development Scenarios in Europe*, research report 169, Lappeenranta University of Technology.

IBM (2011), *Rail Liberalisation Index 2011*, Berlin: IBM Global Business Services.

Ivaldi, M. (2005), 'Restructuring railways: in theory and practice', *CESifo DICE Report Journal for Institutional Comparisons*, **3** (4), 3–9.

Jupe, R. and Crompton, G. (2006), '"A deficient performance": the regulation of the train operating companies in Britain's privatised railway system', *Critical Perspectives on Accounting*, **17** (8), 1035–65.

Knieps, G. (2005), 'Railway (de-)regulation in Germany', *CESifo DICE Report Journal for Institutional Comparisons*, **3** (4), 21–5.

McCraw, T.K. (1984), *Prophets of Regulation: Charles Francis Adams, Louis D. Brandeis, James M. Landis, Alfred E. Kahn*, Cambridge, MA: Belknap Press of Harvard University Press.

Mizutani, F. (2005), 'Regulation and deregulation in the Japanese rail industry', *CESifo DICE Report Journal for Institutional Comparisons*, **3** (4), 10–15.

Mizutani, F. and Nakamura, K. (2004), 'The Japanese experience with railway restructuring', in T. Ito and A. O. Krueger (eds), *Governance, Regulation, and Privatization in the Asia-Pacific Region*, Chicago, IL: University of Chicago Press, pp. 305–36.

Nash, C. (2006), 'Europe: alternative models for restructuring', in J.A. Gomez-Ibanez and G. d. Rus (eds), *Competition in the Railway Industry: An International Comparative Analysis*, Cheltenham, UK, and Northampton, MA, USA: Edward Elgar, pp. 25–48.

Nash, C. (2007), 'Reform of European passenger railways – where do we stand?', in R. Marcario, J. Viegas and D. Hensher (eds), *Competition and Ownership in Land Passenger Transport*, Oxford: Elsevier Science, pp. 95–114.

Nash, C. (2008), 'Passenger railway reform in the last 20 years – European experience reconsidered', *Research in Transportation Economics*, **22** (1), 61–70.

Nash, C. and B. Matthews (2009), *European Transport Policy: Progress and Prospects*, Leeds: Institute for Transport Studies.

Nash, C. and Rivera-Trujillo, C. (2004), 'Railway reform in Europe – principles and practice', paper presented at the Conference on Competition in the Rail Industry', 16–17 September.

Organisation for Economic Co-operation and Development (OECD) (2005), *Railway Reform and Charges for the Use of Infrastructure*, Paris: OECD.

Organisation for Economic Co-operation and Development (OECD) Competition Committee (2006), 'Structural reform in the rail industry', *OECD Journal of Competition Law & Policy*, **8** (2), 67–79.

Seabright, P. and Ivaldi, M. (2003), 'The economics of passenger rail transport: a survey', IDEI Working Paper, no. 163.

Smith, A.S.J. (2005), 'The role of efficiency estimates in UK regulatory price reviews: the case of rail', *Utilities Policy*, **13** (4), 294–301.

Spychalski, J.C. and Swan, P.F. (2004), 'US rail freight performance under downsized regulation', *Utilities Policy*, **12** (3), 165–79.

Wallsten, S.J. (2002), 'Does sequencing matter? Regulation and privatization in telecommunications reforms', Working Paper No. 2817, Washington, DC: World Bank.

Waters II, W.G. (2007), 'Evolution of railroad economics', *Research in Transportation Economics*, **20**, 11–67.

Weidmann, U. and Nash, A. (2008), 'Open access to railway networks: hidden discrimination potential in an integrated railway organisation', paper presented at the 1st Annual Conference on Competition and Regulation in Network Industries, 28 November.

Yu, M.-M. (2008), 'Assessing the technical efficiency, service effectiveness, and technical effectiveness of the world's railways through NDEA analysis', *Transportation Research Part A: Policy and Practice*, **42** (10), 1283–94.

Yvrande-Billon, A. and Ménard, C. (2005), 'Institutional constraints and organizational changes: the case of the British Rail reform', *Journal of Economic Behavior and Organization*, **56** (4), 675–99.

14 Liberalization of air transport

Sveinn Vidar Gudmundsson

INTRODUCTION

Several countries[1] and regions[2] have liberalised air transport markets, spearheaded by the US Deregulation Act of 1978.[3] In this chapter we examine the historic and political drivers and expectations associated with liberalization and how the regulatory environment developed during the post deregulation years both in the USA and Europe.

There have been four principal areas of regulatory reforms in air transport: (1) domestic markets; (2) air services agreements; (3) associated services; and (4) inter-regional open aviation areas. Deregulation in the USA and liberalization in the European Union (EU) are an example of the first; open skies air services agreements[4] of the second; ground-handling, charter and air cargo liberalization of the third; and European Common Aviation Area (ECAA) and the Open Aviation Area (OAA) of the fourth.[5] We discuss these aspects of economic regulatory reform and conclude by reflecting on the accomplishments and failures.

US REGULATORY REFORM OF AIR TRANSPORT

As early as 1946 the state of California had had liberal views on new entry into its intrastate market, leading to 18 new airlines starting service over a 20-year period from 1946. Some of the proponents of deregulation, Levine (1965) and Jordan (1970), reported that liberalized policies in California resulted in the lowest air fares in the world, with large increases in passenger volume. The state of Texas similarly allowed price competition, but not free entry. Airlines like Air California (1967) and Southwest Airlines (1967[6]) had, therefore, experience with discount pricing in a competitive market before the US domestic market was deregulated in 1978.[7]

The political atmosphere before the 1978 Deregulation was one of disillusion with government, fuelled by the Watergate scandal, the Vietnam War and increases in taxation without general perception of improvement in services. These years marked a break with the post-Great Depression years and the Second World War, when strong government was seen as a protector of the people (Lowenfeld, 1981).

The advent of the oil embargo (1973) marked a hike in oil prices and furthered the decline in confidence, shifting political focus away from protectionism to consumer advocacy. It was in this environment that deregulation gained momentum. The economist Alfred E. Kahn[8] played a pivotal role in the efforts to deregulate air transport markets. When taking up a position as chair of the Civil Aeronautics Board (CAB), his agenda was to end its powers to regulate the economic aspects or air transport (Lowenfeld, 1981). The airlines had never been profitable according to Kahn (PBS, 2008); regulation prevented them from competing on price, yet they competed on frequency often flying half-empty planes that could be filled if discounts had been approved by the CAB. With recession in the mid-1970s, airlines had too much capacity. Even if the industry was opposed to deregulation, there was strong motivation to offer discount fares to fill extra capacity. According to Kahn, government had to get out of the way to open up the industry's ability to do so. Another factor was the need to curtail rising inflation by increasing competition and stimulating downward pressure on prices: deregulation was seen as counter-inflationary.

The Kennedy Hearings in 1975 drew attention to the 'route moratorium'. From 1969 the CAB[9] had not granted any new route rights; applications were processed but put on hold. The CAB acted by refusing to add further capacity by granting new route rights, unless there was evidence of inadequate service to the public, which was hardly ever the case. The 'route moratorium' was an effective barrier to entry, thus eliminating competition in the industry (Bailey, 2008; Lowenfeld, 1981). Consequently, the Kennedy Hearings concluded that the CAB's route policies were thwarting low-cost air transport offerings to the public. It was argued that costs and prices were negatively affected by lack of innovation, renewal and competition (Binder, 2007).

While CAB's policies may have stabilized the market share of major airlines, expectations had changed and stable market share of airlines was no longer considered sufficient for the public good. Permitting the airlines to set their own fares, it was argued, would push prices down toward the marginal cost of providing the service at the same time that profitability would go up through better resource utilisation and market segmentation (Kahn, in Oversight of Civil Aeronautics Board Practices and Procedures, 1975, pp. 97–9).

EU AIR TRANSPORT LIBERALIZATION, 1987–97

The most important stepping stones to unified air transport policy in the EU were: (1) the 1985 decision of the European Court of Justice which ruled that Article 81 was applicable, rendering approval of airfares by a member state unlawful (Joint Cases 209-213/84 [1986] ECR 1425);[10] (2) the 1985 Schengen Agreement and the 1990 Convention implementing the Schengen Agreement, enabling free border crossings (implemented in 1995); (3) the Commission's first White Paper on future development of common transport policy;[11] (4) the Bosman ruling[12] on free movement of people in 1995; and (5) the 'open skies' judgments in 2002, stating that member states cannot act in isolation when negotiating international air services agreements (ASA),[13] which meant that air services had to be treated as a subject of Community interest (COM, 2005). These events facilitated liberalization in the EU, by opening up markets.

Given the different legal, economic and political systems in Europe, liberalization is the most straightforward path for regulatory integration. Air transport was exempt from Article 81 of the Treaty, and it was not until the 1980s that the industry entered the jurisdiction of the European Commission. After the resolution of the so called Nouvelles Frontières case, the Commission had powers to attack price fixing, paving the way towards liberalization of air transport (Button, 2007). Given the prospect of full powers transferred to the Commission, the member states had no option but to agree on the Council to embark on a gradual liberalization of air transport. What had seemed impossible became a reality within a short period of time (Bernard, 2006).

The aim of EU air transport liberalization was to increase competition and increase consumer welfare through lower fares following the same economic arguments as in the US. The key difference was that the EU took a route of gradual liberalization, as opposed to full deregulation in one go. Liberalization actually came in three distinct packages: the *1st package* (1987) gave the airlines some flexibility to increase capacity and to adjust fares on EU cross border routes without bilateral negotiations; the *2nd package* (1990) allowed any airline to carry passengers to and from any other member state (third and fourth freedoms),[14] and to carry passengers between any third countries, with origin and destination in the home country (fifth freedom); and the *3rd package* (1993) opened market access, freedom to set air fares, eliminated capacity restrictions and set rules governing the licensing of air carriers. However, stand-alone domestic services in other member states (ninth freedom: cabotage) were protected until July 1997.

It was expected that liberalization would stimulate competition between new carriers and established full-service carriers as well as among the established carriers. The former has taken place on a large scale with low-cost airlines rapidly building market share.[15] However, the latter did not occur to any large extent. The incumbent airlines have more or less kept to their home base and made few attempts to gain foothold in other EU countries[16] or to merge.[17] One of the reasons was that extra-EU flights were still bound by air services agreements having nationality clauses that prevented a carrier operating flights to third countries from another EU country. Such agreements also limited carrier's operating flexibility and cross-border investment in the airlines. If non-discriminatory access to third countries was to be assured and the single air transport market was to function as a fully integrated entity, nationality restrictions in air services agreements had to be removed. The Commission argued for a mandate to conclude air services agreements from the Council. The Council mounted resistance to such a mandate, leaving the Commission to a legal process (Bernard, 2006). This process started with infringement actions against eight member states who had negotiated 'open skies' agreements with the USA, the first being signed with the Netherlands in 1992.[18] In the opinion of the Commission, based on the court ruling mentioned before, these countries, by reserving traffic rights for national carriers prevented free competition in the provision of international air transport between the EU and the USA, to the detriment of consumer welfare.[19]

As a step towards an integrated air transport policy through the infringement actions the Commission began reforms beyond the scope of traditional air services agreements. It thus announced an external air transport agenda with three pillars (COM, 2005): (1) legal certainty of existing bilateral air services agreements between the EU and third countries; (2) a European Common Aviation Area (ECAA[20]) developed for European countries that wish to integrate into the European Common Aviation Market and fit into the 'Neighbourhood Policy' of the Commission"[21]; and (3) comprehensive agreements with third countries.[22]

To ensure legal certainty after the open skies judgment, existing bilateral air services agreements could be amended in two ways: either through *bilateral negotiations* between each Member State and its partners, or through negotiation of *horizontal agreements* with the Commission acting on the basis of a mandate. Each horizontal agreement covers existing bilateral air services agreements in the context of a single negotiation with a third country. From 2003 to 2008 negotiations led to changes in air services agreements of 60 partner states, representing 132 bilateral agreements. During the same time horizontal agreements were signed with 37 states and one regional organization with eight states, that is, a total of 651 bilateral amendments.

The Common Aviation Area (CAA) with neighbouring countries promotes the broader objectives of the European Neighbourhood Policy (ENP) of stability and sustainable development. The CAA covers the eastern and southern borders of the EU.[23] The aim is to create an enlarged air transport market based on a common set of rules. The total size of this market would be in excess of 1 billion people (COM, 2008). The first comprehensive agreement with third countries was the Open Aviation Area (OAA) signed with the USA, that took effect on 30 March 2008. The European flag-carriers will, however, remain prudent due to national political interests curtailing aggressive competitive behaviour. The most likely outcome will be attempts to merge or operate flights out of alliance partner's airports.

INTERNATIONAL LIBERALIZATION OF AIR TRANSPORT

Air transport is unusual compared to most industries, being one of the most internationalized, yet, one of the least globalized. Alliances have been a tool to get around this limitation, but still, international air transport is governed by the so called 'Chicago regime' of restrictive bilateral agreements. The Chicago Convention was signed on 7 December 1944 and was to foster the development of international civil aviation 'in a safe and orderly manner'. Given the rapid development of aircraft during the Second World War, it was clear that air transport would grow rapidly because of vast improvements in technical development, and the introduction of long-haul aircraft capable of flying over multiple borders and countries without landing (Lowenfeld, 1981). Thus, the allied forces foresaw the need for future control of passenger and freight services. Rather than negotiating each new route, the USA and some other nations were keen to set a common basis on which agreements could be made.[24] This aim was not achieved, and all aspects of air transport commercial relations between nations still today have to be negotiated though bilateral agreements.

Bilateral agreements are losing some of their dominance in inter-country negotiations as other methods come into play, such as 'open skies' agreements, with the first such agreement being signed between the USA and the Netherlands in 1992. Open skies agreements are grounded in free-market principles. Thus, there are usually no restrictions on international route rights, number of designated airlines, capacity, frequencies and types of aircraft used. Pricing is determined by the market. A fare can be barred only if both governments coincide, so called 'double-disapproval pricing', but only for certain specified reasons. Furthermore, the 'Fair and Equal Opportunity to Compete' is upheld so carriers may establish sales offices in each other's countries and convert earnings and remit them without restrictions. Carriers are free to provide their own ground-handling services, or choose among competing providers. User charges are non-discriminatory and based on costs. Designated airlines may enter into 'cooperative marketing arrangements', that is, code-sharing or leasing arrangements with airlines of either country, or with those of third countries, subject to usual regulations.

LIBERALIZATION OF THE SUPPORTING INDUSTRY

Increased economic freedom in the air transport industry necessitated access to resources that were usually the grandfather right of incumbent carriers, namely, slots. The general approach both in the USA and the EU was to liberalize these supporting industries and in other cases to make sure that regulation provided access for competitors. Some of the competitive tools created for the new environment also became an issue of rule-making.

Ground Handling

In the USA, ground handling is performed almost entirely by the airlines themselves. Consequently, little was needed to level the market between different airlines. This was, however, not the case in Europe where the airports had almost a monopoly in the market with few self-handling airlines. In the early 1990s competition had been introduced at few airports, most notably Heathrow and Schiphol. At the dawn of full liberalization in the EU complaints about ground handling became prevalent (SH&E, 2002). Thus, it was soon realized that ground handling needed to be air-carrier neutral, in other words, no discriminatory advantage should occur through the control of ground handling in airports, either by carriers or through ground handling operated by the airports themselves.

In the arguments preceding the Directive, the Commission argued that full liberalization should cover all services that constituted contact with the customer and was integral to airline's brand image, that is, passenger handling, flight operations and crew administration, while partial liberalization was applicable to other services such as ramp and baggage handling (Soames, 1997).

The objective of the Ground Handling Directive of 1996[25] was to provide for the competitive provision of services, reduce airline costs, improve quality of service and provide choice of handling agents to airlines. The directive was implemented in stages over the period 1998 to 2001. All airports, regardless of size, were obliged to allow airlines to self-handle for land-side operations, ground administration, aircraft maintenance and surface transport. Airports with annual traffic of more than 1 million passengers or 25 000 tons of freight were obliged to allow air-side self-handling (refuelling, baggage handling, ramp operations, freight and mail handling). Third-party handling was allowed from January 1999 for airports with 3 million passengers or 75 000 tons of freight. From 2001 it was opened for those between 2 and 3 million passengers and 50 000 and 75 000 tons of freight. Restriction to two companies was allowed but one company, according to the directive, must be independent of the airport and the dominant carrier.

Airport Slot Allocation

A 'slot' is defined in the current EC Regulation 95/93 as 'the permission given by a coordinator . . . to use the full range of airport infrastructure necessary to operate an air service at a coordinated airport on a specific date and time for the purpose of landing or take-off as allocated by a coordinator'.

The key concern in allocating slots is optimum use, which often assumes more available seat kilometres (ASKs) per slot as criteria, and the allocation of slots to new entrants to open up competition on new routes (De Wit and Burghouwt, 2008). Congested airports have often used the voluntary International Air Transport Association (IATA) guidelines for slot allocation. These were developed to facilitate schedule coordination in the era of restrictive bilateral agreements, thus posing problems for new entrant airlines. Although, under the EC Regulation 95/93 and as amended by EC Regulation 793/2004, slots can only be allocated to and held by air carriers, the system was taken out of the hands of airlines and given to *slot co-ordinators*. Under the new system, primary allocation is made from the slot pool by the airport coordinator but subject to the recognition of historical precedence in respect of 'series' of slots. The coordinator is also required to take into account additional rules or guidelines established by the airline industry world wide or community-wide, as well as local guidelines proposed by the coordination committee. The regulation also assumes that such rules and guidelines do not affect the independent status of the coordinator and complies with the efficient use of airport capacity. Subject to historical precedence, up to 50 per cent of the slots in the pool shall be allocated to new entrants depending on demand. Since the regulation was put into place, the efficiency in use, as expressed in available seat kilometres per slot, has been greatly improved through replacing of small aircraft on short-haul routes by large aircraft on long-haul routes, thus stepping up the value generation of each slot and the impact on consumer welfare (UK CAA, 2001).

However, it was increasingly evident that the slot allocation regulation was interpreted too tightly. Consequently, an 'artificial' market of secondary slot trades formed, for example, at London Heathrow and London Gatwick. In 2008 the European Commission declared that the existing European Union legislation on airport slot allocation did not prohibit secondary slot trading (PR, 2008a). Taking the London airports as an example the Commission found that the system had shown its value,[26] where airlines have, through secondary trading of slots, been able to mount competition and, for instance, take advantage of opportunities provided by the EU-US Open Aviation Area.

In the USA the issues with slots were different. A US airline can stockpile airport facilities in order to block possible entry of other carriers. This can sometimes be circumvented by the entrant by establishing own facilities, at a cost disadvantage. At some airports this is not even possible as the incumbent may have in its lease a clause giving it ability to block any further construction at the airport. In view of the risk for the airport to have one dominant large carrier that may exit, many airports started to attract carriers by including a 'preferential use' clause in their leases. Such clause gives the airport authority the ability to lease to other airlines airport facilities not used by the leaseholder.

Under the High Density Rule, slots were distributed by Committees representing airlines and airport authorities. As new entrants demanded access to slot-controlled airports, the system crumbled as almost all slots were controlled by the incumbents. Due to this problem and the resulting deadlocks in allocating slots, the Department of Transportation (DoT) amended the High Density Rule in December 1985 by allowing slot trading (GAO, 1990).[27] Under the new system existing slots were allocated according to slot holdings in December 1985. However, in April 1986 the airlines could trade them subject to prior approval from the Federal Aviation Administration (FAA).

Airline Ticket Advertising

The US regulator soon recognized that advertising had become an issue as some airlines tended to emphasize the short-term benefit from capturing the customer's attention through deceptive advertising knowing that a certain per centage of those acting on advertisements will accept the 'true' value of the product even though it was described differently in the advertisement. The DoT in judging airline's adherence to Section 411 of the Federal Aviation Act follows a 'Statement of General Policy' Part 399.80, that lists the following as a breach of the code of conduct in airline advertising: (a) misrepresentation of the quality of service type or size of aircraft, departure times, points served, number of slots, and total trip time; (b) misrepresentation of fares and charges; and (c) misrepresentation of discounts stating that they are available when they are not. Furthermore, Part 399.84 declares advertisements which contain a lower fare than is actually charged to the customer, an unfair deceptive practice, but this practice was often used by the incumbents when a new entrant entered a market, offering a lower fare than prevailed. In addition, non-inclusion in the advertising of code-sharing, when present, is considered a deceptive or unfair practice as the passenger may not realize, unless notified, that there may be a carrier or equipment switch en route. Such practices were found to discriminate effectively against direct service new-entrant airlines.

To address these and other issues, the DoT administered airline advertising standards that constitute allowed *practices*: (a) taxes can be listed separately as long as they are included in the advertisement; (b) fares can be advertised as one-way even though a round-trip ticket is required to receive the advertised fare; (c) restrictions on fares must be listed in the advertisement; and (d) if a low fare is advertised a 'reasonable' number of seats must be made available at that fare. The DoT has found two main complaints on behalf of airline customers regarding advertising (AARA, 1991, pp. IX–X): (a) lack of reasonable number of seats at an advertised low fare (see practice d above); and (b) lack of adequacy of disclosure about restrictions on fares (see practice c above).

Looking towards Europe, regulation was issued in 2008 governing the operation of airline services,[28] laying down rules for the granting of licences, control of airlines and market access to ensure competition in air transport and better quality for the consumer. The regulation aims to improve transparency and enable passengers throughout the European Union to be better informed about prices and to compare offers with all taxes and charges disclosed in published ticket prices in a transparent way. For instance, price discrimination based on place of residence was banned (PR, 2008b). Airlines are consequently obliged to advertise final prices of flights throughout the entire advertising and reservation process, and uniformly over all member states.

Computer Reservation Systems

Computer reservation systems (CRS) could be biased in favour of the owners' airlines (hosts), especially on the actual display itself. To level the playing field, the CAB issued rules in 1984 that corrected this display biases to a certain extent.[29] Nevertheless, the systems continued to generate incremental revenues by a combination of 'architectural bias' and travel agents' (TA) identification with the CRS's host airline, called 'halo effect'. Certain CRS owners included, in their CRS agreements with TAs, clauses that 'tie' the owner's commission levels to the CRS usage. Airlines and travel agents felt increasingly uneasy towards the CRS hosts' ability to control competition, leading to a lawsuit brought against American and United. The plaintiffs alleged that the CRS's hosts were taking advantage of their dominant position. The USA's Department of Transportation

(DoT, 1988) analysed the substance of incremental revenues and found that (ACEA, 1992, p. 147) 'airline revenues in 1986 were about 14 per cent higher for United and 15 per cent higher for American because of incremental effects. These were only moderately lower than the estimates of incremental revenues for the pre-rule period'.

When profits and incremental revenues were added together, it was clear that the CRS's contribution to American and United Airlines had major importance in their growth and competitive position in the market (ACEA, 1992). The whole question of CRS's bias revolved first and foremost around the possible effects on the competitiveness of new entrants and non-owning airlines. One must point out that the 1984 ruling on CRS's bias in the USA was not beneficial to all new entrants, because many smaller carriers relied on connecting traffic (interlining) from the majors. What happened was that connecting flights fell into the third category of screen priority. This caused a serious drop in bookings as TAs had a tendency to book a high proportion of flights from the first screen instead of scrolling through all the screens before making a selection.

In Europe two competing reservations systems emerged: Amadeus[30] and Galileo.[31] All the same concerns as those in the USA were addressed by the regulator through Regulation 2299/1989, which is a Code of Conduct for Computer Reservation Systems. The EC was concerned about the uneven distribution of access to CRSs as regional dominance of the systems differed substantially globally. This had impact on pricing and underlined the importance of competition between the systems. However, the most effective competition was not from within the CRS industry but rather from the Internet that started to replace the CRS as primary distributor of travel products, especially for low-cost carriers that were able to mount effective competition entirely through internet bookings. Because of the Internet and because of de-hosting (separated from owning airlines) of the systems the CRS issue is no longer the threat to competition it used to be. Furthermore, the US DoT eliminated most of the rules as of 31 July 2004. In 2007 the European Commission attempted to follow suit by proposing a revision of the Code of Conduct for computerised reservation systems (COM, 2007).

INDUSTRY STRUCTURE

Starting in the 1950s evidence accumulated showing that government policies did not facilitate the efficiency and income distribution principles popularized in the decades after the Great Depression (Joskow and Noll, 1981). Consequently, voices questioning the need for economic regulation got more and more prominent. The absence of scale economies, industry concentration, market entry, predatory behaviour, subsidies and government ownership are six interlinked issues (the last two primarily a problem in the EU) of the deregulation debate. In the following section we examine these in more detail.

Firm-level Economies

According to Caves (1962), *economies of firm scale* are not considered to be present in the airline industry. After industry consolidation, which peaked in 1987, questions arose if there were constant returns to scale in the industry (Dempsey and Goetz, 1992; Kahn, 1988). Oum and Zhang (1997) used the same data as Caves et al. (1984, 1985) and found only mildly increasing returns to scale in the airlines and constant returns to scale in the railways, thus supporting the general premise of limited economies of scale in the industry. The importance of economies of scale lies principally in the rationale behind industry consolidation, while if there are limited scale-economies, consolidation is rather explained on the basis of market power (Oum and Zhang, 1997). Furthermore, limited economies of scale would lower barriers to entry, especially given the fact that the aircraft, the production plant of the airline, is a mobile asset.

Concentration

Many mergers and bankruptcies in the USA led to controversy as to the benefit of deregulation for the consumer and the airlines. In general, the view was that the industry had become concentrated with respect to the incumbent carriers, while waves of new entry have occurred since deregulation, causing varying overall concentration in the market. Although mergers have been highly prevalent as a result of the deregulated US market, it has been less so in the liberalized EU market, for reasons already stated. However, increasingly it appears that the hurdles for effective airline mergers in the EU are being removed as nationality clauses in air services agreements are being amended. The Air France–KLM merger, through a holding of the two respective airlines, has paved the way for such unions that are likely to occur in the years to come, most likely stimulated by economic downturns and growth constraints.

Market Entry

New-entrant airlines started to appear soon after deregulation in the USA and liberalization in the EU, stimulating segmentation in the markets. However, what many did not foresee in the post-deregulation era was the scale of failures among both incumbent and new carriers; low cost was not, on its own, sufficient to secure sustainable competitive advantage. The higher cost incumbents came up with ingenious innovations, such as hub and spoke networks, yield management systems, computer reservations systems, and frequent flyer programmes, all contributing to raise competition barriers and to reduce the singular impact of a cost advantage in the fight for customers. In addition, infrastructure barriers, due to fast industry growth, limited the availability of valuable scarce resources like terminal access and slots, making competition less viable or simply impossible in some markets.

Predatory Practices

The US Department of Justice, in 1999, issued a complaint against AMR Corporation and its subsidiary American Airlines, of predatory pricing against three low-cost carriers. The complaint alleged that capacity and pricing on four routes were below cost and that the carrier intended to recuperate its losses by monopolizing these four routes. Having lost the case, the US Department of Justice filed an appeal. In 2003, the US Court of

Appeals upheld the District Court's summary judgment, issuing the following in the process: 'Because it is uncontested that American did not price below AVC (average variable cost) for any route as a whole, we agree with the district court's conclusion that the government has not succeeded in establishing the first element of Brooke Group, pricing below an appropriate measure of cost.'[32]

According to Brady and Cunningham (2001) the courts have predominantly departed from a focus on 'intent' to 'economic analysis'. The appropriate measure of cost has consequently led to substantial debate and a stream of research (Baumol, 1996; Dempsey, 1997; Guiltinan, 1996) on how to determine costs. The most adopted approach in academia is the marginal cost measure, while the courts apply various approaches (Brady and Cunningham, 2001). The airline industry, being a services industry with a perishable product, poses specific problems due to the application of yield management systems. Court cases on alleged predatory pricing are consequently difficult to pursue and rarely, if ever, has predation been proven in the industry.

Subsidies and Privatization

One of the key biasing factors in the European liberalization process was the strong inclination by some member states to subsidize their carriers. Unlike the USA where airlines had never been state owned, most European airlines were or had been in state ownership. Consequently, the playing ground had to be levelled by blocking state aid, on the one hand, and increasing the financial viability of the carriers, on the other. State aid obviously has the capacity of blocking competition by raising barriers for more efficient carriers. In 1993 only British Airways was fully privatized, which privatization was preceded by a publicly financed restructuring in 1987 (Graham, 1997).

To shape policy on the issue of subsidies, the EC appointed the Comite des Sages in 1993, which advised one final injection of capital to be allowed, given a number of conditions. Consequently, a number of ailing carriers at the time received substantial subsidies, and in the period from 1991 until 1997, a total of €7.7 billion was approved to various airlines by the Commission (Lawton, 1999). State aid has been defined according to article 87 of the Treaty as state grants, interest relief, tax relief or relief of airport charges, state guarantee or holding, and provision by the state of goods and services on preferential terms.[33] However, if an airline is owned by the state, it is still possible for the 'owner' to provide an 'investment' if a private investor would have behaved in the same way. In most cases arguments have risen over state-owned carriers receiving state aid.

THE BENEFITS

A question that is on top of the minds of deregulation advocates and critics is if there has really been a benefit to the public. With increased efficiency, airlines can offer lower prices and with relaxed entry barriers more efficient firms enter the industry leading to increased competition. Many studies have attempted to examine these factors from different perspectives.

Increased Efficiency and Growth

Doganis (2006) estimated that the cumulated cost advantage of the new carriers was 49 per cent, with the best performing carrier, Ryanair, having 62 per cent cost advantage per seat-km over British Airways, and fares to match.

Seabright and Ng (2001) studied European carriers, and found cost reduction had occurred leading to dissipating difference between the costs of US and EU carriers. According to their findings, employees did comparatively better in Europe than in the USA, pilots earning 37 per cent, and cabin staff 58 per cent more. They found that 10 per cent reduction in public ownership led to 6.5 per cent reduction in airline costs. What is more, airlines facing competition from low-cost airlines experienced cost reduction if effective competition took place. Otherwise costs rose in the fallout of reduced market share.

New Entry and Exit

The market share of new entrants in the USA increased from 1978 to 1985 from 3 to 10.4 per cent of overall market share (Gudmundsson, 1998). During the same period, the former local carriers went from 9.1 per cent share in 1978 to 13.1 per cent in 1985.[34] This was reflected in a General Accountability Office (GAO, 1996) study reporting a 50 per cent increase in flights to small communities looking at 87 small to mid-sized communities in the USA.

In Europe, between 2003 and 2006, low-cost carriers' market share increased by 40 per cent to 20 per cent overall share, but the incumbent airlines lost 10 per cent market share bringing the total share down to 55 per cent. Another parallel development, stimulated by the advent of low-cost airlines, was the reduction in premium intra-EU passenger traffic (33 per cent between 2001 and 2005) (AEA, 2007). Inter-city connectivity in Europe has improved with gains between 1996 and 2004 of 40 per cent in cities served, and 91 per cent in city-pairs served, something which is mostly attributed to low-cost carriers (Fan, 2006). The number of city pairs served by low-cost airlines went up from 17 to 266, while full service carriers increased city-pairs served from 216 to 233. Fan (2006) reports that of 195 city-pairs served exclusively by low-cost carriers in 2004, only 30 were served before 1996.

Lower Fares

Trend in yields for the Association of European Airlines (AEA) carriers has been steadily declining over the past decade. According to the AEA (2007), EU domestic yields ('intra-State') have declined less over the years, while international ('extra-EU') yields have declined the most. A US Government Accountability Office (GAO, 2006) study reports a 40 per cent reduction in median fares from 1980 to 2005. The reduction is proportionally higher on the long-distance routes and heavily travelled hub-to-hub markets, compared with shorter-distance, less travelled markets. An earlier research (Borenstein and Rose, 1994) found that fares dispersion, measured as GINI Coefficient,[35] varied from 3.6 per cent to 83 per cent on the same airline and same route; while dispersion varied highly between airlines and on routes. They found that on routes served by more than one carrier there was lower dispersion in fares between carriers, if frequency was a constant, otherwise dispersion increased.

Since deregulation in the USA, real yields per passenger mile[36] (RPM) have declined from 8.49 in 1978 to 4.0 cents in 2006 (in 1978 prices), a 53 per cent reduction (it was 10 cents in 1971). The causes were primarily increased productivity of aircraft fleets and the influence of increased competition after 1978. Although in the 1981–82 and 1990–91 recessions, airline costs increased due to rising fuel costs with associated fare increases (Morrison, 1995). Morrison (1995) showed that decline in yields attributed to deregulation, over the first fifteen years, was about 22 per cent greater than if regulation had continued. A consumer welfare gain averaging about $12.4 billion (1993 dollars) annually. Gaynor and Trapani (1994) calculated consumer surplus for 70 city-pair markets and found that consumers were better off in 49 markets and worse off in 21 markets, with average welfare gains per market of $4.3 million.

Increased Competition

The GAO reports that markets became more competitive with deregulation, with average number of competitors increasing from 2.2 per market in 1980 to 3.5 in 2005. Borenstein (1992), on the other hand, in an earlier study, found that from 1977 until 1990, the industry became more concentrated with the Herfindahl-Hirschman Index[37] (HHI) rising from 0.106 (1977) to 0.121 (1990). Maldutis (1993) reports similar findings: airlines with more than 1 per cent market share at US airports were 25 in 1979 but only 15 in 1991; at the same time, total enplanements increased from 310.8 million to 346.5 million with the largest peak in 1986, at 389.7 million. The HHI shows that the weighted average for the 50 largest airports increased from 0.222 in 1979 to 0.390 in 1991 (Maldutis, 1993). In city-pair markets (all trips) 0–500 miles there appears to have been increased concentration, while in markets 501–1500 or more miles competition has increased (Borenstein, 1992). Overall, the average city-pair HHI went from 0.531 (1984) to 0.506 (1990).

Airline competition in Europe has increased under liberalization, but the proportion of monopoly routes has remained the same between 1994 and 2006, that is, 71 per cent. However, the increase in the number of routes served from 1994 until 2006 showed that one-airline-routes increased by 60 per cent, two-airline-routes by 20 per cent, three-airline-routes by 270 per cent and four-airline routes or more by 220 per cent (EC, 2006). These figures demonstrate that competition has increased, although growth was heavily attributed to the low-cost airlines that offered affordable fares to new destinations, a similar story to that in the USA.

CONCLUSIONS

Many observers of liberalized air transport markets expect stability, that is, some sort of equilibrium to form over time. Airline executives often view industry concentration as inevitable where only few players will survive. However, air transport markets are expected to have relatively low entry and exit costs (Kahn, 1971). Button (1996, p.277) reasons the non-existence of a natural market equilibrium in air transport, so called empty core 'whenever there is short-run excess capacity, there is unlikely to be a competitive equilibrium'. In such a situation, the resources are allocated inefficiently and actors will attempt to cooperate in their activities or merge (Button, 1996; Gillen, 2006).[38] Concentration means opportunities for entry as remaining players pursue cost restructuring, while periods of high rivalry mean loss of profitability. Evidence points to deregulated air transport markets having cycles of rivalry, stimulated by economic upturns and downturns in addition to extraordinary events. Is regulation the only way to secure stability or will airlines in their decisions counter market instabilities? Mergers and alliances are such stabilizing forces and the regulator has in most cases approved those with some competition enhancing concessions. Although rivalry has been reduced at times through alliances and regulation raising barriers to entry (Gudmundsson and Kranenburg, 2002), a full-scale re-regulation will achieve neither financial stability nor raise consumer benefits. A regulated industry may react to increased fuel prices by arguing for higher prices from the regulator, but if the regulator refuses, the firms will not have the capacity to respond effectively and financial health may deteriorate (Joskow, 1974; Winston, 1998).

Competitive markets are messy and unstable (Kahn, 1971, 1988). Kahn (1988) talks about more entrants, labour disputes, price wars, bankruptcies, consolidation and losses, than he had expected at the outset of deregulation. What was clear was that reorganization was necessary of almost every aspect of the industry to shed inefficiencies associated with regulation (Winston, 1998). Management practice had to change from a set of relatively fixed focal points to a very dynamic environment (Gudmundsson, 1997). Many successes, small and large, were made and many failures occurred when trying to adjust to the new environment. The one carrier that can be labelled a 'deregulation success' is Southwest Airlines (1971), which carved out a clear understanding of principal success factors, one of which has been controlled growth to retain financial stability. Almost all other airlines have pursued uncontrolled growth and consequential contraction or collapse when the external environment changed. Why have these airlines not called for reregulation in the pursuit of stability? According to Levine (2006, p. 2):

> The explanation appears to lie in the fact that comprehensive regulation tended to make firms alike and align their political interests, and deregulation allowed many different players and strategies to emerge, benefiting some and harming others. Accordingly, the industry was unable to achieve the level of cohesion necessary to undo a popular legislative program.

Airline deregulation was designed to benefit consumers that tend to be less organized than other industry stakeholders (Levine, 2006). However, the majority of airlines foresaw benefit in breaking the industry stalemate that was created through CAB policies at the time. Hence, the airlines provided only a lukewarm opposition to deregulation. The homogeneity that comes with regulated industries is lost and, consequently, no one group is likely to pursue re-regulation for the benefit of all (Levine, 2006). What is more, reregulation is unlikely to stabilize the external forces such as oil prices or economic cyclicality or even service levels that are a frequent focus of complaints. The air transport industry has grown rapidly in past decades causing increased strain on infrastructure. Service disruptions associated with congestion will become more frequent and so will environmental constraints. How the airlines deal with ad hoc events and crises certainly is a service issue. None of these primarily external factors, however, justifies economic reregulation of the industry in the interest of consumers. Consumers predominantly want affordable prices and convenient departures. On these two aspects deregulation has scored high for most, as both the USA and European cases show. Overall (GAO, 2006, p. 1) '[t]he evidence suggests that reregulation of airline entry and fares would likely reverse much of the benefits that consumers have gained'. Morrison (1995), in an earlier account, admits that deregulation has its imperfections, but is still preferable to regulation with its flaws. Analysts will continue to disagree over implementation and events, but (Morrison, 1995, p. 33) 'returning the industry to traditional rate and entry regulation would at best be premature and at worst a mistake'.

NOTES

1. Australia, New Zealand, South-Africa, Chile (1979), Costa Rica, Egypt, Tanzania, Uganda, United Arab Emirates.
2. European Union, European Aviation Area, Open Aviation Area [EU-US], Air Transport Agreement among the members states and associate members of the Association of Caribbean States.
3. Public Law No. 95-504.
4. In 1992 The Netherlands signed the first open skies agreement with the USA, despite objections by the European Commission, giving both countries unrestricted landing rights on each others' soil.
5. Eight states have ratified the multilateral air transport agreement among Caribbean states that entered into force in November 1998, marking an important step towards the establishment of a single market for air transport services in the Caribbean region.
6. Started operations 18 June 1971. The idea behind Southwest Airline's operations, as an exemplary post deregulation success story, was not unique, Pacific Southwest Airlines (PSA), which began operations in 1949, probably initiated the philosophy behind the low-fare, no-frills concept (Meyer et al., 1984).
7. Although, PSA (founded 1949; merged with USAir in 1988) and Air California had monopoly status on some intrastate routes that were not served by the interstate carriers, they were in competition on many intrastate routes. Pacific Southwest Airlines had experienced success against the large regulated interstate carriers in California, but with the advent of Air California its success dissipated. The California Public Utilities Commission had been inclined to approve only rates providing 'reasonable' rate of return. Moreover, the Commission had divided markets between PSA and Air California on routes not served by the interstate carriers. Thus, PSA was forced by the Public Utilities Commission to maintain its low fare structure prior to deregulation, but after deregulation it immediately used the opportunity to raise fares to match its high costs. Thus, PSA saw deregulation primarily as an instrument for raising fares and a way to abandon its 'forced' fare strategy; while Southwest Airlines and Air California kept its fare strategy as operating philosophy (Lloyd's Aviation Economist, 1984, p. 29).
8. Alfred E. Kahn (1917–) is the Robert Julius Thorne Professor Emeritus of Political Economy at Cornell University.
9. The Civil Aeronautics Board (CAB) was established through the Civil Aeronautics Act passed in 1938. The Board's main function was the granting of 'license' to airlines to operate specific routes. Once a route licence was granted the CAB had limited powers to revoke the licence. The Act gave the CAB authority to set fares in domestic air transportation in response to proposals from the airlines. It was not allowed to restrict schedules nor prescribe or control equipment used. All airlines in existence were given the right to operate after the Act was passed through the so called 'grandfather clause'.
10. In the case the French authorities attempted to prevent a travel agent to offer discount air tickets.
11. The Commission's first White Paper on the future development of the common transport policy was published in December 1992. The document's aim was the opening up of the transport market, which has been generally achieved, except perhaps for rail sector.
12. In 1995 European Court of Justice ruled on the direct effect of article 39[1] of the EC Treaty concerning worker's freedom of movement and association. The case had a profound effect on the transfers of football players within the EU, by banning restrictions of foreign EU members within the national leagues and allowed professional football players to move freely to another club at the end of their contract [Union Royal Belges des Sociétés de Football Association ASBL & others v. Jean-Marc Bosman; Case C-415/93, ECR I-4921].
13. The 'open sky' ruling delivered in November 2002, invalidated agreements concluded between the USA and Austria, Belgium, Denmark, Finland, Germany, Luxembourg, the UK and Sweden (cases C-466/98, C-467/98, C-468/98, C-469/98, C-471/98, C-472/98, C-475/98 and C-476/98). The EU Court of Justice on 24 April 2007 declared the open sky agreement concluded in 1992 between the USA and the Netherlands

also illegal (Case C-523/04). In November last year the European Court of Justice delivered its ruling in the so-called 'open skies' cases concerning the bilateral air services agreements between a number of EU member states and the USA. The Court found that the 'nationality clauses' in bilateral agreements, which allows the USA to refuse to accept the designation by any airline in which substantial ownership and effective control is not vested in nationals, was in breach of Article 43 of the EC Treaty, which allows nationals of a member state to operate a business in another member state under the same conditions as its own nationals.

14. The Freedoms of the air are aviation rights granting airlines the privilege to enter and land in another country either for technical (Freedoms 1 and 2) or commercial reasons (Freedoms 3 and 4): formulated in the Convention on International Civil Aviation of 1944, (known as the Chicago Convention). The convention succeeded in drawing up an agreement for the technical freedoms but the commercial freedoms became an issue of negotiation between states in bilateral agreements.
15. In 2007 low-cost airlines had 23 per cent market share in Europe, up 92 per cent from 2003 (AEA, 2007)
16. British Airways started its inroads to mainland Europe through a 49 per cent stake in Delta Air, a regional airline based at Friedrichshafen, and renamed it Deutsche BA in 1992. In April 1997 BA acquired the remaining shares. It was acquired by Air Berlin in August 2006, but continued to operate independently, marketed as Air Berlin until being disolved by its parent company Air Berlin on 30 November 2008. In 1997 British Airways acquired a 70 per cent stake in Air Liberté (ceased 2001) and TAT that was gradually taken over in the period 1996 to 1997 and merged with Air Liberté. The merged entity was sold on to the Swissair (SAirGroup) in 2001, which in turn merged it with AOM.
17. The exemption, for large carriers, was the 1995 investment of Swissair (bankrupt in 2001) in SABENA (bankrupt in 2001) for a 49 per cent stake; and Air France-KLM created by a merger between Air France and KLM on 5 May 2004. The French government's share of Air France was reduced from 54.4 per cent to 44 per cent of Air France-KLM. Later its share was reduced to 25 per cent, and then to 18.6 per cent.
18. On 5 November 2002 the European Court of Justice issued a judgment in the 'open skies' cases against eight member states (Austria, Belgium, Denmark, Finland, Germany, Luxembourg, Sweden and the UK). The Open Sky agreement concluded in 1992 between the USA and the Netherlands was, however, not declared illegal by the EU Court of Justice until 24 April, 2007 (Case C-523/04).
19. Press release no. 116/04, 20 July 2004.
20. The ECAA was first signed in June 2006 with Norway, Iceland and the Western Balkan States. To be fully implemented by 2010.
21. As of 2007, 35 countries belonged to the ECAA, covering an estimated 500 million people. See: http://ec.europa.eu/. . ./2006_05_30_annexe1_ecaa_en.pdf.
22. The first such agreement was the Open Aviation Area (OAA) agreement between the EU and the USA which was signed on 30 April 2007.
23. Albania, Algeria, Armenia, Azerbaijan, Belarus, Bosnia and Herzegovina, Croatia, Egypt, Georgia, Iceland, Israel, Jordan, Kazakhstan, Kyrgizstan, Lebanon, Libya, Republic of Moldova, Montenegro, Morocco, Norway, Palestine, Russia, Switzerland, Syria, Tajikistan, Tunisia, Turkey, Turkmenistan, Ukraine, Uzbekistan, Macedonia and Serbia, and Kosovo.
24. All allies and neutral countries attended the Chicago meeting although Russia was not participating in discussions involving the sovereignty of their airspace and left the meeting. Thus, 54 'allied' and neutral nations attended.
25. Directive 96/67.
26. The experience of secondary trading at London airports mirrors that of the USA (1) leads to a liquid and flexible market in slots; (2) is effective in fostering new entry; (3) is supported by the industry, and has proved to be an active market; (4) enables slot leasing; (5) competitors prepared to trade slots with each other freely; (6) the use of slots as security in financing has not developed, nor has there been the development of different regulations for each congested (UK) airport
27. Slot sales have decreased since trading was activated but leasing increased, which indicates that the incumbents will retain control in order to maintain this barrier to entry, as reported in the report from GAO (1990).
28. Regulation 1008/2008.
29. The CRS regulation of 1984, addressed four main issues (ACEA, 1992, p. 2): (1) Prohibiting bias in the listing of airline schedules on CRS's screens; (2) Requiring CRS owners to charge all airlines the same 'booking fee' for listing their schedules and issuing tickets through the CRS; (3) Prohibiting CRS owners from leasing CRSs to travel agents for terms of more than five years; and (4) Prohibiting CRS owners from requiring that a travel agent use a particular CRS system exclusively.
30. Founded in 1987 by Air France, Iberia, Lufthansa, and SAS.
31. Founded in 1993 by Aer Lingus, Air Canada, Alitalia, Austrian Airlines, British Airways, KLM Royal Dutch Airlines, Olympic Airlines, Swissair, TAP Air Portugal, United Airlines, and US Airways.
32. U.S. v. AMR Corp., 10th Cir., No. 01-3202, 7/3/03.

33. See Council Regulation (EC) NO 659/1999 of March, 1999 [Art. 3, para. 1. (iii) of Annex III].
34. America West was the first new airline to reach one billion dollars in sales, achieved in 1990, just before its first Chapter 11 bankruptcy, while People Express was just about to reach the same mark ($928 million in 1985) when it was acquired in 1987 by Continental Airlines. The new carriers were generally struggling financially although many showed spectacular revenue growth.
35. The Gini coefficient is a measure of statistical dispersion, defined as a ratio with values between 0 and 1. A low Gini coefficient indicates more equal price distribution, while a high Gini coefficient indicates more unequal distribution. 0 corresponds to perfect equality (everyone offering exactly the same price) and 1 corresponds to perfect inequality (where one person has all the revenue, while everyone else has zero revenue).
36. The data reflects the activity of US passenger airlines as defined by the US Department of Transportation under Chapter 411 of Title 49 of the US Code (Air Transport Association, 2008).
37. HHI below 0.1 (or 1,000) indicates an unconcentrated index; HHI between 0.1 to 0.18 (or 1,000 to 1,800) indicates moderate concentration; HHI above 0.18 (above 1800) indicates high concentration.
38. Gillen (2006) offers the view that the exogeneous market structure is imposed on the empty core as opposed to being derived from decisions being made. So, for the empty core to perpetuate entrants must be uninformed or optimistic.

REFERENCES

Air Transport Association of America (ATA) (2008), 'Airline bankruptcy and service cessations': www.airlines.org/economics/specialtopics.

Airline Advertising Reform Act (AARA) (1991), (H.R.5124) Hearing before the Subcommittee on Aviation of the Comm. on Publ. Works and Transp., House of Representatives, 102nd Congr., 2nd session, Government Printing Office, Washington, DC.

Airline Competition Enhancement Act of 1992 (ACEA) (1992). Hearing before the Subcommittee on Aviation of the Committee on Public Works and Transportation House of Representatives, U.S. Government Printing Office, Washington, DC, 18 June.

Association of European Airlines (AEA) (2007), *Operating Economy of AEA Airlines*, December, available at: http://files.aea.be/RIG/Economics/DL/SumRep07.pdf (accessed 13 January 2009).

Bailey, E. (2008), 'Air transportation deregulation', American Economic Association, annual conference, January: www.aeaweb.org/annual_mtg_papers/2008/2008_264.pdf (accessed 17 January 2009).

Bailey, E., Graham, D.R. and Kaplan, D.P. (1985), *Deregulating the Airlines*, Cambridge, MA: MIT Press.

Baumol, W.J. (1996), 'Predation and the logic of the average variable cost test', *Journal of Law and Economics*, **39** (1), 49–72.

Bernard, N. (2006), 'Internal market governance in a globalised marketplace: the case of air transport', in N.N. Shuibhne (ed.), *Regulating the Internal Market*, Cheltenham, UK and Northampton, MA, USA: Edward Elgar Publishing, pp. 302–21.

Binder, D. (2007), 'Meaty, beaty, big and monopolistic: the story of General Motors', http://ssrn.com/abstract=989823 (accessed 3 February 2009).

Borenstein, S. (1992), 'The evolution of U.S. airline competition', *Journal of Economic Perspectives*, **6** (2), 45–73.

Borenstein, S. and N.L. Rose (1994), 'Competition and price dispersion in the US airline industry', *Journal of Political Economic*, **102** (4), 653–83.

Brady, S.P. and Cunningham, W.A. (2001), 'Exploring predatory pricing in the airline industry', *Transportation Journal*, **41** (1), 5–15.

Button, K. (1996), 'Liberalising European aviation: is there an empty core problem', *Journal of Transport Economics and Policy*, **30** (3), 275–91.

Button, K. (2007), 'Transport policy', in A.M. El-Agraa (ed.), *The European Union: Economics and Policies*, Cambridge: Cambridge University Press, pp. 298–313.

Caves, D.W. (1962), *Air Transport and its Regulators*, Cambridge, MA: Harvard University Press.

Caves, D.W., Christensen, L.R., and Thretheway, M.W. (1984), 'Economies of density versus economies of scale: why trunk and local service airline costs differ', *Rand Journal of Economics*, **15**, 471–89.

Caves, D.W., Christensen, L.R., Thretheway, M.W. and Windle, R.J. (1985), 'Network effects and the measurement of returns to scale and density for US railroads', in A.F. Daugherty (ed.), *Analytical Studies in Transport Economics*, Cambridge: Cambridge University Press, pp. 97–120.

Communication from the Commission (COM) (2005), 'Developing the agenda for the Community's external aviation policy', Communication from the Commission of 11 March 2005 – Developing the agenda for the Community's external aviation policy, no. 79.

Communication from the Commission (COM) (2008), 'Common aviation area with the neighbouring countries by 2010 – progress report', no. 596.
Communication of the Community (COM) (2007), 'Proposal for regulation of the European Parliament and the Council on a Code of Conduct for computerised reservation systems', no. 709, Brussels, 15 November 2007.
De Wit, J. and Burghouwt, G. (2008), 'Slot allocation and use at hub airports, perspectives for secondary trading', *European Journal of Transport and Infrastructure Research*, **8** (2), 147–64.
Dempsey, P.S. (1997), 'Testimony before the Full Committee on the Judiciary, U.S. House of Representatives', Washington, DC: GPO.
Dempsey, P.S. and Goetz, A.R. (1992), *Airline Deregulation and Laissez-faire Mythology*, Westport, CT: Greenwood.
Department of Transportation (DoT) (1988), 'Study of airline computer reservation systems' (DOT-P-37-88-2), May.
Doganis, R. (2006), *The Airline Business*, 2nd edn, London: Routledge.
Done, K. (2008), 'Air France to drop Heathrow-LA service', *Financial News*, 9 October.
European Commission – DG Energy and Transport (EC) (2006), 'Analysis of the EU air transport industry: final report 2006', Contract no. TREN/05/MDI/S07.52077.
Fan, T. (2006), 'Improvements in intra-European inter-city flight connectivity: 1996–2004', *Journal of Transport Geography*, **14**, 273–86.
Gaynor, M. and Trapani, J.M. (1994), 'Quantity, quality and the welfare effects of US airline deregulation', *Applied Economics*, **26**, 543–50.
General Accountability Office (GAO) (1990), 'Airline competition: industry operating and marketing practices limit market entry', GAO/RCED-90-147.
General Accountability Office (GAO) (1996), 'Certification of new airlines', GAO/RCED-96-8.
General Accountability Office (GAO) (2006), 'Reregulating the airline industry would likely reverse consumer benefits and not save pensions', GAO-06-630.
Gillen, D. (2006), 'Airline business models and networks: regulation, competition and evolution in airline markets', *Review of Network Economics*, **5** (4), 366–85.
Graham, B. (1997), 'Air transport liberalization in the European Union: an assessment', *Regional Studies*, **31** (8), 807–22.
Gudmundsson, S.V. (1997), 'The difference between European and U.S. airline management practice', *Journal of Air Transport Management*, **2** (3), 75–82.
Gudmundsson, S.V. (1998), *Flying too Close to the Sun: The Success and Failure of the New-Entrant Airlines*, Aldershot: Ashgate.
Gudmundsson, S.V. and Kranenburg, H.L. van (2002), 'New airline entry rates in deregulated air transport markets', *Transportation Research Part E*, **38**, 205–19.
Guiltinan, J.P. (1996), 'Aggressive and predatory pricing: a framework for analysis', *Journal of Marketing*, **60** (3), 87–102.
Jordan, W. (1970), *Airline Regulation in America: Effections and Imperfections*, Baltimore, MA: Johns Hopkins Press.
Joskow, P.L. (1974), 'Inflation and environmental concern: structural change in the process of public utility price regulation', *Journal of Law and Economics*, **17** (2), 291–327.
Joskow, P.L. and Noll, R.C. (1981), 'Regulation in theory and practice: an overview', in G. Fromm (ed.), *Studies in Public Regulation*, Cambridge MA: MIT Press, pp. 1–65.
Kahn, A. (1971), *The Economics of Regulation, II: Institutional Issues*, New York: Wiley and Sons.
Kahn, A. (1988), 'Surprises of airline deregulation', *American Economic Review. Papers and Proceedings*, May, 316–22.
Lawton, T.C. (1999), 'Governing the skies: conditions for the Europeanisation of airline policy', *Journal of Public Policy*, **19** (1), 91–112.
Levine, M.E. (1965), 'Is regulation necessary? California air transportation and national regulatory policy', *Yale Law Journal*, **74**, 1416–47.
Levine, M.E. (2006), 'Why weren't the airlines reregulated?', NYU, Law and Economics Research Paper no. 06-18: http://ssrn.com/abstract=896568.
Lloyd's Aviation Economist (1984), 'Southwest Airlines: the alternative route to low costs', October, pp. 28–9.
Lowenfeld, A.F. (1981), *Aviation Law: Cases and Materials*, 2nd edn, New York: Matthew Bender.
Maldutis, J. (1993), In: Financial Condition of the Airline Industry Committee on Public Works and Transportation, Additions to the record: February.
Meyer, J.R., Clinton, V. and Oster, C.V. (1984), *Deregulation and the New Airline Entrepreneurs*, Cambridge, MA: MIT Press.
Morrison, S.A. (1995), 'The effects of airline deregulation in the United States', *Economic Analysis*, **143**, 15–35. Economic Research Institute Economic Planning Agency, Tokyo, Japan.

Oum, T.H. and Zhang, Y. (1997), 'A note on economies of scale in transportation', *Journal of Transport Economics and Policy*, **31** (3), 309–15.
Oversight of Civil Aeronautics Board Practices and Procedures (1975), Hearings before Sub-comm. On Administrative Practice and Procedure of Senate Comm. On the Judiciary, 94th Congress, 1st Session (Kennedy Hearings).
Press release (PR) (2008a), 'Airport slot allocation: The Commission clarifies the existing rules', European Commission, Brussels, 30 April, no. 672.
Press release (PR) (2008b), 'Price transparency and improved safety: more competition and better quality for the air transport sector', European Commission, Brussels, 30 October, no. 1603.
Public Broadcasting Service (PBS) (2008), Alfred E. Kahn interview, http://www.pbs.org/fmc/interviews/kahn.htm
Seabright, P. and Ng, C. (2001), 'Competition, privatisation and productive efficiency: evidence from the airline industry', *Economic Journal*, **111** (473), 591–619.
SH&E (2002), Study on the quality and efficiency of ground handling services at EU airports as a result of the implementation of Council Directive 96/67/EC. Report to European Commission, 7 October.
Soames, T. (1997), 'Ground handling liberalisation', *Journal of Air Transport Management*, **3** (2), 83–94.
UK Civil Aviation Administration (CAA) (2001), 'The implementation of secondary slot trading', report, November.
Winston, C. (1998), 'US Industry adjustment to economic deregulation', *Journal of Economic Perspectives*, **12** (3), 89–110.

15 Road infrastructure and institutional reform: tolling and pricing

David A. Hensher and Demi Chung

INTRODUCTION

The provision of road infrastructure globally is undergoing major institutional change as governments increasingly put budgetary constraints on traditional sources of public sector financing. The funding of roads from common sources (for example, fuel taxes) is becoming more problematic than ever before. While in some jurisdictions it is a matter of allocation, in other contexts it is a more serious concern about the total amount available to allocate. In California, for example, a 2007 Road User Fee Task Force Program issued a White Paper from the Californian Performance Review. Initiated by the Californian Governor and reporting back directly to the Governor's Office, it has been promoted in response to the diminished proceeds from fuel taxes. Funding 'by the gallon' from fuel-use taxes no longer suffices, despite some additional Federal Government contributions. Paying by the kilometre is now on the political agendas of many countries.

The 'new' funding sources are primarily private capital and road user charges. The former is being increasingly mobilized, with the European Commission suggesting that 20 per cent of the trans-European network should be financed from private capital. The growth in congestion in many urban areas accompanied by increased emissions (air pollution and enhanced greenhouse gas) from cars and trucks has refocused interest on user charging, to both reflect efficient internal cost recovery as well as to internalize the increasingly damaging negative externalities of road transport activity. Kilometre fee development (all networks) with or without fuel tax compensation, is an example of the future direction. In California, as little as 0.1 of a US1 cent fee per use-mile, it is suggested, would generate annually US$310 million towards the road funding bill.

The German lorry road users charging (LRUC) scheme has paved the way in Europe for a distance-based charging regime, applied to all heavy good vehicles with a total permissible weight of 12 tonnes or more that use the 12000 km German autobahn network. Of the over 1.2 million heavy good vehicles using the network in Germany, it is estimated that over 470000 of these vehicles are registered outside Germany. The location of Germany makes it a major transit 'corridor' connecting trade between other European countries. This is the first large scale operation of a road user charge scheme to use satellite-based electronic fee collection technology (ITSinternational, 2005). An on-board unit identifies the (pre-registered) vehicle's location and calculates the charge for a specific vehicle class (based on number of axles and emission rating). This data is transmitted to the Toll Collect control centre.

Regardless of what regime currently exists, and what the future holds for existing funding sources, there is a growing recognition that we need to realign the revenue streams to better reflect the use of the road system. While the debate is more than just how to reduce traffic congestion, the latter has been the stimulus to seek out other ways of improving the performance of the road network (see Kalmanje and Kockelman, 2004). Pricing car and heavy vehicle use is being taken more seriously – tolling of roads constructed under public–private partnerships (PPPs), experiments with distance-based user charges (for example, Oregon, see Forkenbrock, 2004; Hug et al., 1997; O'Mahony et al., 2000; for example, Melbourne, see Lay and Daley, 2002) and the introduction of cordon-based congestion charges in London, Stockholm and Durham, have sent a signal that *widespread* pricing linked to tolling and congestion is on its way – it is not *if* but when and *where*.

The biggest challenge over the past 30 years has been in getting pricing on the agenda of the political process. Now it is there, to some extent,[1] and technological support has come a long way, we have many more options in how we introduce charging and hence financing regimes than ever before. In many senses the 'popular' cordon-based charging regimes like that in London (Transport for London, 2003) are not the best way of implementing a charging regime; reflecting in part the fact that, technologically, London (and most of the UK) is behind countries that have invested extensively in electronic tolling systems that are interoperable across the system. For example, Sydney, Melbourne and Brisbane have a growing number of tolled roads that are fully interoperable, requiring a single electronic tag (ETAG). Payment disbursement amongst owners of each toll road is undertaken through a clearing house, just as banks do. So what is the future likely to be?

As an illustration of a dramatic institutional reform linked to the provision of road infrastructure, the Dutch Transport Minister, Camiel Eurlings, announced in early 2008 that satellite-based road-user charging will be implemented throughout the Netherlands to reduce congestion and finance future road infrastructure. The 'kilometre price' proposed is to be differentiated by location, environmental properties of the vehicle, and time of day (effectively a peak/off-peak or congestion charge).[2] It is to be introduced for all vehicles on all roads in the entire country, starting with trucks in 2011 and phasing in a scheme for cars from 2012 to 2016. The Dutch government plans to scrap road tax as well as purchase tax on new cars when the system is introduced. Eurlings states this will provide a fairer system which taxes vehicle use, rather than ownership. Indeed, the minister says that more than half of Dutch road users will actually pay less under the road-user charging scheme. According to calculations by motoring organizations, only motorists who drive more than 18 000 kms a year are likely to be worse off under the new scheme. Importantly, the Dutch government has determined that the costs of operating the national road-user charge will not exceed 5 per cent of the proceeds.

The Netherlands is not alone in this sweeping change; there is legislation in one state in the US. House Bill 3946 passed in 2001 in Oregon State Legislative Assembly has a mandated mission, 'to develop a revenue collection design, funded through user pay methods, acceptable and visible to the public, that ensures a flow of revenue sufficient to annually maintain, preserve and improve Oregon's state, county and city highway and road system'.

Oregon Department of Transportation is required to develop alternatives to current highway taxing use through fuel taxes. The bill does not stipulate distance based charges, however. It is feasible from financial and administrative perspectives and also has compliance with Privacy Acts. This is ultimate economic application of intelligent transport and logistics systems and technology geographic information system ([GPS], [GIS] and so on).[3]

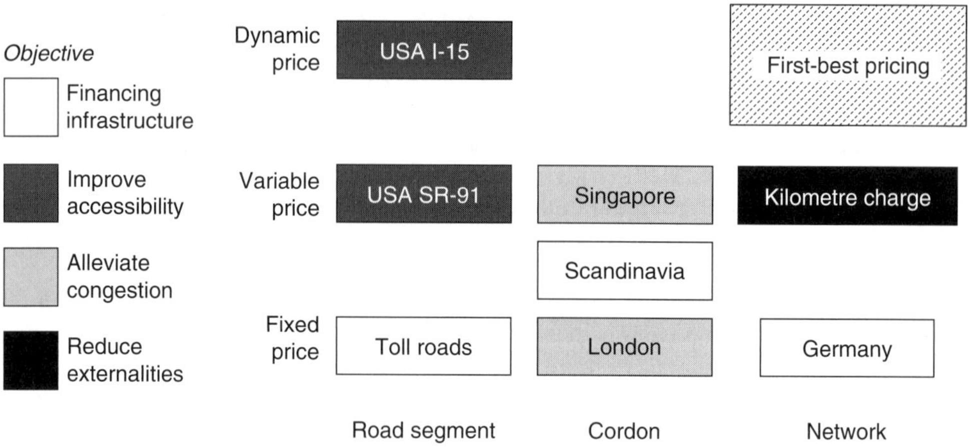

Figure 15.1 An overview of road-pricing measures and linked objectives

Despite the growing interest in a more efficient variable road-user charging regime as a source of revenue for funding road infrastructure, many governments are supporting a politically less risky strategy via the provision of selective high-quality infrastructure that is tolled, and in many cases built and operated by, private consortia under a long-tem concession franchise. Not only does private involvement provide an additional source of funding, it also shields the government from political risk of imposing road charges on users.[4] The remaining sections of this chapter take a closer look at global developments in road infrastructure tolling.

Figure 15.1 provides a useful overview of the road-pricing measures and linked objectives that are currently in place, and which are representative of the mix of measures that are likely to be introduced in new jurisdictions. Tolling of specific roads appears to be more attractive to most politicians, especially in the USA, Australia and South America, than network-based user charging; however, the world will be watching with interest the Dutch initiative. The current reticence with network-based charging appears to be linked to political concerns about votes and fairness, especially where existing charges such as fuel taxes will remain in place. Furthermore a growing number of governments see engagement with the private sector through public–private partnerships as a way of sharing risk but in the context of relatively unambiguous projects such as a very specific road investment that appeal to the private sector. Although a number of banks with experience in infrastructure projects are expressing interest in participating in network-based road-user charging schemes (primarily as a clearing house for the collection and distribution of revenues), the possibility of network-based PPPs should not be discarded.

PRIVATE TOLLING AS A PRECURSOR TO SYSTEM WIDE CHARGING

Private tolling is growing in popularity in many parts of the world (see Larsen and Ostmoe, 2001). The rationale for tolling appears to be a desire of government to bring forward the provision of freeway-quality infrastructure in a budgetary environment where government claims that they could not afford these investments. That is, the focus is on financing as distinct from efficient charging. However, as time passes, we start to gain a better understanding of the real motivation for private participation under what are commonly referred to as public–private partnerships, namely, government funding priorities and not the cost per se. The involvement of the private sector is a means to deliver more road infrastructure in a context where government sees its relevance, but is not prepared to spend on this, relative to commitments to other areas such as health, education and police, and so on. Hemming and Ter-Minassian (2004, p. 31) succinctly express the views of many:

> It is by no means certain that they [the private sector] will be more efficient than traditional public investment. Moreover PPPs can be used to move investment off budget and debt off government balance sheet while the government still bears most of the risk and faces potentially large costs that will eventually be borne by taxpayers.

If PPPs are to deliver high-quality and cost-effective services, there must be equitable risk sharing between the government and the private sector. The quality of services has to be contractible so that payments to service providers can be linked to performance, and the risk of costly contract renegotiations can be minimized. There has to be either competition or incentive-based regulation. An appropriate institutional framework characterized by political commitment, good governance and clear supporting legislation is needed (Hemming and Ter-Minassian, 2004). The government will have to refine its project appraisal and prioritization skills so it is able to manage a complex PPP programme.[5]

Regardless of the true motivation underlying PPPs, severe under-investments in road infrastructure and governments' rigid adherence to surplus budgets have precipitated private tolling of roads. Tollways have been one of the most active PPP markets. For example, in Australia – one of the countries pioneering PPPs – private provision has delivered 11 toll roads, equivalent to A$12 billion investment, since the 1980s (Ernst and Young, 2007, p. 1). Over the years, PPP toll roads have evolved to a stage where greater benefit is being delivered to the public sector. Brown succinctly describes the status quo of the Australian market of private toll roads (2005, p. 437):

> The structure of early toll road agreements seemed to be tilted in favour of the private sector, with the existence of [material adverse events] clauses and the ability to significantly delay rent payments to the government. In more recent examples the private sector assumes more of the downside traffic risk while the government shares in excess toll revenue.

Despite such positive views on PPPs in the transport sector, evidence on the long-term success of the relatively recent PPP phenomenon is not yet available. With an average life of 60 years (Broadbent and Laughlin, 2005, p. 75), most projects have to date only reached their operational phase, and quite a number of them were either sold (for example, the Cross City Tunnel in Sydney, and Skye Bridge in Scotland[6]), in financial turmoil (for example, Sydney Airport Link and Lane Cove Tunnel, and Melbourne Scoresby Freeway in Australia), terminated (for example, metropolitan tram services in Melbourne) or required substantial financial support from government (for example, the

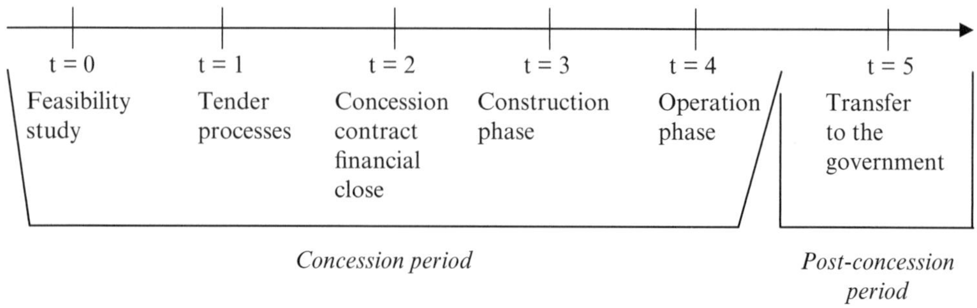

Source: Adapted from Debande (2002: Fig. 2).

Figure 15.2 Sequence of events during the life of a PPP contract

Channel Tunnel Rail Link in the UK and the Guangzhou-Shenzhen Superhighway in China) soon after the construction was completed.

There are many variants on PPPs. Broadly, PPPs[7] are public procurement policies involving the private sector providing asset-based services that are traditionally the responsibility of the government (Broadbent and Laughlin, 2004, p. 4). They are seen as an umbrella term that encompasses a range of financial and organisational relationships between the public and private sectors (Edwards et al., 2004, p. 17). These relationships are established by a concession contract which enables a commercial organization to finance, build and operate an asset for an agreed period. The collaboration between the two sectors is divided into the concession and post-concession periods.[8] These two periods are designed to cover the asset's whole life cycle.[9] A number of events during these two periods set the milestones for contract execution. Figure 15.2 describes the sequence of events across these two periods.

Because these contracts are related to infrastructure-based products, they require, from the private partner(s), a substantial amount of financial capital at the start-up stage,[10] thus the willingness of a profit-seeking private-sector provider to accept the project is conditional on the long-term commitment on the part of the public purchaser. The concession period locks the public sector into a financial contractual relationship with the private party who remains the owner of the property during this fixed term. The length of the concession period is determined on the basis that the sale of the asset-based services is sufficient to discharge construction, financing, operation and maintenance costs, plus a reasonable profit for private investors (Duffield, 2001, p. 27). At the conclusion of the concession, the ownership of the property will normally revert back to the public sector at no charge.[11]

The concession can take many different forms. The typical forms are shown in Table 15.1. The most popular form (and the focus of this study) is 'design, build, finance and operate' (DBFO),[12] where the private sector is contracted to supply a bundling product. This product comprises two distinct elements: the creation of an asset, that is, the construction of a physical infrastructure, and the whole-of-life asset management (WWG, 2006, p. 8). In the DBFO framework, the public sector purchases a service instead of an asset, with pre-defined payment levels, which are payable only when the service meets required standards (Debande, 2002, p. 359).

Table 15.1 Forms of PPPs

Abbreviation	Detail
DBFO	Design, build, finance and operate
DBOM	Design, build, operate and maintain
DBOT	Design, build, operate and transfer
DOD	Design, operate and deliver
BOO	Build, own and operate
BOL	Build, operate and lease
BOOST	Build, own, operate, subsidise and transfer
BOOT	Build, own, operate and transfer
BOT	Build, operate and transfer
BRT	Build, rent and transfer
BTO	Build, transfer and operate
FBOOT	Finance, build, own, operate and transfer
PFP	Privately financed project
Semi-public companies	Government and private enterprise/s jointly own facility
ROT	Rehabilitate, operate and transfer

Source: Duffield (2001, Table 2-7).

The payment mechanism is linked to the requirements set out in the output specification and the results of the risk assessment (Akbiyikli et al., 2006, p. 72). Its objectives are to provide private proponents a number of incentives to deliver value for money. The recoupment of costs and future profit rely on a flow of suitable quality services from the asset, thus it encourages the private proponent to build the required asset on cost, and to use efficient technology (Debande, 2002, p. 360). Further, the revenue receipts flow to the private operator only when the construction of the asset has been completed and the service is fully operational; thus it also motivates the private consortium to finish the construction element as early as possible. Strong evidence suggests that the PPP contractual mechanism has facilitated the integration between the asset creation and its ongoing management compared with contracts delivered under the traditional method (NAO, 2003). Figure 15.3 depicts the incentive scheme established through the interdependence of these core elements in an archetypical DBFO contract. The dashed line connecting the 'payment mechanism' and the 'asset' iterates the principle of DBFO, that is, the purchase of service not the asset itself.

The role the private sector plays in the second element of this bundling product varies between social infrastructure and economic infrastructure projects. Roads are deemed as economic infrastructure projects, where the private sector bears the market risk (NSW Treasury, 2007, p. 1) and are generally funded by user charges or dedicated taxes (English and Guthrie, 2003, p. 503). In these capital-intensive projects, the creation of assets is likely to dominate (WWG, 2006, p. 8). In DBFO roads, after the construction is complete, the provision of the associated service, for example, toll collection, roadwork and lighting maintenance, is a relatively minor component of the arrangement (Walker, 2005). The public sector's involvement is limited to monitoring the adherence to the contract and renegotiation of changes to services supplied (Debande, 2002, p. 367). In exchange, the private operator negotiates a concession right with the government

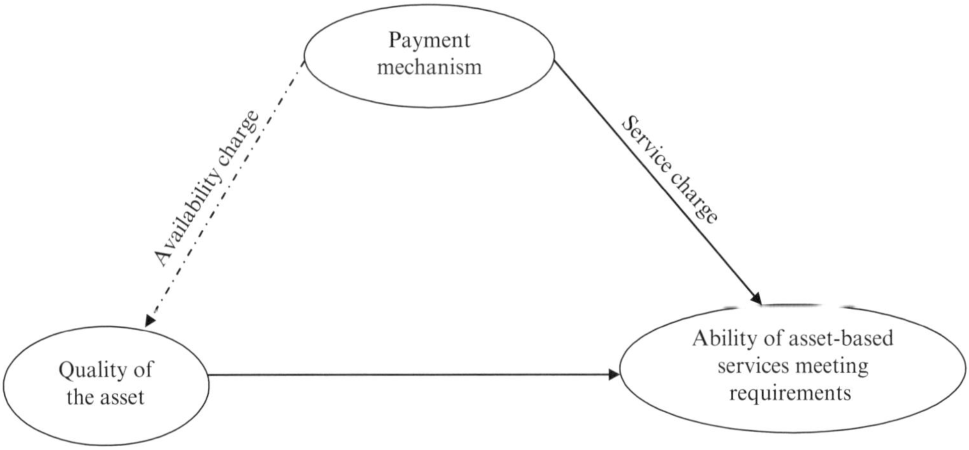

Source: Adapted from Chung (2009, Fig. 1).

Figure 15.3 Relationship between the payment mechanism, quality of the asset and asset-based services

for a period (English, 2005) that warrants the rate of return to private equity (Arndt, 1998; Glaister et al., 2000). These projects are also known as stand-alone projects (see Akbiyikli et al., 2006; WWG, 2006, p. 55).

The DBFO model is typically 'non-recourse' to government. Cash flows generated from the project are the main source of return on equity and debt repayment. Each project is organized in a separate legal entity in the form of a 'Special Purpose Vehicle (SPV)' created under the private ownership. The SPV is also the legal owner of the project related assets during the concession term (Kozarovski, 2006, p. 309). Complex relationships between the SPV and the contracting government agency are intertwined by two primary documents: (1) the lease agreement that grants the SPV a leasehold estate for a specified period; and (2) the Project Deed which specifies the financial arrangements and respective responsibilities between the various parties. The distinct quality of the DBFO package is that there is a minimum interface between the government body and other parties in the relationship cobweb. Once the project reaches financial close, most aspects of the contract's execution and management are facilitated directly by the SPV. The contractual interfaces of the DBFO road model are shown in Figure 15.4.

The proliferation of PPP toll roads has attracted considerable debate in the literature on risk sharing between the public and private sectors and their effectiveness in achieving the objectives of governments. The next section takes a closer look at this crucial topic.

PRIVATE PROVISION IN ROADS: SOURCE OF FINANCE AND RISK-SHARING

In their 1996 study, Fishbein and Babbar analysed key attributes of eight private toll roads from four continents. They found that the financing structure of the road concessions was

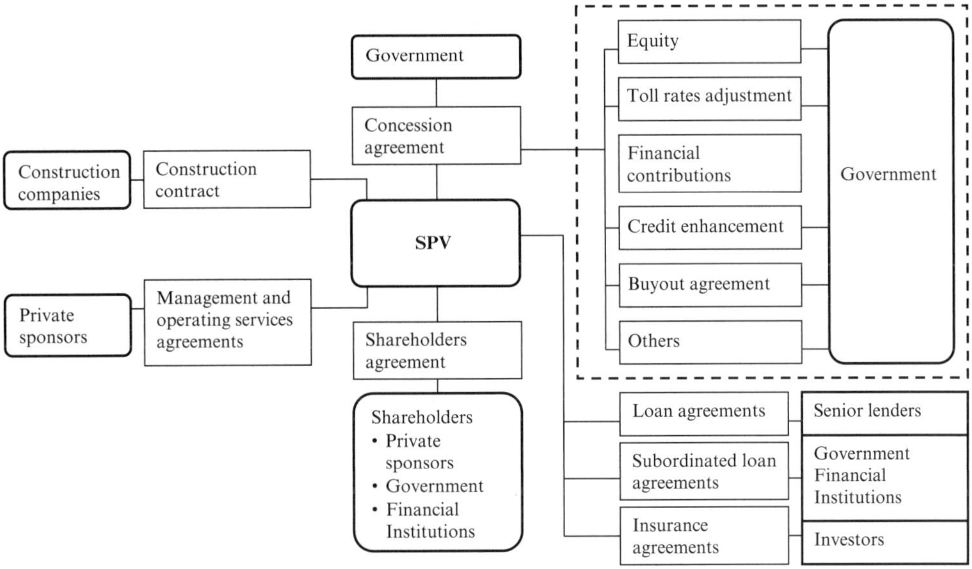

Source: Adapted from Lockwood et al. (2000, Exhibit).

Figure 15.4 Interfaces of DBFO road model

primarily in the form of private equity and debt mix. Except for one, all projects were highly leveraged, with debt to equity ratios ranging from 50 to 100 per cent. Although it is inconclusive about the extent to which the right mix of equity and debt would balance the risk sharing, there is discernable linkage between the sources of funding and specific government objectives. In one case, in order to limit the required toll rates and accelerate the return of ownership, the British government chose to gear the Dartford Bridge with 100 per cent private debt (Fishbein and Babbar, 1996, p. 16). In other cases, recourse to private equity was considered as the impetus to motivate the private proponent to maximize the road's long-term financial performance, rather than maximizing earnings from constructing the property. Fishbein and Babbar alluded to the interdependence between the project's attractiveness to private equity and governments' willingness to allow higher returns on equity.

Both Fishbein and Babbar (1996) and Lockwood et al. (2000) made similar remarks on a project's ability to attract private capital. Traffic demand, public–private risk-sharing arrangements and the level of government financial support played important roles in shaping a project's financial attractiveness. The degree of government support varied according to the project's internal rate of return. A project showing promising returns could be privately financed with minimum government support. One implication of their findings on later developments is that, when a government's approach to road concessions changes to focusing on cost savings, the private proponent is pressured to produce estimates demonstrating its proposal has the greatest potential to generate the highest project return fast. Fishbein and Babbar (1996) also warned that public resistance to tolling would hinder future growth, whereas Lockwood et al. (2000) cautioned that unresolved issues related to risk sharing and conflict reconciliation between the two sectors would deter future cooperative efforts. Both papers favour private finance. While they provide useful insights into the development of partnership toll roads at the international scale, they fail to investigate the causes of common errors repeatedly found in numerous projects, such as erroneous traffic forecasts and misallocation of risk.

In a similar vein, Debande's 2002 study examined privately financed toll infrastructure in the UK. While he was also supportive of the scheme, he went beyond a simple comparison of key attributes across projects and flagged the potential distortion of recourse to private finance. He argued that the premium of private capital was indeed costly. For projects worth investing, they needed to generate sufficient social benefits to compensate these additional costs. Debande took a more integrated approach to examine issues relating to risk transfer. Urban toll roads generally constituted parts of an integrated road network; thus the extent to which demand and volume risks were transferable depended on users' accessibility to alternative free services. If access was relatively convenient, the price of transferring demand and volume risks to the private proponent would outweigh the potential benefit because the private sector would impose prohibitive measures to mitigate the risk of losing to low-cost competing services. Another related matter concerned with social significance is, if the toll road were a vital component of the network, the effectiveness of market discipline imposed on the private sector firm would be greatly reduced (Debande, 2002, p. 370) since government could not afford it to fail. The study concluded that gains from relying on private finance were limited to the design and construction phase, and there was no real transfer of demand or volume risks in road concessions.

Results from Debande's findings are not unique to the UK. This is confirmed by two Australian studies, Mills (1991) and Arndt (1998). Focusing on the *ex ante* contractual mechanisms, both studies scrutinized the risk-sharing arrangements between the two sectors. Mills found that the contractual arrangements of the Sydney Harbour Tunnel (SHT), the first PPP road in Australia, left the government fully exposed to all traffic risk. The tunnel company only bore limited construction risk; other risks were retained by the government and hence taxpayers, and some of the operating risks were even shifted to users. Mills was not convinced that the project could be justified on the grounds of economic efficiency because, in his opinion, no net welfare gain would be secured through the private provision.

An often used argument for private provision in public infrastructure is that private capital adds a welfare increment through the creation of new assets which would otherwise be delayed. This was the justification for the Sydney Harbour Tunnel as well as for the Melbourne Citylink (MCL). The MCL risk profile showed some evidence of risk-sharing between the two sectors; however, through the conditions embedded in the payment mechanism, the state government remained indirectly exposed to traffic flow and other risks that it had tried to transfer to the private company (Arndt, 1998). Nonetheless, the MCL agreement showed promising advancements over any previous Australian toll road. It shifted more risk to the private company and created more symmetry in the acceptance of risk by the project parties. Arndt was optimistic with the growing trend over time as he saw MCL as a step toward a clearer, fairer and more rational risk allocation approach.

Neither Mills (1991) nor Arndt (1998) examined the *ex post* operational performance of the toll roads, therefore the discrepancy (if any) between the actual outcomes and their analyses is yet to be investigated. Shaoul et al. (2006) undertook an *ex post* financial study analysing empirically the value-for-money that the first eight shadow toll roads developed in the UK proposed to deliver. The result confirmed Debande's finding that the cost of transferring risk to the private sector was excessive. The authors were sceptical as to whether these projects could deliver the anticipated risk transfer. While the authors were highly cynical of the private finance initiative (PFI) scheme as a whole, their conclusion was mainly speculative, and did not offer any explanation for the significant gap existing between the financial costs of privately and publicly financed public infrastructure. These limitations were largely due to the expedient opaqueness of estimates in the business cases used to support the PFI in preference to public finance.

A further concern raised in the literature is the effectiveness of the payment mechanisms in achieving government objectives and in implementing the risk allocation strategy. In his analysis of the DBFO transportation projects awarded between 2004 and 2006 in British Columbia in Canada, Aziz (2007) showed that there was a strong association between the design of payment mechanisms and specific government objectives. The payment mechanisms he contemplated ran the gamut from usage-based payment (such as user toll) to a performance-based system (such as availability payment and safety performance payment). Each of these payment systems was designed to achieve various government objectives. The usage payment reflected the transfer of demand risk to the private proponent, and the performance payment represented a low risk premium and better value for money for the public sector. Without analysing the effectiveness of the observed payment mechanisms in achieving government objectives, Aziz extended the overarching principle of 'purchasing services' in PPPs and proposed a hybrid payment system that involved payments for capital.

TYPES OF DBFO IN ROAD TRANSPORT: RISK-SHARING FACILITATION

There are four types of DBFO typically implemented in the roads sector. At the two ends of the spectrum, there are services-based and right-to-toll DBFOs. Services-based DBFOs compensate the concessionaires for the service provision provided to road users. Right-to-toll DBFOs grant the private sector firms a right, after constructing the asset, to operate the facility and to charge final users. On the surface, project financing for these two types relies primarily on the expected cash flow of the project and is typically on a 'non-recourse or limited recourse' basis. This means recourse is limited to the SPV and its assets, and the lenders have no financial recourse for repayment of their loans against the public-sector contracting entity (Debande, 2002, p. 357). A special payment mechanism is often negotiated between the two sectors to ease the burden of debt repayments. Lying in between are return-guaranteed DBFO concessions and 'pseudo' balanced risk–benefit sharing DBFOs. The return-guaranteed DBFOs obligate the government to underwrite all risks for the private proponent. The 'pseudo' balanced risk–benefit sharing DBFOs are masqueraded as free from traffic risk to government, but a closer look reveals that government has committed to letting returns on private equity take precedence over value for money to the general public. Risk-bearing is balanced by way of shifting the burden to road users. Key attributes of these four types of DBFOs identified are highlighted in Table 15.2. Table 15.2 indicates that gradually, the risk-sharing is moving towards an approach in which government shares more upside risk gains with concessionaires, while the downside risk has been passed on to the community; the private sector remains immune from risks.

Table 15.2 DBFO tollway structures

Attribute / Type	Service-based (shadow toll)	Return-guaranteed (real toll)	Pseudo-balanced (real toll)	Right to toll (real toll)
Programme objective	Testing market appetite for transport infrastructure operation	Paving way for private participation in operating transport infrastructure	Moving toward a more balanced risk–benefit sharing partnership	No net cost to government
Project finance	Dependent on project's expected cash flows	Independent of project's expected cash flows[a]	In/dependent of project's expected cash flows	Dependent on project's expected cash flows
Toll variations[b]	n/a	Government regulated	Government regulated	Government regulated
Users' demand elasticity to toll	Inelastic	Median to high[c]	High (when there exist free competing routes)	High (when there exist free competing routes)
Traffic projections	Under-forecasting	Over-forecasting	Over-forecasting	Over-forecasting
Volume risk	Shared	Government	Road users/government	Road users/government
Design and construction risk	Concessionaire	Concessionaire (except for SHT)	Concessionaire	Concessionaire
Maintenance and operation risk	Shared	Primarily with government	Primarily with government/ road users	Government/road users
Network risk	Government	Government	Government/community	Government/community
Government financial contribution	Not directly	Yes (fixed sum[d] payment /interest-free loan/revenue guarantee)	Yes	Not directly
Government guarantee to equity return	No	Yes	Yes	Not directly

Revenue share	No	No	Yes (but highly unlikely due to required return on equity is unrealistically high)	Yes (but highly unlikely due to erroneous traffic forecasts)
Traffic-volume based payment	Government pays for services consumed by travellers	Toll collections pay for project cost and private equity return	Presumably, concessionaire pays land rent for the licence to undertake the project	Concessionaire pays BCF for the licence to charge and retain toll
Payment options for rent	N/a	N/a	Cash or subordinated, non-interest bearing promissory notes	Cash only (no evidence of alternative form)
Cost saving to government	No	No. potentially unlimited cost to government	No, potentially unlimited cost to government	Yes, passed on to users
Examples	shadow toll roads in the UK; A74(M)/M74 in Scotland	Skye Bridge; SHT; Dartford Bridge; M4; M5	M2; ED; MCL	CCT; M7; LCT

Notes:
(a) Toll revenue and government contribution/guarantee to ensure repayment of all project costs plus required return on equity.
(b) Most motorway tolls are set by each motorway operator. Toll price adjustments normally follow CPI movements and must be subject to governments' agreement. There are two exceptions: (i) RTA sets the toll for SHT; (ii) Skye Bridge contained additional condition allowing for movements beyond CPI growth-up to 30 per cent more than CPI should actual toll revenue fall below a certain level.
(c) The competing route of SHT-the Sydney Harbour Bridge charges the same level of toll, price elasticity is expected to be low to medium.
(d) Scottish Office Development Department paid a fixed sum £8.9 million in 1998 prices (NAO, 1997, p. 57).
BCF: Business Consideration Fee; CPI: Consumer Price Index.

Source: Debande (2002); NAO (various years); NSWAGO (2000, 2003, 2006b, 2007); NSW Roads and Traffic Authority (RTA) Contract Summaries (various years).

Two payment structures are used across these four types of DBFOs: the shadow toll programme and the real-toll programme. The largest shadow toll programme in the world was launched by the British Highways Agency (HA) – an executive agency of the Department of Transport (now part of the Department of the Environment, Transport and the Regions) in the early 1990s (Shaoul et al., 2006). The real-toll road programme is also being used in other transport infrastructures in the UK, but its extensive use is seen in Australia. Almost all DBFO tollways in Australia since the early 1990s are real-tolled. The shadow toll programme was implemented to ensure road usage by the public. Initially the UK private sector refused to sign on a road contract if it was to charge consumers a toll (Fourie and Burger, 2000) as it believed that real tolls could 'endanger the policy of creating a private road operating industry' (Shaoul et al., 2006, p. 258). In this programme, a public body pays the private proponent for the use of the tollways by the public (Debande, 2002, p. 376). Because no charge is directly levied on motorists, the programme minimizes traffic risk to the private operator. The costs of the asset are wholly met by the shadow tolls the private company charges the HA (Debande, 2002, p. 372), thus the private company is also insulated from financial risk.

The programme was not rationalized on cost-minimizing, but was policy-driven to test the market's appetite for operating public roads (NAO, 1998; PAC, 1998, paras 48–9); value for money was only a subsidiary goal. In other countries, shadow tolls have become too costly to afford by governments (Bain, 2005). Portugal, the first to initiate the shadow toll, has since dropped the programme because the public body found itself unable to afford payments.

The real-toll programme involves user payment.[13] The initial capital is sourced from private equity and debt. The programme expands the funds available to government, not because of the private finance, but because of subsidies from user charges (NSWIIG, 2005, p. 14). The real-toll programme can be further categorized into two streams. In one stream, government gives no direct payment to the private operator, and the private operator has no financial recourse to government (NSWIIG, 2005, p. 15), hence revenues depend on the volume of traffic and the paying ability of users (Debande, 2002, p. 367). It appears that the demand risk is borne by the private sector alone. However, with user charges as the sole source of rental revenue to government,[14] and with rents payable in cash only when actual traffic exceeds some predetermined threshold, this condition places government as the co-bearer of the demand risk. Nevertheless, the real-toll programme establishes the platform for a system-wide road-charging scheme.

CONCLUSIONS: LIBERALIZING ROAD INFRASTRUCTURE – WHAT DRIVES IT FORWARD?

The initial motive of introducing private provision into road infrastructure is to source additional funds so that proposed road projects will not be pushed back in priority. The whole model has been structured around this concept and to provide incentives to encourage private involvement.

The road infrastructure market has always been an 'unlevelled' playing field. At the time when the concept was fairly new, governments had to offer appealing packages to the market, in which government assumed all project risks in order to entice private capital. The first four shadow toll roads executed in the early 1990s by the UK government are a case in point. They were packaged to align with the government's wish to foster a private-sector road-operating industry (PAC, 1998). To guarantee patronage and insulate the private proponents from traffic and financial risks, no charge will be levied on motorists for using the private tollways. The private proponents are to be compensated by the Highways Agency through shadow tolls that are calculated based on the volume of traffic (NAO, 1998). Recourse to private finance has cost taxpayers in the UK an excessive post-tax risk premium of 63 per cent (Shaoul et al., 2006).

As markets mature and users are more accustomed to paying for the use of specific roads, the playing field has changed markedly in the favour of government. A recent project in Sydney (the contract was signed in 2002), the Cross-City Tunnel, experienced fierce competition in the bidding process (JSCCT, 2006). In order to gain a share in this lucrative market, the private consortia, comprising a foreign property development giant, agreed to the 'no net cost to the government' policy and took on risks that were estimated based on overoptimistic traffic forecasts (NSWAGO, 2006a). The PPP tunnel allowed the government the opportunity to shift the financial risks to motorists. Public resentment over the high-priced toll had sent the tunnel into receivership soon after its opening to public traffic.

Neither extreme offers a stable and conducive environment for an equal partnership in risk-sharing. Government should be mindful that public accountability is a responsibility that cannot be transferred out, and they should be prepared to take on risks that are outside the domain of the private sector. Only a collaborative environment can sustain a long-term working relationship that delivers true value for money.

This discussion on PPPs is very pertinent to broadening interest and commitment to congestion charging. Banks are expressing interest in going beyond the financing, through debt and equity, of toll roads, to area-wide congestion charging schemes. We suggest caution in automatically inviting the private sector, at least until there is a greater transparency in the financial obligations and a fairer sharing of risk and reward between the private and public sectors.

Whatever is promoted, it is important to recognize that infrastructure provision and funding must be flexible enough to accommodate the legacy of the past as well as current prevailing policies and emerging trends. In particular, new approaches must adhere to the following criteria:

- Efficiency – schemes must be analysed from a cost-effectiveness angle taking into account setup and operation costs.
- Relevance – any financial scheme under consideration must be aligned with the broad objectives set out by national policy-makers.
- Fiscal neutrality – new schemes cannot lead to a net fall in the public budget.
- Fairness – new schemes must carefully weigh the benefits and charges as perceived by the majority of citizens.
- Use of resources – infrastructure financing mechanisms must favour an efficient allocation of resources.
- Competition – financing schemes cannot ignore economic laws and market realities nor lead to distortion of competition.

- International compatibility – schemes cannot ignore the globalization of transport nor cross-national frameworks on transport pricing.
- Legacy – contracts negotiated with established concession-owners need to be respected.

Private provision of road infrastructure is an inevitable outcome of the institutional reform in the public sector. It has alleviated the shortage of public funds for large-scale infrastructure investment. Governments need to be wary that: (1) private provision is not a substitute for public accountability; and (2) recourse to private capital should not compromise the public interest. Private provision should only be encouraged as a financing mechanism in accomplishing strategic planning at the broader level that emphasizes allocative efficiency. Overall, that is what the government set out to achieve.

NOTES

1. Not everyone wants congestion charging as London has adopted. More than 74 per cent of residents in Scotland's capital, Edinburgh, have recently voted against City Council plans to introduce congestion charging similar to London's scheme. Donald Anderson, leader of Edinburgh City Council admitted that the planned scheme is now 'dead and buried'.
2. The road-user charge scheme will be facilitated by GPS/speed sensor vehicle tracking, calculated by onboard electronic accumulating odometers, remotely assessing travel from central computers that are capable of applying a range of charging regimes. These include uniform road-use charges and congestion pricing (differential charging according to traffic conditions) including adjusted-upward charges for road use in remote areas (perhaps excluding local residents) where maintenance costs are high and distances travelled are relatively less. Graded distance fees can also be introduced if desired – possibly on equity grounds.
3. These tracking/distance reckoning and road use assessment systems integrate the vehicle and the refuelling setting (Joseph, 2004). Tracking can be defined spatially within a particular jurisdiction such as a state (with an on-board electronic odometer [OEO] shutdown capability when out of state via the GPS system); and within a local jurisdiction. An electronic odometer records distance within jurisdiction. There would be a central computer wirelessly intercommunicating with the OEO, which calculates the road-use charge and applies it at point of sale, at the fuel pump. Once the charge has been paid (through a system such as electronic funds transfer at point of sale [EFTPOS] the system turns OEO back to zero, re-initiating the road-user charge procedure.
4. The concession of the M4 motorway in Sydney expired in 2010. The state government lifted the toll as soon as the ownership was reverted back to the public hands.
5. Currently there are no international accounting standards for reporting PPPs. The lack of such a standard raises concerns about transparency, especially regarding the longer-term implications of such schemes. Discussions with Professor Jean Shaoul (Manchester Business School) have been valuable in clarifying the costs of public involvement.
6. Skye Bridge was bought back by government due to the public objection to toll levies.
7. It is also termed privately financed projects (PFPs) in the procurement policy of NSW, Australia – WWG 2006. The early generation of the British equivalent is named private finance initiative (PFI). In this study, these terms are interchangeable, while PFIs refer specifically to projects undertaken in the UK.
8. To date, only one DBFO project (Sydney M4 motorway) has reached the post-concession period.
9. The life of some transport infrastructures can be well over a century. The Skye Bridge in Scotland has a design life of 120 years (Bain, 2005).
10. Total contract value is usually $50 million or more. But the whole amount of the contract does not entirely come from the private sector party. The public sector supplies physical capital like land and related capital works (WWG, 2006).
11. The zero reversion cost should not be seen as buying a property at no cost. Financial commitments from the public purchaser to the private owner during the concession period, as argued by Heald (2003, p. 359), were in fact paying for the post-concession life of the property.
12. The use of terminology varies between countries. In the UK, DBFO in transport involves the transfer

of ownership at the end of concession period (Glaister et al., 2000) while the similar arrangement in Australia is termed BOOT (Debande, 2002, p. 380).

13. In NSW, Australia, no tolls are to be levied on buses providing regular public transport services or for any other vehicles exempted under the Roads Act or its Regulations (RTA, 1988, 2003a, 2003b, 2004). In Victoria, Australia, no tolls are to be charged on exempt vehicles such as a police vehicle and vehicles performing emergency services under the authority of the State (CityLink, 1995).
14. In recent motorways, the private consortia are required to pay rent to public road authorities for the right to levy tolls and retain toll revenues for its own benefit.

REFERENCES

Akbiyikli, R., D. Eaton, and A. Turner (2006), 'Project finance and the private finance initiative (PFI)', *Journal of Structured Finance*, **12** (2), 67–75.
Arndt, R.H. (1998), 'Risk allocation in the Melbourne City Link Project', *Journal of Project Finance*, **4** (3), 11–24.
Aziz, A.M.A. (2007), 'A survey of the payment mechanisms for transportation DBFO projects in British Columbia', *Construction Management and Economics*, **25** (5), 529–43.
Bain, R. (2005), 'Major road works ahead', *Traffic Engineering & Control*, **46** (7), 248–51.
Broadbent, J. and R. Laughlin (2004), 'PPPs: nature, development and unanswered questions', *Australian Accounting Review*, **14** (2), 4–10.
Broadbent, J. and R. Laughlin (2005), 'The role of PFI in the UK government's modernisation agenda', *Financial Accountability and Management*, **21** (1), 75–97.
Brown, C. (2005), 'Financing transport infrastructure: for whom the road tolls', *Australian Economic Review*, **38** (4), 431–8.
Chung, D. (2009), 'Private provision of road infrastructure: unveiling the inconvenient truth in New South Wales, Australia', *Road and Transport Research*, **18**(1), 68–85.
CityLink (1995), *Agreement for the Melbourne City Link (Act No. 107/1995)*, Melbourne: The state of Victoria.
Debande, O. (2002), 'Private financing of transport infrastructure', *Journal of Transport Economics & Policy*, **36** (3), 355–87.
Duffield, C.F. (2001), 'An evaluation framework for privately funded infrastructure projects in Australia', PhD thesis, University of Melbourne.
Edwards, P., J. Shaoul, A. Stafford and L. Arblaster (2004), 'Evaluating the operation of PFI in roads and hospitals', Certified Accountants Educational Trust (ACCA), London Research Report No. 84.
English, L. (2005), 'Evaluation and monitoring of public private partnerships', Submission to the Legislative Assembly Public Accounts Committee Inquiry into Public Private Partnerships, Sydney, Sub. No. 4.
English, L. and J. Guthrie (2003), 'Driving privately financed projects in Australia: what makes them tick?', *Accounting, Auditing & Accountability Journal*, **16** (3), 493–511.
Ernst & Young (2007), 'The road ahead: future of PPP in Australian road infrastructure', report produced by Project Finance Advisory team, Ernst & Young Australia, Sydney.
Fishbein, G. and S. Babbar (1996), 'Private financing of toll roads', RMC Discussion Paper Series 117, World Bank, Washington, DC.
Forkenbrock, D.J. (2004), 'Mileage-based road user charge concept', *Transportation Research Record*, **1864**, 1–8.
Fourie, C.F. and P. Burger (2000), 'An economic analysis and assessment of public–private partnerships', *South African Journal of Economics*, **68** (4), 305–16.
Glaister, S., R. Scanlon and T. Travers (2000), 'Getting public private partnerships going in transport', *Public Policy and Administration*, **15** (4), 49–70.
Heald, D. (2003), 'Value for money tests and accounting treatment in PFI schemes', *Accounting, Auditing & Accountability Journal*, **16** (3), 342–70.
Hemming, R. and T. Ter-Minassian (2004), 'Making room for public investment', *Finance and Development*, December, 30–33.
Hug, K., R. Mock-Hecker and J. Wurtenberger (1997) 'Transport demand management by electronic fee collection in a zone-based pricing scheme: the Stuttgart MobilPASS field trial', *Transportation Research Record*, **1576**, 69–76.
ITSinternational (2005), 'Ensuring German trucks pay', *ITSinternational*, January/February, 50–51.
Joint Select Committee on the Cross City Tunnel (JSCCT) (2006), *Cross City Tunnel: First Report*, Sydney: JSCCT.
Joseph, J. (2004), 'No distance left to run', *TOLLtrans Technology Review*, 26–31.

Kalmanje, S. and K. Kockelman (2004), 'Credit-based congestion charging', *Transportation Research Record*, **1864**, 45–53.
Kozarovski, D. (2006), 'Public private partnerships – a solution worth pursuing despite their complexity: [paper in: Forum: Public Private Partnerships]', *University of New South Wales Law Journal*, **29** (3), 308–17.
Larsen, O.I, and K. Ostmoe (2001), 'The experience of urban toll cordons in Norway: lessons for the future', *Journal of Transport Economics and Policy*, **35** (3), 457–71.
Lay, M.G. and K.F. Daley (2002), 'The Melbourne City Link Project', *Transport Policy*, **9** (3), 261–7.
Lockwood, S., R. Verma and M. Schneider (2000), 'Public–private partnerships in toll road developments: an overview of global practices', *Transportation Quarterly*, **54** (2), 77–91.
Mills, G. (1991), 'Commercial funding of transport infrastructure', *Journal of Transport Economics & Policy*, **25** (3), 279–98.
National Audit Office (NAO) (1997), 'The Skye Bridge', report by the Comptroller and Auditor General, HC 5, Session 1997–1998, London: The Stationery Office.
National Audit Office (NAO) (1998), 'The private finance initiative: the first four design, build, finance and operate roads contracts', report by the Comptroller and Auditor General, HC 476, Session 1997–1998, London: The Stationery Office.
National Audit Office (NAO) (2003), 'PFI: Construction Performance', report by the Comptroller and Auditor General, HC 371, Session 2002–2003, London: The Stationery Office.
New South Wales Auditor-General's Office (NSWAGO) (2000), 'Review of M2 motorway documentation (reported in Auditor-General's Report to Parliament 2000 Volume 3)', Audit Office of New South Wales, Sydney.
New South Wales Auditor-General's Office (NSWAGO) (2003), 'Roads and traffic authority of New South Wales (reported in Volume 5 of the Auditor-General's 2003 Report to Parliament)', *The Audit Office of New South Wales*, Sydney, pp. 209–18.
New South Wales Auditor-General's Office (NSWAGO) (2006a), 'Performance audit: the Cross City Tunnel project', Audit Office of New South Wales, Sydney.
New South Wales Auditor-General's Office (NSWAGO) (2006b), 'Roads and Traffic Authority of New South Wales (reported in Volume 4 of the Auditor-General's 2006 Report to Parliament)', Audit Office of New South Wales, Sydney, pp. 121–9.
New South Wales Auditor-General's Office (NSWAGO) (2007), 'Roads and Traffic Authority of New South Wales', *Auditor-General's Report to Parliament 2007*, **6**, 69–74.
New South Wales Infrastructure Implementation Group (NSWIIG) (2005), 'Review of future provision of motorways in NSW', The Premier's Department, Sydney.
New South Wales (NSW) Treasury (2007), *Working With Government: Risk Allocation and Commercial Principles*, Sydney: Office of Financial Management.
O'Mahony, M., D. Geraghty and I. Humphreys (2000), 'Distance and time based road pricing trial in Dublin', *Transportation*, **27**, 269–83.
PAC (House of Commons, Committee of Public Accounts) (1998), 'The private finance initiative: the first four design, build, finance and operate roads contracts, 47th Report HC 580 38, London: The Stationery Office.
Roads and Traffic Authority (RTA) (1998), *Eastern Distributor: Summary of Contracts*, Sydney: NSWRTA.
Roads and Traffic Authority (RTA) (2003a), *Cross City Tunnel: Summary of Contracts*, Sydney: NSWRTA.
Roads and Traffic Authority (RTA) (2003b), *Westlink M7 Motorway: Summary of Contracts*, Sydney: NSWRTA.
Roads and Traffic Authority (RTA) (2004), *Lane Cove Tunnel: Summary of Contracts*, Sydney: NSWRTA.
Shaoul, J., A. Stafford and P. Stapleton (2006), 'Highway robbery? A financial analysis of design, build, finance and operate (DBFO) in UK roads', *Transport Reviews*, **26** (3), 257–74.
Transport for London (2003), *Congestion Charging: Six Months On*, October, London: Transport for London.
Walker, B. (2005), '5 things that Auditors-General should be doing about PPPs: [Public Private Partnerships]', *Dissent*, **18**, 50–53.
Working with Government (WWG) (2006), *Working with Government: Guidelines for Privately Financed Projects*, Sydney: New South Wales Government.

16 Public transport liberalization: achievements and future directions

John Stanley

INTRODUCTION

The provision of public transport services in many developed economies has undergone substantial change over the years. With bus services, for example, from an initial situation of private sector provision until about the 1970s, as is still common among developing economies, public sector monopolies became the norm. From the mid-1980s, however, there has been a substantial swing back to private sector service provision, albeit that the enthusiasm that initially accompanied this swing is now eroding somewhat and public provision remains the norm.

Changes in ownership have been so widespread internationally, and their impacts of such interest among public transport stakeholders, that an ongoing international conference has emerged to continually review progress, problems and achievements and suggest directions for improvement. This is the Thredbo Conference series, Competition and Ownership in Land Passenger Transport, which has been held every second year for the past two decades.[1] This conference series and the ongoing research by international bodies such as the Brussels-based International Association of Public Transport (UITP), means that there is a vast array of material reviewing the field.

This chapter draws on that material and on the author's own experience at the front line in negotiating bus service levels and associated service contracts in Melbourne, Australia, and as a director of Melbourne's operator-owned and managed public transport marketing company, Metlink. The presentation is necessarily summary but many references are provided to allow the interested reader to pursue further information.

The chapter starts by providing an overview of the argument for liberalization and types of liberalization (or privatization) that are commonly found in public transport and the benefits that have been expected from such reforms. This includes a discussion on the nature of the expected relationships between the service purchaser ('the authority') and the private provider under various regimes.

A review of some of the outcomes that have been delivered under the new regimes is then presented. This review leads to consideration of the emerging relationships between purchaser and provider in some jurisdictions and contracting regimes to support these relationships. An important emerging focus is on performance of the various parties in the tactical or system planning functions, which are becoming critical to achievement of high-level governmental policy goals from public transport service provision, and linking this performance to operational delivery through the contracting regime. The chapter concludes with some observations about likely future directions.

WHY LIBERALIZE?

As with other sectors in which liberalization became significant from about the mid-1980s, the major motivation for liberalizing public transport service provision has generally been to reduce the growing cost of services to the public purse. The argument is that long-standing public monopolies can get 'bloated', being removed from the market pressures that drive costs down to efficient levels. Transferring operation to the private sector was thus expected to lead to lower costs from more efficient operations. In many cases it was also expected to drive a greater focus on service innovations to better meet customer needs, increasing revenue streams and again cutting the call on the public purse while delivering a better service.

This development was an integral part of the emerging political philosophy of the time, typical of the Thatcher regime in the UK, which saw the role of government as steering rather than rowing. What has been called 'new public management' sought to introduce market-like disciplines into public sector decision-making, to improve service efficiency (reduce costs per unit of output) and effectiveness (for example, responsiveness to customer demands). In the public transport field, the efficiency objective was typically centre stage in the initial rounds of liberalization, with service quality and effectiveness subsequently growing substantially in importance.[2]

The early running on public transport liberalization was made in places such as the UK and parts of Scandinavia, with New Zealand also being an early enthusiast. Australia was an early starter in rail, although some Australian states have had private bus service provision for many decades, uninterrupted by public ownership. European progress outside of the UK more generally has been variable and North America has, perhaps surprisingly, been a slow adopter. Para-transit or community transport services are more likely to be privately provided in North America than urban public transport services. Several South America countries have been relatively active in liberalization (for example, Brazil and Chile).

To illustrate the kinds of objectives pursued by governments in liberalizing public transport service provision, consider the objectives of Melbourne's rail franchising, one of the early international rail liberalizations.[3] The Victorian State Government's 1998 five main objectives in franchising Melbourne's rail services were to (DTF, 2000, p. 9):

1. secure a progressive improvement in the quality of services available to public transport users
2. secure a substantial and sustained increase in the number of passengers using the public transport system
3. minimize the long-term costs of public transport to the taxpayer
4. transfer risk to the private sector
5. ensure that the highest standards of safety were maintained at all times.

While five objectives were listed, it was widely recognized at the time that cost savings and risk transfer were the two key drivers for the franchising. This was typical of early rail and bus privatizations. Whether operators would actually be able to deliver on their promises was not one of the five criteria, although it can be argued to be implicit!

CONCEPTUAL FRAMEWORK

Following the approach developed by van de Velde (1999), it is useful to distinguish between the strategic (S), tactical (T) and operational (O) levels of public transport tasks and competencies. These three levels in the STO framework can be summarized as follows:

- *Strategic* – broad transport policy goals and objectives – what do we want to achieve over the medium to long term? The focus here is on identifying transport policies that will contribute to meeting governmental high level goals to cut traffic congestion, reduce the environmental footprint of travel, reduce social exclusion associated with transport disadvantage and improve transport safety and health outcomes.
- *Tactical* – transport system design and planning to achieve the strategic outcome goals – which services can help to meet these aims and how should they be broadly configured and supported by fare systems, marketing, minimum service levels, and so on?
- *Operational* – delivery of services within the boundaries set by the strategic and operational levels – how should these services be produced?

The strategic (S) or policy role is for government, usually advised by a department of transport. The focus of liberalization has mainly been at the operational level (for example provision of bus or rail services), although there have been forays into liberalization of system/service design activities at the tactical level, especially in the Netherlands (van de Velde et al., 2007), where task allocation has changed hands on some occasions. Stanley and van de Velde (2008) describe this as the tactical level in flux.

Types of Competition

Bayliss (2002) points out that there are basically two types of approach to private sector involvement (or privatization) in public transport service provision: *competition for the market* and *competition in the market*. Competition for the market usually provides the right to operate a service or group of services in an area or region for a defined period, following an open competition (usually by competitive tender, CT). Competition in the market involves operators offering services subject to various rules that govern trading and operating practices. In developed countries, competition for the market is much more common than competition in the market, although UK bus deregulation outside London is a well-known example of the latter, which is considered further below.

A further possibility is private sector service provision under a negotiated arrangement with the authority or purchaser. That negotiation may have been initially based on a winning tender bid that has subsequently been extended by negotiation or it may be based on the authority negotiating with a long-standing private sector provider, without putting services to the broader market. Such negotiated contracts are dealt with separately below.

Table 16.1, drawn from Wallis and Hensher (2007), describes the main types of market arrangements in public transport service provision, ranging from the most liberalized

Table 16.1 Relationship of competitive tendering to regulatory models

Regulatory/market model		Means of determining operator
Provision initiative	Regulatory model	
Market initiative	A. 'Deregulation'	Open (in the market) competition.
	B. 'Quantity licensing'	Authority decision in response to operator application
Authority initiative	C. 'Competition for the market', involving a range of contract/ funding models, e.g. – net cost – gross cost + incentive – gross cost	Typically procured through compulsory competitive tendering (open or restricted competition). May be procured through negotiation with incumbent operator, with CT as potential fall-back ('threatened competition')
	D. Monopoly provision (usually public operator, often open-ended subsidy)	Ongoing provision by monopoly operator

Source: Wallis and Hensher (2007, p.455).

(deregulation) to the least (monopoly provision). The table includes the possibility of tendered and negotiated contracts.

In terms of the most common form of liberalization, competition for the market, the extent of arrangements starts at one extreme with management contracts, in which the authority usually retains ownership of assets, most probably employs staff, keeps revenues and pays all costs, the operator's involvement being limited to operational management in return for a fixed fee. This is the simplest form of gross cost contract.

In most gross cost contracts, however, the authority relinquishes control of assets (vehicles and possibly depots/other infrastructure) and staff to the operator, who is usually required to deliver at least certain specified services for an agreed (tendered) fee. The authority retains revenues. Conditions may be attached to the required state of the assets at completion of the contract, if these are to be returned, under a contractual obligation, to the authority or to a subsequent contractor. A net cost contract goes further and sees the operator retain revenues and receive a fixed sum to support service provision, this sum being the subject of the tender.

Risk Allocation

The major differences between the types of contracts summarized in Table 16.1 relate to risk allocation between the authority and the operator. The greater the degree of risk that an authority seeks to transfer to a private operator, the more operating freedom the operator will require to ensure a return that justifies taking on additional risk.

In Table 16.1, 'Provision initiative' refers to who has responsibility for deciding service levels. Under the 'Market initiative' cases, this responsibility rests with the operator. Under 'Authority initiative', the private provider may have some flexibility in determining service levels, particularly when net cost contracts are used. In practice, service levels are increasingly tending to be determined by the authority and be specified in the tender documents, against which interested providers bid. This is seeing a shift towards gross cost contracts for service provision, in which the operator carries most cost side risks (particularly what are called production risks, rather than risks of cost inflation, which are often dealt with by indexation clauses in operator remuneration) but revenue risks are borne by the authority.

Table 16.2, from Stanley and van de Velde (2008), shows the disposition of risk between various forms of contractual arrangement. It distinguishes production risk, which arises from changes in unit costs and from the way factors of production are combined to deliver a service, from revenue risk, which mainly arises from patronage changes but can also arise from changes in fare levels. The level of risk sought to be shifted to the operator increases as you move from the top left to bottom right of the table. Failures by some authorities to actually achieve risk shifting (for example, some UK rail experience and Melbourne rail, where operators had to be 'bailed out' to maintain service delivery), and a desire by most authorities to maintain control over service levels, tends to move the balance in most cases towards the middle of the table (for example, gross cost contract plus ridership incentive, together with other incentives based on operational performance and customer satisfaction – discussed further below).

As noted in Table 16.1, under 'Means of determining operator', competitive tendering is not the only option available to authorities. Negotiation with an incumbent operator is sometimes used to allocate the rights to operate a service (route or series of routes in an area), especially if an authority is satisfied with an existing operator's performance, and/or is perhaps concerned about a lack of bids if a service is put out to competitive tender. This was the case with refranchising of the Melbourne train and tram contracts in 2002–03, after the failure of the first round of competitive tenders (where one of the winners handed back the keys within three years of winning a major share of service offerings). Negotiated performance-based contracts, as an alternative to competitive tenders, are discussed further below.

AN EFFECTIVE COMPETITIVE TENDERING PROCESS

In a particularly comprehensive and valuable review of UK experience with competitive tendering, Kain (2007, pp. 4–5) sets out a number of matters that need to be considered by an authority in seeking a successful competition for the provision of rail services. These matters are also relevant for bus contracts, although some of the answers to the issues raised will be different because of the relative service attributes (for example, bus has low capital requirements compared to rail, which means different barriers to entry on the asset side). The key matters raised by Kain encompass:

- the specific approach adopted in auctioning
- establishing a robust set of criteria to assess bids

Table 16.2 Risk allocation: examples of contract types

Production cost risk borne by	Revenue risk borne by		
	Authority	Shared	Operator
Authority	Management contract – Operator paid management fee for running service – All production costs and Passenger revenues accrue to authority	Other combinations of incentives	
Shared	Gross-cost contract with shared production cost risk Example 1: indexation transfers some input price risk to the Authority. Operator remains responsible for productive efficiency. Example 2: operator bears production cost risk up to a specific level, after which risk is shared with authority	Other combinations of incentives	Other combinations of incentives
Operator	Gross-cost contract: – The operator is responsible for all costs of running the system. – The authority remains responsible for all revenues – Various clauses can incentivize	Gross-cost contract with ridership incentive: – e.g. variable payment related to ridership gains	Operator responsible for all costs of running the system and for all passenger revenues generated.
	The operator to deliver improved operational quality	Net cost contract with shared revenue risk: – e.g. operator carries revenue risk up to a specific level of divergence with forecast ridership. Risk then shared with authority	Net cost contract: – Operator receives lump sum payment to cover expected deficit on service (or pays a fee for the right to operate a profitable service) Super incentive contract: – Operator receives no lump sum payment but only a variable payment fully dependent on specific performances (e.g. ridership, etc.)

Source: Stanley and van de Velde (2008, p. 24).

- attracting and retaining the competitive market 'for the ground' for future competitions
- ensuring that bidding transaction (competition) costs are not so large as to offset the anticipated franchise benefits
- being able to specify the required output and to monitor and enforce adherence to the committed output
- structuring the franchise contract to handle risk and unanticipated events (that is, uncertainty).

To this list one should add the capacity of the authority to be able to decipher a deliverable bid. The failure of some franchises is evidence that this capacity is not universally present. The possibility of predatory bidding is another issue in this regard, as identified by Alexandersson and Hulten (2003). The last two Thredbo International Conferences on Competition and Ownership in Land Passenger Transport (2005 and 2007) have both agreed that reforming the authority (raising skill levels) is now more pressing than further reforming the operator! It is suggested that the privatization process itself may have contributed to a hollowing out of skills in authorities, for reasons such as some skilled staff moving from the authority to join operators.

Without discussing all these important points in detail, a few are highlighted below, for reasons such as the way the author sees the allocation of the rights to provide public transport services moving.

Selection Criteria

As service expectations of the authority increase their emphasis on quality (for example, on-time running, reliability, cleanliness, driver behaviour, and so on), as well as cost, the complexity of tender document preparation and evaluation increases. As Kain (2007) emphasizes, selection criteria need to be transparent when the bidder is chosen for reasons other than cost, to ensure accountability of the authority. Also, it is not possible for bidders to put forward properly customized bids if the authority has not spelt out the full set of criteria that will be used to evaluate bids, including both:

- the relative weights that will be attached to particular evaluation criteria, and
- how option performance against each criterion will be assessed.

In some cases, authorities seek to select a winning bidder on a narrow base (for example, cost) and then seek to negotiate quality aspects of the service. This at least provides bidders with a common competitive benchmark for bidding but, as quality criteria grow in importance, the focus should not be narrowly constrained to cost as the primary basis for selection.

Encouraging Competition in the Bid Process

Achieving competition in the bidding process is clearly fundamental to liberalization. This can be encouraged by methods such as:

- making the size of the service offerings such as will allow multiple potential bidders, provided this does not lead to uneconomic service sizes. The UITP Regional and Suburban Railway Committee, for example, has argued that a minimum critical mass of business volume for a competitive bid for a regional European rail operation is about 2.5 million train kilometres (UITP, 2006)
- requiring transfer of existing staff to the winning bidder, to assist with availability of expertise
- limiting the need for the bidder to own expensive rolling stock (although Hensher et. al. (2007) argue that, for urban bus services, operator ownership of the assets is most likely to support efficient service provision)
- keeping bid costs at reasonable levels, which may involve paying a fee to the second ranked (and perhaps other) bidder
- keeping the contract length sufficiently long to allow the winner to improve assets and earn a return but not so long as to discourage future bidders.

Notwithstanding the possible use of such measures, there are a number of possible further barriers to be confronted in the process of seeking to introduce greater levels of competition in public transport service provision. Yvrande-Billon (2007) outlines how the French Competition Commission denounced the existence of a cartel between the three leading bidders in that country between 1994 and 1999, showing the importance of effective anti-competitive measures.

Mathiesen and Solvoll (2008) identify how the progress of competitive tendering of bus services in Norway has been accompanied by many takeovers and increasing ownership links among transport companies. They argue that the introduction of auctions gives companies a profit incentive for increasing cross-ownership to reduce the downside risks of losing an auction. Extensive consolidation of providers has also occurred in markets such as the UK and Sweden.[4] This may weaken the competitive element of tendering. Mathiesen and Solvoll suggest transport authorities should differentiate their procurement strategy and use competitive tendering together with other procurement strategies, such as negotiated performance-based contracting, to manage this tendency.

Boitani and Cambini (2006) discuss competitive tendering experience for Italian local bus service contracts and note that the numbers of participants were few and the incumbents gained almost all the franchises, with negligible financial savings. Their analysis suggested that Italy had high unit costs and low labour productivity but that the authorities were unwilling 'to see their own firms thrown out of the market if less efficient than potential entrants'. They link this to an unwillingness to give up the political rents accruing from ownership of local public enterprises. They also note an unwillingness to see labour costs cut and suggest federal intervention to strengthen the propensity to competition and performance-based contracts as part of a possible way out of such circumstances.

Kain (2007) discusses some of the moral hazard risks associated with an authority seeking to attract bids and warns (for example) about authorities offering incentives that may drive tactical bidding processes by potential operators. Cap-and-collar risk-sharing arrangements, which may limit (for example) an operator's downside risk and provide for sharing upside (and modest downside) risks with the authority, may encourage such behaviour.

Key Performance Indicators

With gross cost contracts becoming more common, the inclusion of key performance indicators (KPIs), with associated bonus/penalty provisions, is critical to sustaining performance pressures during the course of the contract. The key to good KPIs is that they should:

- relate to matters over which the operator has some control
- be timely and measureable by the authority, in an unambiguous way
- be few in number and focused on critical variables that are of concern to the authority, which will usually mean they are central drivers of patronage (especially if there is no separate patronage incentive) or of environmental performance.

Key performance indicators typically are of three types:

1. *A patronage incentive*, whereby the operator receives a bonus for patronage growth exceeding some hurdle level and may be penalized if growth is less than this benchmark. The hurdle may be increased if the authority provides funding for additional services or supportive measures such as bus priority lanes, to place the emphasis in the KPI on what the operator is doing to increase patronage. In Melbourne bus contracts, for example, a service elasticity is applied to additional funded service kilometres to increase a base patronage growth hurdle from 2 per cent annual growth before a bonus is payable.
2. *An operational performance regime* – commonly including measures such as cancellations and on-time running, although traffic congestion reduces the operator's control over these variables.
3. *A qualitative performance regime* – where the focus is usually on customer satisfaction rating of the operator's performance across a range of service attributes. Copenhagen is a leader in this assessment method.

Key performance indicators in the second and third categories are sometimes a response to specific customer complaints about poor service level (for example, delays; complaints about poor driving practices, leading to quality criteria related to driver training and behaviour).

The total payable as bonus or penalty under a KPI regime is commonly about 2 per cent of the value of the gross cost contract but may be larger or smaller (for example, Sinisalo (2007) notes that about 5 per cent of contract value is tied to performance and quality incentives and penalties in Helsinki region).

Network Integration

There is increasing focus in public transport service provision on the authority being able to present an integrated network to customers and potential customers. This is part of making the journey experience as simple and seamless as possible, from gathering information about travel options to undertaking the journey.

A weakness with a deregulated system is the difficulty of reconciling the service offerings from various separate and competing suppliers with the requirements of a service network. With services liable to change quickly and operators only dealing with partial networks, the authority faces real problems in providing network information and achieving seamless connections. Service subsidies can be used to fill gaps in the offerings of private deregulated providers but this does not deal with the broader problem of a lack of certainty about what is being offered system-wide, making network marketing problematic. Authorities such as the Greater Manchester Passenger Transport Executive have made progress in this area by developing partnerships and trust with operators outside formal contracting frameworks, to consciously develop an integrated network. However, the relationship has no contractual base and can be fragile (Hensher et al., 2007).

Similar problems arise in tendering for the market if there are multiple tenders in a given area or region. By way of example, Melbourne has one train operator, one tram operator and 26 bus operators. However, the Metlink Melbourne solution has been an effective way to resolve the passenger information and network marketing aspects of the city's network of services (but not network integration). Metlink is a company owned and managed by the operators as a condition of their service contracts. Business and marketing strategies and plans must be approved by the State Director of Public Transport, (the authority), who funds the public transport service contracts and supports the information and marketing services provided through Metlink. The city is experiencing double-digit public transport patronage growth, with the Metlink role in integrating information and marketing services a valuable contribution towards this outcome.

Incomplete Contracts and Relational Contracting

With bus contracts typically of five to seven years' duration and fixed rail longer, there is a challenge for authorities and operators in writing contracts that deliver a sharp edge to operator performance over that period, while retaining the flexibility to respond to changing circumstances. The rapid rise and then fall in fuel prices (and expected future rises), rapidly growing concerns about climate change and the global financial crisis, all suggest that uncertainty will characterize public transport on an increasing scale in coming years. It will simply not be possible to anticipate all possibilities and write complete (detailed) contracts to respond. The notion of even trying invites years spent in litigation!

Hensher and Stanley (2008) focus on such problems and on the development and implementation of service delivery regimes that sustain performance pressures during the life of the contract, not just at the time when contracts are up for bidding in a tender. This approach focuses on building relationships of trust between the authority and operator and placing the contract in this setting, such that change during the course of a contract is normal, expected and easily handled. This is similar to aspects of alliance contracting in construction. Hensher and Stanley see incompleteness as making the case for combining trusting partnerships with legal contracting obligations.

The present author's experience is that this approach works well when the authority and operators, at an industry level, have agreed about the strategic goals being pursued by public transport services and about the tactical level requirements to meet these goals. The contract then sets a negotiated process-driven approach to delivering the system, based on authority–operator relationships. Duncan (2007) and Stanley et al. (2007) outline some of the requirements to make this approach work:

- *common objectives* tied to public policy purposes – where the focus is on shared agreement about the public policy purposes being pursued by service provision, while recognizing that the respective parties are also entitled to their own objectives, whose legitimacy must be respected
- *agreed governance arrangements* – covering matters such as roles and responsibilities of each party, risk sharing arrangements and remuneration arrangements, including incentives, and relationship management arrangements
- *trust* – which is based on many factors, from the broader societal setting within which a culture of trust may or may not figure prominently to matters such as selection of the right partners, confidence in a partner's capacity to deliver, demonstrated good faith, accountability and transparency (useful to support a trusting relationship between the authority and operator but also to assure the wider community that there are safeguards against regulatory capture) and written agreements capturing the above.

In line with this thinking, van de Velde et al. (2007) have reported that a recent review of Dutch tendering experience has called for more relational contracting, focusing more on agreement about the contracting process than on attempting to write down complete contracts. As noted by Stanley and van de Velde (2008), in part this reflects recognition that overly detailed prescriptions of contractual obligations (micro management), with associated KPIs, may distract operators from pursuing the high level goals that are the real purpose of service provision. It also reflects the recognition that it is not possible to fully specify contractual obligations and that, over time, flexibility and a relationship built upon trust are more likely to be effective in high level goal achievement.

Negotiated Performance-based Contracts

As noted above, some privately provided public transport services are not subject to a tender but are negotiated directly between the authority and an operator. This may be, for example, to renew an existing contract, where the authority is very happy with the performance of the incumbent and sees little point in incurring the transaction costs to repeat a bidding process that might be expected to deliver the same operator. This argument is gaining credence as transactions costs increase. Alternatively, it may be that the jurisdiction has a history of private sector provision and is prepared to negotiate with existing operators to continue delivery (the Melbourne experience). Equally, the provider may be an existing in-house (public sector) team, operating under an arm's length service agreement that requires the provider to meet certain expectations.

The authority can take steps to help ensure efficient and effective operator performance in such a negotiated relationship by benchmarking service costs against similar jurisdictions elsewhere, to indicate efficient costs, and then use a contract that applies performance pressure to the operator (a negotiated performance based contract). This could involve:

- the use of KPIs as outlined above
- a rollover incentive in the event of good performance against KPIs
- the threat of competitive tendering in the event of poor performance (including outsourcing if the existing provider is 'in-house').

Negotiated performance-based contracts are particularly suited to an environment where the authority and operator are seeking to pursue relational contracting, based on trust. In such circumstances, the conditions likely to encourage such a relationship, as outlined above, are applicable but the requirement for transparency and accountability is particularly important, to guard against regulatory capture. An open book approach on finances as well as benchmarking against similar services (virtual competition) elsewhere are useful devices in this regard, to support the measures listed above.

ACHIEVEMENTS

It is useful to consider the impacts of liberalization against the initial intended outcomes, particularly cost reduction, risk transfer and, for some jurisdictions, patronage growth. Patronage growth is important because most developed countries are pursuing a policy level ('S') objective of increasing the relative mode share performed by public transport compared to the car, to contribute to the policy goals of reducing traffic congestion, lowering greenhouse gas emissions, increasing social inclusion and improving health and safety.

Cost Impacts

Wallis and Hensher (2007) have produced a valuable assessment of the impacts of a large number of competitive tenders (over 20) for urban bus services, across a wide range of countries (ten), since 1985. Their cost assessment focuses on changes in unit gross costs, to normalize for changes in service levels and to avoid problems of normalising subsidies that are calculated on non-comparable bases (for example, gross cost versus net cost). In some cases, it includes data outlining the impacts from initial tenders and shows how these have changed with subsequent tender rounds. In summary, the cost reductions they identified from the *initial* round of tenders were:

- Great Britain: 50–55 per cent
- Scandinavia: most in the range of 20–30 per cent but a range from 5 to 34 per cent
- USA: 30–46 per cent
- Australia: Perth 22 per cent; Adelaide 38 per cent
- New Zealand: *c.* 40 per cent ex-public operators; *c.* 5 per cent private operators.

Bekken et al. (2006) concluded that competitive tendering produced savings of about 10 per cent in Norway, which was less than most countries achieved. They attribute this to a local bus industry that had improved its efficiency over a long period of time before CT was introduced. This mirrors the New Zealand experience identified by Wallis and Hensher above. Bekken et al. suggest that the threat of competitive tendering was important in ensuring that cost savings were achievable in the Norwegian examples they review.

Cost/subsidy savings are used by government or the authority (depending on to whom these savings accrue) for a variety of purposes. In some cases it is to simply cut the level of funding support that is provided, as has largely occurred in Norway (Bekken et al. 2006). In the Helsinki region, in contrast, savings have largely gone into reducing ticket prices and increasing service levels (Sinisalo 2007).

Wallis and Hensher (2007) also summarize the cost trends in second and subsequent tender rounds for urban bus services in five European countries where CT has reached this stage. They find that the evidence suggests significant real cost increases from the initial tender round figures. They acknowledge that, in some cases, this is attributable to more demanding contract specifications but they also suggest that it is due to several aspects of the CT model, including:

- the greater experience of bidders, resulting in more informed bidding, with less likelihood of bid errors or mis-estimates (sometimes described as the problem of the winner's curse). The implication is that a number of initial bids were unsustainable
- less emphasis by incumbent bidders on retaining market share at all costs
- bidders taking a longer-term perspective, more often providing for asset replacement, with associated higher profit margins
- lesser levels of competition for tenders (in some cases). Wallis and Hensher note, for example, that the average number of bidders fell from six in the 1995 London bids to between two and three by 1997 and that, in New Zealand, the average number of bidders in tenders in the two largest centres had fallen to one in the early 2000s). This tendency reflects benefits of incumbency.

Their research also suggests that the windfall gains in round 1 are primarily in going from public to private operators and that there is little more to save after that. Interestingly, Wallis and Hensher note a convergence on cost levels by the third tender round in the Adelaide urban bus contract tenders, suggesting that efficient cost levels had probably been achieved. The case for negotiated performance-based contracts, rather than re-tender, is relatively strong in such cases if the incumbents are delivering on quality as well as cost.

Turning to rail and to the expected call on the public purse, Melbourne's rail (train and tram) privatization process in the late 1990s promised significant savings to government, based on the winning bids. However, these net cost savings never materialized, partly because of the implementation of corporatization processes before franchising. These corporatization processes included significant cost cutting, which led the State's Auditor General to conclude (VAG, 1998, p. 8): 'After 6 years of cost-cutting and rationalisation of operations, there appears to be limited scope for further large savings to be achieved in an environment where a substantial proportion of existing rolling stock will need replacement in the next few years.' As Stanley (2005) notes, this note of caution did not deter the winning bidders in the 1999 franchising. Over-optimistic bidding was a key factor in that franchising, resulting in the major winner (National Express Group) ceasing operations only three years into the franchising period.

The Melbourne experience is notable because it is a reminder to authorities to clearly understand whether the cost savings they expect will actually be achievable and sustainable. Large cost savings from tenders may reflect opportunistic bidding, in which winners count on being able to subsequently renegotiate payments upwards, believing that governments are risk averse when it comes to the prospect of service cancellations (sometimes called playing 'capture the regulator'). This is a particular concern with rail, where barriers to entry are high.

In reviewing UK passenger train services, Kain (2007) concludes that there is no definitive answer as to whether franchising reduced the call on the public purse. For the broader railway that was British Rail he concludes that costs have increased, although Welby and Nichols (1999, p. 75) argue that: 'Privatisation has fundamentally changed the financial flows within the system, so that in many ways they are now a closer approximation to long run economic costs, in that subsidies should now be providing for the opportunity cost of capital in a way that was previously absent.'

Smith et al. (2009) also review UK rail experience and conclude that franchising does not seem to have succeeded in driving down train operating company costs. They find that initial cost reductions were important but that these have subsequently been lost to rising costs, a part of which they see as related to franchising performance (for example, competition between franchisees for staff serving to bid up labour costs). Alexandersson and Hulten (2003), reviewing Swedish rail franchising experiences, suggested that both gross and net cost contracts had delivered first round savings of about 20 per cent. Expanding the roles performed by drivers and changing schedules were suggested as being important contributors to these savings. Even larger savings were subsequently found in some subsequent tender rounds. Operator costs exceeding bid costs in some gross cost contract tenders were noted in some cases.

Reflecting on the evidence about CT of bus and rail services, participants at the Thredbo 9 Conference concluded that (Stanley 2007): 'CT is usually a sufficient but not necessary condition for cutting PT service delivery costs, compared to costs achieved under public monopoly provision, but is not sufficient to achieve cost reductions under re-tendering.'

Patronage Impacts

The most striking illustration of patronage changes following liberalization is that for urban bus patronage from the UK, where the difference between trends in London (where a competition for the market model was adopted) and the rest of the country (where deregulation was the order of the day) are remarkable. Deregulation took place in late 1986. In London, tendering was introduced over a longer period. The counterfactual of what patronage might have been if liberalization had not taken place is not readily apparent but the simple patronage trends are starkly different and suggestive.

Figure 16.1 shows UK local bus patronage trends for London and outside London, in index form (1985–86 = 100). Whereas patronage in London grew almost 80 per cent over about two decades, that in the rest of the UK fell by over 30 per cent. Vehicle kilometres in London grew roughly in line with patronage, though exceeding patronage growth for much of the period. Patronage growth in London has been very strong since

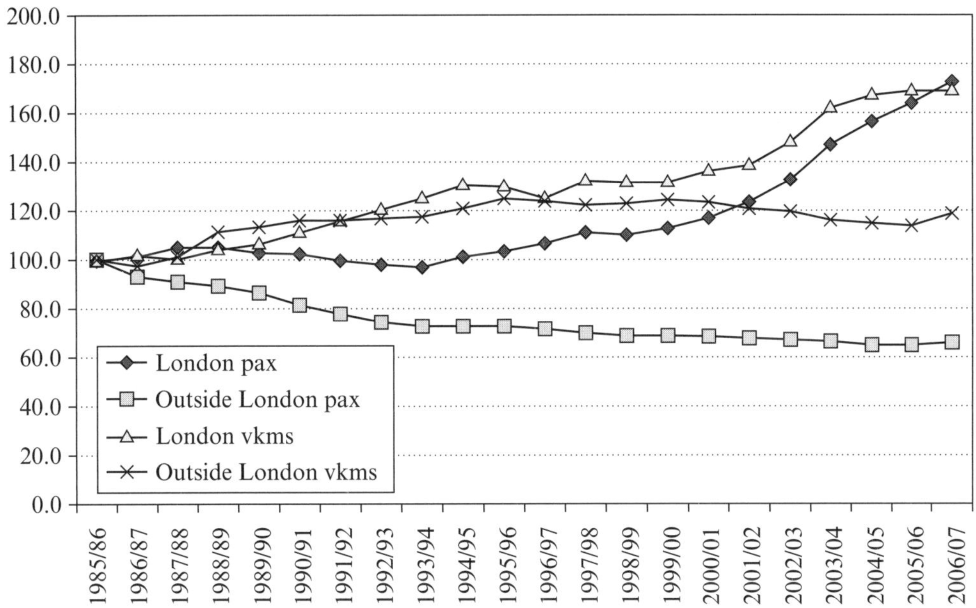

Source: Department for Transport (2007).

Figure 16.1 *UK local bus patronage and vehicle kilometre indices, 1985/86 to 2006/07 (1985/86m = 100)*

the late 1990s. In the rest of the UK, kilometres increased by about 20 per cent between 1985–86 and the early 1990s but have only fluctuated since that time, as patronage has declined.

ICLEI Local Governments for Sustainability (2003) summarizes European experience with patronage growth in a range of service delivery environments. They found that patronage growth was highest in the case of competition for the market, at 1.8 per cent per annum, negative in cities without competition (–0.7 per cent per annum) and poorest in cities with deregulated systems without significant control by public authorities (−3.1 per cent per annum).

The inference is that expecting local bus services to survive in the free market place is unlikely to provide much encouragement to patronage growth. The economic wisdom of deregulation of bus services in a broader context, where there are substantial negative externalities of motor vehicle use, is dubious. Substantial real fare increases outside London reflect this dilemma. While local passenger transport executives outside London can financially support some services (about 15 per cent of total), this is only a minor part of the total picture.

The patronage trends outside London could easily lead a cynic to suggest that, because of deregulation, it is no surprise that the UK has been a leader in researching social exclusion related to transport! Preston's (2004) more measured conclusion is that on-the-road competition can lead to too much service at too high fares with too low service quality. If external costs were factored in to the analysis, the case for a greater role for

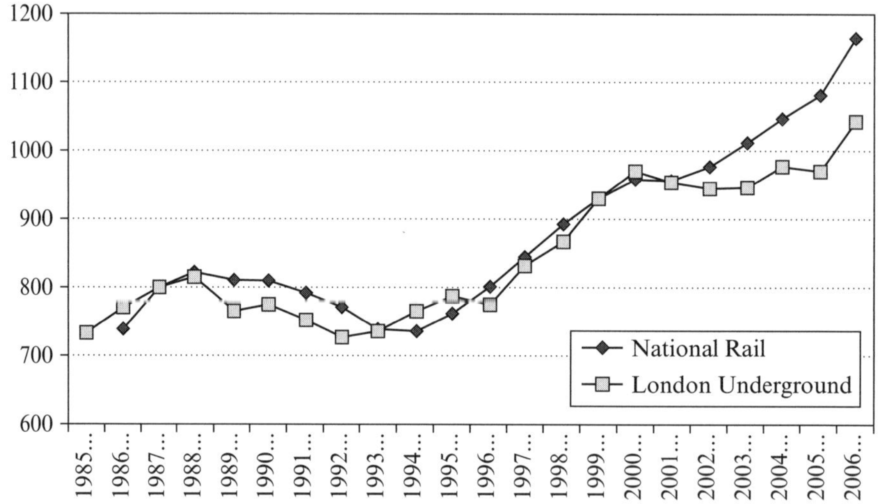

Source: Department for Transport (2007).

Figure 16.2 *UK rail patronage, 1985/86 to 2006/07 (million passenger journeys)*

subsidized services (for example, through passenger transport executives) has merit. Preston (2004, p. 12) concludes that:

> Competition for the market . . . can in theory provide the optimal mix of service quantity, quality and fares at technically efficient cost. But there may be incentives for planners to also provide too much service at too high fares, or for politicians to demand that planners provide too much service because service is highly visible – low fares are less visible.

Rail privatization in the UK took place from 1996. Figure 16.2 shows that patronage growth has been very strong from just before that time. However, the close alignment between patronage growth and liberalization is suggestive that liberalization has played a part in helping to stimulate growth. Smith et al. (2009) have examined this question in some detail, seeking to separately identify the role of influences such as changes in real fare levels and service kilometres provided on patronage. Their conclusion is that liberalization might account for up to about one-fifth of patronage growth in the early period after franchising, but that precise estimation of the relevant contribution is very difficult.

IS THERE A REGULATORY CYCLE?

Looking very briefly across some continents in terms of the progress of liberalization, we observe:

- In recent years Europe has been spinning its wheels somewhat in its push for public passenger transport liberalization. Passenger railway liberalization has not made any progress within some major jurisdictions. Bus liberalization has been tried by many European countries but fully implemented by few. Where implemented, it has delivered some important cost savings and service improvements but the cost gains have proven to be largely one-off. On the supply side, a tendency towards increasing concentration is apparent.

- Perhaps surprisingly, North America has never been as excited as Europe by liberalization and local authorities continue to dominate the market. Slow extensions of private provision are occurring. South America is still in a growth phase of liberalized services but the experience of Santiago, where seeking to change too many system variables at the same time (including a wholesale re-restructure and tendering of services[5]) caused significant delivery problems, is salutary.
- Australia remains solidly privatized in its bus services (public provision is the exception), with both negotiated and competitively tendered provision in operation, but has only limited private passenger rail provision. This is unlikely to be extended.
- South Africa has made little progress on competitive tendering, largely because of labour concerns about employment and conditions (associated with black economic empowerment).

In a very insightful contribution, Gwilliam (2008) has posed the question of whether there is a regulatory cycle for bus transport. This question emerges, inter alia, from observation of trends in liberalisation in both developed and developing countries, including the tendency for increasing industry concentration in countries where liberalization has taken place. Gwilliam depicts an industrialized country regulatory cycle as shown in Figure 16.3, consisting of a cycling through private competitive supply, unregulated private monopoly, regulation of the private monopoly, nationalization and then back, through further regulatory reform to some type of private competitive supply.

Gwilliam notes that not all countries follow precisely the same route and that the savings which flow from the introduction of competition depend on the starting point in the cycle. This cycle very much reflects the current author's perspective on the emerging liberalization process in public transport, not just bus.

Professor Chris Nash (personal communication) suggests that the UK bus experience is also reflective of this regulatory cycle. Nash sees the failure of the completely free market, plus increased concentration (in many areas more than 90 per cent of services are provided by one operator) means increased local government involvement. He indicates that a number of UK Passenger Transport Executives are keen to go for area-wide franchising but questions whether there will be enough competition for the franchise. Nash asks whether the UK might simply end up re-regulating.

There is also some evidence of this process in rail. Following the first round franchising of Victoria's (Australia) metropolitan train, tram and country passenger services (rail and some bus services) in 1999, the major franchisee National Express handed back the keys within three years. While re-franchising took place for the metropolitan train and tram services, the country rail passenger services have returned to public ownership. This was partly due to a decision to implement a major infrastructure and service upgrade but, after seven years, re-franchising is increasingly unlikely.

In the UK, National Express East Coast has recently ceased providing the East Coast main line rail service from London to Edinburgh (among other destinations in Scotland),

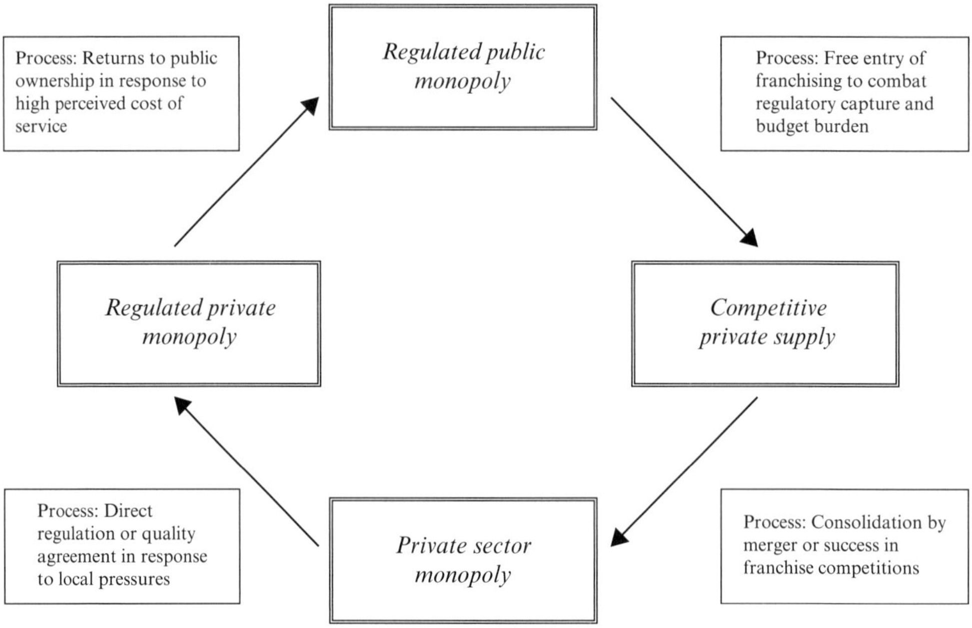

Source: Gwilliam (2008).

Figure 16.3 The industrialized country regulatory cycle

which will revert to public provision. It has been indicated that this will only be for a period of two years, after which time the service, will again be offered to the private sector, since it is essentially regarded as a profitable service (that is, National Express overbid for the rights to provide the service, as in Victoria). The reversion to public provision provides an opportunity to improve broader rail network integration, which has been problematic under private operation. This, in turn, may increase the prospects of wider public provision, reflecting Gwilliam's regulatory cycle.

FUTURE DIRECTIONS

What, then, might an 'ideal' tender and contract look like? The general answer must be that there probably is no such thing. Local circumstances (history, culture, political environment, and so on) will determine what works best. Sometimes this will mean no tender at all but sustaining performance pressure on an incumbent public or private operator. However, some desirable qualities can be indicated.

Government is responsible for the long-term strategic or policy setting process and will expect demonstrated value for money in service delivery, to achieve goals that are related to such outcomes as congestion reduction, low environment impact, social inclusion, safe travel and healthy travel. Clarity in defining desired high-level outcomes is a vital task that demands focus and detail if an effective system and services are to be provided.

The process of awarding the rights to provide service, and the contractual conditions attached to them, flow from the intended high level outcomes. Increasingly, the authority is taking full responsibility for the tactical or system design function, narrowing the service innovations that are open to the operator (often based on unsatisfactory outcomes from early tendering ventures). This is sometimes characterised as regulation re-entering the liberalization process.

Kain (2007), drawing on his review of UK and Melbourne rail franchising, suggests a tender competition requires:

- franchising design that does not undermine underlying network economics
- winning firms to be those that are capable of delivering the services most efficiently
- the anticipated gains from the competition having a high probability of being realized.

He sees these conditions as being most likely to be achieved if: gross cost contracts are adopted – with large service bundles for rail service; service specifications are tightly defined; bid assessment criteria are specific; entry barriers are low; transaction costs are low; state-owned operators are able to bid; contracts are short and complete; high-performance bonds are set; government avoids being a sleeping partner with the provider; a risk-averse, sceptical approach is taken to bid assessments; and cost risk and endogenous revenue risks are completely transferred.

The current author agrees with the three requirements but sees some alternative opportunities to aspects of their realization. The growing interest in maintaining performance pressure during the course of the contract, in an environment of increasing uncertainty (incomplete contracts), suggests a greater possible role for relational contracting with a solid foundation in trust. Transparency and accountability are central ingredients, to guard against regulatory capture, but a relationship-based approach, where the roles, rules and boundaries are clear and understood can release synergies that provide benefits at both the tactical and operational levels. Kain's (2007) warning about avoiding 'cosy' relationships with the franchisee is timely but can be avoided in a principled approach to partnering. KPIs on all three of the operator, the authority and the partnership are an appropriate way to subject this new approach to scrutiny.

In terms of partnering arrangements between purchaser and provider, guiding principles are required on matters such as (Stanley 2007):

- the terms of, and the process for, conducting a tender (if there is to be one)
- how adjustment of public policy over time will be managed
- commercial and financial arrangements – in seeking to achieve a balance between unleashing operator service innovation skills and meeting government/authority requirements for some form of universal service availability, on social grounds, the authority specifying minimum service standards and then bundling remaining service opportunities (kilometres or hours) for the operator to allocate to maximize patronage may provide an efficient way to pursue economic, environmental and social outcomes
- risk allocation and sharing – where risk allocation is increasingly likely to be consistent with gross cost contracts with incentives
- relationship management

- shared governance – Duncan (2007) argues that this is about sharing responsibility for the management of the relationship, based on common principles, not about one party seeking to manage the other
- contractual flexibility
- principled behaviour.

In a situation where an existing public operator can meet private sector benchmarks, continued public provision is defensible. If an incumbent private operator can also show that performance is at efficient levels and is effective, the case for negotiation with that operator has merit (with the threat of tendering in the event of a lapse and/or a limit on the scale of such renegotiation that is undertaken before market testing).

Public transport liberalization is still in its infancy. Research opportunities abound, as illustrated through the continuing Thredbo Conference series. As the process matures, a blend of performance pressure, partnerships based on trust, hard-edged accountability and a growing market, especially from climate change opportunities, will produce significant new understanding.

As a final observation, industry structure will be a point of future interest. There is a tendency to increasing concentration. It would be an ironic outcome of liberalization if a long-term result was the replacement of public monopoly by private monopoly! Drawing on Gwilliam's (2008) idea of a regulatory cycle, liberalization experience to date is suggestive of the probability for some future renationalization of rail operations in jurisdictions where aspiration on liberalization has far exceeded achievement.

NOTES

1. http://www.thredbo.itls.usyd.edu.au
2. Efficiency is often misinterpreted as minimizing costs rather than delivering a given level of service at the minimum cost.
3. The franchise essentially allows an operator to provide services on routes, or groups of routes over a contiguous area, for a specified period and under conditions set out in a contract, where the operator has some freedoms to plan services to deliver operational and passenger service efficiency.
4. Norway has a much lower share of tendered contracts than its Scandinavian neighbours.
5. See, for example, Munoz and Gschwender (2007).

REFERENCES

Alexandersson, G. and Hulten, S. (2003), 'The problem of predatory bidding in competitive tenders – a Swedish case study', paper presented to 8th International Conference on Competition and Ownership in Land Passenger Transport, Rio de Janeiro, Brazil, 14–18 September.

Bayliss, D. (2002), 'Review: urban public transport competition', *Public Transport International*, **3**, 4–9.

Bekken, J., Longva, F., Fearnley, N. and Osland, O. (2006), 'Norwegian experiences with tendered bus services', *European Transport*, **33**, 29–40.

Boitani, A. and Cambini, C. (2006), 'To bid or not to bid, this is the question: the Italian experience in competitive tendering for local bus services', *European Transport*, **33**, 41–53.

Department for Transport (2007), *Transport Statistics Great Britain 2007*, November, London: TSO.

Department of Treasury and Finance (DTF) (2000), *Passenger Rail Franchising in Victoria: An Overview*, April, Sydney: State Government of Victoria,

Duncan, K.C.H. (2007), 'Tactical level partnerships: legal perspectives and challenges', in R. Macario, J. Viegas

and D. A. Hensher (eds), *Competition and Ownership in Land Passenger Transport, Selected Papers from the 9th International Conference (Thredbo 9)*, Lisbon, September 2005, Amsterdam: Elsevier, pp. 437–52.
Gwilliam, K. (2008), 'Bus transport: is there a regulatory cycle?', *Transportation Research A*, **42**, 1183–94.
Hensher, D.A. and Stanley, J.K. (2008). 'Transacting under a performance-based contract: the role of negotiation and competitive tendering', *Transportation Research A*, **42**, 1143–1151.
Hensher, D.A., Yvrand-Billon, A., Macario, R., Preston, J., White, P., Tyson, W., van de Velde, D., van Wee B., de Aragao, J.J.G., Santos, EM., Filho, R.D.O. and Hensher, D.A. (2007) 'Delivering value for money to government through efficient and effective public transit service continuity: some thoughts', *Transport Reviews*, **27**(4), 411–48.
ICLEI Local Government for Sustainability (2003), *Better Public Transport for Europe Through Competitive Tendering: A Good Practice Guide*, funded by the European Commission Directorate General for Environment, Germany, Freiberg: ICLEI.
Kain, P. (2007), 'The pitfalls in competitive tendering: addressing the risks revealed by experience in Australia and Britain', Bureau of Transport and Economics Staff Paper, Australian Government, September.
Mathiesen, T.A. and Solvoll, G. (2008), 'Competitive tendering and structural changes: an example from the bus industry', *Transport Policy*, **15**, 1–11.
Munoz, J.C. and Gschwender, A. (2007), 'Transantiago: a tale of two cities', paper presented to the Tenth International Conference on Competition and Ownership in Land Passenger Transport, Hamilton Island, Australia, September.
Preston, J. (2004), 'The deregulation and privatisation of public transport in Britain: twenty years on', Transport Studies Unit, Oxford University lecture.
Sinisalo, K. (2007), 'Tendering and contracts in the Helsinki region', *Public Transport International*, **1**, 16–17.
Smith, A., Nash, C. and Wheat, P. (2009), 'Passenger rail franchising in Britain: has it been a success?', *International Journal of Transport Economics*, **36**(1), 33–62.
Stanley, J.K. (2005), 'Performance-based contracts in public transportation: the Melbourne experience', in D. Hensher, (ed.), *Competition and Ownership in Land Passenger Transport, Selected Refereed Papers from the 8th International Conference (Thredbo 8)*, Rio de Janeiro, September 2003, Amsterdam: Elsevier.
Stanley, J.K. (2007). 'Workshop B report: creating and maintaining trusting partnerships', in R. Macario, J. Viegas and D.A. Hensher (eds), *Competition and Ownership in Land Passenger Transport, Selected Papers from the 9th International Conference (Thredbo 9)*, Lisbon, September 2005, Amsterdam: Elsevier, pp. 423–36.
Stanley, J.K. and van de Velde, D.M. (2008), 'Risk and reward in public transport contracting', *Research in Transportation Economics*, **22**, 20–25.
Stanley, J.K., Betts, J. and Lucas, S. (2007), 'Tactical level partnerships: a context of trust for successful operation', in R. Macario, J. Viegas, and D.A. Hensher (eds), *Competition and Ownership in Land Passenger Transport, Selected Papers from the 9th International Conference (Thredbo 9)*, Lisbon, September 2005, Amsterdam: Elsevier, pp. 189–212.
UITP (International Association of Public Transport) Regional and Suburban Railway Committee (2006), 'Regional railways contracts for passenger transport', *Public Transport International*, **2**, 11–13.
Van de Velde, D.M. (1999), 'Organisational forms and entrepreneurship in public transport: classifying organisational forms', *Transport Policy*, **6**, 147–57.
Van de Velde D.M., Lutje Schipholt, L. and Veeneman, W. (2007), 'Competitive tendering in the Netherlands: central planning or functional specifications', paper presented to Tenth International Conference on Competition and Ownership in Land Passenger Transport, Hamilton Island, Australia, September.
Victorian Auditor General (VAG) (1998), 'Public transport reforms: moving from a service to a system', Report No. 5, Melbourne.
Wallis, I.P. and Hensher, D.A. (2007), 'Competitive tendering for urban bus services – cost impacts: international experience and issues', in R. Macario, J. Viegas and D.A. Hensher (eds), *Competition and Ownership in Land Passenger Transport, Selected Papers from the 9th International Conference (Thredbo 9)*, Lisbon, September 2005, Amsterdam: Elsevier, pp. 453–88.
Welsby, J. and Nichols, A. (1999), 'The privatisation of Britain's railways. An inside view', *Journal of Transport Economics and Policy*, **33**(1), 55–76.
Yvrande-Billon, A. (2007), 'The attribution process of delegation contracts in the French urban public transport sector: why is competitive tendering a myth', in R. Macario, J. Viegas and D.A. Hensher (eds), *Competition and Ownership in Land Passenger Transport, Selected Papers from the 9th International Conference (Thredbo 9)*, Lisbon, September 2005, Elsevier, Amsterdam: pp. 501–16.

17 Maritime ports

Lourdes Trujillo and Marianela Gonzalez

INTRODUCTION

In the majority of countries, most international trade (export/import), and in some cases also large shares of domestic trade, is done through maritime transport. This means that about 90 per cent of goods exchanged through international trade in the world go through ports since ports act as *interfaces* between maritime and inland modes of transport. This is why also they are at the centre of all intermodal policy decisions. There are 2814 international ports catering to freight traffic registered in the world. Port traffic increases at an average rate of about 3 per cent per year. All this defines ports as economic and service units of notable importance in the global economy.

The role of a modern seaport can be summarized in the following United Nations Conference on Trade and Development (UNCTAD) definition:

> Seaports are interfaces between several modes of transport, and thus they are centers for combined transport. Furthermore, they are multi-functional markets and industrial areas where goods are not only in transit, but they are also sorted, manufactured and distributed. As a matter of fact, seaports are multi-dimensional systems, which must be integrated within logistic chains to fulfill properly their functions. An efficient seaport requires, besides infrastructure, superstructure and equipment, adequate connections to other transport modes, a motivated management, and sufficiently qualified employees. (UNCTAD, 1996)

In the above definition, one of the main characteristics of seaports is the diversity of its activities. The opportunities to introduce competition in the delivery of port services and the optimal way of doing so depends on the size of the port and the service considered. The calibration of the actual opportunities to liberalize depends on the detailed knowledge of the structure and nature of the various activities. Moreover, since all port activities take place in a limited area, it is also relevant to study how they are coordinated, and what is the role that *port authorities* – or any other responsible institutions – must play in regulating seaports' infrastructure and activities.

Traditionally, in most countries ports have been owned and managed by public institutions. Public ownership in the seaport industry has usually been justified by the argument that seaports play a key role for national economies. Moreover, in some countries seaports are regarded as focal points for regional development, and accordingly, they justify the outlay of subsidies from governments for the building and improvement of port facilities.

However, it is not evident that public organization of this industry is necessarily the best option. In particular, tighter public budgets and increasing fiscal needs have led many countries to seek private participation in seaports. Private firms' involvement in ports is not new for the provision of services, since many firms were already present in ports around the world, but it is quite innovative in the construction of port infrastructures.

On the other hand, the continuous process of change in international transport management in the past ten years, from a segmented modal approach towards a much more integrated transport concept together with the incessant technological changes are resulting in an increasing pressure on ports to adapt their role and function to this more demanding operational environment. This entails the rethinking of national port development strategies, in particular, the need to define new partnerships between the public and private sectors in port operations, investment financing and asset management. This leads to a review of the respective roles of public and private actors, and specifically calls for a clarification of the regulatory aspect of this relationship and a particular attention devoted to transport safety and environmental protection (Juhel, 2001).

This chapter offers a revision of the characteristics of all the different services provided by seaports, and describes the approaches used worldwide to introduce private sector participation in the industry. The challenge that modern ports now face is to design more adequate regulatory mechanisms to guarantee efficient outcomes in a context of tight public budgets, particularly in developing countries. There are no unique answers applicable to every port, therefore we try to provide a panoramic view of the feasible models a port can follow, and the best practices observed worldwide.

The structure of the chapter is as follows: the next section presents a brief description of the economic characteristics of seaport activities. The third section describes the problems and challenges that traditional ports face in adapting to changes in maritime transport. In the fourth section we examine some of the more innovative international experiences. Finally, we present the conclusions.

DESCRIPTION OF SEAPORTS' INFRASTRUCTURE AND SERVICES

Economic Analysis of Seaports' Activity

Defined in broad terms, a seaport can be considered a single organizational unit that provides a service to ships. However, instead of a single unit, a seaport is better characterized in economic terms by considering it to be a multi product and multi-service organization (services to ships, cargo and passengers).

There are two basic economic characteristics that define the organization of seaports' activities. On the one hand, the infrastructure where these activities are performed – berths, quays, storage areas, and so on – is expensive to build and exhibits characteristics of indivisibility, that is, it is not possible to enlarge a port in a continuous way and must be built with determined minimum dimensions. On the other hand, the available areas for performing seaport activities are generally very limited. This space limitation implies that the number of feasible operators that can provide services within a port area is, by definition, reduced. Indeed, market size plus physical restrictions are the factors that typically preclude the possibility of competition at some ports.

Since there are many aspects involved, it is useful to divide seaport activities between: (a) infrastructure, (b) services provided by the port, which require the use of the former, and (c) coordination between the different activities performed at ports. The main characteristics of these three elements are analysed below.

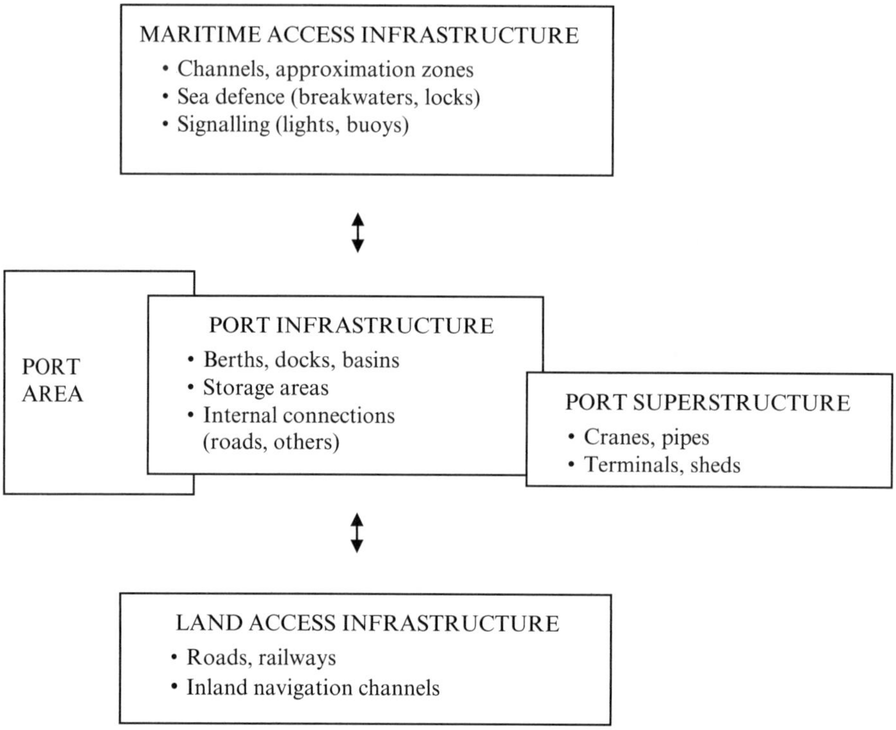

Source: Trujillo and Nombela (2000).

Figure 17.1 Scheme of seaports' structure

Description of Seaports' Infrastructure

The European Union uses an interesting definition of what is and what is not considered a port infrastructure (European Parliament, 1993). First, the *port area* is defined as a complex of berths, docks, and adjacent land where ships and cargoes are served. To reach that area, it is required to have infrastructures related to maritime access (channels, locks, aids to navigation, and so on) and to land access (connections to roads, rail network, and inland navigation). Figure 17.1 shows a scheme of the different types of infrastructures required by a port.

Therefore, the area where seaport activities takes place encompasses both the infrastructure within the port (berths, quays, docks, storage yards, and so on) and the superstructure. Among the elements forming the superstructure, it is possible to distinguish between fixed assets built on the infrastructure (sheds, fuel tanks, office buildings, and so on) and fixed and mobile equipment (cranes, van carriers, transtainers, and so on).

When discussing port infrastructure, it is convenient to define precisely those elements that are included and excluded. As can be observed, there are infrastructures outside the port area, but these are essential for the use of a port (maritime and land accesses). In general, port authorities have responsibility over some of the maritime access infrastructures – breakwaters, lights, buoys – and all elements within the port area. Meanwhile, connections to land networks and the remaining forms of maritime access (channels, locks) are generally owned and maintained by the central, regional or local government.

Description of Port Services

Besides the provision of basic infrastructures for the transfer of goods and passengers between sea and land, there are multiple services provided by different agents at ports, some of whom may even work outside the port area. These services cover all activities linked to the connection between port users and port, from the moment that a ship approaches a port until it ends all its operations (De Rus et al., 1994).

First, there is a group of services related to *berthing*, which include pilotage, towing and tying. All these services can be directly provided by port authorities, or they can be offered by private firms. Pilotage is defined as those operations required for a ship to enter and exit a port safely, and it usually implies the presence in the vessel's bridge (or at least a contact by radio) of an expert with sufficient knowledge of the zone to avoid risks. Pilots can be independent private agents in some ports, licensed by the port authority, while in other cases they are public employees. Towage is the operation of moving a ship using small powerful boats (named tugs) to steer it more easily. Again, it is possible to have private firms providing services for these operations, while in other ports tugs and their operators are directly hired by the port authority.

One of the more important services provided to cargo ships is what is generically labelled as *cargo handling*. This encompasses all activities related to the movement of cargo from/to ships and across port facilities. There is a historic separation between the operations of moving goods from ship's side until they are safely stored within the vessel (*stevedoring*), and those movements from berth to ship's side (*loading*), as a result of these operations traditionally being performed by different workers. Today, however, there are specialized firms that provide all these cargo handling services, using equipment such as cranes and surface transport elements.

The process of cargo handling varies according to the type of goods involved. There is a trend toward the specialization of firms according to the type of cargo, since the equipment required can then be specially designed to be highly cost-efficient. Thus, specialization leads to the formation of *terminals*, defined as specialized berths where all operations are mainly concentrated on a given type of cargo. Container terminals constitute the best example of this trend, since the handling of containers requires large gantry cranes, and land storage is relatively easy with adequate trucks and lifts, but it is highly space-consuming. All these factors make it more convenient for a firm to have a specially designed berth in order to handle containers more efficiently than general cargo berths.

Another type of service demanded by port users are those related to administrative paperwork and permits (sanitary certificates, import/export documents, taxes, and so on). These are usually performed by specialized agents or *consignees*, who are hired by shipping companies to arrange in advance the paperwork and all matters related to the use of port facilities by a ship. Even before a ship calls at a port, consignees start working to arrange that all services required (handling, repairs, supplies, and so on) are contracted for the ship and performed in the shortest feasible period. It is essential for a modern port to have systems to minimize the burden of paperwork for port users, since

BOX 17.1 PORT SERVICES

1. **Infrastructure provision**
2. **Berthing services:**
   - Pilotage
   - Towing
   - Tying
3. **Cargo handling:**
   - Stevedoring
   - Terminals
   - Storage
   - Freezing (fish, others)
4. **Consignees:**
   - Administrative paperwork for ships and cargo
   - Permits (sanitary, customs, and so on)
   - Service hiring
5. **Ancillary services:**
   - Supplies
   - Repairs
   - Cleaning, refuse collection
   - Safety

delays originating in inefficiency in administrative procedures result in large economic losses to shippers.

Finally, there is a series of other ancillary services performed by different agents and firms, working within or even outside the port area. In this group, all supplies to ships must be included, of which fuel and water are probably the most important. There are also services to crew members (for example, medical), and general common services such as cleaning, refuse collection, safety and so on. Some ports can also offer repair facilities to ships, which may involve the use of some special infrastructures.

In summary, there are many different services offered by a port. These services can be performed by a combination of public and private initiatives, and there are several models of ports indicating how private participation is introduced. A summary of all services described in this section is shown in Box 17.1.

Coordination between Seaports' Activities: Port Authorities

There are many different activities being performed simultaneously within the limited space of port areas, with ships constantly entering, being serviced and exiting. Therefore, there is a need for an agent to act as a coordinator to ensure the proper use of common facilities, and to take care of safety and the general design of port facilities. In most seaports, this function is played by an organization called the *port authority*. These are generally public institutions, where local interests are represented, but this configuration is not unique, and it is possible to find examples of purely private port authorities.

There are several organizational modes for seaports, depending on the role that port authorities assume. These are usually labelled as landlord port, tool port and services port (Juhel, 1997):

- *Landlord port*: in this model, basic port infrastructure is owned by the port authority, which is also in charge of its management, and leases it out to operators, mostly on a concession basis. In this way, port services are provided by private firms that own the assets conforming to the port superstructure and all equipment required for service provision (cranes, vans, forklifts, and so on). The port authority retaining all regulatory functions. Examples of this type of port organization are Buenos Aires (Argentina) and Rotterdam (Netherlands). In general, this is the most common form of organization for large ports. A world review of the top 100 container port demonstrates that almost 90 of these ports present this type of organization.

- *Tool port*: as in the landlord model, port authorities are also the owners of infrastructure, but in this mode of organization, they also own the superstructure (buildings and so on) and the equipment (cranes and so on). Private firms provide services by renting port assets, through concessions or licences. The port authority also retaining all regulatory functions. Examples of this category are Antwerp (Belgium) and Seattle (the USA).
- *Services port*: in this model, port authorities are responsible for the port as a whole. They own the infrastructures and superstructures, and they also hire employees to provide services directly. Some British ports have been used as an example to illustrate this type of organization, since its port authority (the Passenger Shipping Association – PSA) is the owner of all assets and it provides all services.

If a connection between the type of port and ownership is to be established, it can be concluded that port authorities of the landlord and tool models are generally public, while the port operators are private firms. Therefore, these two types could be classified as mixed ownership, since although the basic infrastructure is generally public, many elements of the port can be owned by operators. Meanwhile, services ports are more likely to be privately owned, where there is a single private firm operating the port as a single unit.

In principle, the role of port authorities should be exclusively confined to the provision of infrastructure and the coordination of port services. However, in many countries where there is no regulatory institution for seaports, port authorities perform many other tasks, such as investment planning and financing, or regulation of the tariffs that private operators charge to port users.

Moreover as Baird (2002) points out, the evidence suggests significant involvement of the private sector, especially in port operations and services, however, this does not detract from the fact that the public sector, in virtually all instances, takes much more than just a passing interest in its seaport system. Whether through a port authority, marine department, or other body, in the vast majority of countries the public sector retains a central role in seaport planning, regulation, development and investment.

Regarding the efficiency of seaports and the type of ownership, some authors point out that it is possible to find counter-examples in order to argue that there is in fact no relationship between efficiency and the type of port ownership. For example, Cullinane and Song (2002) confirm that, at least in the case of UK ports, it is extremely difficult to conclude that ownership constitutes a significant factor in port performance and efficiency; instead, factors such as geographical location and deregulation seem to have a greater influence on efficiency. In the literature on seaport industry, the comparison between Singapore and Hong Kong is often mentioned. Both are highly efficient ports, with the former being almost completely public and the latter private. As Notteboom and Winkelmans (2001) point out, the general belief that public port organisations are less responsive to demand and supply conditions than private companies is not always confirmed. However even though some remarkable exceptions do exist, there is no consensus about the relationship between efficiency and ownership (Gonzalez and Trujillo, 2009).

SEAPORT INDUSTRY REGULATION

Situation of Seaports before Privatization and Liberalization

In recent years a significant number of countries have implemented policies intended at reforming their port industry, based on the idea that it will improve efficiency and reduce the heavy financial burden placed upon governments. On the other hand, the port infrastructure during the 1960s and 1970s was badly maintained and often poorly managed in most developing countries and in some developed countries. The scenario changed drastically from the early 1980s onward, when technology advances like cargo containerization, induced the maritime transport industry to fundamental restructuring in service networks. Suddenly, there was no longer any distinction possible between ports in developing countries and in industrialized nations. The same problems confronted all ports worldwide (Juhel, 2001).

A traditional seaport, before the introduction of reforms reflecting changes in the maritime transport industry, could be described by the following:

1. National or local government budgets are used to finance the building of most large infrastructure construction costs, but public budgets are becoming tighter.
2. Port authority (generally public) finances the costs of maintenance and repairs for infrastructure.
3. Port authority is financed partly with public funds, and the rest by port tariffs and fees from private firms operating in the port.
4. There is an excess of employees working in the port, who have a high degree of unionization and strong positions at collective bargaining.
5. Port efficiency in terms of costs and waiting times for ships is relatively poor.

This list of stylized facts does not pretend to reflect the exact situation of all seaports in the world, but only to pinpoint the main questions that those seaports that have started introducing reforms have already faced. Moreover technological changes introduced in the last decades, such as containerization of cargo and development of larger and deeper specialized ships, have forced seaports to start a fast renovation of their facilities, in order to be able to provide services to the new needs of shipping companies and, in particular, to a growing demand for container handling services.

Therefore, because of all these reasons, there are now strong capital needs in the seaport industry, in order to make the necessary investments to attend to a growing demand, especially in the context of reduced public subsidies, which are due to tighter fiscal conditions among governments. In this way private financing becomes a very important option.

Thompson and Budin (1997) have identified several reasons that have promoted the introduction of private participation in transport industries. First, the private sector is able to provide services at lower costs than the public sector, since it usually is more productive and efficient. Second, if private capital is used to finance costs, the public sector can devote its scarce resources to other priority areas. Last, the private sector is generally more able to search for business opportunities, and to respond more swiftly than the public sector to changing conditions in competitive markets.

Introducing private participation at seaports appears to be a more attractive alternative for seaports to respond to the challenges pinpointed above. This is a trend observed worldwide, where ports in general are adopting the landlord model. Such arrangements have a long tradition in North American and West European ports. As described above, in this model port authorities keep the ownership of infrastructure in order to avoid the risk of monopolization of essential assets by private firms. These assets are then operated by the private sector. Since 1980 the organization of more than 100 ports in developing and transitional countries has been changed in this way. Concession contracts between port authorities and private firms are the most common instrument to allow private participation at ports.

The role of port authorities is thus transformed from their traditional role as institutions in charge of all port activities to a role in which they are only a coordinator of these activities. When introducing private firms at seaports, there is a new need to design *regulation systems* to monitor the behaviour of private operators. This regulation would usually take place in asymmetric information conditions (firms know their costs and market conditions better than the regulator does). However, it is not strictly required that port authorities perform this regulatory role at seaports. Instead, an independent institution could perform this task.

Forms of Private Participation at Seaports

Increased private participation is found in ports of most countries. However, full privatization of previously public ports is rare, mainly confined to the UK (Heaver et al., 2001). When choosing the best form to introduce private participation in the organization of port services, there are several alternatives depending on port size, initial conditions and the type of service considered. Among the different possibilities, the following can be mentioned (Trujillo and Nombela, 2000):

1. Selling the seaport as a whole (*full privatization*). Using this form, all assets and liabilities are transferred to the private sector, which can be justified by serious fiscal needs from the public sector.
2. Transferring to the private sector parts of the seaport for their development by private operators (*build, operate and own, BOO*). Short-term financial needs justify the use of this form of privatization.
3. Introducing private participation in the port in order to build or renovate facilities required for service provision (*build/rehabilitate, operate and transfer, BOT or ROT*). In this case, the public sector does not lose ownership of the port infrastructure, and even those new facilities built by private firms are transferred to the public sector after a specified period of time. This is the case of classical concessions.

4. Creating a new independent company, from the combination of efforts from two or more firms: *joint ventures*. This type of agreement arises when two parties with common interests join forces. Thus, for example, in some cases a firm can supply technology and know-how, while another might have knowledge of market opportunities and customer contacts. These agreements are not exclusively signed between private firms. There are examples of collaboration between port authorities and private firms.
5. *Leasing*: in some cases, port authorities simply rent port assets to be used by private operators during a fixed period, and thus they obtain income from contract fees. Contrary to concession contracts, in this case private firms are usually not required to make investments, therefore they only assume commercial risks. Some port facilities, such as storage buildings or cranes, are rented by operators under this scheme.
6. *Licensing*: in this case the port authority allows operators to provide some services which only require relatively simple equipment, and thus assets are generally owned by private operators. Infrastructure is provided for these operators to use it, generally for some specified fee, and in some cases they may also use some superstructure elements owned by the port authority. Stevedoring companies, pilots, tug operators or consignees can work under this type of agreement.
7. *Management contract*: a simple form of introducing private participation in a port is by contracting out the port management. In this situation, the port authority is the owner of infrastructure and port facilities, but decisions on its running are taken by a private firm which can provide a more commercial approach to operations. Both investment and commercial risks are in this case faced by the public sector, since managers do not invest their own capital in the port. The port of Bristol (UK) is an example of this type of contract, where facilities are owned by the local government, but the port is managed privately.

Needs of Regulation

When choosing among the options in the list above to determine which is the best alternative for a particular port, the port objectives must be evaluated, and the constraints that the port authority faces must be considered. The type of service may determine the possible degree of private participation. A basic determinant would be to consider whether the service requires the exclusive use of a port's fixed assets.

Services that do not require an exclusive use of infrastructure or superstructure port facilities

Within this group, there are services such as pilotage, towing, consignees and the other ancillary services to ships and crew. In many ports, as a result of safety arguments, there is a tradition that some of these services be provided by the public sector. In particular, all berthing services are in most cases considered a 'public service obligation' – that is, every port user has a right to be provided those services – and are therefore directly provided by port authorities to avoid the possibility of the service being disrupted.

Pilotage is a typical example of a compulsory service organized on a monopolistic basis in many ports. Pilotage is required for ships above some given capacity or length, and for dangerous cargoes. There are exemptions in most ports, such as for regular passenger services (ferries). The degree of public intervention in this service varies across countries. In some of them, pilots are civil servants, and thus they are subject to common state rules. In other cases, they are organized as independent agents more or less self-regulated by their own associations.

Regarding other berthing services (towage and tying), there are also diverse solutions among ports. Both are generally considered a 'public service obligation', and in practice are performed directly by port authorities or indirectly by licensed private operators. Towing services can be provided exclusively by a single firm, or in the case of large ports, it can be feasible to have several companies competing among themselves (De Rus et al., 1995). Some ports do not strictly control these services, and they only require some minimum conditions (technical capacity, safety and environmental standards, and so on) for private operators to have a licence to operate. In this case, towage tariffs can be determined by market conditions and not fixed by the port authority.

Therefore, it is possible to conclude that this set of auxiliary services to ships and crew can work reasonably well through a system of licences by which several operators are authorized to provide services within the port area. The activities of these operators can be regulated in terms of their prices and quality of services. In some cases, it is possible to have several operators competing among themselves (for example, consignees of pilots). It would then not be required to have strict regulation on their charges, unless collusive practices are detected.

For other services, as in the case of towage, it can be more complicated to have more than a reduced number of operators, since their number will be limited by port size. In the case of medium/small ports, it is clear that there is a need to establish some limits on prices and conditions of service, in order to avoid market domination by a few firms who may try to exploit their position to extract rents from port users.

Services that require exclusive use of assets

These services require the use of one of the most scarce resources at seaports: space. Thus, within this group, we would include terminals for cargo handling, storage areas, repairing docks and fuel suppliers. It is more complicated to introduce private participation in these services, since operators need to use assets that are considered to be optimally owned by the port authority. Therefore, concession contracts need to be written carefully in order to reconcile private operators' interests with port authorities' objectives. At the same time, contracts must include incentives for private operators to maintain or enhance assets as required.

The number of operators for these services is by definition extremely limited, although it will vary according to port size. Similarly, the need to establish some regulation over charges and quality depends on what type of port is considered and how many alternative ports are close to it. Thus, for example, in the case of a port within a region with a highly competitive environment, the port authority or the institution in charge of regulation does not need to be extremely concerned about excessively high charges by private operators. In that situation, private firms must self-control their prices, in order to avoid the risk of losing market share in favour of competitors.

A possible characterization of the different sizes and degrees of development that a seaport can reach is presented in Stopford (1997). Depending on port size, it is more or less required to have economic regulation of those ports where private operators use fixed assets. Broadly, we can distinguish two types of situations according to the degree of development reached by a seaport. First, there are those ports with a reduced market size that do not require more than a general cargo terminal, which can serve all kinds of goods and containers, or in other cases, they can have one terminal specialized in dry-bulk goods.

For these ports, it is possible to consider the introduction of some form of competition among those firms that are willing to operate in the port. Thus, it is possible to establish a system of auctions where private firms bid for the right to operate the terminal. Once the bidding process is over and a single operator is chosen, it is necessary to have some regulation over the charges that this firm imposes on port users, since otherwise it would enjoy a monopoly position. Price-cap systems or a rate-of-return type of regulation would constitute alternative options to regulate the behaviour of private operators, depending on the information and the experience that the regulatory institution might have on the type of service subject to regulation.

However, this need for regulation is less strict if there is competition *between* ports. In cases where a region offers alternative ports to shipping companies, there is less need to regulate prices charged at the terminal, since the market mechanism would make the private operator keep prices low or lose traffic. On the contrary, if those alternative ports do not exist, the private operator enjoys some market power that must be controlled by regulation. As an illustrative case, users of the Mexican ports of Veracruz and Manzanillo have repeatedly complained about high tariffs, and they are asking for a regulatory institution to be created in order to limit the monopolistic position of terminal operators.

In the case of large seaports, where there is a large enough volume of traffic, it is probable that competition can exist *within* the port. If a large port is divided into several independent terminals, it is possible to induce competition between operators for the traffic that calls at the port. In this way, regulation of prices is less of an issue, since if the market mechanism works reasonably well, private operators will restrain price rises themselves. However, some form of supervision would still be needed, since the situation is prone to collusion between competitors (due to the usually small number of parties involved).

As an example, the port of Buenos Aires (Argentina) was split into six different concessions to operate its terminals by independent companies. Although there were some initial problems and the market has eventually reduced the number of operators, there have been substantial improvements in port outcomes (Serebrisky and Trujillo, 2005).

Therefore, the conclusion derived from this analysis is that the introduction of private participation in the seaport industry appears as the more attractive option for ports trying to develop and adapt to the new conditions of the maritime transport market. Modern ports are in need of huge investments to enhance their facilities, in order to be able to provide those services demanded by shipping companies. Since in all countries it is increasingly difficult for governments to finance the required investments, the optimal solution is to try to attract private capital for investment in ports, and to also improve efficiency through the liberalization of port practices and the introduction of competition. The role of public sector institutions then changes from being direct providers of services to becoming regulators and supervisors.

The correct design of concession contracts for the collaboration between the public and private sectors is now the cornerstone of port systems. As discussed above, the existence of competition determines the need for the regulation of private operators, but even when competition is present and regulation not strictly required, there is still a need for port authorities to have some degree of control over the infrastructure assets that private firms are using.

PRIVATIZATION AND THE LIBERALIZATION OF SEAPORTS: SOME INTERNATIONAL EXPERIENCES

In general terms, the process of privatisation and liberalization of the main seaports around the world has been characterized primarily by the use of concession contracts introducing private participation, rather than by selling seaports' assets to the private sector. Through this process of concessioning, port authorities reduce their functions and are transformed into landlords responsible for coordinating all activities performed at the port. Consequently, they receive all rents accruing from asset renting.

There are several patterns of privatization and deregulation that the seaport industry is experiencing, depending on the region and the initial situation of the ports.

European ports

In Europe the model seems to be the increasing introduction of private firms in the provision of port services, but in general, infrastructures are kept within the public sectors and, in some countries, governments continue financing investment costs. However, there is a debate within the European Union (EU) on a plan to create a system of self-financing ports, which would not receive subsidies from governments (European Commission, 1997). The idea would be that port authorities should design port tariffs adequately in order to finance expansions of infrastructure, or seek private participation to finance infrastructure. In this region, limitation on the use of public funds is not mainly motivated by a lack of financial capacity from states, but on competitive considerations. It is argued that in order to have a single market where all ports compete on equal conditions, it is not fair that some states subsidize their facilities, while other seaports have to finance their facilities themselves.

The organizational structure of European ports is shown in Table 17.1 which was compiled by the European Commission based on research and questionnaires issued to the relevant of 13 member states of the EU (ports that participate in the survey represented the 50–60 per cent of the European port traffic at the end of 1998).

As can be seen, a variety of models exists. At one extreme, ports are run as departments of the national, regional or local administration, or under the exclusive auspices of a port authority. In both cases, the management body has the obligation to implement policy decisions taken elsewhere. In particular, the port authority has large responsibilities for port development, provision of infrastructures, safety, services and coordination of all activities, public or private, within the port. At the other extreme, ports established as private enterprises with managerial decision-making purely based upon economic considerations with no public influence at all. Between these positions lie the vast majority of EU ports, characterized by public influence in evidence but reserved, for example to planning, safety or land management.

Table 17.1 Organizational structure of European ports

	Public		Private	
	Type I	Type II	Type III	Type IV
Ownership	Public	Public	Mixed	Private
Autonomy of port management	Very restricted	Limited	High	Complete
Responsibility of port management	State-operated/tool port/landlord port	Landlord port (predominant)/ tool port	Full service port	Full service port
External public funding	Extensive	Important	Very limited	No public aid
Cost-recovery practice	Not principal objective	Partial recovery predominant	Full services, some infrastructure investments	Full costs recovery
Access to provision of services	Open tender/ direct agreement	Direct agreement predominant	Direct agreement	Normally closed
Relative importance in terms of traffic (*)	Limited (8%)	Very important (75%)	Limited (7%)	Limited (10%)
EU countries	Dk, Gr, F, P, D, I	B, Dk, Fin, F, D, Gr, NL, P, E, S, I	Dk, Ir, S, UK	Mostly UK, but also in other member states

Note: (*) Traffic estimated based on member states replies and best evidence available.

Source: Commission of the European Communities (2001).

Next we describe three of the European experiences in port reform. Two of them are referred to ports situated in the Le Havre-Hamburg Range, where ports have a significant hinterland and compete fiercely for cargo to destinations such as South Germany and Middle and Eastern Europe. The third example is the privatization that took place in the UK ports. The UK singular approach, characterized by the total sale of some ports, has received much criticism (Baird, 1995; Saundry and Turnbull, 1997; Thomas, 1994).

Port governance in the Netherlands has been characterized by municipality or region public organization that own and control ports with national government being only responsible for the major investment in port infrastructures. The three main Dutch ports (Rotterdam, Amsterdam and Zeeland seaports) are owned by local government and they are autonomously run. Since 1995 the Port of Rotterdam Authority was obtaining more autonomy from the municipality and finally, in 2004, the port authority changed from a municipal department to a public corporation. As a consequence of this transform, the port management is controlled by a supervisory board. Although the municipality of Rotterdam has been the only shareholder, since 2006 the national government has a minority share in order to secure effective spending of public resource (De Langen and ven der Lugt, 2007). The Amsterdam Port Authority is a department of the municipality of Amsterdam that became more independent from the municipality in 1998. In 1970 the Port Authority of Terneuzen was established and a year later was the Vlissingen Port Authority. Since 1998 these two ports are managed under a single port authority (Zeeland Seaports).

In Belgium there are four seaports, all situated in Flanders. Port governance is normally decentralized and these ports are managed at the municipal level or by autonomous bodies. None of these ports is completely privatized (Meersman et al., 2007). At the beginning, the municipal ports were subject to the same municipal regulation as any other municipal service, but they were transformed in autonomous municipal port companies.

The most radical reform in the European region has been introduced in the UK, where most ports have been fully privatized. The method used is extremely uncommon because it concerned the comprehensive transfer of all three core functions of the port: landowner, port operation and regulatory. The process was developed in two steps. The first one took place in the early 1980s by privatizing the Associated British Ports (ABP), an institution that had ownership of all former state ports, and involved the sale of state-owned ports and railway ports. In 1991, with the Ports Act, started a second step consisting of the transfer to the private sector of major trust ports. Apart from privatization a deregulation of the labour market was carried out in 1989 with the abolition of the National Dock Labour Scheme. For the port users this latter fact has been more important than the privatisation. In 2006, the top 20 ports handled the 84.5 per cent of the total UK port traffic. Of these ports, 15 are private ports, three are trust ports and two are municipal ports. Converse to the expected result from privatization, Saundry and Turnbull (1997) point out that privatization has actually constrained certain forms of competition and they warn about a potential overcapacity risk. For Baird and Valentine (2007) when a seaport is state owned, one of their main objectives is facilitate trade and promote economic and social benefit; however, when they asked private port owners about this type of questions, the majority answer has been to make money.

Asian and Pacific Ports

In some Asian and Pacific countries, an intensive process of transformation of public ports into companies with a commercial orientation is taking place. Private sector participation in financing infrastructure building was already begun long ago, as in the cases of Japan and Hong Kong. In both these countries, the main port facilities are financed, developed and operated by private firms under long-term concessions and only regulatory and strategic planning issues and the provision of land and supporting infrastructure continue in the public sector. In Thailand, terminals are leased out to private companies, while the port authority runs all ports in the country. In several countries in the Asian region (Korea, Philippines, Malaysia and China), privatization has led to shipping companies also actively participating in the development of seaports (for example, Maersk and P&O). We show three of the most interesting reform experiences that have taken place in Asia and Pacific region.

Over the past few decades, following the national economic reform, the governance structure of China's ports has considerably changed. Before 1984 the port governance was characterized by a high level of centralization with a complete planned economic system. The central government owned all ports and controlled all port activities and decision-making. In 1984 began a movement towards decentralization, so most of the ports were jointly managed by both central and local governments. The changes have derived in a more decentralized model in making decisions with more business orientation. A very important issue has been the introduction of private participation in the port sector and the promotion of foreign investment in the port industry, with a limit of 49 per cent. In 2004, the Port Law progressed in the decentralization and corporatization process replacing the port authority with both a port administration bureau (provincial or municipal) and a port business enterprise, which can participate in the open market. The central government has no longer been involved in the ownership of China's ports. With respect to the foreign private investors, the ceiling on shares in ports has been eliminated. In spite of the benefits of port reform, Cullinane and Wang (2007) report a possible overcapacity in the container terminals and point out some remaining problems: lack of skilled workers, insufficient length depth, poor road, rail and waterway services, delays and low quality level in some port services such as pilotage.

The Port of Singapore experience is very interesting. It has been one of the largest and most efficient public ports. Before its privatization, the Port of Singapore Authority had both regulatory and terminal operator functions. In 1996, following a privatization tendency in the country, a privatization and corporatization process was initiated. The corporatisation of the Port of Singapore Authority in 1997 transformed the port from its previous status as a government body to one where, although it remains completely government owned, it is independent of the government. This process has made it possible for the port to be more flexible and quicker in responding to other opportunities outside of the country (Tongzon, 2006). At the moment, the Port of Singapore goes on being one of the most efficient hubs in the world. This situation is due to not only the privatization but also the global economic condition and to factors like location and connectivity. In spite of its predominant situation, Cullinane et al. (2007) consider that the port must increase the quality, productivity and efficiency of the port's services because of the intense competitiveness of the ports in the region.

The mechanism for restructuring the Australian ports was also privatization and corporatization. This process took place at state level. Although some ports were privatized (all South Australian ports and Geelong and Portland ports in Victoria) the prevailing model was corporatization. Each state defined its own corporatization model and, in some cases, included elements of privatization. The main objective was transform the ports into commercial business with a market orientation; however, Everett and Robinson (2007) assert that this purpose has not been achieved.

Latin American Ports

Latin America is a very dynamic region in terms of seaport concessioning, building of new facilities, and importantly, rehabilitation of existing ports. The model of mixed public/private financing of seaports is quite successful in countries of this region, for three reasons. First, there is a strong need to seek capital to finance investments, since most governments are highly constrained in their budgets, due to debt servicing payments. Second, rapid economic growth is generating new traffic that demands new facilities and more efficient services. Third, fierce competition makes it necessary for ports to upgrade their facilities or risk being displaced by rivals.

In the 1990s many Latin American countries started a reform process that had some aspects in common: the dismantling of the public agency in charge of the management and operation of the ports, the decentralization of the system and the introduction of private participation. Most of the seaports granted autonomy and were allowed to introduce private participation in cargo-handling services, a practice that was prohibited before. There was also a movement toward the liberalization of port tariffs, with the objective of promoting competition between ports at a regional level. The main ports in each country have been subject to important reforms, and substantial private participation has been introduced through concession of terminals. Most of infrastructure has been passed to the private sector through concessions. Hoffmann (2001) points out that a difference between Latin America and Europe privatisation process is that in Latin America, after the privatization, there is less public involvement in planning, investment and regulation than in Europe.

Given the relevance of seaport reforms in Latin America, we describe three interesting cases: the experiences of Chile, Argentina and Mexico. Even though these particular countries are chosen, there are many others in the region that have already transformed their ports, or have advanced plans for that purpose. Other remarkable experiences in this region are those of Brazil, Panama, Nicaragua, Costa Rica, Colombia, Uruguay and Peru.

Until 1997 the Empresa Portuaria Chilena (Emporchi) was the owner of the 12 state ports. The management of Emporchi can be divided in two phases. From 1960 to 1980, Emporchi dealt as a service port. Seaports in Chile were characterized by the split of cargo handling among two different groups of workers. Stevedoring operations were performed by specialized port workers, while loading/unloading operations were done by Emporchi employees. Both groups enjoyed some monopolistic positions. In 1981 legal changes were introduced by passing a new Seaports' General Law, which effectively eliminated the monopoly of Emporchi in cargo loading operations, allowing private participation in the industry for those services. Almost simultaneously, a different law abolished the system of licences for stevedoring, allowing any worker to perform those services for shipping companies. These regulatory changes permitted the significant entry of new private operators and a competitive market for cargo handling was established. The impact on costs was substantial: for general cargo, cost savings of $17.7 per handled ton were obtained, while for dry-bulk goods the savings were estimated in $1.43 per ton, and for liquid cargoes there were no improvements (Tortello, 1998).

At the end of 1997, a law seeking to modernize state ports was passed, which sought to transform the Chilean port system and adapt it to the new needs of maritime transport. The law intended to introduce more private participation to achieve the objective of modernizing the ports. The law envisages splitting Emporchi into ten autonomous public companies, which will run the 11 state ports. These new companies will act as port authorities, managing ports' infrastructure, and are not allowed to provide cargo-handling or berthing services. The idea is that the new port authorities should contract all those services with private operators, through licences and concessions. All those concession contracts will be left under the supervision of the new autonomous port authorities (Hall, 1997). Although the decentralization and concession of the main ports was finished in 2004, three years later a new privatization wave was proposed but it has been postponed mainly due to the unions' press.

The reform in Argentina also began with a revision of legislation, with significant changes being introduced in 1992. Among these changes, it is remarkable to observe the decentralization introduced by the elimination of the central public agency responsible for ports (Administración General Portuaria, AGP), and the transfer of ports to regional governments. Small ports were transferred to municipal governments, while the large port of Buenos Aires was split into three separate zones: Dock Sud, transferred to the province of Buenos Aires; Puerto Sur, which is still to be developed; and Puerto Nuevo, which remains in the hands of the central state.

Other characteristics of the reform process were the deregulation of all port services, including those related to berthing such as pilotage and towage, and the elimination of restrictive working practices. Most importantly, the reform has introduced significant private participation in building and operating port infrastructure. Given its special role within the maritime transport industry, particular attention has been devoted to the development of container terminals, and the reform has attained large improvements in terms of the volume of traffic and productivity.

The port of Buenos Aires has been one of the more innovative world experiences in port reform. The state part of the infrastructure (Puerto Nuevo) was split into six terminals and concessioned for periods between 18–25 years, with the aim of avoiding the concentration of terminals by forcing bidders to win at most only one terminal each. In 1998, market conditions had changed the situation, and there was only two terminals competing in the container market within Puerto Nuevo. Even though the reduction in the number of operators indicates that the process could have been better designed and was probably implemented too fast, its results in terms of port outcomes have been positive (Serebrisky and Trujillo, 2005).

The success of seaport modernization in Argentina must be assessed by the increasing participation of private capital in the sector, but most importantly, by the changes that have been introduced in working practices. However, some issues remain to be improved, for example, customs, dredging, coastal navigation, dry ports, labour issues.

The reform of the Mexican port system started in 1993 and followed a pattern similar to that of many other reforming countries. Until then, the port system was managed centrally by a network of public firms. As part of the strategy used to reform the port system, two type of actions were followed. First, an adequate legal framework was needed to allow private firms to enter the port industry as operators. This was built in the new Ports Law passed in 1993. The second action was the dismantling of the public agency Puertos Mexicanos, responsible up to 1993 for the ports' network and was the only agency in the country authorized to build port infrastructures and to provide port services. The reform rested on three key instruments: decentralization, privatization and introduction of competition in the port system.

Decentralization started in February 1994, when the APIs were created and assumed the functions of planning, building infrastructures and promoting the port, apart from tasks on safety. They act as landlords since the Ports Law precludes them generally to act as port operators and requires them to contract with third parties. However, there are not full port authorities, since that role is legally attributed to Secretaría de

Comunicaciones y Transportes. The main 16 ports created APIs accountable to the federal government mostly, five APIs were controlled by state governments (tourism and fishing ports or local ports), and only an API (Acapulco) was completely private. The last phase considered for the process of ports' reform is the transfer of APIs to the private sector, by selling their shares to investors.

In almost every port, private participation was introduced through auctions for concession contracts between APIs and private firms to provide port services. Private participation has induced significant changes in the port industry, in terms of investments on infrastructure, and improvements in quality of service and tariffs. Thus, total investments in equipment and new terminals were around 6000 million pesos between 1995 and 1998, 60 per cent of which was private investment.

Liberalization and competition, between ports and between operators within ports, resulted from some restriction in the auctions for concessions and required liberalization of tariffs and elimination of cross-subsidies and barriers to entry. An important ruling was the initial restriction for firms not to win more than one concession on each coast (Pacific, Atlantic). This geographical restriction was later modified, and now the only restriction is that a firm must not gain a relevant position in the relevant market (this was applied, for example, for the tourist cruise markets).

The results of port reform in Mexico have been positively assessed by Estache et al. (2002, 2004) and Trujillo and González (2007), in terms of improvements in the average performance of the sector.

CONCLUSIONS

Overall the port reform rested on three key instruments, decentralization, privatization and the introduction of competition.

Decentralization implies that each port must have an autonomous, self-financing port administration, so that the government will have only a supervisory role over the system. This was pursued by creating an independent port administration at each port or group of small ports. Privatization implies that the port industry must be open to the participation of private investors for the operation of terminals and other facilities, and eventually even port administration. Liberalization implies free entrance of operators in the ports, it means the improvement of the competition between ports and between operators within ports, when the competition is not possible in some services, it could be necessary to have some type of regulation.

The process of privatization and liberalization of the main seaports around the world has been generally characterized by the use of concession contracts, rather than by selling seaports' assets to the private sector. The consequence of this process is that port authorities are transformed into landlords responsible for coordinating all activities at the port.

Consequently, international evidence suggests that the most common approach of port administration is a landlord model, with a public port authority which basically develops regulatory functions and controls maritime traffic and private enterprises which operate in cargo-handling and other port services, such as tug and pilotage.

REFERENCES

Baird, A. (1995), 'Privatization of trust ports in the United Kingdom: review and analysis of the first sales', *Transport Policy*, **2**(2), 135–43.

Baird, A. (2002), 'Privatization trends at the world's top-100 container ports', *Maritime Policy and Management*, **29**(3), 271–84.

Baird, A.J. and Valentine, V.F. (2007), 'Port privatisation in the United Kingdom', *Research in Transportation Economics*, **17**, 55–84.

Commission of the European Communities (2001), 'On public financing and charging practices in the Community sea port sector', SEC 234, Commission Staff Working Paper.

Cullinane, K. and Song, D. (2002), 'Port privatization policy and practice', *Transport Reviews*, **22**, 55–75.

Cullinane, K. and Wang, T.-F. (2007), 'Port governance in China', *Research in Transportation Economics*, **17**, 331–56.

Cullinane, K., Yap, W.Y. and Lam, J.S.L. (2007), 'The Port of Singapore and its governance structure', *Research in Transportation Economics*, **17**, 285–310.

De Langen, P.W. and van der Lugt, L.M. (2007), 'Governance structures of port authorities in the Netherlands', *Research in Transportation Economics*, **17**, 109–37.

De Rus, G., M. González, C. Román, B. Tovar and L. Trujillo (1995), *Competitividad de los Puertos Españoles*, Documento de Trabajo, Tribunal de Defensa de la Competencia, Madrid.

De Rus, G., C. Román, and L. Trujillo (1994), *Actividad Económica y Estructura de Costes del Puerto de La Luz y de Las Palmas*, Madrid: Cívitas.

Estache, A., González, M. and Trujillo, L. (2002), 'Efficiency gains from port reform and the potential for yardstick competition: lessons from México', *World Development*, **30**(4), 545–60.

Estache, A., Tovar de la Fé, B. and Trujillo, L. (2004), 'Sources of efficiency gains in port reform: a DEA decomposition of a Malmquist index for México', *Utility Policy*, **30**(4), 221–30.

European Commission (1997), *Green Paper on Sea Ports and Maritime Infrastructure*, Official Publications of the EU Communities, COM (97) 678 final.

European Parliament (1993), *European Sea Port Policy*, Directorate General for Research, Transport Series E-1, 7-1993.

Everett, S. and Robinson, R. (2007), 'Port reform: the Australian experience', *Research in Transportation Economics*, **17**, 259–84.

Gonzalez, M.M. and Trujillo, L. (2009), 'Efficiency measurement in the port industry: a survey of the empirical evidence', *Journal of Transport Economic and Policy*, **42**(2), 157–92.

Hall, K.G. (1997), 'Latin American ports proceed to privatize? Trend paces development of other sectors', *The Journal of Commerce*, 26 June.

Heaver, T., Meersman H. and Van de Voorde, E. (2001), 'Co-operation and competition in international container transport: strategies for ports', *Maritime Policy and Management,* **28**(3), 293–305.

Hoffmann, J. (2001), 'Latin American ports: results and determinants of private sector participation', *International Journal of Maritime Economics*, **3**(2), 221–41.

Juhel, M. (1997), 'Government regulation of port activities: what balance between public and private sectors?', *International Course on Privatization and Regulation of Transport Services*, Washington, DC: World Bank.

Juhel, M. (2001), 'Globalisation, privatisation and restructuring of ports', *International Journal of Maritime Economics*, **3**, 139–74.

Meersman, H., Van de Voorde, E. and Vanelslander, T. (2007), 'Fighting for money, investments and capacity: port governance and devolution in Belgium', *Research in Transportation Economics*, **17**, 85–107.

Notteboom, T. and Winkelmans, W. (2001), 'Reassessing public sector involvement in European seaports' *International Journal of Maritime Economics*, **3**, 242–59.

Saundry, R. and Turnbull, P. (1997), 'Private profit, public loss: the financial and economic performance of the U.K. ports', *Maritime Policy and Management*, **24**, 319–34.

Serebrisky, T. and Trujillo, L. (2005), 'An assessment of port reform in Argentina: outcomes and challenges ahead', *Maritime Policy and Management*, **32**(3), 191–207.

Stopford, M. (1997), *Maritime Economics*, London: Routledge.

Thomas, B.J. (1994), 'The privatization of United Kingdom seaports', *Maritime Policy and Management*, **21**, 134–48.

Thompson, L.S. and Budin, K.J. (1997), 'Global trend to railway concessions delivering positive results', Public Policy for the Private Sector, Note no. 134, December, World Bank, Washington, DC.

Tongzon J. (2006), 'Privatization: the Port of Singapore experience', Asia Research Center, Murdoch University, Australia.

Tortello, M. (1998), 'La organización de la estiba en los países de Mercosur y Chile: El modelo portuario de Chile'. I Congreso Internacional de Estiba, Puertos del Estado, Las Palmas de Gran Canaria.

Trujillo, L. and González, MM. (2007), 'Evolution of Mexico's port efficiency in a multiproduct context: a distance function approach', in O. Chisari (ed.), *Regulatory Economics and Quantitative Methods*, Cheltenham, UK and Northampton, MA, USA Edward Elgar.
Trujillo, L. and Nombela, G. (2000), 'Seaports', in A. Estache and G. De Rus (eds), *Privatization and Regulation of Infrastructures: Guidelines for Policymakers and Regulators*, Washington, DC: WBI Development Studies.
United Nations Conference on Trade and Development (UNCTAD) (1996), *Report of the Intergovernmental Group of Exports on Ports on its Second Session*, TD/B/CN.4/57, Geneva.

18 Liberalization in the water sector: three leading models

Claude Ménard and Aleksandra Peeroo

INTRODUCTION

Over the past 15 to 20 years, problems in the water sector have become more visible and pronounced. Developed countries struggle to provide the much needed investments to maintain and update their water systems. The main concern in less developed countries on the other side is the lack of access to safe drinking water which combined with inadequate sanitation causes more than 5 million deaths every year.[1] Being aware of these difficulties and the pervasive changes in other infrastructure sectors, one must note the remarkably slow pace of reform in the water sector. Moreover, the most systematic reforms until now have been implemented in developed countries although the situation of the water sector seems less urgent than in the developing world.[2]

By reform, we mean substantial changes in decision rights, changes that modify the governance and in many cases the allocation of property rights of the existing operator. Thus understood, reform can take many different forms. Since the early 1980s, its epitomized model has been that of liberalization, identified with the introduction of competition and, most of the time, with privatization. However, it might be relevant to retain a broader definition of liberalization, particularly in the water sector, in which competition in the market barely exists while competition for the market remains a challenge. In this chapter we consider liberalization in the continuation of Baumol et al. (1982) as reforms that introduce contestable markets. Beside full privatization, private sector participation in its different forms is therefore part of that movement. It can even be argued that corporatization, and the market pressure it intends to introduce on public entities, is a step in that direction since it makes public corporations more responsible and more responsive to incentives.[3]

This extended concept of liberalization might be useful to capture what has been and is still going on in the water sector. Indeed, full privatization remains an exception in the sector. In Western Europe, notwithstanding the isolated case of England and Wales, liberalization has mainly taken the form of private sector participation.[4] Correspondingly for less developed countries Gassner et al. (2009) show in their extensive survey of reforms of urban water and sanitation systems that state-owned enterprises (SOE)[5] still represent the standard case. Only 141 reforms translated into significant changes in ownership of the entities involved, under the form of private sector participation, of which only 10 per cent is divestiture (ibid., p. 24), with these changes mostly concentrated in Latin America and the Caribbean (Gassner et al. 2007, p. 29).[6]

The mixed results achieved by the reform movement make things even more complex. Indeed, although the literature on this issue is still very limited, data available show ambiguous outcomes when comparing the performance of reformed SOE with solutions involving private sector participation; and, similarly, there is no straightforward answer when comparing the performance of the various forms of private participation (Gassner et al., 2009; Ménard and Saussier, 2002; Wallsten and Kosec, 2008).[7]

In what follows, we focus on organizational alternatives involved in liberalization processes as defined above, in order to emphasize the challenging institutional changes at stake and the various difficulties these solutions face. Since most changes have occurred in urban water systems so far, we leave aside the problems of liberalization in the provision of water and irrigation in rural regions, and the difficult trade-offs it involves.

Our chapter is organized as follows: the next section comes back to the characteristics of the water sector. It describes some key features that may explain the slow pace of reform and that may also help better understanding characteristics and limits of that process. Then, a historical and global overview of the liberalization movement and the actual state of the water sector is presented. The third section examines more specifically the main drivers towards and factors of resistance to the liberalization process in the water sector. The fourth section looks at how these factors operate in three models of liberalization, illustrated with stylized examples from European countries.[8] The fifth section takes stock of this examination to point out challenges of liberalization in the water sector, which have to do with guaranteeing integrity and coherence of water systems. The conclusion then summarizes our main arguments and raises some important issues not discussed in the chapter.

CHARACTERISTICS OF THE WATER SECTOR

In order to understand the difficulties faced by reforms of urban water systems and the very limited steps towards liberalization, even in countries that championed this type of reform, we have to keep in mind some key characteristics of water systems.

Inherent Attributes in Comparison with Other Infrastructures

On the one hand, water systems share common features with other 'critical infrastructures' that have been liberalized.[9] It is a network industry that looks like the epitome of a natural monopoly. Urban water systems benefit from substantial economies of scale and recent technological developments facilitating the installation of insulated cables within water networks allow economies of scope.

On the other hand, there are substantial differences with other network industries, such as electricity or gas, not to mention telecommunications, which make urban water systems less prone to liberalization. First, the proportion of sunk costs in urban water systems has no parallel in other sectors. Second, urban water systems benefit from very substantial economies of density, so that competing entities would have little incentives to extend the network to areas with small and dispersed population. Third, because of the high costs of long-distance transportation, urban water systems are decentralized. Fourth, the implementation and maintenance of water systems involve important externalities: they can disrupt economic activities, as when repairs disturb urban traffic; they can benefit from urban planning; and they can have an immediate impact on health. Fifth, water has no substitute and concerns all human beings, with no exception. As such, it is perceived as a 'social' good that imposes strict constraints on operators, whether they are public or private, particularly when it comes to pricing and operation. Last, because there is no substitute, competing usages are a continuous and increasing source of tension.

As a consequence, urban water systems combine two sets of difficulties that hamper 'liberalization'. First, the need to maintain the technical integrity of these systems under the conditions described above imposes a mode of organization that locally remains almost everywhere a 'natural monopoly'. Second, because these monopolistic properties combined with a 'good' absolutely essential to human survival, water systems require institutions that guarantee the coherence between technical-organizational requirements and needs to be satisfied; this could well explain why water systems remain so tightly regulated, whatever the type of reforms implemented.[10]

Organization of the Water Sector

Indeed, the slow pace of reform in the water sector contrasts with the substantial changes that have affected other network industries. The push towards liberalization in the network industries was initiated in the developed countries and rapidly took the form of deregulation. It started in the airline industry in the USA, with the deregulation process in the late 1970s. The telecommunication sector in the USA followed shortly after. This movement accelerated throughout the 1980s with other countries endorsing similar reforms, first with telecommunications, then with energy, urban and rail transportation and, later on, water. What was acknowledged to work in the developed countries was then exported towards the developing world, particularly at the initiative and under the pressure of the World Bank in conformity with the so-called 'Washington Consensus'.[11]

Liberalization then became identified with privatization. However, the two concepts differ. Strictly speaking, privatization involves the transfer of property rights on assets from public to private entities. Liberalization refers to a broader and in a sense less specific policy, which is to introduce competitive pressure, without necessarily involving privatization. Whereas full privatization is often viewed as a way to reform infrastructures, with telecommunications as the paramount example, energy or, more recently, railways, it was and remains a very uncommon solution in the water sector due to its specificities, England and Wales being the main and most significant exception. However, rules designed to introduce competitive pressure have been adopted in many countries.

This trend towards liberalization in the network industries has been accompanied by substantial changes in the mode of regulation. The organization of the market is increasingly switching from a 'command-and-control' to a contractual approach. In the network industries, deregulation rarely meant full competition. It rather focused on introducing 'competition for the market' wherever implementation of 'competition in the market' was not possible or too costly, and often did so through contracts. 'Corporatization' of public entities, 'private sector participation' and 'public–private partnership' became various tools in the hands of market-oriented reformers. Over the past 20 years the number of people worldwide provided with drinking water by the private sector rose to 753 million[12] (Pinsent Masons, 2008, p. 30). The actual situation and its recent evolution for the different regions in the world are summarized in Table 18.1.

Although the exactitude of such aggregated data always remains debatable, the mag-

Table 18.1 Number of persons served by the private sector per region

Region	% people	
	2004	2008
Western Europe	45	44
C & E Europe	5	10
ME & Africa	4	6
South & Central Asia	0	1
South East Asia	6	12
Oceania	10	25
North America	19	21
Latin America	21	17
World total	9	11

Source: Adapted from Pinsent Masons (2004, p. 19, 2008, p. 40).

nitude is unambiguous; only about one tenth of the world's population receives its drinking water from a private entity, often under the form of partnerships in which private operators have a rather limited role. Public management is the prevailing mode of organizing the water sector, with an overwhelming majority of 89 per cent. Keeping the perception of water as a social good in mind, this result might not be that surprising. The USA and Canada, but also some European countries like the Netherlands and Germany, most of the Middle Eastern countries (for example, Egypt, Iran, and Lebanon), all South Asian states and most of Africa are examples for this type of water service provision. However, public management takes many different forms. There are major differences between direct public management, in which the provision of water services depends on the general municipal budget or an independent department of the local administration, and delegated public management, in which the municipality 'delegates' the provision to a communal enterprise which operates under company law. The switching from direct to delegated public management during a reform process is known as corporatization, and can be viewed as a step towards liberalization.[13]

A second interesting finding, as shown by Table 18.1, concerns the geographic concentration of private sector participation in the provision of water. Indeed, substantial private involvement exists primarily in Western Europe (mainly in the UK and France and to a much lesser extent in Spain, Greece and Italy), which confirms that so far most reforms involving the private sector have been implemented in developed countries. Table 18.1 also shows the state of private sector participation in Latin America, for the most part in Chile, but also in Columbia and Brazil.[14]

However, these numbers can be misleading in so far as providing an umbrella to many different forms of private participation. Full privatization, with the sale of the physical assets (pipes and so on) and therefore the transfer of the main property and decision rights to a private owner, is a mode of organizing the provision of drinking water that has found nation-wide application only in England and Wales and is otherwise merely very sporadically used (for instance in the Czech Republic or in Chile). The concept of public–private partnerships (PPPs) is much broader, including a variety of solutions,

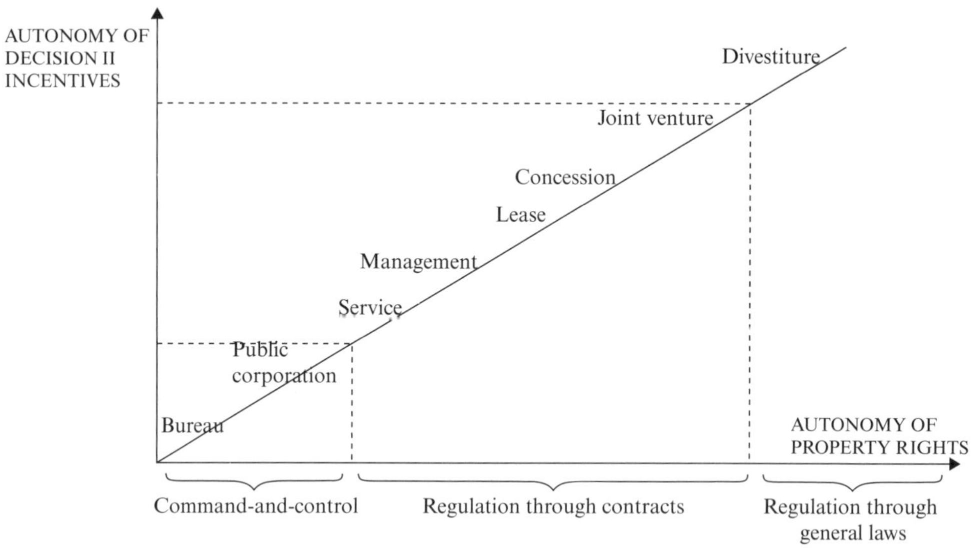

Source: Adapted from Ménard (2009, p. 97).

Figure 18.1 Different arrangements for providing water (or other public utilities in that respect)

which depend on the distribution of decision rights between the public and the private party. At one end of the spectrum of PPPs, commercial service contracts call the private entity to provide specific services on behalf of the municipality, such as billing or collecting invoices. Management and lease contracts attribute further responsibilities to the private operator, for example, the operation and maintenance of the water system. This form of PPP is often used when the necessary infrastructure already exists. Concessions are at the other end of the spectrum; the private partner has not only to operate and maintain the system, but also to provide the necessary investments. Figure 18.1 summarizes the main arrangements in the provision of water as classified by the degree of control over decision and property rights and the key variables at stake.

DRIVERS AND CHALLENGES TOWARDS LIBERALIZATION IN THE WATER SECTOR

The movement towards liberalization of urban water systems was mainly driven by financial and ideological factors. Yet, it turned out to be far less 'liberal' than in other sectors.

The Main Drivers Towards Reform

The most important driver towards water sector reform has been and remains finance (Finger et al., 2007, p. 3). While developing countries face the huge costs of expanding systems for an explosive urban population, public entities in developed countries have to deal with ageing networks that need heavy investments. Even in well-maintained systems, municipalities have to face investments required by tougher quality standards and new legislative frameworks. According to Hall and Lobina (2007) the urban wastewater directive of 1991 (91/271/EEC) provides one illustration of a regulation having a major impact by significantly increasing the costs of water services. In order to implement the standards required and to avoid increased taxes or responsibilities for higher prices of water, municipalities have a strong incentive to delegate the provision of water services to private partners, or to corporatize public entities in the hope of realizing efficiency gains.[15]

The inclination to do so has been amplified by the increasing technical complexity of water systems, largely due to the implementation of stringent quality standards (Massarutto et al., 2007, p. 198). Even in developed countries, municipalities or regional authorities often lack the necessary expertise, while private water companies have acquired technical know-how during long years of business. This may also help to understand why a handful of big companies dominate this market.

Moreover, the massive investments required, either because of higher quality standards, or because of the need to renovate ageing systems, or because of the demographic explosion in developing cities, are confronted everywhere with major budget constraints. For example, in the case of the European Union, the so-called 'stability pact' of 1992 imposed tougher rules on the budget of the member states, with a cascading effect even when water services fell under the responsibility of municipalities, since subsidies and/or financial support were increasingly restricted (Schouten and van Dijk, 2007, p. 30). Similarly, in developing countries, pressures towards liberalization and private sector participation took place in contexts in which the water sector crisis combined with major macroeconomic constraints (Shirley and Ménard, 2002).

These factors provided support to the belief that became dominant in the 1990s that liberalization would solve these problems at a lower cost than if water remained in the hands of public authorities. One must keep in mind that 'liberalization paired with privatization' was the great slogan of the so-called 'Washington Consensus', with the hope that process would feed growth.

Meeting Resistance

However, for reasons partially embedded in the specific characteristics of the water sector, coalitions of interests – including users – tend to develop resistance to that process almost everywhere. One reason might be that users are unwilling to see the provision of a good essential to their survival and without substitute in the hands of private monopolies. Most interestingly, this resistance seems to be less rooted in the fear of higher tariffs – although this might become an issue when pricing rules are perceived as unfair (Shirley and Ménard, 2002) – than in the perception that water is a critical infrastructure providing an essential public good that should remain in public hands or, at least, under extremely tight public control. A review of recent surveys clearly exhibits this trend when it comes to water (Bonnet et al., 2006).

Of course this resistance is fed by rent seeking groups. With liberalization identified to privatization or significant private participation and with the perspective that 'rationalization' will follow with major job cuts, unions tend to oppose these plans in order to maintain employment for their members. In several cases, such resistance led to changes in the initial plans. For example, in Munich unions successfully prevented the corporatization of sanitation services (Lanz, 2005). In some cases, public authorities and/or international donors have tried to anticipate these problems, accompanying reforms with 'sweetness' for employees (Pérard, 2007; Shirley, 2002.).

One last obstacle to liberalization comes from government opportunism (Savedoff and Spiller, 1999). This problem is rooted in the different lengths of the life cycles of a politician and a public utility, particularly of water systems where water pipes can last roughly 80 years if properly maintained. As a result, public officials do not feel too committed to the long-term effects on the network of the solutions they adopt. In the long run, providers of water services need to recover the high costs of their sunken assets, but in the short run they can survive with prices that cover only operating costs, which are well under average costs. This provides strong incentives for government opportunism – keeping prices low for electoral purposes. Consequently, private sector participation might be discouraged, particularly if public commitments are not credible, which depends on institutional endowments, notably political institutions. Note that political risks might also hamper public corporations, challenging reforms even when there is no change in formal ownership.

THREE MODELS EXEMPLIFIED BY EUROPEAN CASES

The distinct forms that liberalization can take in the water sector and the problems they face are well illustrated by the recent evolution of several European countries. As suggested above, the need to guarantee the technical integrity of water systems and the coherence between a system and the needs it must meet could well explain resistance to radical changes, the fact that liberalization has secured local integrated monopolies everywhere and that it happened at a slow pace.

In order to substantiate our argument, we focus on the urban water sector in three countries: England and Wales, France and Germany. In our view, they offer three alternative models of liberalization in the water sector: England and Wales have privatized their water sector as early as 1989; France, which has a long tradition of PPPs in that sector, has introduced significant changes to increase competitive pressure; and Germany has basically maintained the public management of its system, although with a trend towards corporatization. In what follows we emphasize structural characteristics typifying each form, but we also exhibit commonalities that, in our view, come from the need to align the modes of organization to socio-technical requirements of urban water systems.

England and Wales: Full Privatization, but Close Monitoring[16]

The example of England and Wales is unique in that no other country has completely privatized its water sector. This case provides a stylized example of how financial considerations backed by a strong ideology fed the most radical form of liberalization: full divestiture.

However, this 'natural experiment' has at least two distinguishing features with respect to the reform of other utilities. First, because of the characteristics of water systems emphasized above, the sector remains tightly monitored through the very visible hand of an independent regulator, the Office of Water Services (OFWAT). Second, and this makes the case even more interesting, while liberalization of the water sector in most countries goes hand in hand with a decentralization process, it went in the opposite direction in England and Wales, having been accompanied by an increased centralization. As early as 1973, municipalities started losing their grip over water services, with the creation of ten regional water authorities based on river basins. These authorities were not only responsible for water resource management, but also for the provision of water and sanitation services.

In 1989, these regional water authorities were privatized and their original tasks split. Water resource management fell into the hands of the newly created National Rivers Authority, a public entity that was replaced by the Environmental Agency in the mid-1990s. The newly created water companies, on the other hand, were legally bound to provide water and sanitation services. They hold both property and decision rights, and assume all the risks of a private company. At the same time, they operate under severe constraints imposed by several national regulators. Therefore, liberalization operates under a quite complex institutional arrangement.

Of particular interest for our purpose is the role of OFWAT, which is to promote competition and to protect consumer interests. Users actively participate in the regulation process through the ten consumer councils for water. These consumer councils are in direct contact with both the water companies and with OFWAT. By dealing with consumer complaints, the councils play an important role in supporting OFWAT's monitoring of the private water companies. Furthermore, the consumer councils for water and OFWAT are legally bound, by the Water Act 2003, to cooperate and exchange information. This can be of particular importance when major decisions like a merger of water companies are debated. The Office of Water Services operates as a classical regulator, not that different from the traditional regulatory commissions of public utilities, setting performance standards and service targets and, above all, fixing tariffs. The innovation is that OFWAT uses a price cap mechanism that, in principle, is reviewed every five years and automatically takes into account inflation, performance and efficiency. Through benchmarking, OFWAT is also in charge of ensuring (virtual) competition.

However, the five-year term was not fully respected. Owing to a rapid increase in prices paralleled with jumps in profits of operators that generated a public outcry, the regulator intervened and changed the rules of the game.[17] Moreover, there is a continuing debate on the effectiveness of a solution with almost no competition *in* the market. Therefore, next to the already existing yardstick competition which is a major tool in the hands of the regulator, new instruments have been introduced or are under consideration (Cave, 2009; Euromarket, 2004a, p. 341f;). One way is the so-called *common carriage* which consists in the sharing of one operator's physical assets (pipes). Another approach is to enhance *competition on the vertical supply markets* meaning that the water companies delegate intermediate activities to other party(ies). Furthermore, the introduction of *cross-border supplies* is meant to allow consumers to get a connection from another supplier. There are also self-contained water supplies, with privately owned wells providing water to one or several sites. These independent supplies do not fall under the authority of OFWAT and are therefore called *unregulated supplies*. Finally, within an *inset appointment* one water company replaces another in a specific area. It is the most important of these new instruments; nevertheless, use of it has been made in only 18 cases so far (Cave, 2009, p. 6).

However, 40 demands for inset appointments are currently waiting for approval (Cave, 2009, p. 98). It is possible that the implementation of these measures will amplify in the future since the government recently signalled its willingness to further increase competition in the water retail market – in line with the underlying wish to achieve a completely competitive retail market in the future, with customers freely choosing their provider, alike the electricity sector. Nevertheless, the movement into a real competitive water market is slow and loaded with barriers as two recent reports show (Cave, 2008, 2009).

To sum up, the need to guarantee the alignment of the new mode of organization to the monopolistic properties of water systems and the institutional design in which it is embedded, with the key role of the Parliament and of the central regulator as its 'arm', has ended in a complex arrangement in the UK case. Difficulties encountered in the implementation of a truly competitive environment, and the tight and finicky regulation that frame the entire urban water system command a moderate appraisal of the liberalization implemented through the UK privatization model.

France: More Competitive Pressure in PPPs, but Limited Risk-taking[18]

The French way towards liberalization is different. It is rooted in a long tradition of public–private partnerships. The first delegation contract in the French water system goes back to 1856. However, PPPs have not always been as widespread as they are now, owing to recent legal changes designed to favour a more open, competitive sector.[19] The main drivers behind the French liberalization movement are financial constraints, as in the UK, with ideological forces playing a much more subdued role. However, and this differs from the UK, it never reached the point of full privatization; and it was accompanied by a movement towards decentralization.

A turning point in that respect are two laws of decentralization passed in 1982, initiating a radical movement of transfer of competences from the central government to regional, departmental and municipal levels. Following these laws, the population supplied with water services under public management dropped from 50 per cent to 25 per cent within a few years.

The movement towards opening water systems to *competition for the market* was also fed by two other factors. Most French political leaders shared the widespread view conveyed by the worldwide push towards liberalization that PPPs would enhance efficiency. A snowball effect thus developed: impregnated with the new power associated to their extended responsibilities and simultaneously constrained by increasingly tight budgets, local public authorities succumbed to imitation. They found in widely publicized successful PPPs a source of inspiration for reforming their own water services, with the promise of efficiency gains and an alleviated financial burden. Another law adopted in 1999 (the so-called 'Chevènement law'), on inter-municipal cooperation, strengthened this trend by encouraging administrative unitization among small and medium size communities. Nowadays, only 21 per cent of the population gets their water through public management (Aubin and Varone, 2007, p. 47). Hence, 79 per cent of the population is provided by one of the three major French water operators: Veolia, Suez-Lyonnaise des Eaux or SAUR. This means that, while liberalization accompanied decentralization in France whereas it went side by side with administrative centralization in the UK, liberalization produced similar effects when it comes to the organization of the industry: it favoured big operators. This trend towards concentration was amplified by the increasingly demanding EU standards for drinking water and wastewater treatment. Being absurdly small sized, most French municipalities lacked the necessary know-how to comply with the tougher regulation and had to turn towards the expertise of the leading firms (Pezon, 2008).

However, liberalization remains limited. First, as almost everywhere in water systems, there is no competition in the market, while competition for the market attracts at best two or three competitors (most of the time from the big three). Moreover, 88 per cent of delegation agreements in the sector are lease contracts (*affermage* in French), which means that the responsibility to invest (and the associated risks) stays with the municipality while the operator maintains and operates the network and manages the services. In almost all cases, the private operator collects water bills, takes a share contractually agreed upon, and transfers the remaining part to the municipality for reinvestment in the system.

Hence, unlike in the British model, risks are not fully shouldered by the private operator but shared, so that there is a significant gap between property rights and decision rights. Another difference is that the system is monitored through contractual agreements, not through an independent regulator. However, contracts are tightly bounded. It was already so for PPPs before the push towards liberalization. In the 1990s, the eruption of corruption scandals linked to bribery for obtaining delegation contracts and other irregularities put the model under pressure (Domanski, 2006). In order to better monitor the contractual regime, partly through intensified competition for the market, new laws were introduced. The so-called 'Sapin Act' from 1993 intended to prevent corruption and to increase transparency, among other measures by making public tendering compulsory. Not long after, the so-called 'Barnier Act' (1995) specified the rules for PPPs, clarifying conditions of entry, restricting contract duration, and forcing public authorities to publish annual reports on performance. It was rapidly followed by another law, the 'Mazcaud Act', also from 1995, reducing the possibilities to extend contracts and obliging private operators to deliver annual reports to the delegating authorities.

How much these measures have enhanced competition in the French water sector is debatable. In a 1997 report, the French National Audit Office already highlighted the large number of exceptions to these laws, particularly with respect to the duration,[20] and the problems resulting from a highly oligopolistic market structure, with three players leading the ball. Moreover, the propensity to keep risks in the hands of public authorities might facilitate private participation, but it also severely restricts responsibilities for private operators. Whether liberalization has favoured competition therefore remains an issue.

Germany: Corporatization as a Step towards Liberalization?[21]

Of our three liberalization models for the water sector, the German model is clearly the less striking. Among the many factors that may explain this situation, a leading one might be embedded in the very nature of German political institutions. Germany is a federal state. Since each of the 16 states (*Länder*) has its specific law, the water sector is extremely fragmented, with thousands of small, local operators and many different modes of organization.

Consequently, no big operators dominate the national market, contrarily to the French or British situation. Only about 30 per cent of the German population obtains their water under arrangements involving private sector participation, and this represents only about 9 per cent of the total number of operating entities. The private sector is mostly active in big cities and in the eastern part of Germany (where, after the reunification, deteriorated water networks required heavy investments at a time of tight constraints on public finance).

In fact, PPPs remain the exception. Most of the water services are run under public management. Strong resistance towards liberalization developed in Germany, not only due to opposition by unions, but also because of persistent reluctance among the population to see water 'privatized'. A 2008 public opinion poll shows that 84 per cent of the Germans are against further private sector involvement in the supply of stately provided services.[22] And regularly, when German municipalities' citizens are able to decide in a referendum, there is a very strong reaction against private sector involvement in local public utilities. Illustrations are provided by the case of Münster, a city of 270000 people in the west of Germany, whose citizens voted 'no' to the privatization of the municipal multi-utility[23] in 2001 with a two-third majority and, more recently, by the case of Leipzig, a city of 511000 people in the east of Germany, where 87 per cent of votes were against a partial sale of 49.9 per cent of the stock to Gaz de France.

However, pressures to 'liberalize' the system have had an impact. Germany is moving away from the traditional *Regiebetriebe*, that is, from direct public management with water services run within municipal departments and financed by the general municipal budget. Severely criticized by the European Union (EU) and the Organisation for Economic Co-operation and Development (OECD) (Palaniappan et al., 2006, p.12f.), among others, as lacking transparency and being opaque, the old form of direct public management is progressively leaving way to 'corporatization'.

Due to the federal structure of Germany and its extremely fragmented water sector, it is difficult to find consistent data concerning the different modes of organization. However, a study of the German Association of Energy and Water Industries (BGW, 2005, pp.14 ff.) indicates an interesting development. The data on the water utilities show clearly that the German water sector has known a process of corporatization from 1986 to 2003.[24] Whereas direct public management in the form of *Eigenbetriebe* accounted for 63.3 per cent of the water volume in 1986, it was only 14.9 per cent in 2003 (and even only 4 per cent in 2005). During the same period, corporatized utilities have grown from 12.7 per cent to 30.2 per cent.

The main reason for this liberalization movement through corporatization lies in the growing financial pressure on municipal budgets. Municipalities hope to gain from corporatization in two ways. First, once the utility is corporatized, it is more difficult for users to point the finger at the city council and blame the municipal administration for a service that is not perceived as functioning properly. Secondly, there is the hope to benefit from efficiency gains that can be realized within a more disciplining mode of organization which also leaves room for more powerful incentives.

RESULTS AND TRENDS: COMFORTING, PUZZLING OR DISTURBING?

The three cases of England and Wales, France and Germany illustrate the many different forms that the introduction of market pressures can take in the water sector. A key argument behind the liberalization movement is that the introduction of private interest or at least the very possibility of private sector participation would enhance performance.[25] However, this remains much debated in empirical works.

Mixed Impact of Changes in the Allocation of Property and Decision Rights

At the empirical level, several studies on the effect of the reallocation of ownership on efficiency exhibit ambiguous results in the water sector. Studies from Ménard and Saussier (2002), Pérard (2007), Gassner et al. (2007, 2009) and Wallsten and Kosec (2008), among others, suggest that ownership may not be the key issue, which does not mean that it can be ignored. At the same time, Cave (2008, 2009) and a report from the World Bank (2009) emphasize the advantages of private participation, although not necessarily full privatization in the case of the World Bank. And a challenging fact is that major operators at the international level do not lobby that much in favour of full privatization.

Whereas the standard economic theory predicts inefficient public ownership, case-based evidence raises questions about such a general statement. Pérard (2007, p. 31) suggests that public water services can be managed very efficiently when the institutional setting is right, as he illustrates in the case of Tunisia, with efficient water services provided by two centralized agencies monitored by the government.[26] In this country, water is reasonably priced and represents 0.65 per cent of the per capita income per month.[27] At the same time, unaccounted-for-water (UFW) of 18 per cent keeps up with the 19.2 per cent of the completely privatized water services of England and Wales and is even better when compared with the 26.4 per cent of the French PPPs (BGW, 2005, p. 24).

Wallsten and Kosec (2008) confirm the view that ownership alone does not explain the performance of water utilities. It is rather the competitive pressure that makes the difference: benchmarking leads to a better performance. This might be compatible with public ownership, if benchmarking is implemented. Our three liberalization models may well substantiate this argument. The German water sector, with thousands of small, (mainly) public entities that allow citizens/consumers to compare price and quality, outperforms the two other models. According to the BGW study (2005, p. 24), which looked at several key indicators, water loss is at a high in France, with 26.4 per cent UFW , followed by England and Wales (19.2 per cent), while Germany exhibits a low 7.3 per cent of UFW.[28] Consistently, investments per cubic meter of water are highest in Germany (€0.55), followed by England and Wales (€0.47), with France significantly behind (€0.37). Last, the per capita water bill per year for a standardized water consumption accounts for €100 in England & Wales and €85 in France, while it accounts for €82 in Germany. Of course, these indicators are very incomplete and do not provide clear-cut 'proof' of the performance of the alternative models for liberalization. However they converge with some other troubling factors.

When it comes to full privatization, one could point out flaws in the institutional design of innovative solutions. For example, in the England and Wales case, the regulator (OFWAT) plays a major role with respect to the performance of the sector. Changes in its mission over time have generated uncertainties. While its initial role was to facilitate competition, it has progressively shifted towards actively promoting it, leading to tighter price caps over recent years. As a result, tariffs for water services in England and Wales are quite low. But at the same time this pressure on the operators leads to under-investment and the prospect of deterioration of the networks in the medium term. Furthermore, OFWAT's credibility has been undermined when the regulator did not enforce the fines it imposed on some operators. Moreover, new instruments have not proven to significantly improve competition. Whether or not this will change in the future, due to recent developments, such as the promised movement towards some retail competition, remains to be seen (Cave, 2009, p. 77).

If we now turn to the second model of liberalization, PPPs, we are immediately confronted with standard problems of tariff increases, under-investment, especially towards the end period of contracts, risk-averse strategies of operators so that public authorities tend to bear most of the uncertainties, and the very high rate of renegotiations, all of which questions the presumed efficiency of this mode of organization. In his extensive study of PPPs in Latin America, Guasch (2004) showed that 74.4 per cent of water contracts were renegotiated, on average only after 1.6 years! From a transaction cost perspective, these frequent renegotiations are very expensive: directly, because they bind the parties in a process that takes 3 to 12 months, needing experts, lawyers and so forth; but also indirectly since they undermine the credibility of the operators, the public authority and, ultimately, the supporting institutions. More recently, Gassner et al. (2009) concluded their review of the private sector participation in electricity and water distribution with some scepticism about the overall impact of PPPs.

Finally, if we turn to corporatization, it also confronts problems, particularly because of political interferences on tariffs and employment. A major issue with respect to the performance of public corporations concerns overstaffing, as is well illustrated by the German case of Munich. The ratio of employees per 1000 connections of the city's utility provider amounts to 19.5, albeit a ratio of 4 or 5 is usually considered enough to deliver satisfactory performance.

Expectations: Showdown . . . or Slowdown?

Taking these difficulties into account, what future can be expected for liberalization in the water sector, and particularly for private sector participation?

First, we have to bear in mind that the implication of the private sector in water services remains low. It represents only 11 per cent of the provision of drinkable water worldwide, although the percentage is higher in urban areas (25 per cent), since operators can take advantage of density effects, particularly in urban areas of developed countries (33 per cent). Moreover, the push towards PPPs, especially in developing countries has been quite bumpy (Gassner et al., 2007; Pérard, 2007).[29] Nevertheless, PPPs are still viewed as an important option for developing countries since they should allow access to capital and the necessary expertise. At the same time, data suggest that private operators are increasingly reluctant to assume risks of significant investments, which finds an expression in the growing role of management and lease contracts (Palaniappan et al., 2006), in which the main risks remain with the public party. In many cases, the trend is rather towards corporatization, with the end of mega-concessions (Saghir, 2006).

This revision of the initial optimism might very well affect developed countries as well. In the European Union, the Commission has passed a procurement directive intending to favour the delegation of public services and to intensify competitive pressures. However, the same authority has imposed tougher regulation as well as full cost recovery for externalities, thus introducing elements of uncertainty that might make private operators reluctant to assume too many risks in long-run investments (European Commission, 2008).

This may partly explain some significant evolutions regarding the liberalization of the water sector in the EU.[30] First, the English model of complete divestiture remains unique and has not been imitated.

Second, private sector participation seems to have been increasingly associated with the role of large groups, feeding an oligopolistic structure. Even within the English model, there has been a significant movement towards vertically integrated companies. As a result, the English authorities have amended the Water Industry Act from 1991 with the Enterprise Act (2002), which imposes tighter monitoring of future merger and acquisition projects.[31] Nevertheless, the trend towards integration remains powerful everywhere, including among public corporations. Pezon (2008) highlights the increasing role of supra-communal authorities integrating several water and sewerage services in France, particularly in rural areas.[32]

Third, several water companies facing increased financial pressures have proposed arrangements in which networks would be run by private non-profit organizations that would 'delegate' the service provision to for-profit companies (Euromarket, 2004a). Liberalization would then involve switching from public–private participation to private–private participation, with a quasi-public entity assuming almost all risks.

Hence, liberalization in the water sector is no longer identified with full privatization and is even less identified than before with substantial private sector participation. In France, private operators remain welcome, particularly in areas with a low population density since public authorities are faced with heavy investments required by tougher regulations that they can hardly face. Unfortunately these are also areas less attractive to private operators. Simultaneously, there are signs of pressure towards re-municipalization in larger and more profitable cities (Grenoble, already in the 1990s, and now Paris, and possibly other major cities).[33] In Germany, the Federal Agency for Hydrology has explicitly opposed the liberalization movement (Bundesanstalt für Gewässerkunde, 2002, p. 28), arguing that under the existing publicly monitored regime the quality of drinking water is exceptionally high in international comparison, with a highly secure supply. However, the need to improve the efficiency of the system has been acknowledged, and a working group has defined a modernization strategy that would rely on benchmarking (including among corporatized utilities) as well as on economies of scale and scope through integration of water and sewerage services, a sensitive issue in Germany because of the diversity of fiscal rules among municipalities and *Länder* as well as between these two sectors. Over all, there seems to be an agreement that the trend towards more concentration is widespread in Europe (Euromarket, 2004b). An issue then arises as to whether this will favour a handful of very large companies, be they private or public, thus creating a very oligopolistic market structure at the European level.

CONCLUSION

In this chapter, we have developed two main points. First, we have argued that 'liberalization' is a relatively vague concept, providing an umbrella for many different types of reform. For a time, it has been identified with privatization. However, full privatization in the drinkable water sector, on which we focused here, remains a very exceptional case, so far. If we relax the concept of liberalization in order to focus on the idea that it corresponds to the introduction of competitive pressures, then there has been unambiguously a powerful movement towards liberalization in the water sector, with the prevalence of two main models that are distinct from full privatization: PPPs, which take different forms of contractual arrangements that basically introduce more or less intense 'competition for the market'; and 'corporatization', that is, the transformation of services providing water into legally and economically autonomous entities accountable for their performance very much like a private firm would be, so that benchmarking becomes possible.

Second, referring to three 'stylized' examples from Europe, with comparable and converging environments in terms of standards of living, consumption of water, technologies in use, and so on, we have exhibited mixed results for the three "models" under review. One reason for this state of affairs is that modes of organization for providing water evolve slowly and are deeply embedded in specific institutions that combine national traditions, as with the French PPPs, and micro-institutions, as with the German local water utilities. But even more importantly, we have argued that there are specific characteristics of the water sector that may well explain resistance to liberalization, perceived as a transfer of property rights and/or of decision rights that would deprive citizens from their control over a non-standard good. These specific characteristics affect both the supply side, since water provision obeys conditions that are very close to that of a pure natural monopoly, and the demand side, since water is so crucial for human survival and therefore so politically sensitive.

As a result, drinking water provision is a highly exposed activity, that can hardly escape opportunistic pressures that take many different forms: political opportunism from public authorities, opportunistic behaviour from operators (whether they are private or public), and third party opportunism from interest groups (including consumer associations). Consequently, liberalization in the water sector is a mode of reform that has a dense and omnipresent institutional dimension. It is at risk of generating high economic and political transaction costs, thus creating uncertainties or even having dramatic consequences that feed suspicion regarding liberalization. Shirley and Ménard (2002) argued that this might contribute to explain the slow pace of reforms in the water sector, and the dramatic events it often requires before these reforms are initiated. It is most of the time only when major macro tensions, such as very tight financial constraints, combine with sector crisis, as when drinkable water does not reach significant segments of the population, that restructuring water systems comes high on political agendas.

Ultimately, for all these reasons, one can predict with a quite high probability that liberalization in drinking water systems and the introduction of competitive pressures will remain on the agenda and, at the same time, with limited effects. One may even suspect that with the increasing scarcity of the resource and the growing conflicts of interest among its users, public control will increase rather than decrease. We are entering a world in which strategic control over water becomes also a geostrategic issue.

NOTES

1. The literature on water issues is much more substantial than that which is available on sanitation, although epidemiologists have shown that safe water without adequate sanitation does not do the job (Esrey, 1996).
2. This is beginning to change, however: on the one hand, reforms become more far-reaching in less developed countries and, on the other hand, the provision of drinkable water becomes a more critical issue in high-income countries, its spectre already showing up in the southern belt of the European Union for instance.
3. The public debate about corporatization is ambiguous, with some viewing corporatization as a step towards full privatization while others consider it is a cover for public entities to escape liberalization.
4. According to Pinsent Masons (2008), it represents 40 per cent of the provision of water in this area. However, this might be an exception for the time being. In the USA, reforming the water sector remains very low on the agenda of policy-makers. Even in Europe, despite pervasive private sector participation in some countries, the liberalization of the water sector keeps being limited in scope and scale, as well illustrated in Finger et al., 2007.
5. By state-owned enterprises we mean all entities owned and monitored by public authorities, central, regional or local.
6. However, trends toward private sector participation seem to have developed in Asia over the last years. See http://ppi.worldbank.org/explore/ppi_exploreSector.aspx?SectorID=2 (accessed 7 April 2011).
7. For recent and challenging views about the successes of private sector participation, see World Bank (2009) and Cave (2009).
8. There are several reasons why we focus on these examples. First, the examples provide a relatively wide spectrum of alternative solutions for liberalization. Second, the institutions, the level of development of these countries and the characteristics of their water systems make comparisons meaningful and allow the argument to focus on the main issues at stake. Last, but not least, complex matters in developing countries could hardly be encompassed in the limits here.
9. By critical infrastructures, we mean infrastructures that are fundamental for the functioning of our societies, so that they must be reliable, safe and secured (Kunneke et al., 2010).
10. For a more detailed analysis of these two set of difficulties, see Ménard (2009).
11. This programme is explicit in World Bank (1994). For a summary and discussion of the 'Washington Consensus', see http://en.wikipedia.org/wiki/Washington_Consensus (accessed 7 April 2011).
12. This number being an estimate (the same author proposes 721 million, p. 40).
13. However, as already mentioned, it could also be viewed as a way to avoid privatization and real competition.
14. Recent data suggest important breakthrough of private sector participation in Asia (http://ppi.worldbank.org/explore/ppi_exploreRegion.aspx?regionID=2, accessed 7 April 2011).
15. See García et al. (2007, p. 57f).
16. For a relatively detailed overview of the UK water sector, see Euromarket (2004a: *Case Study UK*).
17. With the Water Act of 2003, OFWAT's mission officially switched from facilitating to effectively promoting competition. Since then, OFWAT's price caps have become tighter, squeezing the profits of the private water companies.
18. If not otherwise mentioned, the numerical data is from Euromarket (2004a).
19. The movement is still going on. A new law was adopted on 28 July 2008, to facilitate public–private partnerships in the provision of public infrastructures and services.
20. There is a possibility to extend contracts if initially unplanned investments have been made by an operator.
21. For an analysis of the German water sector see Euromarket (2004a: *Case Study Germany*) and Domanski (2006).
22. Public opinion poll undertaken by Forsa, January 2008.
23. In Germany, multi-utilities are a very common way of providing local public services like water, gas, and electricity through one provider, the Stadtwerke.
24. This trend is further confirmed by a more recent study (BGW, 2008, p. 14).
25. This is tying up to the considerations about contestable markets, where the simple threat of some other company chasing away the incumbent is sufficient to improve the latter's performance.

26. See also Touzi (2009). A similar conclusion can be derived from the case study on Santiago-de-Chile in Shirley (2002, ch.6).
27. Own calculations based on Baietti et al. 2006, p. 5 and SONEDE (http://www.sonede.com.tu/fra/E_SC.html (accessed 1 December 2009).
28. These numbers are consistent with data provided by the French Environment Institute (IFEN) and the economic regulator for the English and Welsh water sector (OFWAT), see http://www.ifen.fr/donnees-essentielles/eau/gestion-de-l-eau-potable-et-des-eaux-usees/la-distribution-de-l-eau-potable.html?taille= and http://www.ifen.fr/actualites/presse/detail-d-un-communique-ou-dossier/article/eau-potable.html?tx_ttnews[backPid]=879&cHash=012221ae42 for the French case and http://themes.eea.europa.eu/Specific_media/water/indicators/WQ06,2003.1001 for England and Wales (accessed 3 June 2009).
29. For detailed data on the evolution of PPPs, see the World Bank Private Participation Infrastructures database at: http://ppi.worldbank.org/explore/ppi_exploreSector.aspx?sectorID=4 (accessed 7 April 2011).
30. For an overview of a similar trend in the US, see Warner (2007).
31. However, as a reaction to the Cave report from 2008, the government considers reviewing the current merger regime, which has been criticized as being too rigid (Cave, 2009, p. 77).
32. The trend towards integration in the sector has also been noticed in developing countries (see Pérard, 2007).
33. A similar trend for water and other utilities might also be on its way in the US (see Warner, 2007).

REFERENCES

Aubin, D. and F. Varone (2007), 'Policies regulating the water supply and sanitation sector in nine European countries', in M. Finger, J. Allouche and P. Luis-Manso (eds), *Water and Liberalization European Water Scenarios*, London: IWA. pp. 34–53.

Baietti, A., W. Kingdom and M. van Ginneken (2006), 'Characteristics of well-performing public water Utilities', *Water Supply and Sanitation Notes*, note no. 9, May, World Bank.

Baumol, W.J., Panzar, J. and Willig, R.D. (1982), *Contestable Markets and the Theory of Industry Structure*, San Diego, CA: Harcourt Brace Jovanovich.

Bonnet, C., P. Dubois, D. Martimort and S. Straub (2006), 'Empirical evidence on satisfaction with Privatization in Latin America: Welfare Effects and Beliefs', Working Paper, IDEI, Toulouse, October.

Bundesanstalt für Gewässerkunde (BfG) (2002), *Water Resources Management – Country Profile Germany*, Report 27, Koblenz: BfG.

Bundesverband der deutschen Gas- und Wasserwirtschaft (BGW) (2005), *Branchenbild der deutschen Wasserwirtschaft 2005*, Bann: wvgw Wirtschafts- and Verlagsgesellschaft Gas and Wasser Mbth.

Bundesverband der deutschen Energie- und Wasserwirtschaft (BGW) (2008), *Branchenbild der deutschen Wasserwirtschaft 2008*, Bann: wvgw Wirtschafts- and Verlagsgesellschaft Gas and Wasser Mbth.

Cave, M. (2008), *Cave Review of Competition and Innovation in Water Markets. Preliminary Report*, http://www.defra.gov.uk/environment/water/industry/cavereview (accessed 23 February 2009).

Cave, M. (2009) *Cave Review of Competition and Innovation in Water Markets. Final Report*, http://www.defra.gov.uk/environment/water/industry/cavereview (accessed 3 June 2009).

Domanski, A. (2006), 'Corruption et Services Publics – Le Cas du Secteur de l'Eau', Master's thesis, Université de Paris (Panthéon – Sorbonne).

Esrey S.A. (1996) 'Water, waste and well-being: a multi-country study', *American Journal of Epidemiology*, **143**, 608–23.

Euromarket (2004a), 'Work package 4: analysis of the legislation and emerging regulation at the EU country level', in *Research Project on Water Liberalisation Scenarios: An Empirical Analysis of the Evolution of the European Water Supply and Sanitation Sectors, Community Research Project by the European Commission*, http://mir.epfl.ch/webdav/site/mir/shared/import/migration/EUROMARKET_D4_Final_Report.pdf (accessed 3 June 2009).

Euromarket (2004b), 'Work package 3: analysis of the strategy of the water supply and sanitation operators', in *Research Project on Water Liberalisation Scenarios: An Empirical Analysis of the Evolution of the European Water Supply and Sanitation Sectors, Community Research Project by the European Commission*, http://mir.epfl.ch/webdav/site/mir/shared/import/migration/D3%20-%20Work%20Package%203%20final%20version%20July%202004_1.pdf (accessed 3 June 2009).

European Commission (2008), 'Economics in water policy: the value of Europe's water', *Water Information System for Europe – Water Note on the Implementation of the Water Framework Directive 5*, Brussels: European Commission.

Finger, M., J. Allouche and P. Luis-Manso (eds) (2007), *Water and Liberalization. European Water Scenarios*, London: IWA.
Garcia, S., L. Guérin-Schneider and L. Breuil (2007), 'Analysis of the European Union explicit and implicit policies and approaches in the water supply and sanitation sectors', in M. Finger, J. Allouche and P. Luis-Manso (eds), *Water and Liberalization European Water Scenarios*, London: IWA. pp. 54–81.
Gassner, K., A. Popov and N. Pushak (2007), *An Empirical Assessment of Private Sector Participation in Electricity and Water Distribution in Developing and Transition Countries*, working paper, Washington, DC: World Bank: http://siteresources.worldbank.org/INTSDNETWORK/Resources/2007_June_Impact_of_PSP_in_elec_and_water.pdf (accessed 3 June 2009).
Gassner, K., A. Popov and N. Pushak (2009), 'Does private sector participation improve performance in electricity and water distribution?', *Trends and Policy Options*, **6**, Washington, DC: World Bank.
Guasch, J. L. (2004), *Renegotiating Infrastructure Concessions: Doing It Right*, Washington, DC: World Bank.
Hall, D. and E. Lobina (2007), 'International actors and multinational water company strategies in Europe, 1990–2003', *Utilities Policy*, **15**, 64–77.
Kunneke, R.W., J. Groenewegen and C. Ménard (2010), 'Aligning institutions with technology: critical transactions in the reform of infrastructures', *Journal of Economic Behavior and Organization*, **75**, 494–505.
Lanz, K. (2005), *Case study – Munich, Germany*, Watertime, Research project financed by the European Commission: http://www.watertime.net/docs/WP2/D34_Munich.doc (3 June2009).
Massarutto, A., E. Linares and V. Paccagnan (2007), 'Liberalisation and private sector involvement in WSS: the European experience', in M. Finger, J. Allouche and P. Luis Manso (eds), *Water and Liberalization, European Water Scenarios*, London: IWA. pp. 196–216.
Ménard, C. (2009), 'From technical integrity to institutional coherence: regulatory challenges in the water sector', in C., Ménard and M. Ghertman (eds) *Regulation, Deregulation, Reregulation, Institutional Perspectives*, Cheltenham, UK and Northampton, MA, USA: Edward Elgar, pp. 83–108.
Ménard, C. and S. Saussier (2002), 'Contractual choice and performance', in E. Brousseau and J.M. Glachant (eds.), *The Economic of Contracts. Theory and Applications*, Cambridge: Cambridge University Press. pp. 440–62.
Palaniappan, M., H. Cooley, P. Gleick and G. Wolff (2006), 'Assessing the long-term outlook for current business models in the construction and provision of water infrastructure and services', paper for the Global Forum on Sustainable Development, OECD, Paris, November.
Pérard, E. (2007), *Private Sector Participation and Regulatory Reform in Water Supply: The Southern Mediterranean Experience*, Development Centre Working Papers, Paris: OECD.
Pezon, Ch. (2008), 'Intercommunalité et Durabilité des Services d'Eau Potable et d'Assainissement', Rapport, Ministère de l'Environnement et du Développement Durable (MEDD).
Pinsent Masons (2004), *Water Yearbook 2004–2005*, London: Pinsent Masons.
Pinsent Masons (2008), *Water Yearbook 2008–2009*, London: Pinsent Masons.
Saghir, J. (2006), 'Public–private partnerships in water supply and sanitation – recent trends and new opportunities', presentation for the Global Forum on Sustainable Development, November, OECD; Paris.
Savedoff, W. and P. Spiller (1999), *Spilled Water – Institutional Commitment in the Provision of Water Services*, Washington, DC: Inter-American Development Bank.
Schouten, M. and M.P. van Dijk, (2007) 'The European Water Supply and Sanitation Markets', in M. Finger, J. Allouche and P. Luis-Manso (eds), *Water and Liberalization European Water Scenarios*, London: IWA pp. 11–33.
Shirley, M.M. (ed.) (2002), *Thirsting for Efficiency – the Economics and Politics of Urban Water System Reform*, Amsterdam: World Bank.
Shirley, M.M. and C. Ménard (2002), 'Cities awash: a synthesis of the country cases', in M.M. Shirley (ed.) *Thirsting for Efficiency – the Economics and Politics of Urban Water System Reform*, Amsterdam: World Bank, pp. 1–41.
Touzi, S. (2009), 'La nécessaire évolution des impératifs d'équité et d'efficacité dans la gestion de l'eau potable en Tunisie', PhD dissertation, University of Corsica.
Wallsten, S. and K. Kosec (2008), 'The effects of ownership and benchmark competition: an empirical analysis of U.S. water systems', *International Journal of Industrial Organization*, **26**, 186–205.
Warner, M. and A. Hefetz (2007), 'Beyond the market versus planning dichotomy: understanding privatization and its reverse in US cities', *Local Government Studies*, **33**(4), 555–72.
World Bank (1994), *World Development Report 1994: Infrastructure for Development*, Washington, DC: World Bank and Oxford: Oxford University Press.
World Bank (2009), *2008 Annual Report*, Washington, DC: World Bank.

19 Liberalization in the postal and delivery sector

Michael A. Crew and Paul R. Kleindorfer

INTRODUCTION AND ORIGINS OF LIBERALIZATION IN THE POSTAL SECTOR

Modern postal service began with the reforms originating with the Penny Post proposed by Sir Rowland Hill in his famous 1837 pamphlet 'Post Office Reform: Its Importance and Practicability'.[1] Prepayment by means of a stamp was a significant innovation in 1840. For the price of a stamp, delivery anywhere in the country was provided. The twin notions of ubiquitous delivery and uniform price have been essential aspects of the modern postal service since the 'Penny Black' was first issued in 1840. Major steps in improving and automating mail and parcel processing operations occurred during the nineteenth and twentieth centuries. However, the major revolution in postal and delivery services did not occur until the second half of the twentieth century.

At the end of the twentieth century, a wave of technological advances based on innovations in optical fiber and information technologies began to transform infrastructure industries. The importance of integrating the hitherto isolated postal sector with other communication technologies led to the recognition of the necessity to rethink the governance and regulation of postal service providers. Some would suggest that this process began with the laws that changed national post offices into more independent, 'business-like' government agencies, for example, the Post Office Act of 1969 in the UK and the Postal Reorganization Act of 1970 in the USA. Others would point to the division of ministries for post, telephone, and telegraph (PTTs) into separate postal and telecommunications agencies, for example, the British Telecommunications Act of 1981 or the German Postreform I of 1989. In our view, however, the most convenient 'start date' for the present wave of fundamental postal reform and market liberalization is 1988, when the European Union began work on a regional postal policy and the government of New Zealand first broached the idea of repealing the postal monopoly.

From these stirrings a wave of postal reform has spread across the globe. Landmarks include publication of the European Postal Green Paper (1992), publication of the Australian Industry Commission Paper (1992), repeal of the Swedish postal monopoly (1993), substantial privatization of the Dutch Post Office (1994), adoption of the European Postal Directive (1997), setting an end date for the German postal monopoly and the creation of a German regulator in Postreform III (1997), repeal of the New Zealand postal monopoly (1998), creation of a British postal regulator with authority to end the postal monopoly (2000), substantial privatization of Deutsche Post (2000), privatization of the Japanese Post Office (2005) and adoption of a third European postal directive setting an end date for all postal monopolies in Europe (2008).[2]

Postal reform may be considered part of a broader movement worldwide, which included privatization of network industries. The intellectual foundations of the movement might be traced to an uprising in economic thinking in the 1960s and 1970s, which argued for the greater use of market forces and greater commercial orientation rather than bureaucratic structures in major sectors of the economy. This uprising was visible in different countries at different times, with sectors such as telecommunications and energy leading the way, with privatization, new regulatory forms appearing, lowering of entry barriers, and other forms of market liberalization. The postal sector seemed immune to these currents for a long time. However, the increasing need to integrate postal operations with other means of communications, and the increasing need to come to grips with electronic substitution and technology change have pushed the postal sector inexorably in the direction of commercialization and markets. Finding an appropriate confluence with these trends has come to be seen as the primary objective of postal reform.

This chapter examines a number of aspects of postal reform and associated market liberalization that have developed to meet the challenges described above. The next section reviews the major drivers of postal reform in a number of countries and the following section discusses examples of countries that are currently undertaking such reforms. The final section concludes with some thoughts on open research questions in this area.

POSTAL REFORM WORLDWIDE

Postal reform is taking firm hold around the world. The driving force is undoubtedly the changes that have been occasioned by increased competition, first, by the emergence of substantial private express companies in 1980s and, second and more importantly, by the evolution of the Internet-based services in the 1990s. Electronic competition, in particular, means that national postal operators (hereafter POs) are facing declining demand for the letter mail that not only provides higher margins than, for example, newspapers and advertising mail, but has traditionally been the *raison of d'être* of the government post. For most governments, this increasingly competitive environment clearly implies the need to allow POs to operate more efficiently and more like commercial companies as opposed to government bureaucracies. This transformation requires not only operational changes but also the freedom to price in a more commercial manner. At the same time, governments are unwilling to allow POs to follow commercial logic to the point of cutting back on politically popular services in areas of the country or for portions of the population. Therefore, in granting some commercial freedom, governments have also found it necessary to define with precision the standards of universal postal service which must be maintained, that is, the universal service obligation (USO). In addition, governments, or POs, still have to finance the USO as traditional postal markets are being eroded. Yet, at the same time, it is generally agreed that the need to finance the USO or expand into new markets neither justifies the extraction of monopoly rents from captured mailer nor the use of predatory pricing against competitors. This three-way balancing between commercial freedom, the USO, and fair competition has been a continuing theme in the postal reform area and will continue to be for the foreseeable future.

At the start it should be recognized that 'liberalization' and 'reform' have very different meanings, depending on the country. The term is used here rather broadly to represent legal and institutional changes that are focused on two policy dimensions: (1) the governance of the national PO; and (2) the structure of the market. On the first dimension, POs have undergone changes in many countries from being government organizations to corporate structures, to more commercial undertakings and even to privatized entities. On the second dimension, changes have occurred both upstream and downstream to allow or encourage competition. Upstream, these changes have been to promote 'worksharing' (mail preparation, collection and pre-sorting activities provided by customers or third parties in direct competition with the incumbent PO). Downstream these changes have been to lower progressively the weight or value threshold above which organizations other than the PO are allowed to deliver mail items to homes or businesses. Liberalization efforts began with reducing the reserved area, namely, the weight or price threshold below which the PO has a statutory monopoly. The range from relatively minor upstream changes, as in the USA, all the way to full market opening (FMO) and the complete removal of the reserved area, as in the EU. The term 'liberalization' is used to cover this whole spectrum of activity, but it is obvious that every country has its own story and needs to be evaluated in specific terms. Nonetheless, the drivers that have led to liberalization have been often the same, namely, declining or stagnating mail volumes resulting in part from electronic substitution (often called e-substitution) of First Class and transactions mail, and the lack of productivity improvement and product innovation in the PO in many countries.

Concerning e-substitution, postal service, especially the delivery of letters, is faring badly as a result of technological change in the economy that has improved the prospects of other sectors at the expense of mail. Optical fiber and computers have had a dramatic impact on the economy. The Internet has grown from a tool used by scientists to a ubiquitous phenomenon. The marginal costs of employing the Internet, for example, are near zero, in contrast to the marginal cost of a traditional letter. The impact of the Internet is truly monumental and it will be a significant and continuing threat for the traditional postal service. Mail has faced threats in the past from new technologies, notably telephone and facsimile (fax). The difference between the threat from the Internet derives from its ubiquity relative to fax, much greater scope, multimedia features, email and effectively zero marginal cost. In addition, while the Internet and associated developments in information technology have had a considerable impact on cost reduction, quality and new product offerings in other sectors, the same cannot be said for the postal sector. Partly this is the result of the labor-intensive nature of postal operations, but it also reflects in many countries the lack of focus of public organizations on customer needs and technology-driven innovation.[3]

Adding to the problem of stagnating mail volumes is the continuing central importance of the USO. The USO has been a dominant feature of postal service for many years. Its origins lie in the Roland Hill reforms 1840 which introduced ubiquity and pricing uniformity. Since then the USO has been expanded to include post office density and collection access requirements, and quality standards. The USO has traditionally been maintained by restricting entry to the postal market by means of a 'reserved area' (RA). The reserved area was defined in terms of weight or value thresholds. The prohibition of entry in the reserved area prevents entrants from cream skimming by entering the low-cost markets and undercutting the PO. On the other hand, prohibiting entry leads to a weakening stimulus for innovation and optimizing customer value. So, there has been continuing pressure in the postal sector and in other previously monopolized businesses (for example, energy and telecommunications) toward liberalization. The fundamental tension in the liberalization debate is the presence of large fixed costs from the USO, and the need to maintain volumes and the scale economies that are a traditional feature of postal service. At the same time customers, and especially larger mailers, have pushed for reforms in order to promote productivity and product innovations. The tension is made all the greater since most of the business of large mailers is in low-cost urban areas, which would see significant decreases in prices under entry if the PO is required to maintain a costly USO financed in part by margins on mails delivered in the low-cost areas.

The European postal sector has been the showcase for the tension just noted. The debate on increased competition has focused primarily on the desire to establish an internal European market and on integrating postal services with other communications markets. Other forms of communication have seen great technological advances (telecommunications, mobile telecommunications and the Internet) and these often have important implications for changes in postal products as well. The emerging substitutes have gradually increased the competitive pressure on letters and have resulted in a decline of overall mail demand in many countries, as several recent country studies have noted.[4]

Other central factors driving postal reform and liberalization have included shifting consumption patterns, at the level of corporations as well as households and small and medium-sized businesses. These shifts have resulted in an increased need for refined market and product segmentation with corresponding pricing policies and service performance. Improved and new technologies allow the POs, but also other actors in the postal market such as large mailers and consolidators, to increase the efficiency of their logistics processes, leading to higher flexibility, more capacity to treat large volumes, and improved performance at lower costs. In addition to these economic and market factors, increasing budgetary restrictions have made national governments less willing to finance the deficits of public service organizations, including POs. Thus, increasingly, policy makers are feeling the need to review (and perhaps adapt) the principles of the postal market model to allow POs to be self-sustaining.

Given these complexities of interacting markets, changing consumer demands, and the financial and social importance of the postal sector, it is not surprising that market-based solutions, that is, liberalization, have become the focus of attention for postal reform. At the same time, it has been recognized that if there is not a reasonable balance between the burdens placed on the PO and its USO, competition could result in inefficient cream skimming and even financial insolvency of the PO. Recent literature in postal economics has pointed to several general themes and principles underlying the balance of the burden of the USO and commercial freedom allowed the PO.[5] The four basic themes that have emerged as important are: (1) the postal market characteristics in the country; (2) the structure and efficiency of the incumbent PO/universal service provider (USP); (3) the scope of the USO in the country; and (4) the nature of regulatory regimes imposed. These four interdependent themes are summarized in Figure 19.1, together with potential impacts associated with these that could arise from liberalization.

As shown in Figure 19.1, the economic outcomes at stake are, chiefly, the viability of the PO (and potential financing needs to support the USO), the effects on consumers as measured by price levels and the maintenance of the USO, and employment effects in the postal sector. In addition, the question of whether workable competition can be established and can provide the benefits of market discipline and innovation remains of central interest in the liberalization and postal reform debate.

Considering the factors presented in Figure 19.1, d'Alcantara and Amerlynck (2006) and PwC (2006) analysed these in detail for a number of European countries. Typical values for these parameters are presented in Table 19.1.

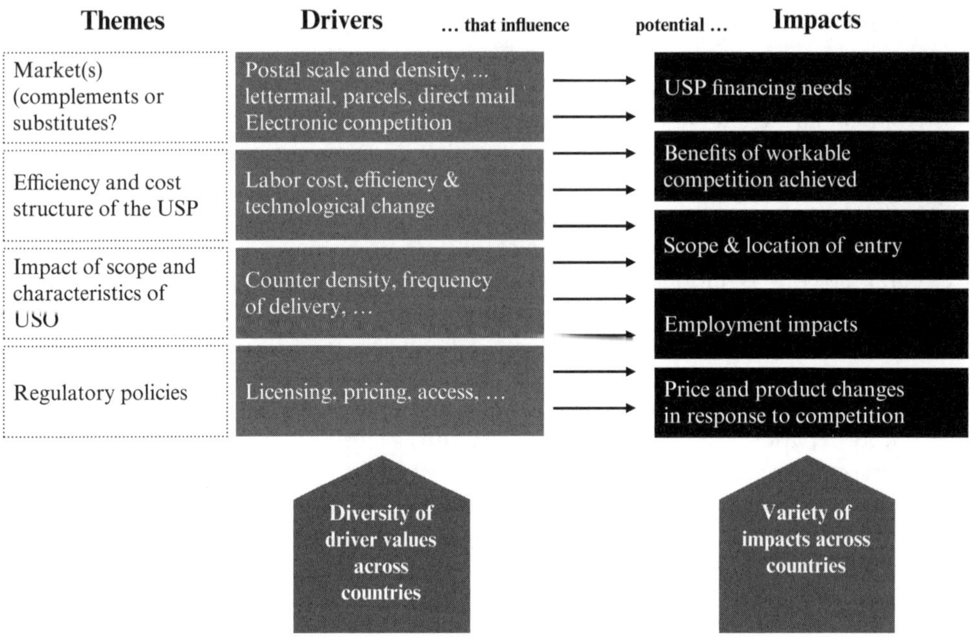

Source: PwC (2006).

Figure 19.1 Themes, drivers and country-specific impacts of liberalization

Table 19.1 Representative ranges of values for postal liberalization factors

Name	Definition	Range of values
Scale	Mail items per person per year	10–700
Urbanization rate (one element of postal density)	Percentage of population in urban area (UN standard)	40–95%
Labor cost ratio	Ratio of freelancer vs employee FTE cost	50–100%
Demand elasticity households	Absolute value of price elasticity of demand	0.1–0.8
Demand elasticity business mailers	Absolute value of Price elasticity of demand	0.4–1.0
Automation % in sorting	Percentage of mail item automatically processed	10–90%
Automation % in sequencing	Percentage of mail item automatically processed	10–80%
Flexible USP workforce	% of part-time or flexible workers (FTE) hired by USP	0–20%
Franchised counters	Percentage of counters franchised	0–50%

Note: * FTE = full-time equivalent; see d'Alcantara and Amerlynck (2006) and PwC (2006) for details.

A further key factor affecting the outcomes of market opening is the efficiency of the USP. Some of the drivers of efficiency are noted above. The wave of restructuring activity already launched in EU in anticipation of the competition enabled by the Third Postal Directive provides some good examples. There are also some interesting contrasting case studies outside of the EU. Let us consider the details.

ILLUSTRATIVE COUNTRY CASE STUDIES

The Netherlands and New Zealand were among the first of the POs to recognize the importance of becoming more commercialized (Toime, 1991), followed closely by Sweden, which opened its markets to competition in 1993. The Netherlands and Germany embraced privatization, while New Zealand Post and Sweden Post adopted a corporate structure and have remained wholly owned by the government.

Across Europe, a major restructuring of the postal sector across all 27 member states of the European Union (EU) is well underway. Shortly after the formal opening of the European internal market in 1992, the postal and logistics sector was recognized as a sector in need of reform. The Postal Directive of 1997, as amended in 2002 and 2008, is moving the EU to fully opening the postal market to competition by 1 January 2011. FMO implies, among other things, no statutory monopoly privileges (that is, no 'reserved area') for any of the components of the postal value chain or for any postal products. Thus, competitors will be free to collect, sort and deliver all types of documents and parcels whenever and wherever they wish. The impending market opening is transforming the European postal sector from slow-moving government bureaucracies, with a limited range of traditional products and excess employment, into a vibrant new industry. Crew and Kleindorfer (2008b) and Crew et al. (2008) provide details of the ongoing transformation of European postal markets.

Some of the countries that are implementing wide-ranging postal reform have done so through changes in regulation and corporate structure directed at promoting commercialization of the national PO. A common approach has been to promote strong management incentives for performance through executive profit targets as, for example, in Australia, Canada, Sweden and New Zealand, where POs are organized as corporations, with a for-profit, commercial culture, but with the government still the sole owner. Other countries are headed to full privatization. As of the beginning of 2008, 69 per cent of equity in the postal service is privately owned in Germany and 100 per cent in the Netherlands and Malta.

In addition, Post Danmark A/S recently transformed itself into a public limited company (that is, a corporation), and then began a process of privatization; 25 per cent of its shares are now owned by a private investor. In 2005, Post Danmark and its investor, in turn, bought 50 per cent of the Belgian Post Office (La Poste). The Austrian Post Office (Österreicherische Post AG) began the process of privatization with an initial public offering in 2006, in line with the changes that are occurring in Europe in communications markets as well as in parcel and express markets. The first tranche of shares in the Austrian Post Office, amounting to 49 per cent of the business, was sold in an oversubscribed Initial Public Offering, and shares began trading on the Vienna Exchange on 31 May 2006. The Italian Post Office (Poste Italiane) has also announced the possibility of an IPO. Japan Post is a further interesting case of privatization, which was the central issue in the Japanese parliamentary elections in 2005 (see Campbell and Porges, 2008).

The above represents some highlights of the fast-moving developments in the postal arena outside the USA. To give a sense of some of the important details that have occurred, the experiences of five countries are summarized: New Zealand and Sweden, which were early entrants to postal reform and have remained corporatized but not privatized operators; Germany and Netherlands, the two largest privatizers; and the United States, with its new postal reform act of 2006.[6] This review of the experiences of these salient examples in the postal reform area aims at setting out some of the main issues that are at stake in the postal reform movement.

Commercialization: New Zealand Post and Swedish Post

New Zealand Post and Sweden Post were early adopters of a commercialization approach. New Zealand Post is chartered and organized as a corporate entity and is expected to be run as a business: specifically, it is expected to give its primary shareholder, the government, a reasonable return on invested capital. It is also expected to maintain a specified minimum level of national postal services agreed with government; since April 1998 it has done so with an open market without a reserved area. Executives face performance-based executive compensation systems, modeled after the private sector with which they compete for managerial talent. Shares in the corporation are owned by the Minister of Finance and the Minister of State Owned Enterprises. There is virtually no sector-specific regulation of postal operators except a registration procedure and consumer-protection measures administered by the Ministry of Economic Development. The combination of corporate structure, a charter that underlines the commercial nature of their activities, proper incentives for management, and light-handed regulation have all combined to make New Zealand Post a successful business venture. Reflecting the efficiency improvements associated with commercialization, revenue per employee has more than doubled in nominal terms, while the real price of a single-piece stamp has fallen by almost a third during the period 1987–2004.

In Sweden, the market was fully opened to competition in 1994, with Sweden Post (known as Sverige Posten AB) maintaining responsibility for the USO. The price of a single-piece stamp was increased dramatically, by 38 per cent between 1993 and 1997, but subsequently moderated. In addition, Sweden Post took action to change the way it provided universal service: The number of full-service postal counters was decreased dramatically after liberalization, with extensive franchising taking place. Similarly, in high-cost rural areas, fixed counter services were replaced by rural postmen and women who acted to deliver the mail and to sell postal services. These and other activities by Sweden Post have led to significant cost savings and improved quality of service (measured by the percentage of mail meeting its delivery requirements). Faced with competition in major urban areas because of the open market, Sweden Post has undertaken significant restructuring to better align the cost, value and prices of its services with customer demand. Although fully owned by the government, Sweden Post's status and culture as a for-profit corporation and the competition it faces have made Sweden Post an often cited example to illustrate the benefits of corporatization and competition.

Sweden Post turned out to be a formidable competitor, so formidable that it quickly faced antitrust action from its major competitor, CityMail. Ultimately a settlement was reached. Interestingly, CityMail has changed ownership a number of times. It has rarely reported profits, though it did so in 2005 and may now have finally reached a stage of profitability under the ownership of Norway Post. The Swedish case raises interesting issues of the conflicts of interest created by public enterprise. Where a PO acts in a commercial manner and prices aggressively, it may be subject to antitrust action. If the PO is publicly owned, the government has a conflict of interest. A private company finds itself suing a company owned by the government and relying on the government to adjudicate.

Commercialization without privatization thus introduces inevitable anomalies. While legal institutions may provide adequate safeguards, at a minimum public ownership makes the application of competition law less straightforward than when two privately owned companies are parties to an antitrust dispute. A further difficulty with commercialization relative to privatization is that the role of incentive compensation for executives is limited because it is not related to an exogenous indicator of value like stock price. So, attempts to link compensation to performance are subject to additional problems of both measurement and legitimation. Perhaps most significantly, absence of private ownership means that public postal operators are not subject to the scrutiny and discipline of capital markets.

Privatization: The Netherlands and Germany

The Netherlands changed the structure of its post office in 1989 from a state enterprise to a private company whose stock was entirely owned by the government. As it turned out, this change was the first step towards full privatization. A close commercial relationship began to develop between the Dutch post office and TNT, an international express carrier with its origins in Australia. In 1994, the flotation of the Dutch Post as part of KPN (the combined postal and telecommunications operator) took place on the Amsterdam Stock Exchange. In 1996, KPN acquired TNT. In 1998, KPN 'demerged' into two companies, the new KPN, a telecommunications company, and TNT Post Group (TPG), a post and transportation company. In 2005, the various strands of the relevant business were regrouped and renamed as TNT. TNT is now a large company with 161 582 employees worldwide and global revenues of 10 885 billion euros in 2007. Its success has arisen from a solid base of mail revenue upon which to build, but its more recent activity is clearly global in scope, reaching out in both the express area and in letter mail activities in various countries. TNT has used its privatized status, and its improved knowledge of investor relations and markets, to gain an early mover advantage in many of the growing areas of business-to-business and business-to-consumer commerce. It is now a force to be reckoned with in traditional mail, express and parcel markets, and supporting infrastructure, as well as in international trade.

The German privatization story begins with 'Postreform I' adopted in 1989. Postreform I created separate departments within the Ministry for Posts and Telecommunications for postal services, postal banking and telecommunications. The new board of directors of the postal agency, Deutsche Bundespost Postdienst (German Federal Postal Service), introduced a greater degree of private sector management and accounting practices. Postdienst was reorganized into four business sectors: letters, parcels/express, post offices, and new business segments. At the time, Postdienst was in poor shape and facing huge deficits. Its situation was not made any better by having to absorb 86000 workers from the East German Post Office after reunification of East and West Germany in 1990. Transforming it into a modern and efficient company looked like a very difficult task. Germany hired a successful manager and businessman, Klaus Zumwinkel, to commercialize the operation.

In 1994, 'Postreform II' transformed Postdienst into Deutsche Post A.G. (German Post), a company organized under normal corporate law with government as the sole shareholder. At the same time, the German Constitution was amended to declare that government, rather than Postdienst, was obliged to ensure 'appropriate and adequate' universal postal services and that such services would henceforth be provided on the basis of private enterprise, with the universal service provider initially specified as the newly formed Deutsche Post A.G.

In 1997, 'Postreform III' adopted a new regulatory framework for postal services. Regulatory authority over the postal sector was committed to the telecommunications regulator, which was renamed 'Reg TP' (Regulierungsbehörde für Telekommunikation und Post). The postal monopoly was replaced by a more narrowly drawn 'licensed area'. The regulator was entrusted with ensuring universal service and regulating postage rates. As a transitional measure, Deutsche Post was granted an 'exclusive license' for five years, that is, until the end of 2002.

Legislative and operational reforms culminated in the sale of around a third of the equity in 2000 in an initial public offering. In 2001, Germany went so far as to change its Constitution, so that it could sell more than 50 per cent of the postal system. At the beginning of 2008, private investors own 69 per cent of what is now called Deutsche Post World Net (DPWN). The 1 January 2008 also saw the complete elimination of the reserved area in Germany, joining Sweden and the UK in opening its markets to end-to-end competition.[7]

DPWN has become a force of change in the postal market worldwide; privatization has transformed a once sleepy bureaucracy into a major multinational corporation. Maschke (2002) describes the transformation process of the German Post Office into the corporate giant DPWN over the course of the 1990s and early 2000s. The compensation structure was transformed and managers imported from other industries, the mail and parcels network were modernized within Germany, and new products (like hybrid mail and e-commerce) were developed. DPWN now has interests not only in the traditional mail and parcels businesses, but also in express mail, logistics, banking, and more. Indeed, DPWN has grown to a company with sales of 63512 billion euros in 2007 and total employment of 520012, making it one of the world's largest corporations. The growth of the German Post Office from a company with total postal employees of some 230000 in 1989 to the mega-corporation DPWN, with over a half million employees in 2008, is a remarkable history of transformation, privatization and expansion into global markets.[8] DPWN is both an interesting case study of the transformation from public bureaucracy to profit-driven corporation as well as an indication of the strategic possibilities that liberalized markets can enable. DPWN's major transformation has faced some significant challenges as well, as the recent $1 billion loss from its DHL unit in the USA illustrates.

Separating Market Dominant and Competitive Products (the USA)

The 2006 Postal Accountability and Enhancement Act (PAEA) enacted a number of significant changes to the Postal Reorganization Act (PRA) of 1970. Many of these changes relate to the modernization of the operational, financial, and management structures of the United States Postal Service (USPS) in the light of increased competition and USPS's continuing requirement to meet its universal service obligation. In addition, PAEA includes a number of provisions that require separate accounts and papers for USPS's competitive products, that is, those products that compete with similar products offered by commercial entities. Following the suit of many other countries, PAEA also introduced a form of price-cap regulation for USPS market dominant products. Given the size and importance of USPS for the postal industry, the history and structure of the postal reform movement in the USA is described in more detail later in this chapter. The approach taken in the USA stands out in contrast to that of the four countries above, which are clearly further along the postal reform path than the USA. While the developments envisaged in the USA under PAEA do not go as far, they must still be seen in the context of major secular trends affecting postal service, particularly competition from electronic media, postal regulatory reform and the opening up of postal markets worldwide to increased competition.

Summary Comments on International Postal Reform

Nations across the world are opening up their postal systems to commercialization and in a growing number of cases to privatization. Unionized workforces are common in POs before and after reform. In addition, workers in some countries enjoyed civil servant status. The combination of these two factors gave rise to wage and benefit terms that were often in excess of competitive levels, often exacerbated by the legacy costs of pensions or healthcare commitments made to earlier generations of postal workers now or soon to be retired. The task of designing and implementing changes in governance and regulation for POs thus involves a number of major problems, in addition to the environmental changes encompassing electronic substitution. It is no surprise, therefore, that postal reform and commercialization of POs remains a work in progress in many countries. Nonetheless, the direction of change is clear, and there is growing evidence of the benefits from change. What is emerging from international efforts to date is that, when coupled with appropriate changes in regulation and commercial restructuring of the PO, allowing greater competition into the market for postal services can provide an important spur to efficiency and innovation for both corporatized and privatized post offices.

CONCLUSIONS: THE WAY FORWARD

This section summarizes the main findings and open research questions suggested by worldwide postal reform and liberalization. In considering these findings, it is important to recall that postal reform and liberalization has taken many forms, often quite specific to individual countries. Most importantly, the scope of liberalization itself has varied widely, from only very minor restructuring of the charter of the PO, to greater opening of upstream worksharing to full market opening. The general dimensions on which liberalization can be classified are the degree of commercialization of the PO, on the one hand, and the level of legal restrictions imposed on potential entrants on the other. Liberalization and reform has resulted in a greater reliance upon commercial and profit incentives, reinforced by competition, to drive productivity, customer innovation and customer orientation of the PO. The expected results of such liberalization also include a better alignment of cost and value of services. Liberalization and competition in the market place are anticipated to provide the pressure for improved alignment, while commercialization and privatization are the critical enabling features; they provide a profit-oriented rationale for market segmentation, value-based pricing and the design of products to meet customer needs rather than simply following regulatory rulings. This, in theory, has been the vision or mental model behind postal reform and liberalization.

Balancing USO, Commercial Freedom and Liberalization

The first and most critical lesson is the issue of balance between: (1) the obligations faced by the PO in terms of the USO and various associated constraints (for example, a uniformly priced single-piece product, delivery frequency constraints, post office locations and services); (2) the degree of liberalization and competition allowed in the market; and (3) the commercial freedom and profit-orientation allowed or promoted within the PO to support its obligations in the face of allowed competition. Failure to address this issue of balance can lead to embarrassing and costly policy outcomes, some of which may be irreversible.

Similarly, liberalization has proceeded against a perspective that the market will be complementing and replacing the traditional (profitable) functions of the PO/USP for business customers and that pricing, uniformity requirements and other market-impeding restrictions on business products need to be relaxed. These increases in pricing flexibility for the incumbent PO would then allow greater development of competition in the business segment, with potential effects of increased innovation, new products and services. For example, in the EU, where allowed, entrants are developing new products, particularly in direct mail, along with new business models. To the extent that these have a better fit with the needs and willingness-to-pay of business customers, these services can also generate additional traffic and revenue for the incumbent if upstream and downstream access are promoted and used properly. However, this must be seen in the context of balancing the three factors noted above so as to allow the PO, as USP, an opportunity to achieve financial viability and play an effective role as a focal point for innovation and continuing stability of the USO.

Several factors make postal sector liberalization unique: first, unlike most other network based industries, the postal market is not likely to continue to expand significantly, at least in its traditional mail products. Second, the postal sector is less technologically intensive and more labor intensive than other network based industries.

Third, more than in other network based industries, the USPs were and are still to a large extent viewed as affected by public interests, imbued historically with important public roles such as the USO as well as their increasing role as a participant in communications markets. These factors, together, spell out both interesting challenges as well as the potential benefits of market-oriented solutions.

Regulation, the USO and PO Governance under Liberalization

The current evolution in the EU and elsewhere allow a number of conclusions concerning the role of regulation, the USO and restructuring of PO activities in promoting the balance described above in the presence of liberalization. Only the most important points are reviewed here.

As competition will be focused on the business segment, the PO/USP must be allowed commercial freedom in pricing and product definition, including applying non-uniform prices for bulk mails, including access tariffs. Failure to do so will be a recipe for forced loss of market share to less efficient entrants. Finding the right balance is a central research and policy issue. Research questions on this matter are discussed in Crew et al. (2008).

Liberalization is intended to promote a shift to commercial and competitive conditions, dictated by customer value. Doing so while assuring the PO the ability to fund its USO and remain financially viable has seen the broad adoption of Price-cap Regulation (PCR) to stimulate efficiency and innovation by the PO. Many open research questions exist in this area (for example, Crew and Kleindorfer, 2008a).

Access pricing (including worksharing discounts) is a critical element of liberalized markets, and should be subject to a transparent and non-discriminatory negotiated access regime. In such a regime, price-cap constraints and structural constraints on the PO's access pricing should be implemented at an aggregate level, and not subject to micro-management by the regulator of the myriad details or varieties of access products. Open research questions are discussed for this issue in Crew and Kleindorfer (2008d).

Pricing stability for smaller customers remains important. Smaller customers, including households, devote a very small part of their overall budget to postal services. Notwithstanding this fact, postal services in general and the USO in particular remain important elements of the overall communications market. Therefore, continuing stability of prices for small customers is likely to remain an important element of 'affordability' discussions concerning the USO and other postal products. For open research questions here, see De Donder et al. (2006).

Proper accounting of the USO/CSO burden, audited annually, has become an important focus in liberalized postal markets. This has included identifying the specific drivers of this burden and is stimulating in the EU a discussion of the value of the constraints and attributes associated with the USO. For research questions on this issue, see Bergum (2008), Panzar (2008) and Buser et al. (2008).

With the costs of the USO identified, POs and their regulators are beginning to see the fundamental importance of addressing, motivating and improving the differential in the human resource costs between USPs and new entrants, for example, through the introduction of restructuring plans, outsourcing schemes, collective working agreements and settlement of historical liabilities such as pension liabilities. More generally, restructuring has made significant inroads in the postal sector in liberalized markets. This observation is extremely important as the experience of others in restructuring is important in creating credibility for its benefits. In the postal sector, the experience of privatized as well as corporatized operators is reviewed in this chapter and suggests fundamental changes in culture, financial performance and market orientation of restructured postal organizations. These changes rely on POs having the appropriate regulatory and governance structures. For research and policy issues on this matter, see PwC (2006), d'Alcantara and Gautier (2008) and Reisner (2008).

Focusing now on the USA, the experience there shows the huge benefits garnered by USPS through liberalization of its upstream worksharing market. These benefits for the US economy have been estimated by Cohen et al. (2006) as being over $11 billion in 2004 alone. Moreover, Pearsall (2005) has shown how important these have been to stimulating volume growth in transactions mail and in direct mail. Thus, a key lesson here is that upstream liberalization alone can provide a significant stimulus to innovation, volume growth and cost savings, both for customers and for the incumbent PO. Research questions on this point are spelled out in Bradley et al. (2008) and Crew and Kleindorfer (2008c).

Finally, the importance of regulation in promoting the balance required between the USO, commercialization objectives and liberalization is the key issue here. Fortunately, these matters have attracted the attention of economists and indicated that there now exists considerable knowledge about several of the key issues involved in postal reform and liberalization. These include the structure and nature of access pricing; the scope and financing of the USO, the structure of price, profit and service quality regulation in the postal sector, and the nature and benefits of potential restructuring alternatives for the PO.

NOTES

1. See Crew and Kleindorfer (1991) for a discussion of the economic impact of Roland Hill's Penny Post and its continuing impact on modern postal services.
2. The reader interested in these developments should see, for example, Crew and Kleindorfer (2008b, 2009) and Crew et al. (2008).
3. For a discussion of product innovation in the postal sector, and its relationship to liberalization, see Felisberto (2008) and Pintsov and Obrea (2008).
4. A number of these have been published in Crew and Kleindorfer (2006a, 2008b) and Crew et al. (2008).
5. See, for example, Ambrosini et al. (2006), Boldron et al. (2006), Crew and Kleindorfer (2006b) and PwC (2006) for a discussion of the extensive literature on balancing the costs of the USO, market power of the PO and commercial freedom allowed it.
6. Sweden is examined in detail in Andersson (2008), and major aspects of the reforms in the other three countries are covered in Crew et al. (2008). The US case is discussed in detail in Crew and Kleindorfer (2008c). The reader should note that this review was written in 2009, and further changes have occurred since then in all of the case study countries.
7. As noted in Dieke and Wojtek (2008), there are some important innovations in the opening of the German market, including the requirement, also imposed in January 2008, that all competitors must pay above a specified minimum wage to any employees involved in products in the USO sector of the market.
8. The German Post Office absorbed some 86000 employees from the previous East German Post Office in 1990 upon reunification of East and West Germany. The expansion in employment that followed saw two different trends. One was the reduction in employees involved in postal work (to around 125000 presently) and the other the rapid expansion of what has become DPWN in courier and logistics operations. Maschke (2002) provides details on these changes and other aspects of the DPWN restructuring process.

REFERENCES

Ambrosini, Xavier, François Boldron and Bernard Roy (2006), 'Universal service obligations in the postal sector: economic learnings from cross-country comparisons', in M.A. Crew and P.R. Kleindorfer (eds), *Progress toward Liberalization of the Postal and Delivery Sector*, New York: Springer pp. 23–38.

Andersson, Peter (2008), 'Liberalization and market performance: towards higher efficiency in Sweden,' in

M.A. Crew, P.R. Kleindorfer and J.I. Campbell (eds), *Handbook of Worldwide Postal Reform*, Cheltenham, UK and Northampton, MA, USA: Edward Elgar, pp. 298–315.

Bergum, Kristin (2008), 'Calculating the net cost of the USO: a practical example from Norway' in M.A. Crew, P.R. Kleindorfer and J.I. Campbell (eds), *Handbook of Worldwide Postal Reform*, Cheltenham, UK and Northampton, MA, USA: Edward Elgar, pp. 110–22.

Boldron, Francois, Denis Joram, Lise Martin and Bernard Roy (2006), 'From the size of the box to the costs of universal service obligation: a cross-country comparison', in M.A. Crew and P.R. Kleindorfer (eds), *Liberalization of the Postal and Delivery Sector*, Cheltenham, UK and Northampton, MA, USA: Edward Elgar, pp. 37–52.

Bradley, Michael D., Jeff Colvin and Mary K. Perkins (2008), 'An economic model of the regulatory structure created by the Postal Accountability and Enhancement Act of 2006', in M.A. Crew, P.R. Kleindorfer and J.I. Campbell (eds), *Handbook of Worldwide Postal Reform*, Cheltenham, UK and Northampton, MA, USA: Edward Elgar, pp. 67–79.

Buser, Martin, Christian Jaag and Urs Trinkner (2008), 'Economics of post office networks: strategic issues and the impact on mail demand', in M.A. Crew, P.R. Kleindorfer and J.I. Campbell (eds), *Handbook of Worldwide Postal Reform*, Cheltenham, UK and Northampton, MA, USA: Edward Elgar, pp. 80–97.

Campbell, James I. Jr and A. Porges (2008), 'How much postal reform in Japanese postal privatization?', in M.A. Crew, P.R. Kleindorfer and J.I. Cambell Jr (eds), *Handbook of Worldwide Postal Reform*, Cheltenham, UK and Northampton, MA, USA: Edward Elgar, pp. 373–87.

Cohen, Robert H., William W. Ferguson, John D. Waller and Spyros S. Xenakis (2006), 'Worksharing: how much productive efficiency, at what cost and at what price, universal service without a monopoly', in M.A. Crew and P.R. Kleindorfer (eds), *Progress toward Liberalization of the Postal and Delivery Sector*, New York: Springer, pp. 141–52.

Crew, Michael A. and Paul R. Kleindorfer (1991), 'Rowland Hill's contribution as an economist', in M.A. Crew and P.R. Kleindorfer (eds), *Competition and Innovation in Postal Services*, Boston, MA: Kluwer Academic, pp. 1–12.

Crew, Michael A. and Paul R. Kleindorfer (2006a), *Progress Toward Liberalization of the Postal and Delivery Sector*, New York: Springer Verlag.

Crew, Michael A. and Paul R. Kleindorfer (2006b), 'Approaches to the USO under Entry', in M.A. Crew and P.R. Kleindorfer (eds), *Liberalization of the Postal and Delivery Sector*, Cheltenham, UK and Northampton, MA, USA: Edward Elgar, pp. 1–18.

Crew, Michael A. and Paul R. Kleindorfer (2008a), 'Regulation and the USO under Entry', in M.A. Crew and P.R. Kleindorfer (eds), *Competition and Regulation in the Postal and Delivery Sector*, Cheltenham, UK and Northampton, MA, USA: Edward Elgar, pp.13–22.

Crew, Michael A. and Paul R. Kleindorfer (eds) (2008b), *Competition and Regulation in the Postal and Delivery Sector*, Cheltenham, UK and Northampton, MA, USA: Edward Elgar.

Crew, Michael A. and Paul R. Kleindorfer (2008c), 'Reform of the United States Postal Service: an unfinished task', in M.A. Crew, P.R. Kleindorfer and J.I. Campbell (eds), *Handbook of Worldwide Postal Reform*, Cheltenham, UK and Northampton, MA, USA: Edward Elgar, pp. 123–42.

Crew, Michael A. and Paul R. Kleindorfer (2008d), 'Pricing for postal access and worksharing,' in M.A. Crew, P.R. Kleindorfer and J.I. Campbell (eds), *Handbook of Worldwide Postal Reform*, Cheltenham, UK and Northampton, MA, USA: Edward Elgar, pp.32–66.

Crew, Michael A. and Paul R. Kleindorfer (eds) (2009), *Progress in the Competitive Agenda in the Postal and Delivery Sector*, Cheltenham, UK and Northampton, MA, USA: Edward Elgar.

Crew, Michael A., Paul R. Kleindorfer and James I. Campbell (eds) (2008), *Handbook of Worldwide Postal Reform*, Cheltenham, UK and Northampton, MA, USA: Edward Elgar.

D'Alcantara, Gonzales and Bernard Amerlynck (2006), 'Profitability of the universal service postal provider under entry with economies of scale in collection and delivery', in M.A. Crew and P.R. Kleindorfer (eds), *Progress toward Liberalization of the Postal and Delivery Sector*, New York: Springer, pp. 39–58.

D'Alcantara, Gonzales and Alex Gautier (2008), 'Route cost structure and the USO', in M.A. Crew and P.R. Kleindorfer (eds), *Competition and Regulation in the Postal and Delivery Sector*, Cheltenham, UK and Northampton, MA, USA: Edward Elgar, pp. 125–35.

De Donder, Philippe, H. Cremer, P. Dudley and F. Rodriguez (2006), 'A welfare analysis of price controls with end-to-end mail and access services', in M.A. Crew and P.R. Kleindorfer (eds), *Liberalization of the Postal and Delivery Sector*, Cheltenham, UK and Northampton, MA, USA: Edward Elgar, pp. 53–72.

Dieke, Alex Kalevi and Ralf Wojtek (2008), 'Competition, wages and politics in the delivery sector: the case of postal minimum wages in Germany', in M.A. Crew, P.R. Kleindorfer and J.I. Campbell (eds), *Handbook of Worldwide Postal Reform*, Cheltenham, UK and Northampton, MA, USA: Edward Elgar, pp. 282–97.

Felisberto, C. (2008), 'Evidence on the effect of liberalisation and competition on innovation in the postal sector', CDM Working Papers, École Polytechnique Fédérale de Lausanne.

Maschke, Walter (2002), 'Transformation at Deutsche Post World Net: using the example of socially compatible workforce adjustment', in M.A. Crew and P.R. Kleindorfer (eds), *Postal and Delivery Services: Delivering on Competition*, Boston, MA: Kluwer Academic, pp. 303–20.

Panzar, John C. (2008), 'Funding universal service obligations', in M.A. Crew, P.R. Kleindorfer and J.I. Campbell (eds), *Handbook of Worldwide Postal Reform*, Cheltenham, UK and Northampton, MA, USA: Edward Elgar, pp. 98–109.

Pearsall, Edward S. (2005), 'The effects of worksharing and other product innovations on U.S. postal volumes and revenues', in M.A. Crew and P.R. Kleindorfer (eds), *Regulatory and Economic Challenges in the Postal and Delivery Sector*, Boston, MA: Kluwer Academic, pp. 213–42.

Pintsov, Leon A. and Andrei Obrea (2008), 'Postal reform and product innovation', in M.A. Crew, P.R. Kleindorfer and J.I. Campbell (eds), *Handbook of Worldwide Postal Reform*, Cheltenham, UK and Northampton, MA, USA: Edward Elgar, pp. 176–94.

PricewaterhouseCoopers (PwC) (2006), 'The impact on universal service of the full market accomplishment of the postal internal market in 2009', European Commission, Brussels, May. See: http://ec.europa.eu/internal_market/post/studies_en.htm (accessed 9 September 2009).

Reisner, Robert A.F. (2008), 'Postal transformation at USPS', in M.A. Crew, P.R. Kleindorfer and J.I. Campbell (eds), *Handbook of Worldwide Postal Reform*, Cheltenham, UK and Northampton, MA, USA: Edward Elgar, pp. 143–59.

Toime, Elmar (1991), 'Competitive strategy for New Zealand Post,' in M.A. Crew and P.R. Kleindorfer (eds), *Competition and Innovation in Postal Services*, Boston, MA: Kluwer Academic, pp. 275–82.

PART III

COUNTRY AND REGIONAL STUDIES

20 Liberalization of network industries in the European Union: evolving policy issues[1]

Matthias Finger and Marc Laperrouza

INTRODUCTION

Europe has a long tradition of infrastructures, even though this tradition differs from country to country. What makes Europe particularly interesting is the process of liberalizing infrastructures which started around the 1980s. More than a process of economic liberalization, it is a process of integrating Europe politically. In fact, what makes the European experience unique worldwide is the fact that the liberalization of the infrastructures is conceived as, and becomes a means of, integrating Europe economically and politically, so as to reposition it globally. Thus, infrastructure liberalization must always be seen, in Europe, as part of broader policy objectives and problems, such as European social cohesion, European competitiveness, as well as the protection of the environment, among others. Consequently, the way the infrastructure sectors in Europe are liberalized evolves over time in light of new challenges that emerge – such as fiscal and environmental challenges or the integration of new member states – as well as in light of the problems that arise in the very process of infrastructure liberalization (Geradin, 2006).

In this chapter, we focus on this last aspect, namely, on the evolution of Europe's approach to infrastructure liberalization over time. There is a clear 'old liberalization paradigm' with which the European Commission started the very process of liberalization: its goal was the *creation* of markets in most network industries. Indeed, in many infrastructure sectors there were no markets at all. The entire value chain was monopolistic and in many countries the firms providing the services were owned and operated by the respective nation states.

Contrary to expectations, the creation of markets caused additional institutional problems (for example, interplay between competition law versus sector-specific regulation, non-discriminatory market access). In other words, whereas the market solved efficiency and productivity issues in certain segments (for example, those open to competition) the entire system became much more complex to manage. Not only do freer markets require more rules (Vogel, 1996), but finding an equilibrium between competitive and non-competitive elements often requires a delicate act of political, financial and organizational balancing. In most network industries at least one element of the value chain still retains its natural monopoly characteristic. Financing and managing this element – along with ensuring the functioning of the entire infrastructure system – often causes headaches for member states.[2]

However, after two decades of liberalization in Europe a new paradigm is starting to emerge. It now focuses rather on the *management* of markets and of competition. This is not to say that institutional issues related to market creation have disappeared. Member-state heterogeneity, tensions between the European Commission (EC) and member states, as well as tensions between liberalization and integration objectives all dictate ongoing policy adjustments, influencing both the direction and the pace of liberalization.

By borrowing examples from different infrastructure sectors we will show that the transition between the two paradigms – namely, the transition from the market creation to the market management paradigm – not only takes into consideration the specificities of each sector and their respective evolutionary path, but also the political nature of the European Union (EU). We focus in this chapter in particular on three problematic policy issues in this transition, that is, market access, management of the monopolistic element, and the interface between competitive and non-competitive segments.

Our chapter is thus structured as follows: the first section sets the stage by recalling the state of the network industries prior to liberalization and the reasons behind the EC's decision to liberalize them. Here, we in particular show how the network industries are related to other policy areas in Europe. The next section presents an intermediary balance sheet of the current state of liberalization in Europe in the main sectors, focusing in particular on the main achievements and policy issues emerging from the liberalization process. The third section examines how these policy issues, as they emerged during the initial process of liberalization, compared with the current policy issues. It places the evolution of policy issues in the framework of the shift from market creation to market management.

SETTING THE STAGE

The purpose of this section is to explain the particular importance of the various infrastructures in the context of broader European policy issues, especially in the perspective of political integration. Infrastructures are indeed important to achieving certain policies (for example, competitiveness of Europe, environmental protection, consumer benefits, integration, and so on). They also affect many policies, such as social cohesion, employment, budget and fiscal policies or transport research policies. Creating a market in the various infrastructure sectors is generally considered to be a tool to achieve at least some of these policies. This first section is structured into three parts, namely the historical background, the explanation as to why infrastructures are policy related, and a discussion of how infrastructure liberalization leads to tensions with these policy areas.

Historical Background

The European Commission has used the liberalization of infrastructures as a stimulus to further integrate the common market and to enhance the competitiveness of the Union (European Commission, 1999). Allowing for free trade of these essential goods and services across national boundaries was seen as a central driver towards the creation of a truly common market. In parallel, consumers were to benefit from lower prices and better services, and the competitiveness of the EU would increase.

This approach was protracted and certainly not uncontested. Historically, the regulation and industrial organization of infrastructures were considered as an issue of national sovereignty and hence fell under the sole authority of the member states. For instance, an initiative of the EC to harmonize the public procurement of the post and telecommunication administration in 1968 was strongly opposed by the Council of Ministers (Schneider, 2001, p. 62) and thus failed. The first successful initiatives of the EC aiming at a structural reform of infrastructures go back to the late 1970s for air transportation (European Communities, 1979), the 1980s for telecommunications (European Commission, 1987) and the early 1990s for postal services (European Commission, 1991).

Nowadays the geopolitical positioning of the EU and its ambitions keep playing an important role in the ongoing reform process. As the European economy strongly builds on international trade, it is threatened to lose its competitiveness vis-à-vis the USA and even more so vis-à-vis the emerging Asian economies. In this respect, well-functioning infrastructures are considered particularly important, especially in the areas of transportation, energy (oil and gas), and information and communication technologies (ICT). The latter are particularly important for Europe's aspiration to become a knowledge society.[3]

Initiating and framing an EU policy on infrastructure liberalization have proven extremely challenging since the member states have quite different stakes in the different sectors, as well as different political ideologies. In the UK, for instance, the Thatcher government provided important incentives to reduce the role of government, privatize infrastructures and introduce competition in order to increase efficiency and provide better service to customers. Consequently, the UK was far ahead of the EU liberalization curve. On the other extreme, countries like France had reasons to stick to their traditional approach of state-controlled infrastructures. Somewhat surprisingly, citizens have generally remained quite satisfied with the legacy organization of the infrastructure sectors and tend to view liberalization efforts with scepticism. Thus, the historical infrastructure providers were not really challenged even though they were not always performing optimally. On the contrary, whenever liberalized network industries suffered from incidents (for example, railways in the UK), this prompted a backlash and called for a return to the previous organizational mode.

Over time, and given the wide variety of vested interests, the EU developed quite a broad framework for infrastructure liberalization, allowing member states to adapt the reforms and restructuring to their domestic styles of regulation. Initial fears that the reforms might proceed very slowly and with marginal results gave nevertheless way to quite unexpected results. In most cases, the process developed its own dynamics, going far beyond initial expectations, as the case of telecommunication demonstrates (Schneider, 2001, pp. 63–4). In this sector there was quite a struggle in the late 1980s about the liberalization of the terminal markets. The proposal of the EC provoked strong opposition from most EU governments. Some countries, led by France, even appealed to the European Court of Justice, claiming that the Commission exceeded its powers. However another group of countries gathered around the UK, the Netherlands and Germany, supported the policy objectives of the Commission in favour of a more liberal telecom policy. Although this political conflict was settled with a compromise in 1989, the economic forces of the liberalization process developed their own dynamics.[4] Similar examples can be found in other infrastructure sectors, as these experiences in the telecommunication sector catalysed the restructuring of other infrastructures, most prominently the postal sector[5] and the electricity and gas sectors (Arentsen and Künneke, 2003; Haas and Glachant, 2006), as we will see below.

While the necessity to restructure the infrastructures is no longer disputed in Europe, one can still find significant differences throughout the member states with respect to the degree and the pace of the restructuring of these essential sectors. Several reasons explain such discrepancies:

Legislation
European Union directives only prescribe minimum requirements and ultimate deadlines. As a result, member states are free to implement the directives faster while allowing for a greater degree of liberalization, or they might 'drag their feet' and look for exemptions in the EU legislation. In this context, one must also mention the overarching principle of subsidiarity, which leaves some diversity in the design and implementation of the directives.

Economic policy towards market structure and conduct
Some countries, such as Germany and France, foster the Rhineland model of economic policy (Albert, 1993), granting 'national champions' a dominant market position in order to safeguard national interests, for instance, with respect to secure energy supply or reliable railroad transportation services. Under such conditions, there is only limited room for competition. On the other hand, countries with an Anglo-Saxon approach, such as the UK and most of the new member states, promote strong competition between a variety of different firms, ideally none of them having a dominant position.

Technical barriers
Competition might be hampered by a lack of network capacity, especially with respect to long distance international transport and transmission lines. This case is well known from European electricity and gas markets. Typically, countries on the fringe of the EU are less interconnected as compared to north-western continental Europe, which significantly restricts opportunities for market entrance and competition. Next to these physical limitations there are also different technical norms and standards which create 'invisible' market barriers. For instance, in the rail transport sector, different technical systems for signalling and monitoring of rail traffic were used throughout Europe for most of the twentieth century. Trains needed and often still need different technical equipment to meet heterogeneous national standards, and hence cannot be easily moved between countries or regions without enduring significant costs related to duplicating equipment. To resolve this interoperability barrier, the EC established a European-wide standard – the European Rail Traffic Management System (ERTMS). Despite notable progress, its deployment across European networks struggles to overcome technological path dependencies (Laperrouza, 2008).

Socio-political specificities
Some outcomes of infrastructure liberalization might not always meet socio-political expectations. The liberalization of the postal sector is a good example in this respect. In some countries like Germany and the Netherlands, there was a significant public discussion on the employment conditions of postmen and the number of deliveries per week. Traditionally, they had a fixed payment for a predefined number of working hours. Newcomers in this sector introduced a new salary scheme, based on the pieces of mail delivered. This scheme proved to be much cheaper compared with the fixed payments of

the incumbent firms. However, for the postmen the employment conditions were inferior, since the overall salaries decreased and the employment conditions were suddenly far less privileged. The newcomers not only outperformed the traditional mail companies in this respect. They also introduced another delivery scheme. Instead of five or six deliveries per week, they had a significant lower frequency (two or three deliveries per week). Labour unions and some political parties claimed 'unfair' competition. Newcomers were forced to have similar employment conditions and delivery schemes as the traditional mail providers. In fact, Jordana et al. (2006) have shown that member states were able to shape their sectors according to the preferences of their national policy communities.

In spite of the differences observed between the member states and the infrastructure sectors, one can find an ongoing process of gradual integration and convergence of these sectors into the common EU market. The European Directives play an important role in providing a common ground for this policy across all sectors and in enforcing it throughout the EU. Liberalization of infrastructures in the EU is, however, more than transposing community directives into national laws. Indeed, once transposed, the national laws still need to be implemented.

Overall, the liberalization of the infrastructures in Europe is a complex interplay between Directives from the Commission, on the one hand, and changing international market structures as a consequence of globalization, technological innovation resulting in new services and production processes, changing consumer needs and expectations, and innovative opportunities for the regulation and organization of these complex sectors, on the other. These aspects are exemplified in more detail in the next section.

Infrastructures and European Public Policy

As noted above, the primary rationale behind the liberalization of network industries in Europe was to improve their sectoral performance and thereby to generate wide-ranging macroeconomic benefits, including an increase in the competitiveness of Europe as a whole. To be more precise, market opening ('liberalization') was to lead to a change in industry/market structure which would in turn modify conduct, hence improve industry/market performance (for example, through lower prices) with spillovers on macroeconomic performance. As such, the liberalization of infrastructures can be seen as an important instrument of public policy.

In its report on the liberalization of network industries (European Commission, 1999, p. 37), the EC defined the main objectives of public policy as (1) the introduction of competition wherever this would enhance welfare, (2) the protection of customers and suppliers from abuse of dominant positions, (3) optimal use of costly infrastructure, (4) ensuring adequate incentives to invest, reduce costs, improve quality and innovate, (5) the prevention of serious disruptions of service or supply, (6) ensuring the provision of defined levels of certain services at 'affordable' prices and (7) ensuring an adequate level of consumer protection.

From an institutional perspective, public policies for the network industries were often 'bundled' together, such as the Common Transport Policy (CTP) which groups road, rail, air and water or the Energy Policy (EP) for electricity and gas. One could go back as far as to the 1956 Spaak Report to find some public policy themes related to transport, that is, non-discrimination in charging, development and financing of infrastructure and a common transport policy (Comité intergouvernemental créé par la Conférence de Messine, 1956). While initially coined in the mid-1950s, the CTP gained momentum in the early 1980s and resulted in the liberalization of the four different modes of transport. Nowadays, the CTP includes strategies to complete the internal market, further develop a trans-European network of major transport corridors and better connect Europe internationally. In line with EU social and environmental goals, it also aims to promote the efficient use of the different transport modes, to improve safety and sustainability while protecting the interests of employees, passengers and citizens.[6] As to the latest iteration of the EP, its objectives deal with the security of energy supply, the availability of affordable energy as well as the promotion of environmental sustainability.[7]

It is however fair to say that all the public policy objectives did not receive the same level of attention either from the EC or from the member states.[8] Some of these objectives had to face more resistance than others from certain stakeholders. For instance, a heavy emphasis had been placed on the introduction of competition across all network industries wherever possible but the type and extent of competition achieved varies largely across Europe. Some objectives were derailed or postponed by a changing economic environment as incentives to invest had to meet with the harsh reality of the financial crisis. Somewhat surprisingly, services of general economic interest benefitted from ongoing policy attention.[9] At the same time some of the initial policy objectives shifted over years and new ones (for example, preservation of the environment) emerged, displacing older ones before they were actually dealt with.

Conflicting Priorities

While common policy objectives/frameworks are drafted and give a general direction for a policy area, the liberalization of the network industries is mostly taking place through sector-specific measures/policies. Such a 'vertical' approach makes sense as there are specificities to each sector which cannot be easily transferred to other sectors. This is particularly noticeable in the transport sector where issues of access to essential facilities are found both in airlines and railroads but where the levels of standardization and /or harmonization differ radically. In the first case almost any plane can land on almost any runaway in Europe. In the second case, only a minority of trains can run on non-domestic railway networks. At the same time, a number of horizontal issues (for example, interdependencies) make the coordination of the policies in the network industries a necessity.[10]

The liberalization of the network industries thus leads to conflicts in policy priorities. These can include trade-offs between short-term versus longer-term objectives or efficiency versus equity objectives (Doyle, 1998, p. 59).[11] In fact, a multitude of policy areas (for example, industrial, economic, regulatory or social) are directly or indirectly impacted by the liberalization process. Finding a balance between the objectives and across the policy areas demands permanent trade-offs. For example, liberalization leads to cream-skimming by new entrants, which in turn poses a problem to public services. Indeed, the historical operator will lose its most profitable clients and thus will increasingly have problems cross-subsidizing its public service obligations which generally have not been changed. As a consequence, policy intervention is necessary so as to define and guarantee the public services, as well as the security of supply. Similarly, liberalization creates all kind of non-anticipated technical problems, namely, in the areas of interoperability or system management. Again, policy intervention becomes necessary so as to make infrastructure systems work from a purely technical point of view. Finally, liberalization quite naturally triggers the development of new markets, whose negative environmental consequences call again for policy intervention.

INTERMEDIARY BALANCE SHEET

In this section, we provide a short summary of the evolution of liberalization in some important sectors and conclude with an overview of achievements in terms of market opening and competition.

Electricity

The electricity sector was historically vertically integrated at the national level and the different companies cooperated at the international level so as to ensure the interconnection of the networks and the transport of electricity across Europe. Also, there was an overcapacity in electricity generation facilities in Europe prior to the sector's liberalization. It is in this context that the European Commission started the process of liberalizing the electricity market in 1996 by issuing a first Directive on third party access to the network (Directive 96/92) in order to allow for competition in electricity generation. A second Directive (03/54) seven years later aimed at accelerating the process of liberalization and opened the market for industrial customers as of July 2004 and for households as of July 2007. More precisely, this second Directive defines four key principles of electricity market liberalization, namely (1) the regulated access to the network at transparent and non-discriminatory prices, (2) the legal unbundling of the grid (transmission and distribution) from the producers, (3) the creation of national independent regulatory authorities, whose main role is in the definition of the grid tariffs, and (4) the definition of a certain public service in terms of accessibility, quality, and affordable prices. At the same time, the Commission ruled about cross-border exchange of electricity, as such cross-border exchanges are essential for the creation of a single European market.[12]

In 2005 the Commission evaluated the progress of electricity liberalization and concluded that the pace of European integration was too slow, as characterized by significant differences in the price levels across the member countries. The main reason for such lack of integration were said to be market access barriers, insufficient unbundling, as well as insufficient interconnection capacities. Subsequently, the Commission issued a Third Directive which came into force in June 2009. This Directive focused, in particular, on the following three elements: (1) the Directive seeks a more effective separation of the grid from production but offers nevertheless alternatives to full ownership unbundling in the form of the so-called ISO (or Independent System Operator[13]); (2) the Directive strengthens the independent national regulatory authorities by giving them more powers (such as the ability to fine operators); and (3) the Directive creates a European Agency for the Cooperation of Energy Regulators (ACER), which will mainly play a role in transboundary matters (cooperation among regulators, transboundary energy exchanges, transboundary grid development).

After an initial decline of the wholesale electricity prices in Europe, these prices have started to increase again after 2005 (Percebois, 2008). This is mostly the result of the fact that consumption continued to increase while investments in production capacities was lagging. In many European countries incumbents maintain significant market shares in most of the countries, whereas competitors cannot access markets of neighbouring countries because of lack of interconnection capacity. One can also observe a certain consolidation process in electricity production across Europe with French EDF, for example, acquiring companies across Europe. In this respect, the acquisition of nuclear power plants, for example, has raised questions of sovereignty over security of supply.

Over the past ten years there have been significant investments in renewable energies such as off-shore and on-shore wind production (Denmark, Germany), as well solar production (Spain). At the same time the development of small-scale decentralized solar and gas production facilities can also be observed. These developments in turn pose significant challenges to the electricity grid which has to be reinforced, equipped for two-way transmission and overall be made smarter (European Commission, 2006). Further challenges are, besides the development of electricity production to meet growing demand, grid congestions and subsequent developments of the grid, especially the interconnection capacities across Europe (Giesbertz and Mulder, 2008).

Besides liberalization policies which continue – for example, furthering market access, stronger regulation of grid tariffs – policies in the electricity sector pertain mainly to grid development (the so-called copper-plate Europe) and carbon dioxide (CO_2) reduction, in other words the promotion of renewable energies, but also energy savings. The underlying policy issues pertain to security of supply on the one hand and sustainability on the other.

Telecommunications

Telecommunications was liberalized at an early stage in Europe. Liberalization has been made possible mainly thanks to the technological dynamics that took place in the telecommunications sector during the past three decades. One can distinguish between two main steps. The first liberalization step led to the full opening of the telecommunications market.[14] In this first step policy-making focused on establishing common technical standards. Starting from 1984, the policy sought to develop common standards and specifications, shared research programmes between operators and industry and special programmes for the least developed regions (using structural funds). Since 1987 a second policy phase saw liberalization as the main focus and culminated with the liberalization of all telecommunications services and networks by 1 January 1998.

The second step was characterized by a different approach to liberalization, namely one based on competition law and aimed at identifying market power among operators, in particular incumbents. In 1999, the European Commission initiated a major review of existing EU telecommunications law that resulted in the adoption in 2002 of a new regulatory framework for electronic communications, enacting seven directives and related documents in connection with ICT regulation.[15] The fast pace of technological change (for example, growing convergence between telecoms, broadcasting and information technology) and service innovation in the ICT sector further prompted the European Commission to undertake a review of the framework in 2002.[16] The EU revised its approach in 2003 to the sector specific regulation of telecommunications by aligning it with competition law (Cawley, 2008). In November 2007, the Commission issued a number of proposals for reform focusing on four main areas: more competition, better regulation, strengthening the internal market, and protecting consumers better. The electronic communications directives were developed in response to a dynamic and increasingly unpredictable market in which a growing number of competitors are participating.[17]

The dynamics in the telecommunications sector have been important as a result of combined liberalization and technological progress. Competition has, at least in the beginning of the process, been impressive. Prices have fallen significantly (for example, consumers spent about 30 per cent less of their income in 2002 than they did in 1996), product and service innovation took place at a rapid pace (Geradin, 2006). At the same time, only few pan-European operators (for example, Vodafone, Deutsche Telekom, France Telecom-Orange) have emerged. Therefore, the market remains fragmented, in part due to the different regulatory requirements operators face in the member states.

In this fast-developing sector, the regulatory framework needs to be revised on a regular basis to ensure it continues to serve the best interests of consumers and industry. An agreement on the EU Telecoms Reform was reached by the European Parliament and Council of Ministers in November 2009 (European Commission, 2009). While formally adopted by the Council and European Parliament, the reform package still needs to be transposed into national legislation by the 27 members states (by June 2011). The rules governing telecommunication in Europe can be divided into five groups: consumer rights (for example, universal service), market access (for example, interoperability), radio spectrum, broadcasting (for example, mobile television) and pan-European services (for example, single European emergency numbers).

Over the years, the main policy theme has been the move towards a single market for telecommunication services and equipment to progressively remove barriers to pan-European operation and supply. This policy has seen an evolving interaction between four elements: liberalization, harmonization, competition and public service. In contrast with other network industries, the telecommunications sector undergoes almost permanent technological change. As a result, it requires constant policy adaptation. Whereas universal service obligations used to be limited to the provision of fixed lines, the question is now whether to also include broadband services.

Information and communication technologies are considered a key driver of European competitiveness. Like in the other sectors, industry concentration, and thus a slowdown in market dynamics looms. Competition between market players needs therefore to be maintained and strengthened. At the same time, significant private investments need to be realized (for example, in broadband infrastructure). As a result, the European telecommunication policy will constantly need to balance the need for competition with investment incentives that ensure a dynamic development of the sector.

Postal Services

The postal sector is actually composed of several businesses: typically, there is the mail/letter business, the parcel business and the express business. Mail is basically a domestic activity which has recently come under pressure from the information and communication technologies (called 'substitution', see below), whereas parcels and express are increasingly an international activity. Historically, postal financial services were part of the postal sector but have been privatized over the past 20 years in many countries. Unlike the other network industries, liberalization in the postal sector was much less problematic, at least at first sight (see below). As a matter of fact, it only took three Directives in order to fully liberalize the postal market. This apparently smooth liberalization process can be explained by the fact that the postal industry is not really a network industry, that is, by the fact that there is no monopolistic physical network to which access has to be regulated. In most of the countries, therefore, the entire value chain is being liberalized (end-to-end liberalization).[18] On the other hand, the postal sector displays one particularity in the form of a historically highly developed public service, reaching all the households of a country with a regular service delivery.

The liberalization in the European postal sector takes the form of a gradual reduction of the monopoly protection of the historical operator (also called incumbent), while not changing much in the Public Service obligation imposed upon this same historical operator. In order do so, the European Commission proceeded along four steps: it published in 1992 a Green Book in which this very logic of the liberalization of the European Postal Services was outlined. Directive 97/67 (1997) limited the monopoly to 350 grams, that is, thus liberalized parcels and express services. It also defined the so-called universal service obligation, that is, the minimum public service the incumbents had to offer. Directive 02/39 (2002) further reduced the monopoly protection to 100 grams and as of 2006 to 50 grams. Directive 08/06 (2008) finally decided that the monopoly protection would be totally abolished by 2011 (for certain countries by 2013).

In 2008 the European Commission presented a study by Ecorys (Ecorys, 2008) , which evaluated the state of postal market liberalization. Its findings are still valid today. The study shows, among others, that the market was sufficiently competitive in the express and parcels segments, and that it started to become competitive in the international and unaddressed letter segments. However, the market did not function in the domestic addressed letter segment, where generally less than 10 per cent of market share were held by competitors, something which has not changed until today. As a matter of fact, the incumbents generally conserve some competitive advantages in this market segment, which take the form of market entry barriers such as minimal wages imposed upon competitors, VAT exemptions, or licensing conditions for competitors. Generally, the historical operators have prevailed but are losing market share. A few of them – for example, Deutsche Post/DHL and Dutch Post/TNT – have become global players and buying even private operators. They remain entrusted with the universal service obligation in their home country, which generally has not changed over the past 20 or more years. The incumbents however try to gain more freedom in the way they have to meet these universal service obligations (for example, by transforming post offices into so-called agencies). Most of them have corporatized and some even privatized. Overall, however, the letter market seems to stagnate and even to decline, something which is mostly due to the substitution effects by the information and communication technologies.

Historically, the main challenge for the historical operators in the postal sector was its liberalization (loss of market share) combined with the fact that they have to provide universal services. Today, the universal services remain a challenge in many countries inasmuch as they remain defined in a quite traditional (that is, physical and thus costly) way (for example, last-mile delivery, maintenance of post offices), whereas the needs and demands even of the household customers have substantially changed in the communication age. This leads to the fact that especially letter mail is increasingly being substituted by electronic communications, while the incumbent has to maintain a distribution network which is increasingly difficult to finance. It is thus not the end of the letter monopoly that causes problems for the incumbent but rather the changing communication behaviour.

In terms of public policy, this means that it is increasingly difficult, worldwide, to have purely sectoral, that is, postal, public policy. At least in the case of the public service, policy-makers must reassess the needs and protection of their citizens in broad communication terms. This, in turn, means that public service in the future encompasses postal and telecommunications simultaneously, meaning that the public communications need to be designed together. Another policy challenge pertains to the historical operator: indeed, postal markets are increasingly segmented and many of these markets are consolidating and concentrating. Policy-makers therefore need to think about their strategy vis-à-vis their historical operator and evaluate the possibility to at least partially privatize them. This, of course, can only be done, once the question of guaranteeing a universal service in the communication age has been solved.

Railways

Railways are a typical network sector, whose liberalization is made difficult by the technical nature of the industry and a predominant national focus. Over the years, the declining competitiveness of rail in comparison with other transport modes led to a significant reduction in modal share. In an effort to revitalize the sector, the European Community and the member states have initiated reforms at the end of the 1980s through a number of policy initiatives. The first essential initiative was the White Paper *A Strategy for Revitalising the Community's Railways* (European Commission, 1996). It was followed by the White Paper *European Transport Policy for 2010* (European Commission, 2001) which included ambitious political targets for the development of the sector. Several legislative steps were taken to liberalize the rail service markets in Europe and to develop a common European railway area (Eisenkopf et al., 2006). From 2001 onwards, three packages progressively opened up the rail market with the aim to create a Single European Railway Market. The first package consisted of three directives (2001/12/EC, 2001/13/EC and 2001/14/EC) which were substantially amended in 2004 (second package) and 2007 (third package). The idea was to gradually open rail to competition at the European level.

Liberalization of the European railway sector is a protracted and particularly slow process. While all member states (with rail networks) have formally implemented the Directives in the first package, the current status of reform shows mixed results. On the one hand, the process of formally transposing legislation has almost been completed and the institutions required by law already exist. On the other hand the level of implementation differs considerably from one member state to another and one can hardly speak of a single and liberalized European railway market.

In a nutshell, the European railway sector currently faces three problems: first, there is a low level of competition. Except for a few countries like the UK, there is very little competition in the passenger segment. New entrants have made more inroads in the freight segment but the historical operators remain dominant in their home markets. Secondly, there is insufficient regulatory independence. In many countries regulatory institutions are not yet fully operational or devoid of real powers. There is little doubt that fair competition can only be ensured if the regulatory bodies are free from political interference and given the means to reign in the power of the incumbent operators and secure non-discriminatory access for new entrants. Finally, the sector suffers from insufficient investment. There, member states and the European Commission will need to work hand-in-hand to agree on long-term investment strategies and multiannual financing contracts to provide greater predictability for infrastructure development.

These issues, as well as a lack of full technical interoperability, a lack of a common approach to rail safety between national railway networks, significant administrative expenses for rolling stock acceptance and licensing procedures, pose significant barriers to the objective of market integration. In this respect, a strengthening of the European Railway Agency's competencies is being sought from various stakeholders (for example, Deutsche Bahn) in order to minimize discrimination on the technical side.

To address these issues and move forward the reform process, the European Commission has proposed to 'recast' the first railway package – in an effort to simplify, clarify and consolidate the existing legislative body. For instance, uncertainty remains as how to interpret the directive related to the separation of infrastructure management from operations. Germany and Belgium have opted for a structure in which the infrastructure manager is part of a holding company that also includes one or more railway companies.

Among the objectives of the Commission is also to guarantee accessibility for all rail operators to the so-called ancillary services – railroad yards, terminals, diesel re-fuelling stations, ticketing facilities – as they often pose an obstacle to the liberalization of rail networks. Finally, the Commission still aims for a complete liberalization of the rail passenger market. As of 1 January 2010, international passenger services are open to competition. As to the opening of competition at the domestic level, the Commission plans to reflect on the liberalization process before proposing an additional legislative package in 2012 leaving still more time before full competition is achieved. In France, for instance, the incumbent railway operator (SNCF) prepares for competition by 2016.

Historically, railways have benefited from strong union support and a capacity to mobilize the society at large towards defending the notion of public service. As a result, reform of the sector is politically delicate and requires a permanent dialogue between the various stakeholders. Some of the most pressing policy issues include the organizational model allowed under EU legislation, ensuring sufficient funding for rail infrastructure or creating a level playing field between different modes of transport (Di Pietrantonio and Pelkmans, 2004).

Air Transport

In air transport liberalization, the European Commission follows in fact the liberalization that has taken place since the early 1990s in the USA. Initially, its main focus was on the airlines. To recall, here the liberalization basically takes form of abolishing the bilateral agreements among countries which define the so-called 'air transport rights'. It is the market, rather than governments which decide upon the air transport offers, the frequencies, the capacity, and the prices. The liberalization of airlines took place by way of three Directives: the first in 1987 (Council Decision 601/87 and 602/87, Council Regulation 3975/87 and 3976/87) liberalized tariffs and allowed codesharing among airlines. The second directive was issued in 1990 (Regulations 2342/90, 2343/90 and 2344/90) and further removed restrictions on international passenger and cargo transport. The third dated 1993 (Regulation 25407/92 on licensing of air carriers, 2408/92 on market access and 2409/92 on fares and rates) led to the effective opening of the European airlines market by authorizing as of 1997 the so-called ninth freedom.[19]

However, it appeared that the liberalization of the bilateral transportation rights was only one of the conditions for creating a functioning European air transport market. A next critical issue is the access to the *airports*, as airports are typical monopolistic bottlenecks, especially when it comes to the attribution of airport slots. Airport slots were first regulated in 1995 (Council Regulation 95/93). However, in 2009 the Commission significantly strengthened its regulations and instituted independent airport coordinators at congested airports. Yet, in the future, the Commission would like to go even further and allocate airport slots by way of a more market based approach. Similarly, so-called groundhandling (maintenance, preparation, catering, and so on) was also identified as a monopolistic bottleneck. Consequently, the Commission issued a Directive already back in 1996 (reference) to liberalize this market.

The technical nature of air transport, as well as the safety concerns, are also addressed by the Commission, especially in the light of liberalization. As of 2003 the Commission has created the European Aviation Safety Agency (EASA). This agency is competent in matters of licensing of aircrafts, engines and instruments, and its competency was extended in 2009 to operations and personnel. The purpose of the agency is not only to ensure high safety standards throughout Europe, but also to promote technical harmonization. It constitutes probably a first step towards a full-fledged European air transport regulator, as it is planned in the future to also regulate airports and air navigation services.

Finally, air traffic control (ATC) also needs to be mentioned. Indeed, European airspace reaches its capacity limits. Considering this, the Commission started its project entitled European Single Sky back in 1999. The ultimate aim of this project is to promote a more efficient and environmentally more friendly use of airspace, but also to enhance safety by restructuring the current airspace into 11 functional airspace blocks. While initially the focus was mainly on safety and capacity problems, the subsequent project Single Sky II also focused on noise and CO_2 emissions. A high level group is coordinating all these activities since 2006 and increasingly also looks at how airports and how EASA can be integrated into this Single Sky project.

Without doubt, these various efforts of the European Commission have led to very substantial changes in the European Air transport sector. To be mentioned, in particular, is the development of the low-cost airlines which have profoundly changed the European air transport landscape. Subsequently, restructuring among the traditional network airlines has taken place. Some of them have gone bankrupt, while most others integrated in three major airline alliances. In this context, one may also mention the fact that the European Commission is increasingly watching out on subsidies for airlines (state aid).

While air transport is indeed liberalized in Europe, the main next challenges are indeed airports and in particular airport slots, access to groundhandling services, and the creation of a single European sky. Underlying to these challenges is of course the gradual transfer of regulatory powers from the national regulatory authorities to EASA, the European Aviation Agency which will, sooner or later become a EU-wide air transport regulator.

Beyond liberalization, many policy issues remain in the air transport sector. The most important such policy issue is without doubt CO_2 emissions. Indeed, the European Commission launched its Emissions Trading Scheme (ETS) in 2005. From 2012 aviation will also be integrated into the scheme. Other policy issues pertain to safety and to state aid.

Overview of Sectoral Achievements

The approach of the European Commission is, as one can see, quite piecemeal and sector-specific. Nevertheless, one can distinguish between the liberalization in the following three areas, transport, energy and communication.

In transport – rail, road and air – the Commission is mainly motivated by efficiency and environmental concerns. It seeks to achieve a better modal split, trying in particular to transfer cargo traffic from the road to rail. Liberalization, for the Commission, is the main tool to achieving this goal, but it is far from being reached as freight has a hard time competing with road traffic.

In energy, the objectives of the Commission are more ambiguous; on the one hand, it seeks to achieve more efficiency by opening the market to competition while, on the other hand, the Commission seeks to achieve a certain security of supply thanks to the free flow of energy. While the first objective was achieved at least in the beginning of operations, the second objective was not. Indeed, the issue of energy security becomes ever more political and actually risks threatening the very liberalization process.

In the area of communications, the Commission has basically abandoned the postal services and bet exclusively on telecommunications as the means to achieve the information society. It is indeed the promotion of the information society that has become the political objective and all telecommunications policies are now subordinated to it. However, progress in this matter is again quite slow with a few success stories, for example, in the area of roaming charges.

Competition and market integration have progressed at a different pace across sectors since the initiation of the liberalization process (Table 20.1). Regarding competition, rail transport and postal services competition remained concentrated. Likewise, the energy sector is dominated by domestic incumbents and exhibits a lack of real competitive behaviour. Two sectors have made notable progress towards instilling competition. In telecommunications major players have grown at the expense of incumbents. In air transport the arrival of new carriers ('low-cost') has contributed to reshape the competitive environment. Some bottlenecks, however, remain and impede competition in a number of sectors. A serious problem comes from the inadequate implementation of liberalization directives in some of the member states. The absence of EU-wide regulatory authorities negatively affects the creation of truly integrated markets as national authorities are poorly adapted to deal with cross-border issues – their scope of action is typically limited to their member-state borders. Competition in network industries is further

Table 20.1 European sectoral achievements so far

Sector	Current status of competition	Current status of market integration
Telecommunications	Extensive	Relatively limited
Postal service	Limited	Relatively limited
Electricity	Limited	Significant
Air transport	Extensive	Extensive
Railways	Limited	Relatively limited

impeded by the fact that incumbents tend to rely on their market power to prevent new entry on their markets. Finally, the creation of EU-wide markets in network industries is negatively affected by various forms of economic patriotism and government interference. As to market integration, much remains to be done. A set of factors has slowed down the process: (1) physical barriers (for example, interconnection, interoperability), (2) legal barriers for example, inconsistencies in sector-specific regulation, suspensory effects, and mutual recognition), (3) inherited advantages of incumbents (for example 'grandfather rights' and long-term contracts).

EVOLUTION AND ANALYSIS OF EU'S INFRASTRUCTURE POLICY ISSUES

In this section we first revisit some of the main policy issues as they have emerged from the process of liberalization as described in the section above. We then identify the main current policy issues and compare the two situations briefly. We finally discuss the transition from the old paradigm (market *creation*) to the new one (market *management*), as this evolution can be observed in each of the above presented sectors.

By-and-large policy issues during the market creation era revolved mainly around issues of proper unbundling, as well as questions of access and access pricing (European Commission, 1999, p. 76), as can be seen from the first column of Table 20.2.

As one can see from the second column of Table 20.2, policy issues in 2010 are quite different, even though some issues persist over 20 years. But after 20 years of liberalization, the European Commission seems to be much more concerned with issues of market concentration, as well with issues of security of supply, even though the traditional issues of access, access pricing and unbundling linger. This evolution of the issues does however not mean that the process of the liberalization of the network industries is being questioned by the European Commission or even halted. Rather, these issues arise because of the continued pursuit of the liberalization process.

Indeed, even though policy issues seem to evolve over the past 20 years, liberalization of the network industries is far from being achieved. As a matter of fact, the main underlying issues of such liberalization remain as relevant as ever, namely, questions of access, of remaining monopolistic bottlenecks and of course issues pertaining to the interface between competitive and non-competitive elements in each sector.

Table 20.2 Evolution of EU's infrastructure policy issues, 1990–2010

Sector	Policy issues in 1990s	Policy issues in 2010
Telecommunications	Determination of modes of access and access prices Universal service obligation (definition and financing) Network interconnection (pricing principles, collusive agreements)	Financing and rollout of Next Generation Networks Prices of services
Postal service	Nature, degree and speed of liberalization	Viability of the historical operators after market opening and corresponding viability of the public service
Electricity	Competitive behaviour in generation Unbundling, grid ownership	Grid capacity and allocation of scarce grid capacity Security of supply Generation prices
Air transport	Allocation of slots at airports Necessity for strong competition policy for airlines	Airport slot allocation Single European Sky
Railways	Introduction of competition Authority in charge of regulation Access pricing and slot allocation Social impact of organizational change	Non-discrimination, interoperability, environment, investment

Source: Adapted from European Commission (1999).

Access

Far from having gone away, the issue of access still remains at the centre of the liberalization process. However, the notion of access has been refined over the years. For instance in the railway sector, debates around essential facilities have shifted from the network access to access to locomotives and maintenance yards. In the electricity sector, the access question has also shifted to scarce trans-boundary capacities, as in the gas the access question has now shifted to storage, for example. Access remains relevant also in the air transport sector in the case of airport slots and airport facilities, and in the postal sector some regulators want to open up access to the network of the historical operators. In other words, as access to the basic infrastructures is being solved, the question of access is increasingly focused on particular bottlenecks, some of which can perhaps be addressed, in the future, by competition law, rather than by sector specific regulation.

Monopolistic element

In most network industries, at least one but generally several elements of the value chain remain monopolistic. In the postal sector for example, the distribution network, even though contestable, remains in the hand of the historical operator. In the telecommunications sector the local loop, along with mobile termination remain equally monopolistic and there are concerns that local fibre will become equally monopolistic. In the electricity

Table 20.3 Remaining non-competitive elements in network industries

Sector	Non-competitive element
Electricity	High-voltage transmission, local distribution
Telecommunications	Local loop and termination of calls on other network
Postal services	Delivery network
Air transport	Access to runways, such as take-off and landing slots
Railways	Track and signalling infrastructure

Source: Adapted from OECD (2001).

sector, both the high-voltage and the local distribution grids are of course monopolies. Similarly, in the railway sector the infrastructure, including the railway stations, remain monopolies. Finally, in the air transport sector, airport slots and facilities, along with the air traffic control will remain monopolistic long into the future. Table 20.3 summarizes this observation.

Interface between competitive and non-competitive elements

As significant monopolistic bottlenecks remain in almost all the network industries, the question of the interface between the competitive and the non-competitive elements remains as important as ever. This is the question of who is taking on the responsibility to coordinate the overall infrastructure system. While historically this was done by the incumbent and at the European level by the collaboration between the incumbents, this coordination role is no longer assumed by any of the operators. In a first step, regulators at the national level often take on the role of overall system coordination. But at a certain point of market integration, the coordination at the national level appears insufficient and supra-national actors need to play a more active role.

This is again where the European Commission comes in; indeed, after being unbundled and deregulated, the network industries develop a dynamic of their own, which in turn leads to new policy issues (for example, concentration, security of supply) on top of the traditional issues of proper unbundling, non-discriminatory access, and state subsidies. As a result, the European Commission finds itself increasingly in the role of having to coordinate between the competitive and the non-competitive aspects of the infrastructure systems, an evolution which will, more or less inevitably, lead to some form of European regulatory agencies. Ultimately, the Commission will end becoming a regulatory body itself, an evolution which was foreseen by Giandomenico Majone already back in 1996 when he coined the term 'regulatory Europe' (Majone, 1996).

However, not everything needs to be regulated at the European level; there is a good case for a complex system of multilevel governance, whereby the basic infrastructure networks in Europe are regulated at a European level, while regional networks can be regulated at a national or regional level and distribution networks at a local level. The question remains, which role the nation state will be left to play in such a multi-governance system.

CONCLUSION

In this chapter we have discussed the shift of paradigm that took place in the European Union. Over the last two decades the emphasis on market creation has given way to the management of these markets and by extension of competition. In most sectors, and with the exception of the remaining monopolistic element (for example, management of tracks in railway or high-voltage transmission in electricity), markets have now been created and are functioning reasonably well. Part of the success can be attributed to the strong tendency in European policy-making to favour competition and market integration (Héritier, 2001, p. 848).

Markets and policy-making have evolved hand-in-hand during the liberalization process. Policy-makers, infrastructure providers and customers have, at times, learnt the hard way that markets only work up to a certain point and that the idiosyncrasies of network industries put them in a different category than, say, manufacturing companies. For instance, fair market access needs to be ensured, investments cannot be foregone without running risks of security of supply.

Over time the 'market only' paradigm has been replaced by a more integrative approach seeking to balance internal market mechanisms with other goals. This 'managed competition' raises the question of how further the liberalization of infrastructures can go. Indeed, the initial assumption that sector-specific rules would give way to general competition law has not held way. In addition, both technological and institutional characteristics pertaining to the infrastructure sectors do not bode well for the disappearance of sector-specific rules.

One can fairly say that the liberalization of the network industries in Europe has notably progressed in recent years, in large part due to the continuous work of the European Commission. In spite of the common policy objectives, carried out via Directives, the picture of liberalization across all member states is still very heterogeneous. In a nutshell, countries are still in very different stages of reform and one can find significant differences in levels of competition across sectors and between countries.

In part because of the question of subsidiarity and in part because of the massive restructuring effort required, the effect of infrastructure restructuring on European integration has not played out as much and as fast as one could have hoped for. Common policy issues have often not found their counterpart in the institutionalization of reforms. Some form of agencification has taken part in almost all the infrastructure sectors but there has yet to emerge one European-wide body regulating a network industry.

In light of the importance of infrastructure in the European economic and political landscape, one can safely expect that the ongoing restructuring of network industries will remain at the centre of public policy concerns. As a result, an important part of the liberalization agenda will still be designed and carried out (or at least driven) by policy-makers rather than by regulatory institutions.

NOTES

1. We thank Rolf Künneke for very helpful comments and suggestions.
2. For instance in railways the infrastructure cannot be profitable unless the full costs are passed onto the train operators. If that were the case, passengers or shippers would put more cars and trucks on the roads.
3. In the so-called Lisbon Agenda the Head of States and Governments agreed to the goal for the EU 'to become the most competitive and dynamic knowledge-based economy in the world' in 2010. Presidency Conclusions of the Lisbon European Council, 23 and 24 March 2000 (http://www.europarl.europa.eu/summits/lis1_en.htm).
4. The globalization and commercialization of this sector fundamentally changed expectations of market actors and accordingly new strategies and alliances developed. The traditional sector organization hampered these ambitions. Even the incumbent operators could not deny these new business opportunities anymore. Big European operators were prepared to enter the global market and engage in new alliances. This was only feasible in a liberalized market environment. Hence, in 1993 the European Commission decided to introduce competition by opening up telecom networks for third parties within a time period of five years. This example demonstrates the dramatic change of liberalization policy in the EU 'from a non-issue to a central area of EC competition and industrial policy' (Schneider, 2001, p. 64).
5. An overview of the liberalization of postal services in the EU is provided by (Geradin, 2002).
6. For Steer Davies Gleave (2009, p. 13), the overarching goal of Europe's Common Transport Policy (CTP) can be described as 'the promotion of an efficient, sustainable, safe and secure transport system capable of enabling trade and mobility whilst minimizing costs for users and society as a whole'. They further divide policy objectives under economy (for example, competitive internal market for transport), society (for example, protection of passenger rights) and environment (for example, sustainability of transport).
7. The European Council's adoption of a comprehensive Energy Action Plan was heralded as a milestone in the creation of the European Energy policy.
8. For some countries transport should be market-led while others see transport as an instrument of public policy.
9. The EC has published annual reports evaluating the performance of network industries in that respect between 2001 and 2007.
10. The EC periodically evaluates the European policies regarding network industries.
11. For instance, the liberalization of the air transport sector has led to the emergence of low-cost carriers, driving a demand for 'week-end city hoping' that may be run counter to policies aiming at safeguarding the environment.
12. Regulation (EC) No 714/2009 of the European Parliament and of the Council of 13 July 2009 on conditions for access to the network for cross-border exchanges in electricity, repealing Regulation (EC) No 1228/2003.
13. The Independent System Operator or ISO-model allows for the transmission grid's assets to be owned by producers, whereas the grid itself is operated by a company that is independent of the owners' assets. Full Ownership unbundling means that the grid company and the energy producing company are legally separate entities.
14. At this first juncture, the local loop was unbundled and household customers were given the freedom to choose their providers.
15. The main instruments of the new EU regulatory framework for the electronic communications sector include: (1) the Framework Directive, which sets out the main principles, objectives and procedures for an EU regulatory policy regarding the provision of electronic communications services and networks; (2) the Access Directive, which stipulates procedures and principles for imposing pro-competitive obligations regarding access to and interconnection of networks on operators with significant market power; (3) the Authorization Directive, which introduces a system of general authorization, instead of individual or class licences, to facilitate entry in the market and reduce administrative burdens on operators; (4) the Universal Service Directive, which requires a minimum level of availability and affordability of basic electronic communications services and guaranteeing a set of basic rights for users and consumers of electronic communications services; (5) the Privacy and Electronic Communications Directive, which sets out rules for the protection of privacy and of personal data processed in relation to communications over public communication networks; (6) the Radio Spectrum Decision, which establishes principles and procedures for the development and implementation of an internal and external EU radio spectrum policy; and (7) the Commission Competition Directive.
16. The regulatory framework that existed in the EU prior to the implementation of the 2002 regulatory framework was aimed primarily at managing the transition from monopoly to competition, and was therefore focused on the creation of a competitive market and the rights of new entrants.

17. A key objective of the directives was to create a flexible regulatory framework that is capable of responding to new technologies, convergence, and an increasingly competitive market.
18. This fact is nevertheless disputed: unlike in the European Union, the USA have chosen a different approach to postal market liberalization, whereby everything but the so-called 'last mile' is being liberalized. Similarly, the UK regulator has chosen an approach which regulates the access of the competitors to the historical operator's (Royal Mail's) distribution network, even though the network is virtual and not physical.
19. The ninth freedom is the total opening of the market, whereby an airline from one European country can offer its services even within another country.

REFERENCES

Albert, M. (1993), *Capitalism against Capitalism*, London: Whurr.

Arentsen, M. and Künneke, R. (eds) (2003), *National Reforms in European Gas*, Amsterdam: Elsevier.

Cawley, R. (2008), 'The new European Union approach to sector regulation in network industries', *Network Industries Quarterly*, **10** (1), 23–5.

Comité intergouvernemental créé par la Conférence de Messine (1956), *The Brussels Report on the General Common Market*, Washington, DC: L.B. Tennyson, Information Representative, European Community for Coal and Steel.

Di Pietrantonio, L. and Pelkmans, J. (2004), 'The economics of EU railway reform', *Competition and Regulation in Network Industries*, **5** (3–4), 295–346.

Doyle, C. (1998), 'Liberalizing Europe's network industries: ten conflicting priorities', *Business Strategy Review*, **9** (3), 55–66.

Ecorys (2008), 'Main developments in the postal sector (2006–2008)', report.

Eisenkopf, A., Kirchner, C., Jarzembowski, G., Ludewig, J., Rothengatter, W. and McCullough, G. (2006), 'The liberalisation of rail transport in the EU', *Intereconomics*, **41** (6), 292–313.

European Commission (1987), *Towards a Dynamic European Economy. Green Paper on the Development of the Common Market for Telecommunications Services and Equipment. Appendices*, Brussels: European Commission.

European Commission (1991), *Green Paper on the Development of the Single Market for Postal Services (Communication from the Commission)*, Brussels: European Commission.

European Commission (1996), *A Strategy for Revitalising the Community's Railways. White Paper*, Brussels: European Commission.

European Commission (1999), *Liberalisation of Network Industries – Economic Implications and Main Policy Issues*, Brussels: European Commission.

European Commission (2001), *White Paper: European Transport Policy for 2010: Time to Decide*, COM(2001) 370, Brussels: European Commission.

European Commission (2006), *European Smart Grids Technology Platform: Vision and Strategy for Europe's Electricity Networks of the Future*, Luxembourg: Directorate-General for Research.

European Commission (2009) *Agreement on EU Telecoms Reform Paves Way for Stronger Consumer Rights, an Open Internet, a Single European Telecoms Market and High-Speed Internet Connections for All Citizens*, Brussels: European Commission.

European Communities (1979, *Contribution of the European Communities to the development of Air Transport Services – Memorandum of the Commission*, Luxembourg: EC.

Geradin, D. (2006), 'The liberalization of network industries in the European Union: where do we come from and where do we go?', contribution to the project 'Globalisation Challenges for Europe and Finland', Economic Council of Finland, pp. 1–21.

Geradin, D. (ed.) (2002), *The Liberalization of Postal Services in the European Union*, The Hague: Kluwer.

Giesbertz, P. and Mulder, M. (2008), 'Economics of interconnection: the case of the northwest European electricity market', *International Association for Energy Economics Newsletter*, 2nd quarter, 17–21.

Haas, R. and Glachant, J.-M. (2006), 'Competition in the continental European electricity market: despair or work in progress?', in F. Sioshansi and W. Pfaffenberger (eds), *Electricity Market Reform: An international Perspective*, Oxford: Elsevier, pp. 265–315.

Héritier, A. (2001), 'Market integration and social cohesion: the politics of public services in European regulation', *Journal of European Public Policy*, **8** (5), 825–52.

Jordana, J., Levi-Faur, D. and Puig, I. (2006), 'The limits of Europeanization: regulatory reforms in the Spanish and Portuguese telecommunications and electricity sectors', *Governance*, **19** (3), 437–64.

Laperrouza, M. (2008), 'Dealing with standardization in liberalized network industries: some lessons from the European railway sector', paper presented at the Second Biennial Conference of the Standing Group on Regulatory Governance (ECPR), 5–7 June 2008, Utrecht.
Majone, G. (1996), *Regulating Europe*, London: Routledge.
Organisation for Economic Co-operation and Development (OECD) (2001), *Restructuring Public Utilities for Competition*, Paris: OECD.
Percebois, J. (2008), 'Electricity liberalization in the European Union: balancing benefits and risks', *The Energy Journal*, **29** (1), 1–20.
Schneider, V. (2001), 'Institutional reform in telecommunications: the European Union in transnational policy diffusion', in M. Green Cowles, J.A. Caporaso and T. Risse-Kappen (eds), *Transforming Europe: Europeanization and Domestic Change*, Ithaca, NY: Cornell University Press, pp. 60–78.
Steer Davies Gleave (2009), *Evaluation of the Common Transport Policy (CTP) of the EU from 2000 to 2008 and Analysis of the Evolution and Structure of the European Transport Sector in the Context of the Long-Term Development of the CTP Final Report*, Brussels: European Commission, Directorate-General Energy and Transport.
Vogel, S.K. (1996), *Freer Markets, More Rules: Regulatory Reform in Advanced Industrial Countries*, Ithaca, NY: Cornell University Press.

21 Liberalization and regulation of telecoms, electricity, and gas in the United States

Mark A. Jamison[1]

INTRODUCTION

The USA has a long tradition of commission-style regulation of privately owned utilities. This fact has both advantages and disadvantages. On the plus side producers, political bodies, regulatory agencies, courts, consumers and other players are fully adapted to the idea and practice of independent regulation. Furthermore there is a ready pool of talented professionals to analyse issues and develop solutions as new issues emerge. But these advantages can also be disadvantages during times of change. Complex, well-developed systems are often slow to recognize new realities and are costly to change because regulatory policies create interest groups that benefit from the status quo. In some instances new technologies and policies cannot be introduced incrementally, but rather strand investment and challenge investor's willingness to provide funds.

In this chapter we examine the development and evolution of utility regulation in the USA, focusing on energy and telecommunications. We begin with the development of these industries, taking as given the traditions, institutions and legal frameworks created through the regulation of transportation and other industries, even though these laid critical foundations for utility regulation. We begin by describing the economic and political context for regulation. We then examine regulation for each sector. We conclude with a brief review of emerging issues.

THE US CONTEXT

In general the economy of the USA is marked by private enterprise and competitive markets. There are swings in political views when the favorable view of liberal markets is less warmly embraced than at other times, but the relative success of the market system, the US Constitution's protection of private property, and an American culture that has for most of its history viewed favorably individual responsibility and liberty have worked together to maintain the economy's market orientation.[2]

One notable exception to this history of liberal markets has been the regulation and provision of public utility services. Although the term 'public utility' is not well defined, the industries treated as such have tended to be monopolistic and are affected with the public interest, meaning that their performance significantly impacts the social and economic functioning of the country and they are subject to legal obligations above and beyond those of other enterprises (Glaeser 1927, pp. 170–71; Phillips, 1993, pp. 1–2, 83-121).

Utility enterprises are treated differently than other businesses in the economy in at least two respects. First, the utilities are generally regulated monopolies. The regulations typically include restrictions on the extent to which a utility can discriminate in its service to similarly situated customers, limits on the providers' prices, and an obligation to serve customers who are willing to pay the regulated price. However, the government's authority to regulate prices is limited in that the regulated prices must allow the operator an opportunity to maintain its financial integrity, that is, the operator is allowed to recover its reasonable costs and cover its costs of equity and debt (Glaeser, 1927, pp. 174–75; Phillips, 1993, pp. 319–31, 376–82). This limit on the government's ability to regulate prices is designed to check opportunistic behavior by the government, which would in turn reduce investment and services (Williamson, 1983). The second way that utilities are different is that they are sometimes government owned, although private ownership is the norm. As we explain in more detail below, power providers such as the Tennessee Valley Authority (TVA) and Bonneville Power Administration (BPA) are federally owned. In addition there are more than 2000 municipally owned electric utilities (Brown, 2005; Liggett, 2000).

It can generally be concluded that, relative to the common alternatives of government provision of utility services or unregulated monopoly, the regulatory system in the USA has worked to the benefit of customers and service providers. However, it has not been without its critics. In 1940 Horace Gray argued that the concept of a private monopoly cannot be harmonized with the idea of public interest (Gray, 1940). He believed that regulation is unable to control monopoly and that social or national goals should supersede concepts of fair profit and cost-based pricing. Sam Peltzman (1976) argued that regulation occurs because of rent seeking on the part of powerful political interests and so could have little to do with a broader and vaguely defined public interest. Harry Trebing (1976, 1984a, 1984b) counters that these and similar views are based on simplifications that exaggerate the problems with government intervention and the adequacy of markets, but he found value in the debates these views generated on the shortcomings of regulation.

Regulation in the USA is an outgrowth of an interaction between economic argument and the US legal system. The federal government's legal authority to regulate comes from the interstate commerce clause in the US Constitution, which provides that the federal government has the authority to regulate commerce among the states. The various state governments' legal authority to regulate comes from their police powers, which the Constitution says are not delegated to the federal government. These powers allow the states to protect the health, safety, morals and general welfare of their citizens. Through a series of legislative actions, court actions, and the like, these police powers came to include the authority to regulate prices and services of businesses that were deemed to be public utilities. Because the interstate can be affected by state regulation and visa versa, there are constant frictions between state and federal regulators (Phillips, 1993, pp. 83–121).

One of the earliest court cases to establish the concept of a regulated business affected with the public interest was *Munn* v. *Illinois*.[3] This case upheld legislation in the State of Illinois to regulate grain elevator prices. The US Supreme Court's decision found that a state has the authority to regulate a private company in the public interest if the private company could be seen as a utility. In *Munn* the court based its argument in part on the private business's market power. Subsequently in *Brass* v. *Stoeser* [4] the court confirmed that the state of North Dakota could regulate private businesses – again grain elevators – even though the individual businesses lacked market power because of the effects the businesses could have on their customers. One of the earliest cases restricting governmental authority over prices was *Smyth* v. *Ames*,[5] which protected utilities' rights to due process and found that regulatory commissions had to base prices on the value of the utilities' assets. The court's asset valuation criteria proved to be problematic, but it was remedied in 1944 in *Federal Power Commission* v. *Hope Natural Gas*[6] where the Supreme Court determined that it should not impose a specific ratemaking methodology on regulatory commissions, but rates had to be just and reasonable, allowing a service provider to operate successfully.

Institutional economists, who emphasized issues of market power and fairness, dominated the economic discourse that helped create and shape regulation.[7] In his studies of the industrialization of the USA, Charles Adams argued that because railroads were capital intensive, their value to the economy would be increased if prices were controlled so as to increase the volume of usage (Trebing, 1987). Martin Glaeser (1927) helped refine the public utility concept and explained the economic and legal rationale for regulation, as well as the emerging lessons from experiments in the regulation of prices. James Bonbright and Gardiner Means (1932) provided insights into problems with utility holding companies and Bonbright (1961) wrote a foundational text on utility ratemaking, emphasizing fairness and concrete regulatory issues (Berg and Tschirhart, 1995). The basic postulates of these and other institutionalists were that regulation was needed to control market power, pursue public and social interest goals, improve efficiency and individual choice, and provide due process for stakeholders and operate with transparent processes (Trebing, 1987).

Neoclassical economics, which emphasizes markets and optimizing behavior subject to technological and regulatory constraints, shaped much of the current thinking on regulation. For example, Alfred Kahn's classic texts, published in 1970–71 and republished in 1988, provided a rigorous, yet non-mathematical explanation of economic efficiency aspects of utility pricing and the impacts of market structure on regulation. Work by William Baumol (1977, 1979, 1982, 1983, 1986), Gerald Faulhaber (1975, 1979) and William Sharkey (1981, 1982a, 1982b) introduced ideas of contestable markets, added rigor to concepts of natural monopoly, explained the cost structures and pricing incentives of multiproduct firms, and developed approaches for allowing price flexibility for multiproduct utilities serving in both competitive and monopoly markets. Their work also laid foundations for cross-subsidy concepts, deregulation, and access pricing applied in telecommunications, and avoided cost concepts applied in energy. Neoclassical economics also introduced other innovations important in utility regulation, including peak-load pricing (Boiteux, 1949; Hirshleifer, 1958; Steiner, 1957; Williamson, 1966), risk assessment in the cost of capital (Sharpe, 1964), option pricing (Black and Scholes, 1973), the use of auctions to allocate scarce resources (Coase, 1959; Vickery, 1961), and Ramsey pricing (Ramsey, 1927). Although first formalized in recent times by the Austrian economist Stephen Littlechild (1983) for the regulation of British Telecom, the advancement of price cap regulation to overcome information asymmetries was also accomplished by application of neoclassical economics (Sappington, 2002).

As we describe in the next section, these legal and economic concepts both laid the foundation for utility regulation in the USA and led to its constant evolution from a system designed to provide stability, control market power, and promote fairness, to one that now focuses largely on efficiency and economic incentives.

REGULATORY MODELS AND APPROACHES IN THE USA

All state and federal regulatory bodies in the USA are permanent agencies, established either by constitution or statute, but mostly by statute.[8] State laws and constitutions create the state agencies and federal agencies exist under federal statutes. In general the state public utility commissions (PUCs) are defined by their respective state laws as independent and not part of any other agency of the government. As a result they are free to make their own final decisions, subject only to court appellate processes, although situations arise where a state legislature will change a PUC decision by revising the state statute. In just a few states, such as Massachusetts and Iowa, and in the case of the Federal Energy Regulatory Commission (FERC), the agencies are within a government department for administrative matters, but not for their regulatory decisions (Brown, 2005; Phillips, 1993, pp. 83–121).

Regulatory agencies in the USA combine legislative, executive and judicial powers of government. Governmental powers in the USA are divided into three categories that form the three branches of government. Basically the legislative branch writes laws (that is, creates basic policy), the executive branch enforces and carries out the laws, and the judicial branch interprets the laws and resolves legal disputes. Regulatory agencies combine these three powers when they, for example, set rates and standards (legislative), enforce statutes and their own administrative rules (executive), and interpret statutes (judicial). Thus the agencies are inherently independent agencies, although accountable to the legislatures, executive authorities, and courts (Brown 2005; Phillips, 1993, pp. 83–121).

Federal regulatory agencies are essentially sector specific: the FERC regulates energy (electricity and natural gas) and the Federal Communications Commission (FCC) regulates telecommunications. State PUCs are generally multisector. In the following sections we focus on the development and evolution of regulation by sector, beginning with telecommunications and then proceeding with energy. We do not elaborate on transportation or water regulation.

Fixed Line Telecommunications

Telecommunications in the USA began with competition between Alexander Graham Bell and Western Union over patent rights to the telephone device and the establishment of telephone exchanges.[9] Bell negotiated an agreement with Western Union that gave Bell, whose company became AT&T, the telephone patents. Bell enjoyed a monopoly on telephone service until the key patents began expiring in the late 1800s, at which time numerous competitors – called independent telephone companies – arose to challenge Bell, which had left so much unmet demand during its first monopoly era that soon after the patents expired the independents served more lines than did Bell. Early telephone technologies could not carry calls between cities, so telephone service was limited to service within a city, which became known as a local exchange (Brock, 1981, pp. 97–99, 104–05). This technology boundary became a regulatory boundary through the telephone franchising process: prior to the development of state and federal regulation, telephone companies obtained franchises from city governments and these franchise agreements regulated prices, although in some instances prices were controlled by state legislatures (Mueller, 1993, 1997, p. 37; Nix and Gabel, 1993; Gabel, 1994).

The rapid rise in competition gave way to industry consolidation, raising concerns about monopolization. This concern coupled with the refusal of some rivals to interconnect their networks fed interest in regulating the growing industry. However, political bodies were ill suited for regulating sophisticated companies that were able to leverage information asymmetries, so state legislatures began creating PUCs in the early 1900s (King, 1912). The rise of long-distance technology, which made interstate calling possible and the development of wireless telecommunications prompted Congress in 1910 to extend the authority of the country's transportation regulator, the Interstate Commerce Commission, to include telecommunications. But neither the Interstate Commerce Commission nor the FCC, which was created by the Communications Act of 1934 and took over telecommunications regulation, had authority to regulate local telephone[10] prices because these services were considered intrastate commerce (Brock, 1981, pp. 158–61, 178–80).

In the early twentieth century AT&T opposed both regulation and competition, but could not avoid one without having the other. As a result the company embraced the idea of regulated monopoly. Unfortunately for AT&T, federal legislation did not actually grant the company a monopoly. So in three decades when the company Hush-A-Phone proposed to attached devices to AT&T phones, something AT&T opposed, the courts and a reluctant FCC permitted Hush-A-Phone to do so, determining that companies and customers could attach devices to the public switched network as long as the devices were privately beneficial and not publically harmful.

Although competition began on the edge of the network, it quickly switched to the network's core. Hush-A-Phone constituted competition at the edges of the network: customers wanted to attach non-AT&T equipment to the AT&T network. Another company, Carterfone, represented a network innovation: Carterfone allowed customers with private radio networks to connect those networks to AT&T. Presumably with proper legal authority AT&T could have offered a radio service, but it did not do so. Furthermore the FCC's *Above 890 Decision*,[11] which set aside a portion of the microwave radio spectrum for use by customers, enabled customers to self supply high capacity circuits, further enabling network competition.

The *Above 890 Decision* also made possible the development of an upstart company that eventually became Microwave Communications Inc. or MCI, which constructed a microwave network and allowed others to lease capacity. MCI soon began leasing its circuits in very creative ways: leasing periods became increasingly short and access to circuits became increasingly automated until, eventually, a customer could 'lease' an MCI circuit on a per minute basis by simply dialing a special code into their AT&T phone. This, of course, was simply long distance service under the guise of circuit leasing.

What MCI demonstrated is that price and service barriers to entry – even if embraced by the sector regulator – can be overcome. However, this was not the lesson that regulators thought they learned from the MCI experience. They concluded that since there was competition at the edges of the network and in the core of the network, but not in the fixed line portion, that it must be because the fixed line is a natural monopoly. This incorrect belief shaped telecommunications policy from the late 1970s to the mid-1990s.

Despite the growing cracks in the AT&T monopoly, the company dominated the industry from the 1920s until the breakup in 1984, controlling approximately 90 per cent of the country's fixed lines and accounting for over 90 per cent of the long distance revenue (Economides, 2004). It was also during this time that the country achieved what became known as universal service. Conventional wisdom is that subsidies from long-distance service to local service, coupled with an AT&T monopoly, led to universal service. However, as Milton Mueller (1993) demonstrated, telecommunications grew faster during the competitive eras than in the monopoly eras and universal service was largely achieved before the FCC adopted an extensive subsidy scheme.

The belief that fixed line was a natural monopoly provided the underpinnings of the breakup of AT&T in 1984. The basic concern was that AT&T had used its control of bottleneck local telephone lines to foreclose competitors and to cross subsidize its potentially competitive markets (Temin, 1990). The breakup required AT&T to divest its ownership of the Regional Bell Operating Companies (RBOCs), which owned the local fixed lines (Hughes, 1996).[12]

Facilitating competition in long distance, making the local telephone subsidy scheme compatible with long-distance competition, and addressing further encroachments by competition into the divested RBOCs' networks occupied regulators' time until around 1993 when it became clear that the monopoly over fixed lines could not be sustained. Fixed line was opened to competition on a state-by-state basis until the passage of the Telecommunications Act 1996 (hereafter the 1996 Act),[13] which made local telephone competition a national policy.

The 1996 Act provides three methods of entry for local telephone service. Entrants can build their own facility-based network, lease portions of an incumbent's network, buy an incumbent's service and resell it, or use some combination of these three approaches. The essential trade-off in the 1996 Act is that the RBOCs and GTE were permitted to offer long-distance service in exchange for effectively opening their local monopolies to competition.[14] GTE was permitted to offer long distance immediately upon passage of the 1996 Act, but the RBOCs had to satisfy a 14 point checklist.[15]

Research following the 1996 Act demonstrates that its entry methods had positive, but modest impacts on the development of fixed line competition. Zolnierek et al. (2001) find that the policies helped competition develop, but primarily in urbanized areas. Roycroft (2005) shows finds that lower prices for leasing network facilities from incumbents led to more entry. Jamison (2004) finds that when regulators require incumbents to receive lower profits on wholesale services sold to rivals than on retail services, incumbents limit entry. Foreman (2002) and Crandall (2001) find that entrants building their own networks are more successful than those that lease facilities or resell incumbent services.

Even though provisions of the 1996 Act had some success, overall the legislation is viewed as confusing, contradictory and disappointing. Crandall (1997) finds much legacy regulation has been left in place, new layers of regulation have been created, problems of non-economic pricing are unresolved, and new opportunities for contentious regulatory games are created because the legislation was inordinately complicated. He also concludes that the 1996 Act created a class of companies that had little chance of long term survival (Crandall, 2005). Nicholas Economides (1998) views the legislation as a positive development, but notes that its implementation was delayed by lengthy legal and regulatory proceedings. The Consumer Federation of America and the Consumers Union (2000) find that the 1996 Act resulted in too little competition.

Wireless Telecommunications

Wireline competition and AT&T's early interest in delaying such competition affected the development of the US wireless industry. AT&T's Bell Labs developed the concept of cellular phones in 1947[16] and proposed to the FCC that it allocate a large number of radio frequencies to cellular service so that the service could be widespread, meaning that AT&T would have an economic incentive to develop the new technology. However, the FCC allocated only a limited amount of frequency at that time; only enough to allow twenty-three simultaneous phone conversations in the same service area.

AT&T continued to press the FCC for additional radio spectrum for mobile services, but it was not until 1968 that the FCC reconsidered its position and formed an advisory committee to determine whether cellular technology could be successful. Following Motorola's invention of the first modern portable telephone handset in 1973, the FCC authorized three tests of cellular, granting experimental licenses to AT&T for Chicago and Baltimore, and a third experimental license to American Radio Telephone Service, Inc., for Washington, DC.

It was not until 1982 that the FCC authorized commercial cellular service, beginning with the first generation (1G) mobile technology known as AMPS a year after the service had become available in Nordic countries. The FCC chose to issue two licenses in each market area. One license was given to a fixed line operator serving the area.[17] The other license was to be issued to a non-fixed line operator based on a process called a beauty contest through which rivals would submit applications to the FCC, each explaining why it was the best qualified for the license. This process overwhelmed the FCC when applicants submitted volumes of documentation, some using multiple semi-trucks to carry their applications to the FCC offices, so part way through the process the FCC decided to assign its remaining non-fixed line licenses using a lottery system. The FCC retained the duopoly market structure until 1995 when it auctioned radio spectrum for use by PCS, which was the second generation (2G) cellular technology. This brought additional service providers into the market, which intensified competition and launched rapid growth in mobile services in the USA.

According to the FCC's 12th report on competition in mobile services, mobile telephony covers almost the entire country.[18] The statistics show that over 99 per cent of the total US population lives in census blocks[19] that have one or more operators. More than 95 per cent of the population lives in areas with at least three mobile carriers. These statistics at least somewhat overstate the actual coverage and competition because, within a census block there may be dead zones, or areas where a customer cannot access a mobile radio signal. However, even if the coverage is overstated, there is general consensus that very few Americans cannot receive cellular service at their homes and that the vast majority have their choice of service providers.

Natural Gas

Natural gas was unimportant in the USA prior to the 1920s, but that changed with the development of new oilfields and improvements in pipeline technology. The popularity of natural gas has grown so much that in 2008 it comprised 24 per cent of the energy consumption in the USA. In contrast, coal provided about 23 per cent of the US power in 2008 and nuclear power provided only 8 per cent. Industry development has features similar to that of fixed line telecommunications development: the gas industry was considered primarily a local or state issue until technology made interstate commerce possible. Also the industry was treated as a natural monopoly until economic forces demonstrated that many functions were better regulated by markets than by government institutions.

Growth in the use of natural gas began around the time of the end of World War II. Net production increased from slightly less than 7 billion cubic feet in 1950 to over 19 billion in 1979, and decreased to about 17 billion in 1990. Meanwhile pipeline transmission increased from 82000 miles to 280000 miles during this period (Phillips, 1993, pp. 694–5, citing *Moody's Public Utility Manual*, 1982, and the American Gas Association's *Gas Facts*, 1990).

Regulation developed because of beliefs in the natural monopoly nature of the industry and regulation was shaped by changes in technology. Regulation began at the city level through franchises and then extended to the state level because cities lacked jurisdiction over intercity pipelines. Prior to the improvements in pipeline technologies that made interstate pipelines possible, 87 per cent of natural gas consumption was concentrated in the six principal gas producing states of Texas, Louisiana, California, Oklahoma, and Texas (Phillips, 1993, p. 693). The new technologies made interstate pipelines possible, which triggered efforts by states to assert jurisdiction in the 1910s and 1920s. The Supreme Court ruled that the states could not regulate these pipelines because they represented interstate commerce, but it wasn't until the passage of the federal Natural Gas Act 1938 (NGA)[20] that Congress filled the regulatory void, giving the Federal Power Commission jurisdiction over interstate gas transmission (Natural Gas Supply Association, 2008).

The natural gas supply chain includes production or extraction at the wellhead, transmission through pipelines, and distribution through local distribution companies (LDCs). The NGA focused on pipeline regulation, not wellhead regulation. That changed in 1954 when the Supreme Court ruled that that natural gas producers selling into natural gas into interstate pipelines were 'natural gas companies' under the NGA, and were subject to Federal Power Commission (FPC) regulation.[21] The FPC adopted rate of return regulation for these prices. Initially the FPC regulated individual producer prices, but the workload was unmanageable for the FPC because of the large number of producers. As a result the agency began a process of setting rates on a regional basis, dividing the country into five separate producing regions. This approach also proved unworkable because of the diversity of costs across producers within a region and after ten years of effort, the FPC had set rates for only two of the five regions. Furthermore the regulated prices were significantly less than market-based prices for natural gas not sold into interstate pipelines, leading to distorted producer incentives. As a result the FPC abandoned its rate of return approach and established a national price ceiling of

$0.42 per million cubic feet. However, this too was below the market price for natural gas (Natural Gas Supply Association, 2008; Phillips, 1993, pp. 698–9).

The FPC's attempts at wellhead price regulation consistently resulted in regulated prices below the market rate, resulting in natural gas shortages in the 1970s. So in 1978 Congress passed the Natural Gas Policy Act (NGPA)[22] to create a single national natural gas market, allow a balancing of supply with demand, and permit market forces to establish wellhead prices. Rather than relying upon the FPC, the NGPA established statutory maximum prices for wellhead sales. Some of these price controls were scheduled to be phased out by 1985, but prices for wells in production before passage of the NGPA were to remain regulated. The underlying theory of this asymmetric treatment of wellhead prices was that the deregulated prices would provide an incentive for new exploration, but existing wellheads would not be allowed a windfall benefit from these potentially higher prices (Natural Gas Supply Association, 2008). But as a result of this dichotomous treatment of old and new natural gas, some existing wellheads were taken out of production so that producers could receive the higher profits allowed for gas from new wellheads.

The NGPA created a single national market by giving FERC, which had replaced the FPC in 1997, authority to approve interstate transmission of natural gas on behalf of intrastate pipelines or LDCs (Natural Gas Supply Association, 2008).

The NGPA took significant steps towards deregulating prices at the wellhead, but left in place regulation of prices for sales from an interstate pipeline. Under the NGA and the NGPA, pipelines purchased natural gas from gas producers, transported it to its customers, and sold the bundled product at a regulated price. This system came under pressure in the early 1980s when a number of industrial customers began switching from using natural gas to other forms of energy. Because the switching was in response to the economic distortions caused by regulation, the FERC tried to allow pipelines to give special pricing arrangements to these large customers, but courts overruled FERC on the grounds that this was discriminatory. So in 1985, FERC issued Order No. 436,[23] also known as the Open Access Order, which established a voluntary framework under which interstate pipelines could serve as transporters of natural gas and not gas merchants, allowing customers the opportunity to purchase gas at the wellhead and pay the pipeline for transport. To ensure that pipelines did not favor their own merchant operations, FERC established minimum and maximum transport rates. Even though almost all pipelines participated in this voluntary program, in 1992 FERC issued Order No. 636,[24] also known as the Final Restructuring Rule, making the separation of transport and sales mandatory. Under this rule, the FERC required that production and marketing arms of interstate pipeline companies be arms-length affiliates (Natural Gas Supply Association, 2008; Phillips, 1993, pp. 704–15).

Meanwhile, the Natural Gas Wellhead Decontrol Act of 1989 (NGWDA)[25] completed the deregulation of wellhead prices begun under the NGPA by directing the deregulation of all 'first sales' of natural gas. The NGWDA defined 'first sales' as sales to a pipeline, an LDC, an end user, or any sale preceding these sales. It excluded from first sale any sales of gas by pipelines and local distribution companies.

Today the restructuring and deregulatory approaches for natural gas are viewed as relatively successful (Costello, 2009). The FERC regulates transmission by pipelines and wholesale sales and transmission for interstate commerce. It also oversees siting and abandonment of interstate natural gas pipelines and storage facilities, and ensures the safe operation and reliability of proposed and operating liquefied natural gas terminals.[26]

Electricity

Similar to the telecommunications and gas industries, the US electricity industry formed organically through the development of local electric systems. New technologies created interstate markets for electricity and the federal government responded with federal regulation.

Electricity was developed largely by private businesses, but also by municipalities. Many industrial customers produced their own electricity in the early twentieth century, but soon began buying their power from vertically integrated electric utilities because of their scale economies and reliability. Soon the industry integrated horizontally through mergers and acquisitions, government programs, legal and regulatory mandates, and various policy decisions (Brown, 2005; Liggett, 2000).

The development of regulatory regimes paralleled the sector development. Initially municipalities granted franchises or licenses to local operators or created their own utilities, which came to be known as public power. Horizontal integration resulted in utilities that extended beyond the jurisdictional reach of local governments. This, plus a need to reduce corruption and professionalize regulation, led to the development of regulation by states (Brown, 2005; Glaeser, 1927; Knittel, 1999). Most states already had regulatory commissions for railroads or other monopoly industries and most states simply gave these commissions authority for electricity regulation, along with regulation of telecommunications, natural gas, and water and wastewater (Brown 2005).

Prompted in part by the rise of multistate holding companies that were beyond the jurisdiction of state regulatory commissions, federal regulation of electricity appeared in the 1930s with the Federal Power Act (FPA) and the Public Utilities Holding Company Act (PUHCA),[27] both enacted as part of President Franklin D. Roosevelt's New Deal reforms along with the formation of the Tennessee Valley Authority (TVA) and Bonneville Power Administration (BPA). The FPA created the FPC and, in filling a gap in state regulation, namely, the regulation of interstate wholesale electricity transactions, gave the FPC jurisdiction over all wholesale power exchanges and the use of the transmission grid for such transactions. The federal government formed TVA in the southeast and BPA in the northeast to take advantage opportunities for hydropower in situations where private investment was not forthcoming. Also as part of the New Deal, the National Rural Electrification Act (REA) led to the development of rural electric cooperatives in areas unserved by private operators and municipalities. These cooperatives are owned by their customers and generally receive federal subsidies in the form of low interest loans initiated by the National Rural Electrification Act (REA).

Today the electricity sector is largely privately owned by publicly traded companies, but about 25 per cent of the US load is served by government owned entities and rural electric cooperatives. The government owned operators include federal entities, such as the TVA and BPA, and more than 2000 public power utilities (Brown, 2005; Liggett, 2000). Research has shown that this mix of private and public ownership has resulted in some inefficiency, owing in part to the inefficiencies of regulating private companies. Rose and Joskow (1990) find that the government owned generation companies are slower to adopt new technologies than are their privately owned counterparts. Berry (1994) finds that investor owned electric companies are more technically efficient than are rural electric cooperatives, which are owned by their customers. Kwoka (2006) finds that privately owned electricity distribution companies are less technically efficient than their publically owned counterparts unless faced with either competition for large customers or benchmark competition.

The division of responsibilities between state and federal regulators often led to turf battles because all market participants are in some way regulated by both the state and federal regulators. States oversee the rates charged by electric distribution companies, but the FERC oversees their energy purchases and their access to the grid. The FERC oversees transmission companies' revenues and access policies, regulates transmission reliability, monitors and investigates energy markets, and licenses and inspects private, municipal, and state hydroelectric projects, but states oversee their siting, licensing, and the like. Independent power producers and other generating companies are regulated in the same ways as transmission companies,[28] but states also affect them through the states' oversight of retail markets and distribution companies (Brown, 2005). Often state PUCs lack authority over prices of public power entities and rural electric cooperatives.

The second major structural reform of the sector found its beginnings in the late 1970s. Prompted in part by rising energy costs in the late 1960s and into the 1970s, Congress passed the Public Utility Regulatory Policies Act of 1978 (PURPA), which allowed nonutility facilities to enter the wholesale market. Generation by non-utilities was further advanced by the Energy Policy Act of 1992 (EPAct), which created a new category of power producers and exempt wholesale generators (EWGs), which are wholesale producers that do not sell electricity in retail market and do not own transmission facilities.

This major structural reform, also called 'dereg', short for deregulation, continued until the debacle with California's electricity system in 2000. Dereg was intended to intensify competition in generation by reforming grid access, divestiture of some generation assets, and the creation of trading institutions (OECD, 1998). Generation was unbundled from regulated utility operations, which continued to provide transmission and distribution. In some cases generation was sold or transferred to a utility affiliate, but in others the generation was actually sold off (OECD, 1998; Lien, 2008). Currently about 40 per cent of the generating capacity is owned by non-utility generators (Lien, 2008, citing the US Energy Information Administration, 2006).

California was one of the first states in the USA to adopt dereg. In the 1990s, California customers routinely paid above the national average for their electricity in a system that was dominated by three vertically integrated utilities, each of which was a monopoly in its own territory.[29] The California Public Utilities Commission (CPUC) recommended restructuring in the early 1990s and the legislature adopted statutes to do just that in 1996. For a time after deregulation began in 1998, the system more or less worked.

Under California's dereg plan, the utilities sold off about 50 per cent of their power plants and the state created an independent system operator (ISO) to manage the transmission grid, and a power exchange, which daily purchased electricity from power plants to supply the needs of the utilities. Wholesale prices were set in the spot market, but the retail prices that utilities could charge were regulated by the CPUC and could not rise with increases in wholesale prices.[30]

In the summer of 2000 wholesale prices for electricity rose sharply. Soon the prices California's utilities paid for electricity reached a crisis level and with retail price caps in place, the utilities were unable to recover their costs. Two utilities approached bankruptcy and suppliers threatened to stop shipments.[31] Insufficient supply led to rolling blackouts. Efforts at FERC and in California to address the crisis devolved into a state–federal battle, with the California governor blaming FERC for not addressing the critical issues and accusing producers and marketers of manipulating the markets. Some FERC commissioners joined in accusing the industry for artificially creating the crisis, but FERC limited its formal response to allowing California's utilities to stop buying exclusively from the spot market and instead negotiate long-term contracts.[32] The California governor and politicians, and many interest groups criticized the FERC response as being tepid. One of the California utilities sued FERC.

By early 2001 electricity demand had outstripped supply and the state continued to experience blackouts. The policy battle was between federal regulators, who did not want to cap wholesale prices to accommodate retail price caps, and state regulators and politicians, who did not want to lift retail price caps to accommodate deregulated wholesale prices. By the summer of 2001 the FERC acquiesced and imposed wholesale price caps, a move that is largely credited with ending the California crisis.

Compared to California, another dereg situation, the Pennsylvania, New Jersey and Maryland (PJM) market has worked well. PJM was created as a power pool in 1927 by three utilities in Pennsylvania and New Jersey, but by the time it became an ISO in 1997 it included eight interconnected utilities (Bowring, 2006). Joseph Bowring attributes PJM's success to features that were missing from the California market, including flexibility for buyers to enter into bilateral contracts, to self-schedule generation, and to supply their own generation. He also credits nodal pricing – a system for determining prices in which prices are calculated for a number of locations on the transmission grid (nodes), each of which represents a location where energy is either injected or drawn. Bowring concludes that these effective systems were possible in the case of PJM because the members had a history of working together, developed a complete set of market rules prior to implementation of the market, adopted an independent governance structure, created an effective monitoring process and formed an adaptive process for changing policies as circumstances changed.

Several studies have concluded that the restructured markets in general have had positive impacts, but issues remain. Bushnell and Wolfram (2006), Fabrizio et al. (2007) and Shanefelter (2006) credit the markets with improving technical efficiency, as does Lien (2008). However, studies have also found that improvements could be made. For example, Bushnell et al. (2008) find that residual vertical integration in restructured markets in higher commodity prices, even though Bowring observes that this feature improves market stability and success.

PRESENT CHALLENGES AND OPPORTUNITIES

Current service providers in the USA vary in size and scope. The largest telecommunications operator, the vertically integrated AT&T, provided about 45 per cent of the country's fixed lines and had $124 billion in annual revenue in 2008, but it is second to Verizon in numbers of mobile customers with 77 million subscribers. Verizon provided 31 per cent of fixed lines. New entrants into the fixed line market held a 17.6 per cent market share in terms of of switched lines in 2007, down from a high of 19 per cent in 2005. Wireless was 49 per cent of industry revenue in 2007, up from 16 per cent 10 years earlier (FCC, 2008; Hoovers, 2009). The gas industry is less concentrated than the telecommunications: industry revenues of $92 billion in 2007 were spread across a large number of companies, with the three largest gas utilities – Atmos, Southern California Gas, and Pacific Gas and Electric – receiving only about 10 per cent (AGA, 2009). The largest electric utility in the US in 2008, Exelon, had revenue of nearly $19 billion, but retail electricity revenue for the country was nearly $344 billion in 2007. The next largest electric providers – Southern Company and FPL Group – had a little more than $15 billion each in revenue in 2008 (DoE, 2009; Standard and Poor's, 2009).

Many of the problems that have been encountered in utility liberalization in the USA can be described as failures to fully adapt to new visions and new circumstances. In telecommunications regulators and policy-makers failed to understand that traditional industry boundaries and pricing schemes would not withstand competitive market forces, that regulatory management of entry and technology would delay innovations and lower efficiency, and that entry is easier and more effective in new markets than in legacy markets with little unmet demand. In natural gas there was a failure to see that regulating prices for some customers, but not all, would lead to arbitrage and that attempting to bundle a potentially competitive commodity (gas) with a monopoly service (distribution) would lead to uneconomic bypass if customers were able to find alternatives. In electricity, policy-makers in California, and probably at FERC, at first failed to appreciate how completely dereg would change electricity pricing – in part because the political process that redesigned the markets lacked expertise – thinking that retail price caps would have no consequence but to limit consumer prices. As the system began to unravel, federal and state officials were slow to recognize the contradictions in their dereg approach. Furthermore, as Jones and Rosenberg (2008) explain, market reform initiatives often fail to foresee that the utilities themselves need to undergo cultural changes and that these changes can be costly, difficult, and prone to failure.

Adaptive challenges also form the underlying theme for the emerging challenges for US utilities and their regulators. In telecommunications the major issues are broadband penetration, universal service, and radio spectrum management. For broadband and universal service the policies of the outgoing Bush administration and the stated policies of the incoming Obama administration both suffer from the application of legacy ideas to new technologies and new markets. The Bush administration deregulated broadband as an information service, which ignores its nature as a transmission service,[33] and kept in place the traditional universal service subsidies that have been ineffective and bias markets towards legacy technologies. The new administration has so far embraced the Organisation for Economic Co-operation and Development (OECD) practice of simple measures of broadband and has stated that it wants to apply the legacy subsidy system to broadband, continuing the long-held assumption, refuted by research, that telecommunications penetration is most effectively advanced through subsidies. A more effective approach might be to recognize that broadband has many dimensions, focus on the development of the broader ICT system of which broadband is a component, and leverage what the US economy is generally good at, innovation and competition. Regarding radio spectrum management, the USA has been a leader in applying market-based management. A primary challenge is to continue that approach and avoid the temptation to substitute political judgment for economic decision-making by businesses and customers regarding how spectrum should be used.[34]

In energy the leading issues are investment and environmental. Regarding investment there is a need to expand generation and transmission capacity and to replace aging facilities. The incoming Obama administration has indicated that constraining carbon emissions and increasing the use of renewable energy are high priorities. Taken together the investment and environmental issues have the potential to create a perfect storm of higher costs and risk while imposing changes in technologies, system design and political roles. That costs for utilities and prices for consumers will be higher is clear. Indeed, higher prices are likely to be an important policy instrument to limit energy consumption. The new administration's stated carbon emissions targets are ambitious and unlikely to be met without new technologies, actual decreases in energy consumption, or both. Technology challenges exist throughout the value chain, from producer to consumer. Indeed some potential technologies – such as plug-in hybrids automobiles, distributed solar and wind generation, and smart grid – imply system changes that could result in two-sided markets for power, experimentation, changing roles for utilities and consumers, and changing roles for regulators as federal regulators take increased authority over facility siting, environmental regulators either increase their say over utility issues or utility regulators increase their jurisdiction over environmental issues, and international bodies exert say over what have traditionally been domestic issues. The uncertainties with respect to policy direction, regarding whether policy-makers have the political will to allow adequate cost recovery, how regulators will treat costs of technology and system experiments, and technology costs are making the financial community and the utilities reluctant to make long-term investments, which limits opportunities for long-term solutions.[35]

CONCLUSION

This chapter examined the development and evolution of utility regulation in the USA, emphasizing the experiences with market reforms. We find that regulation first developed to address practical concerns with utility markets and then evolved to emphasize efficiency concerns and recognize realities of policy-maker, regulator, industry, and consumer incentives. Even though the system adapted as circumstances changed, miscues were common when policy-makers failed to recognize the extent of the changes they were experiencing and the full implications of their policy changes. This is both good news and bad news as the country faces new changes of broadband penetration in telecommunications and growing investment needs and environmental regulations for energy: regulators, policy-makers, and stakeholders are experienced in creating new approaches for solving problems, but are also prone not fully recognizing how bottom-up, individual economic decision-making will interact with top-down economic controls.

NOTES

1. The author would like to thank Lynne Holt, Sanford Berg, Ted Kury, Rolf Kunneke and Matthias Finger for their comments and suggestions on this chapter; and Brenda Buchan for her research on the California electricity crisis. All errors and omissions are the responsibility of the author.
2. Clearly there are many who might disagree with this view of the USA or argue that the country is moving away from its liberal roots. I will not expand upon these issues because they are beyond the scope of this chapter.
3. *Munn* v. *Illinois*, 94 U.S. 113, 131–2 (1877).
4. *Brass* v. *North Dakota* ex rel. *Stoeser*, 153 U.S. 391 (1894).
5. *Smyth* v. *Ames*, 169 U.S. 466 (1898).
6. *Federal Power Commission* v. *Hope Natural Gas Co.*, 320 U.S. 591 (1944).
7. For more complete descriptions of the contributions of institutional and neoclassical economics to utility regulation, see Miller and Samuels (2002), Berg and Tschirhart (1995), and Faulhaber and Baumol (1988).
8. Statutes typically spell out the duties, responsibilities, financing, and authority of the agencies (Brown 2005).
9. We omit the development of the telegraph. Western Union was the primary telegraph provider in the USA.
10. In the USA, line rental and local calling were traditionally combined into a single service called local service.
11. *In re Allocation of Microwave Frequencies Above 890 Mc.*, Dkt. No. 11866, 27 FCC 359 (1959), *aff'd on reh'g*, 29 FCC 825 (1960).
12. Had the authorities not believed that fixed line was a natural monopoly, other viable options might have been considered, including removing barriers to competition in fixed lines, taking steps to ensure that the emerging cellular technology would be a substitute for fixed line, and creating a horizontal breakup into vertically integrated network operators.
13. Pub. L. No. 104-104, 110 Stat.56.
14. The AT&T divestiture agreement restricted the RBOCs from providing long distance service, except in limited areas (called Local Access Transport Areas, or LATAs). LATAs were typically the size of an area code region that would have existed in 1982. The RBOCs were prohibited from carrying long distance calls across LATA boundaries. GTE was prohibited from providing interLATA long distance under a different consent decree.
15. Tomlinson (2000, pp. 320–21) provides details on these preconditions.
16. A cell is an area served by a mobile carrier antenna. Radio frequency refers to the frequency with which electromagnetic wave cycles pass a given point per second. For example, the FCC's A Block license for broadband PCS included frequencies around 1850 MHz, which means 1 850 000 000 cycles per second.
17. If more than one fixed line operator was in the area and wanted the license, the operators negotiated a joint business arrangement.
18. FCC, 'Twelfth Report', http://hraunfoss.fcc.gov/edocs_public/attachmatch/FCC-08-28A1.pdf, accessed 21 November 2008.
19. A census block is the smallest geographic unit used by the USA Census Bureau for its decennial population census. In cities, a census block is often a city block, but in rural areas a census block may be many square miles.
20. Natural Gas Act, 52 Stat. 824 (June 21, 1938).
21. Phillips Petroleum Co. v. Wisconsin (347 U.S. 672 (1954))
22. Natural Gas Policy Act of 1978, Public Law No 95-621, 92 Stat. 3350 (November 9, 1978)
23. *In re Regulatino of Natural Gas Pipelines After Partial Wellhead Decontrol*, Order No. 436, 33 para. 61,007, 50 Fed. Reg. 52,217 (1985), *modified*, Order No. 436-A, 33 FERC para. 61,372 (1985), *modified*, Order No. 436-B, 34 FERC para. 61,204 (1986), *reh'g denied*, Order Nos. 436-C, 436-D, 436-E, 34 FERC para. 61,404 (1986), *rev'd in part and remanded sub nom.*
24. *In re Pipeline Service Obligations and Revisions to Regulatinos Governing Self-Implementing Transportation Under Part 284 of the Commission's Regulations*, 59 FERC para. 61,030 (1992).
25. Pub. Law 101-60, 103 Stat. 157.
26. FERC, 'What FERC does', http://www.ferc.gov/about/ferc-does.asp, accessed 7 September 2008.
27. Public Utility Act of 1935, Public Law No. 333, Sec. 2(a)(7)(1935).
28. FERC, 'What FERC does', http://www.ferc.gov/about/ferc-does.asp, accessed 7 September 2008.
29. William Booth, 'Mutiny on the meter? California races to fix deregulated system as prices, tempers rise', Washington Post, 3 December 2000.
30. For a full analysis of the California case, see Borenstein (2002).

31. Pacific Gas & Electric filed for bankruptcy the following year.
32. The utilities were also no longer forced to sell their generation into the PX, and could sell directly to their customers. See FERC's chronology for 'Addressing the Western energy crisis' at http://www.ferc.gov/industries/electric/indus-act/wec/chron.asp (accessed 20 April 2009).
33. As the FCC uses the terms, an information service provides or modifies content. Transmission services do not modify content.
34. For more extensive discussion, see Hazlett (2008b) and Weiser and Hatflied (2008)
35. For additional discussion of US electricity issues, see Cannell (2009), Chupka et al. (2008), Chupka and Basheda (2007) and Steinhurst (2008).

REFERENCES

American Gas Association (2009), available at http://www.aga.org.

Baumol, William J. (1977), 'On the proper cost tests for natural monopoly in a multiproduct industry', *American Economic Review*, **67** (5), 809–22.

Baumol, William J. (1979), 'Minimum and maximum pricing principles for residual regulation', *Eastern Economic Journal*, **5** (1–2), 235–48.

Baumol, William J. (1982), 'Contestable markets: an uprising in the theory of industry structure', *American Economic Review*, **72** (1), 1–15.

Baumol, William J. (1983), 'Some subtle pricing issues in railroad regulation', *International Journal of Transportation Economics*, **10** (1–2), 341–55.

Baumol, William J. (1986). *Superfairness: Applications and Theory*, Cambridge, MA: MIT Press.

Berg, Sanford V. and John Tschirhart (1995), 'Contributions of neoclassical economics to public utility analysis', *Land Economics*, **71** (3), 310–30.

Berry, Dan Michael. 1994, 'Private ownership form and productive efficiency: electric cooperatives versus investor-owned utilities', *Journal of Regulatory Economics*, **6**, 399–420.

Black, Fisher and Myron Scholes (1973), 'The pricing of options and corporate liabilities', *Journal of Political Economy*, **81** (3), 637–54.

Boiteux, Marcel (1949), 'La tarification des demandes en point: application de la theorie de la vente au couit marginal', *Revue generale de l'electricite*, **58**, 321–40; translated as 'Peak loading pricing', *Journal of Business*, Apr., 1960, **33** (2), 157–79.

Bonbright, James C. (1961), *Principles of Public Utility Rates*, New York: Columbia University Press.

Bonbright, James and Gardiner C. Means (1932), *The Holding Company*, New York: McGraw Hill.

Borenstein, Severin (2002), 'The trouble with electricity markets: understanding California's restructuring disaster', *Journal of Economic Perspectives*, **16** (1), 191–211.

Bowring, Joseph (2006), 'The PJM market', in Fereidoon P. Sioshansi and Wolfgang Pfaffenberger (eds), *Electricity Market Reform: An International Perspective*, Amsterdam: Elsevier, pp. 451–76.

Brock, Gerald W. (1981), *The Telecommunications Industry: The Dynamics of Market Structure*, Cambridge, MA: Harvard University Press.

Brown, Ashley (2005), 'The structure and process for regulation of the electricity industry in the United States', working paper, Harvard Electricity Policy Group, John F. Kennedy School of Government, Harvard University.

Bushnell, James and Catherine Wolfram (2005), 'Ownership change, incentives and plant efficiency: the divestiture of U.S. electric generation plants', University of California Energy Institute, Center for the Study of Energy Markets, Working Paper WP-140.

Bushnell, James B., Erin T. Mansur and Celeste Saravia (2008), 'Vertical arrangements, market structure, and competition: an analysis of restructured US electricity markets', *American Economic Review*, **98** (1), 237–66.

Cannell, Julie (2009), 'The financial crisis and its impact on the electric utility industry', Edison Electric Institute, Washington, DC, available at: http://www.eei.org/ourissues/finance/Documents/Financial_Crisis_and_Industry.pdf (accessed 20 April 2009).

Chupka, Marc W. and Gregory Basheda (2007), 'Rising utility construction costs: sources and impacts', available at: http://www.eei.org/ourissues/finance/Documents/Rising_Utility_Construction_Costs.pdf (accessed 20 April 2009).

Chupka, Marc W., Robert Earle, Peter Fox-Penner and Ryan Hledik (2008), 'Transforming America's power industry: the investment challenge 2010–2030', available at http://www.eei.org/ourissues/finance/Documents/Transforming_Americas_Power_Industry.pdf.

Coase, Ronald H. (1959), 'The Federal Communications Commission', *Journal of Law and Economics*, **2**, 1–40.

Consumer Federation of America and Consumers Union (2000), 'Lessons from 1996 Telecommunications Act: deregulation before meaningful competition spells consumer disaster', available at: http://www.consumersunion.org/pdf/lesson.pdf (accessed 20 April 2009).

Costello, Ken (2009), 'The natural gas industry at a glance', National Regulatory Research Institute, available at: http://nrri.org/matrix/gas/natural_gas_industry_at_a_glance.pdf (accessed 15 September 2009).

Crandall, Robert (1997), 'Are we deregulating telephone services? THINK AGAIN', Brookings Policy Brief Series # 13, Washington, DC: The Brookings Institution.

Crandall, Robert (2001), *An Assessment of the Competitive Local Exchange Carriers Five Years After the Passage of the Telecommunications Act*, Washington, DC: Criterion Economics.

Crandall, Robert (2005), *Competition and Chaos; U.S. Communications Since 1996*, Washington, DC: Brookings Institution.

Department of Energy, Energy Information Administration (DCE) (2009), 'Electric Sales and Revenue 2007', available at: http://www.eia.doe.gov/cneaf/electricity/esr/esr_sum.html (accessed 20 April 2009).

Economides, Nicholas (1998), 'The Telecommunications Act of 1996 and its impact', Working Paper, Stern School of Business, New York University.

Economides, Nicholas (2004), 'Telecommunications regulation: an introduction', Working Paper, Stern School of Business, New York University.

Fabrizio, Kira R., Nancy L. Rose and Catherine D. Wolfram (2007), 'Do markets reduce costs? Assessing the impact of regulatory restructuring on US electric generation efficiency', *American Economic Review*, **97** (4), 1250–77.

Faulhaber, Gerald R. (1975), 'Cross-subsidization: pricing in public enterprises', *American Economic Review*, **65** (5), 966–77.

Faulhaber, Gerald R. (1979). 'Cross-subsidization in public enterprise pricing', in John T. Wenders (ed.), *Pricing in Regulated Industries: Theory and Application, II; Papers Presented at an Economic Seminar*, Denver, CO: Regulatory Matters Division, Mountain States Telephone and Telegraph Co., pp. 76–121.

Faulhaber, Gerald R. and William J. Baumol (1988), 'Economists as innovators: practical products of theoretical research', *Journal of Economic Literature*, **26** (2), 577–600.

Federal Communications Commission (FCC) (2008), 'Trends in telephone service', available at: http://www.fcc.gov/wcb/iatd/trends.html (accessed 20 April 2009).

Foreman, R.D. (2002), 'For whom the bell alternatives toll: demographics of residential facilities-based telecommunications competition in the United States', *Telecommunications Policy*, **26** (9–10), 573–87.

Gabel, David (1994), 'Competition in a network industry: the telephone industry, 1894–1910', *Journal of Economic History*, **54** (3), 543–72.

Glaeser, Martin G. (1927), *Outlines of Public Utility Economics*, New York: Macmillan.

Gray, Horace M. (1940), 'The passing of the public utility concept', *The Journal of Land & Public Utility Economics*, **16** (1), 8–20.

Hazlett, Thomas W. (2008a), 'A law & economics approach to spectrum property rights: a response to Weiser and Hatflied', *George Mason Law Review*, **15** (4), 975–1023.

Hazlett, Thomas W. (2008b), 'Property rights and wireless license values', *Journal of Law and Economics*, **51** (3), 563–98.

Hirshleifer, Jack (1958), 'Peak loads and efficient pricing: comment', *Quarterly Journal of Economics*, **72** (3), 451–62.

Hoovers. (2009), Database available at http://premium.hoovers.com (accessed 20 April 2009).

Hughes, Miles W. (1996), 'Telecommunications reform and the death of the local exchange monopoly', *Florida State University Law Review*, **24** (1), 179–217.

Jamison, Mark A. (2004), 'Effects of price for local network interconnection on market structure in the US', in Erik Bohlin, Stanford L. Levin, Nakil Sung and Chang-Ho Yoon (eds), *Global Economy and Digital Society*, Amsterdam: Elsevier, pp. 301–20.

Jones, Douglass N. and Edwin A. Rosenberg (2008), 'Utility restructuring and its unnatural acts', *Fortnightly's Spark*, (February), 1–2, 7–9.

Kahn, Alfred E. (1988), *The Economics of Regulation: Principles and Institutions*, vols 1 and 2, 2nd edn, Cambridge, MA: MIT Press (New York: Wiley, 1970–71).

King, Clyde Lyndon (1912), 'The need for utility commissions', in Clyde Lyndon King (ed), *The Regulation of Municipal Utilities*, New York: D. Appleton.

Knittel, Christopher (1999), 'The origins of state electricity regulation: revisiting an unsettled topic', working paper, Department of Finance and Economics, Boston University.

Kwoka, John E. (2006), 'The role of competition in natural monopoly: costs, public ownership, and regulation', *Review of Industrial Organization*, **29**, 127–47.

Lien, Jeff (2008), 'Electricity restructuring: what has worked, what has not, and what is next', working paper, United States Department of Justice.

Liggett, William D. (2000), 'Historical overview of the electric power industry', in *The Changing Structure of*

the Electric Power Industry 2000: An Update, Washington, DC: Energy Information Administration, US Department of Energy, pp. 5–8.
Littlechild, Stephen (1983), *Regulation of British Telecommunications' Profitability*, London: HMSO.
Miller, Edythe and Warren J. Samuels (eds) (2002), *The Institutionalist Approach to Public Utilities Regulation*, East Lansing, MI: Michigan State University Press.
Mueller, Milton (1993), 'Universal service in telephone history: a reconstruction', *Telecommunications Policy*, **17** (5), 352–69.
Mueller, Milton (1997), *Universal Service: Competition, Interconnection, and Monopoly in the Making of the American Telephone System*, Washington, DC: AEI Press.
Natural Gas Supply Association (2008), 'The history of regulation', http://www.naturalgas.org/regulation/history.asp, accessed 7 September 2008.
Nix, Joan and David Gabel (1993), 'AT&T's strategic response to competition: why not preempt entry?', *The Journal of Economic History*, **53** (2), 377–87.
Organisation for Economic Co-operation and Development (OECD) (1998), *United States – Regulatory Reform in the Electricity Industry*, Paris: OECD
Peltzman, Sam (1976), 'Toward a more general theory of regulation', *Journal of Law and Economics*, **19** (2), 211–40.
Phillips, Charles F. Jr (1993), *The Regulation of Public Utilities*, Arlington, VA: Public Utilities Reports, Inc.
Ramsey, Frank (1927), 'A contribution to the theory of taxation', *Economics Journal*, **37**, 47–61.
Rose, Nancy L. and Paul L. Joskow (1990), 'The diffusion of new technologies: evidence from the electric utility industry', *The RAND Journal of Economics*, **21** (3), 354–73.
Roycroft, Trevor R. (2005), 'Emperical analysis of entry in the local exchange market: the case of Pacific Bell', *Contemporary Economic Policy*, **23** (1), 107–15.
Sappington, David E.M. (2002), 'Price regulation', in Martin E. Cave, Sumit K. Majumdar and Ingo Vogelsang (eds), *Handbook of Telecommunications Economics, Volume 1: Structure, Regulation, and Competition*, Amsterdam: Elsevier, pp. 225–96.
Shanefelter, Jennifer Kaiser (2006), 'Restructuring, ownership and efficiency: the case of labor in electricity generation', University of California Energy Institute, Center for the Study of Energy Markets, Working Paper WP-161.
Sharkey, William W. (1981), 'Existence of sustainable prices for natural monopoly outputs', *Bell Journal of Economics*, **12** (1), 144–54.
Sharkey, William W. (1982a), 'Suggestions for a game-theoretic approach to public utility pricing and cost allocation', *Bell Journal of Economic*, **13** (1), 57–68.
Sharkey, William W. (1982b), *The Theory of Natural Monopoly*, New York: Cambridge University Press.
Sharpe, William F. (1964), 'Capital asset prices: a theory of market equilibrium under conditions of risk', *Journal of Finance*, **19**, 425–42.
Standard and Poor's (2009), 'Electric utilities comparative company analysis', database.
Steiner, Peter O. (1957), 'Peak loads and efficient pricing', *Quarterly Journal of Economics*, **71** (4), 585–610.
Steinhurst, William (2008), 'The electric industry at a glance', National Regulatory Research Institute, available at: http://nrri.org/pubs/electricity/electricity_at_a_glance.pdf (accessed 20 April 2009).
Temin, Peter (1990), 'Cross subsidies in the telephone network after divestiture', *Journal of Regulatory Economics*, **2**, 349–62.
Tomlinson, Richard G. (2000), *Tele-Revolution: Telephone Competition at the Speed of Light*, Glastonbury, CT: Penobscot Press.
Trebing, Harry M. (1976), 'The Chicago School versus public utility regulation', *Journal of Economic Issues*, **10** (1), 97–126.
Trebing, Harry M. (1984a), 'Public utility regulation: a case study in the debate over the effectiveness of economic regulation', *Journal of Economic Issues*, **18** (1), 223–50.
Trebing, Harry M. (1984b), 'Public control of enterprise: neoclassical assault and neoinstitutionalist reform', *Journal of Economic Issues*, **18** (2), 353–68.
Trebing, Harry M. (1987), 'Regulation of industry: an institutionalist approach'. *Journal of Economic Issues*, **21** (4), 1707–37.
Vickery, William (1961), 'Conterspeculation, auctions, and competitive sealed tenders', *Journal of Finance*, **16**, 8–37.
Weiser, Philip J. and Dale Hatfield (2008), 'Spectrum policy reform and the next frontier of property rights', *George Mason Law Review*, **15** (3), 549–609.
Williamson, Oliver E. (1966), 'Peak load pricing and optimal capacity under indivisibility constraints', *American Economic Review*, **56** (4), 810–27.
Williamson, Oliver E. (1983), 'Credible commitments: using hostages to support exchange', *American Economic Review*, **73** (4), 519–40.
Zolnierek, J., Eisner, J., Burton, E. (2001), 'An empirical examination of entry patterns in local telephone markets', *Journal of Regulatory Economics*, **19** (2), 143–59.

22 The liberalization of infrastructures in Latin America[1]

Francesc Trillas and Miguel A. Montoya

INTRODUCTION

The Latin America and Caribbean region has been a laboratory of liberalization and regulatory reform in the infrastructure sectors for the past two decades. Private investment in infrastructures was seen as part of a reform package designed to take the region out of the *lost decade* after the international economic crisis of the 1970s and the regional debt crisis of the early 1980s. In no other region has the development strategy been based as much on private investment in infrastructures.

As Figure 22.1 shows, most of this private investment has gone to the telecommunications sector, but energy and transport have also received more private investment than any region of the world between 1990 and 2007. Some segments of these sectors have been liberalized in the sense of allowing competition either in the market or for the market (through competitive bidding for concession contracts). Competition was expected to increase output, reduce costs and improve customer service, while privatization was expected to provide better incentives to firm managers and staff, while reducing red tape and politicization. Liberalization and privatization pose significant institutional challenges, however, and there has been high cross-country variation on how these challenges have been addressed. Industry reform based on private investment works best when it is accompanied by an effective and strong regulatory framework.

Infrastructure industries are a key part of welfare[2] (as measured for example through the UN Human Development Index) and a key determinant of economic growth. Besides, their performance depends on policies that are linked to a large extent to the political cycle. It is a key issue then to explore the links between infrastructure policies, social welfare and the political cycle. We show in this chapter that, although reform based on private investment may deliver significantly higher welfare levels, the improvements have been far from universal.

We seek clues to guide what some have called 'second generation' reforms[3] aimed at providing stability and permanence to foreign private investment in the infrastructure sectors in Latin America. As it will be shown, public's sentiment vis-à-vis privatization has been increasingly negative. This 'disillusionment'[4] has been driven, according to Kessides (2004), by 'employee layoffs, price increases, perceived long delays in benefits, and the distributional impacts of privatization'. Recent political unrest in some Latin American countries (such as Argentina, Venezuela and Bolivia) has focused, *inter alia*, on the role of foreign private investment.

Although, historically, the emphasis of regulatory economics has been on efficiency, attention to distributional concerns is also needed, as many scholars increasingly rec-

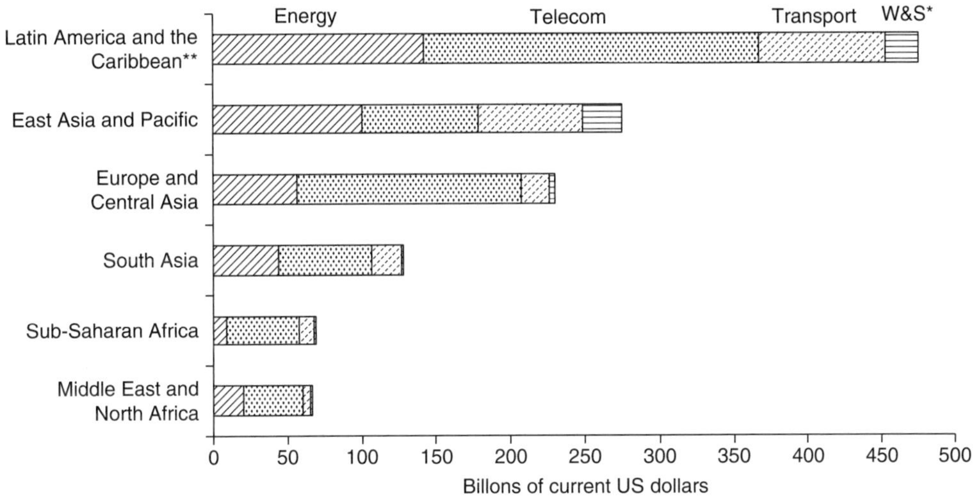

Notes:
* Water and sewerage.
** Includes: Cuba, Dominica, Grenada, Guatemala, Guyana, Haiti, St Kitts and Nevis, St Lucia, St Vincent and the Grenadines.
Investment in infrastructure projects with private participation covers infrastructure projects in telecommunications, energy (electricity and natural gas transmission and distribution), transport, and water and sanitation that have reached financial closure and directly or indirectly serve the public. Incinerators, movable assets, stand-alone solid waste projects, and small projects such as windmills are excluded. The types of projects included are operation and management contracts, operation and management contracts with major capital expenditure, greenfield projects (in which a private entity or a public-private joint venture builds and operates a new facility), and divestiture.

Source: World Bank, Private Participation in Infrastructure Project database.

Figure 22.1 Private investment in infrastructure (1990–2007) by region and type of sector

ognize. Distributive issues are actually at the core of some regulatory dilemmas, such as the trade-off between rent and efficiency (by which to achieve efficient results, such as high investment, some positive rent levels must be guaranteed to regulated firms), and the tension between liberalization and cross subsidies (by which prices are above cost in some regulated segments to fund the deficits in high cost or politically sensitive segments).

If the dilemmas between firms' rents and efficiency take place in democratic societies with high inequality, the existence of political problems when these rents are made more transparent (for example, through well-audited foreign multinationals) should not be surprising. At least since the 1960s, inequality in Latin American countries has been higher than in any other region of the world.

In this chapter, first we summarize the main features of the reform of the infrastructure industries in Latin America and the Caribbean, including a summary of the evidence on infrastructure deployment and investment in the recent past. Next, we analyse the reasons and drivers of reform. We also review the regulatory framework that has accompanied the reform process. After that, we briefly conclude.

THE REFORM OF INFRASTRUCTURES IN LATIN AMERICA (1990–2007)

In this section we focus on some facts about reform in the infrastructure sectors in Latin America and the Caribbean. We provide some data on the evolution of liberalization (allowing competition) and privatization (selling public assets to the private sector). We leave the role of regulation for another section below. A key issue in the infrastructure sectors is that reform typically involved replacing what historically had been public investment by private investment.

The degree to which reform has been able to attract private investment and the impact of such investment on welfare (in the form of access to basic services) are explored here. The public policies to attract and remunerate this private investment depend to some extent on the political process, and on how politics channels the distributive tensions in this region. For these reasons we also present some facts about the interaction between reform and the political process.

Economic and Infrastructure Development Indicators

Basic infrastructures are a key ingredient of a decent life and an important determinant of economic growth. Although the stock of capital does not determine the steady-state rate of growth of income per capita, it does determine the speed of the transition path and some aspects of the stock of capital related to basic living conditions (access to water, electricity, communications) determine the way in which average income per capita is linked to the welfare of the vast majority of families. That is why it is important to see how income per capita, infrastructures expansion and some measures of the quality of life evolved during the reform years.

Table 22.1 gives a snapshot of the living conditions in the region in the recent past. All countries increased income per capita since 2000, but great disparities persist across countries; highest in this particular ranking are Mexico and Trinidad and Tobago, and lowest is Nicaragua (we exclude Haiti, the poorest country, because of lack of data for most of the indicators we focus on in this chapter). The Human Development Index also shows an increasing pattern, and the life expectancy at birth has increased in all countries between 1990 and 2006. However, not all countries improved their figures in issues such as their sanitation facilities, where some countries such as Nicaragua and Bolivia stay at a very low level.

All countries improved their access to distributed water, although only two of them granted access to all citizens. There is only one country in the region, namely Uruguay, that has universal access both to distributed water and sanitation facilities.

Improvements in telephony have been widespread since 1990, benefiting not only from industrial reform but also from technological change. However, some countries such as Bolivia, Nicaragua and Peru still have less than 65 subscribers per 100 inhabitants in 2007. The diffusion of mobile telephony between 1990 and 2006 has been spectacular in the region, reaching levels close or beyond one line per person in several countries. This rapid growth reflects many factors, common to other regions, but probably at work here is a preference for a technology that requires low sunk costs as compared to fixed telephony, given the difficulties of attracting sunk private investments due to the

Table 22.1 Economic development indicators

Country	GNI per capita, PPP (current international $)*		Human Development Index			Life expectancy at birth, total (years)			Improved sanitation facilities, urban (% of urban population with access)			Improved water source (% of population with access)		
	2000	2006	1990	2000	2005	1990	2000	2006	1990	2000	2006	1990	2000	2006
Argentina	8890	11670	0.813	0.862	0.869	72	74	75	86	91	92	94	96	96
Barbados	..	..	0	0	0.892	74	75	77	99	99	99	100	100	100
Belize	4640	6030	0.75	0.795	0.778	72	..	72	..	71	..	..	91	..
Bolivia	3080	3840	0.606	0.677	0.695	59	63	65	47	52	54	72	82	86
Brazil	6840	8700	0.723	0.789	0.8	67	70	72	82	83	84	83	89	91
Chile	8930	11660	0.788	0.845	0.867	74	77	78	91	95	97	91	93	95
Colombia	4620	6100	0.729	0.772	0.791	68	71	73	81	83	85	89	91	93
Costa Rica	6630	9540	0.794	0.83	0.846	76	78	79	96	96	96	..	97	98
Dominican Republic	4160	5750	0.697	0.757	0.779	68	71	72	77	79	81	84	92	95
Ecuador	4440	6810	0.714	...	0.772	69	73	75	88	90	91	73	88	95
El Salvador	4330	5320	0.653	0.716	0.735	66	70	72	88	89	90	69	79	84
Guatemala	3480	4270	0.592	0.667	0.689	63	68	70	87	89	90	79	91	96
Honduras	2510	3380	0.634	0.668	0.7	66	68	70	68	74	78	72	80	84
Jamaica	4900	5960	0.713	0.744	0.736	72	71	71	82	82	82	92	93	93

Table 22.1 (continued)

Country	GNI per capita, PPP (current international $)*		Human Development Index			Life expectancy at birth, total (years)			Improved sanitation facilities, urban (% of urban population with access)			Improved water source (% of population with access)		
	2000	2006	1990	2000	2005	1990	2000	2006	1990	2000	2006	1990	2000	2006
Mexico	8950	11970	0.768	0.814	0.829	71	74	74	74	88	91	88	93	95
Nicaragua	1780	2380	0.61	0.671	0.71	65	70	72	59	57	57	70	77	79
Panama	6850	9370	0.752	0.797	0.812	72	74	75	..	77	78	..	92	92
Paraguay	3310	4040	0.718	0.749	0.755	68	70	72	88	88	89	52	69	77
Peru	4830	6440	0.71	0.763	0.773	66	69	71	73	80	85	75	81	84
Suriname	4650	7110	...	...	0.774	68	69	70	90	90	89	..	91	92
Trinidad and Tobago	10370	20550	0.784	0.796	0.814	70	69	70	93	92	92	88	91	94
Uruguay	7750	9930	0.806	0.842	0.852	73	75	76	100	100	100	100	100	100
Venezuela, RB	8380	11050	0.762	0.776	0.792	71	73	74	90	..	..	89	..	..

Notes: * GNI per capita based on purchasing power parity (PPP). PPP GNI is gross national income (GNI) converted to international dollars using purchasing power parity rates. An international dollar has the same purchasing power over GNI as a US dollar has in the USA.

Source: World Development Indicators Database, World Bank, 2008.

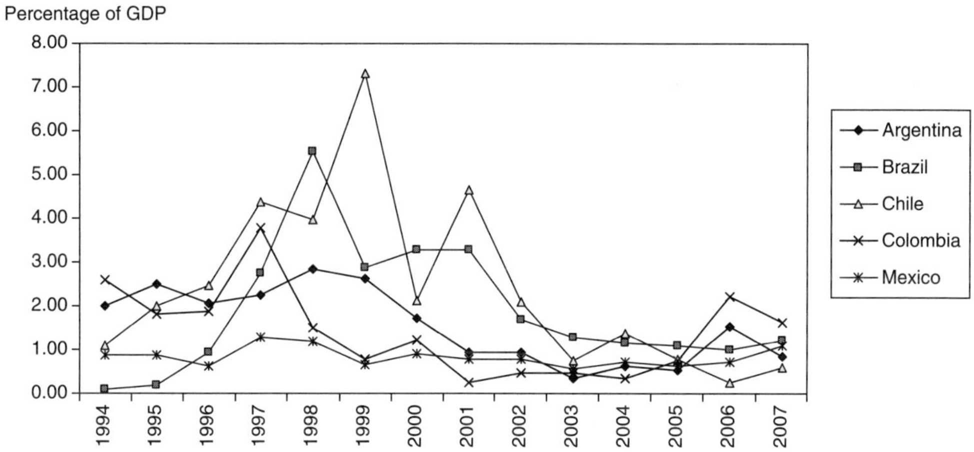

Source: World Bank, Private Participation in Infrastructure Project database.

Figure 22.2 Private investment in infrastructure in select Latin American countries, 1994–2007

regulatory hold up problem, to be explored below (this is most clear in the comparison between fixed and mobile lines in Argentina, for example). Broadband Internet shows a similar potential, but the diffusion curve is still behind mobile telephony. In basic transport, however, the improvements have been minimal, as paved roads and railways are at a similar level than in 1990. Some countries such as Chile have been able to attract high amounts of private investment in toll highway concessions, but in many others unpaved roads are still a clear sign of underdevelopment.

There are general improvements in production of electricity per capita. However, even the improvements in capacity per capita are far from universal, linear or spectacular. This shows the inherent difficulties in reforming this particular industry. On the one hand, electricity is a non-storable good that requires complex coordination between the different industry segments. This implies that if competition is introduced into one competitive segment such as generation, unbundling from the network segments must be possible and regulatory institutions should be in place so that incumbent network firms do not discriminate against entrants. On the other hand, access for many communities was still unavailable or informal, which posed many distributional problems that have proved difficult to solve.

Figure 22.2 focuses on private investment flows. The Latin America and Caribbean region was by far the region of the world that attracted more private investment since 1990. Most of this private investment went into the telecommunications sector, and was concentrated in a few countries, such as Brazil, Mexico, Argentina and Chile. Colombia, Peru and Venezuela, also attracted significant quantities of private capital. The sector where it was most difficult to attract private investors was the water and sewerage sector, where some countries were not able to attract private investors, although they were able to do so in telecommunications, energy or, even, transport. Figure 22.2 shows the evolution over time of investment flows. Private investment flows peaked in the late 1990s to

Table 22.2 Privatization and liberalization of telecommunications in selected Latin American countries

Country	Liberalization					
	Privati-zation	Local	Long distance	Inter-national	Mobile	Value added
Mexico	1990	1997	1997	1997	1989	1989
Nicaragua	1998				1993	
Argentina	1990	2000	1999	2000	1988	1987
Bolivia	1995	1995	2000	2000	1988	1988
Brazil	1998	1998	1998	1998	1997	1991
Chile	1988	1994	1994	1992	1988	1985
Colombia	n.a.	1995	1997	1998	1994	1992
Peru	1994	1999	1999	1999	1990	1994
Venezuela	1992	2000	2000	2000	1991	1991

Source: Sancho and Jordana (2004).

decline later after the financial crisis of 1997–98 and the backlash against privatization that took place around the same time (more on this below). These flows slightly increased again after 2005.

Liberalization and Privatization

The main drivers of industry reform in Latin America were liberalization and privatization. However, this does not mean that all countries immediately privatized all their publicly owned firms in the infrastructures sectors and introduced competition everywhere.

Table 22.2 reports the year in which telecommunications privatization started in a selection of countries. The first country to privatize was Chile, at the end of the Pinochet military dictatorship. Liberalization of local telephony started in general after privatization, although competition in other potentially competitive segments (such as mobile or value added services) could start before privatization. The liberalization of other segments where the potential for competition was less clear took more time to arrive, reflecting the fact that to attract private capital in some of these sectors, investors were assured that they would not face serious competition, at least in the short run.

Table 22.3 summarizes some evidence on the reform of the electricity industry. Sioshansi and Pfaffenberg (2006) provide a series of updated country studies in electricity reform, including Chile, Brazil and Colombia (see also Victor and Heller, 2007). It can be seen that, although most countries introduced some reform to attract private investment and allow new entry, the particular characteristics of reform differed across countries. The degree of vertical integration allowed was high, with only Bolivia and Argentina imposing obligatory vertical separation between competitive and monopolistic segments.

The single buyer (SB) scheme is typical of a situation that permits the incorporation (in an energy production chain) of private enterprises as third parties, in charge of producing and delivering energy products to the integrated state enterprise, which would be acting as the single buyer. This scheme is also known as partial opening up.

Table 22.3 Reforms in electricity systems and size of the markets

Installed generation capacity (MW)	Centralized control		Regulated integrated system	
	Integral part of the state	Certain degree of business autonomy	Single integrated structure	Various business units
0–500			Barbados	
501–1000				
1001–2000	Cuba	Uruguay		Costa Rica
2001–5000		Paraguay		Ecuador
5001–10 000				
10 001–20 000				
> 20 000				Venezuela Brasil

Installed generation capacity (MW)	Single buyer		Open market	
	Integrated distribution	Distribution breakup	Vertical integration permitted	Obligatory vertical breakup
0–500	Suriname Nicaragua			
501–1000	Jamaica, Honduras	Panama	El Salvador	Bolivia
1001–2000	T & Tobago			
2001–5000	Dominician R.			
5001–10 000			Chile	
10 001–20 000			Colombia	Argentina
> 20 000	Mexico			

Source: Bouille and Wamukonya (2003).

As for a regulated integrated structure scheme (RI), there is greater autonomy for the privately owned companies, mixed enterprises, or public entities under provincial or municipal jurisdiction. The state assumes the role of regulator, whereby it can give priority to technical and economic considerations, but always maintaining a certain amount of political considerations. This scheme is characterized by the absence of contestability in the markets. The state effectively participates in investment decisions and in price formation.

Institutional and Political Indicators

Private investment does not operate in a vacuum. The protection of property rights must be credible, and this is achieved by institutions, including the judiciary, that enforce contracts and make the rule of law prevail. More specifically, given the long-lived nature of the specific investments needed in the infrastructure sectors, the institutional framework must make intertemporal cooperation possible and desirable, by which investors trust that their investment efforts will in the future be rewarded with sufficient revenues, and consumers trust that any price increases will be compensated by good access and quality conditions in the foreseeable future.

Related to the political cycle, in Latin America political change has taken place in most countries in the region between 1990 and 2006. Only a minority of countries, such as Chile and El Salvador, have experienced total political stability in terms of the ideological sign of the party or coalition in government. As democracy spread in the region, this tested the stability of the regulatory regimes and reform packages put in place.

REASONS AND DRIVERS OF INDUSTRY REFORM

The transformation of the network industries in most Latin American and Caribbean countries had three main ingredients, each of which had in turn different timing, forms and varieties in different countries (as seen in the previous section). These ingredients were liberalization, private investment and regulatory reform.

Liberalization implied introducing the possibility of competition. This could be either in the market, through rivalry between products sold by different firms with little product differentiation, or for the market, through competitive bidding. The rationale behind liberalization was that output levels would increase, customer service would improve, prices would decrease for given costs, and costs would decline through better incentives and innovation.

Of course, these positive outcomes are achieved when an industry is close to the conditions of the perfect competition model, that is, many firms and free entry and exit. In network industries, even in those segments that are not naturally monopolistic (electricity generation, mobile telecommunications), entry barriers are high and there is usually room for a small number of firms. Then the role of competition policy is key to reduce the market power of firms. Where competition in the market is not possible or desirable, there is the possibility of competition for the market through bidding for concession contracts. This is possible when projects can be individualized and isolated, such as highways, tunnels or generation plants (such as independent power producers). Competitive bidding increases the likelihood that the winning firm will be the most efficient or the one requiring the lowest subsidy or consumer prices, but the design of the competitive process has its challenges. The number of firms competing for the contract has to be as high as possible, and collusion between bidders needs to be avoided. When the contract has a long duration, the possibility that unforeseen contingencies arise is high and renegotiation may lead to costly bargaining and undesirable outcomes. In these cases, concession contracts must be accompanied by sound and effective regulatory frameworks.

Private investment may take the form of entry of new firms in greenfield projects or privatization of state-owned firms. Privatization can take the form of selling the assets to a private firm or consortium with significant participation of foreign operators that contribute both capital and technological expertise (the case, for example, of electricity distribution or water firms in Argentina), or it can take the form of initial public offers, by selling shares to the general public, including possibly the workers of the firm (the case of Chilean electricity). The rationale behind privatization was that the public sector had inherently inefficient management practices and that it lacked the resources to overcome the huge access deficits in infrastructure sectors in these countries. Privatization was expected to provide better incentives to firm managers and staff, while reducing red tape and politicization.

Liberalization and privatization pose significant institutional challenges, however, and there has been high cross-country variation on how these challenges have been addressed. Of course, the realization of the high early expectations depended on private firms developing good corporate governance arrangements and on governments developing accompanying regulatory frameworks that made risky investments possible.

Overall, at the beginning of the 1990s most countries in the region, and international financial institutions such as the International Monetary Fund and the World Bank, believed that the rationale behind liberalization and privatization was strong, and that the advantages would somehow easily materialize. Fiscal (high public deficits and debt levels), ideological (a discredit of economic strategies based on a large role for the public sector) and technological reasons (especially in telecommunications and electricity generation) made many agents put more emphasis on the potential advantages than on the caveats of the strategy. It was not for some years that, in the twenty-first century, more emphasis was put on qualifying the initial strategies and paying more attention to the caveats.

Effective industry reform must take into account the specific characteristics of developing countries, such as Latin American and Caribbean countries: regulatory capacity problems, commitment problems, accountability problems, and fiscal and income inequality problems. Either the transformation strategy must acknowledge these problems and devise a strategy that deals with them, or industry transformation must wait or at least run parallel to an overall reform strategy that fixes capacity, commitment, accountability and fiscal problems.

Laffont (2005) goes in depth into the characteristics of infrastructure sectors in developing countries:

- Higher deadweight cost of public funds due to distorting taxation. The under-development of tax systems implies that collecting one dollar is significantly more costly in developing than in developed economies. World Bank estimates report a deadweight loss exceeding 1 for each unit of funds to be raised against a 0.3 in developed countries. The high cost of the tax system, coupled with enormous social needs and high inequality, entails high pressure to redistribute any rents known to be generated. In particular, there is a high short-term opportunity cost of complying with the commitment not to expropriate the quasi-rents necessary to remunerate the sunk investments in infrastructure sectors.
- Higher cost of auditing and enforcement. The inefficiency of the public sector and absence of well trained bureaucracy are at the basis of under-developed institutions to enforce private contracts.
- Private ordering instead of the rule of law. The behaviour of public servants is often short sighted and dominated by private concerns, translating into relatively low standards of probity in public administration in a number of countries, up to cronyism and corruption.
- Lower ability to enter into long-term contracts. Institutions to protect property rights are usually informal and arenas where agents can engage in credible

long-term commitments are relatively unusual in less developed countries (LDCs). Poorly protected private property rights and under-developed business codes of behaviour are the norm in many of these countries.

The list of special features of developing countries that affect privatization processes should be supplemented, according to Parker and Kirkpatrick (2005), with the following features:

- Imperfectly competitive and incomplete markets. Existing producers often have high market power and there is absence or scarcity of markets (typically, financial markets) where agents with different allocations across states of nature and over time can exchange their endowments.
- Regionalized and sometimes ethnically distinct labour markets, with appointments through connections. Labour markets are thin, and absence of meritocratic appointments reduces the incentives to invest in human capital. This translates into management weaknesses and patronage in appointments at firms and regulatory institutions.

The combination of Laffont's and Parker and Kirkpatrick's features gives an idea of the complications that arise in regulatory reform policies in developing countries, and in Latin America in particular. Laffont further explores how industry and market structures,[5] power of incentives,[6] access prices and universal service should differ in light of such specific characteristics. Although we refer to Laffont (2005) for further details, the general conclusion is that policies that would be efficient in developed countries (such as price caps or other 'incentive regulation' mechanisms) are much more difficult to sustain in developing countries.

What is the link between the impact of regulatory reform on different income groups and political attitudes towards reform? Santhakumar (2005) and Martimort and Straub (2009) provide empirical and theoretical perspectives, respectively, on this issue. Santhakumar (2005) analyses survey data about attitudes towards reform in the electricity sector in India and reports that those states with more connected households and average quality of service are more prone to oppose reforms such as privatization, because in these states subsidies to electricity consumption are the highest and the gains in terms of new connected households derived from privatization are minimal. His is, however, a study of initial opposition to reform and not of backlash against existing reforms. Martimort and Straub (2009) build a model motivated by the backlash against privatization in Latin America. They argue that with public ownership the costs are concentrated on dispersed and disorganized taxpayers who carry the burden of the soft budget constraint problem. With private ownership the soft budget constraint no longer applies in their model, but well-publicized (and higher than before privatization) prices concentrated on the connected middle classes can determine a scenario characterized by a strong political opposition compatible with improvements in efficiency.

Harris (2003) analyses the impact of reforms on the poor, and reports several cases (notably in Chile and Colombia) where private sector involvement has been accompanied by high levels of network expansion in water, sanitation and electricity, as compared with places still under public provision. However, as reported in Easterly and Servén

(2003), infrastructure expansion in Latin America in general has been below the expectations raised at the beginning of the 1990s. Harris (2003) also points out that the connected poor, since they spend a higher proportion of their income on utilities services, have suffered disproportionately from tariff increases that have been necessary to attract private investors, and that subsidies to the poor have in several instances not been well targeted enough.

Reform may also have a desirable distributive feature. Engel et al. (2003) argue that concession contracts to private investors (such as build, operate, transfer, or BOT, contracts, used for example in toll highway franchising) distribute resources from consumers to taxpayers, if franchises perform under a hard budget constraint (costs have to be covered through tariffs, and not through taxation). If the consumer population is not the total population, as is the case with highways, this may have an equalising impact on income distribution.[7] However, Engel et al. (2003) report that many of these franchises in Latin America (they analyse cases from Argentina, Chile and Colombia) were renegotiated and finally did require public funds for cost overruns. They propose that, to avoid such problems in the future, concessions should be accompanied by sectoral regulation separated from ministries to deal with post contractual hazards, and that franchise periods should be variable to reduce demand risk.

Leipziger et al. (1999) find that public provision does not benefit the poor, but conclude that, with private provision, the public sector must keep a role as regulator and as provider of well-targeted subsidies. They argue that the weaker the regulation, the less protected are the interests of the poor.

It is useful to explore the costs and benefits of targeting macroeconomic goals with microeconomic instruments (in this case, regulatory reform). This is precisely what is done in Easterly and Servén (2003). The reduction in investment and public expenditure in infrastructure as part of the economic adjustment of Latin American countries in the 1980s and 1990s is equivalent, according to these authors, to replacing debt with 9 per cent interest with debt with a 20 per cent interest or even more. Instead of new debt, what took place in most Latin American countries was a cut in public sector expenditure on maintenance and construction of infrastructure. Such expenditures are estimated to have a high social rate of return (they give specific examples ranging from 19 per cent to 117 per cent).[8] Cutting expenditures in a project that has a high rate of return is economically equivalent to acquiring debt with a high interest rate: both policies free resources today, in exchange of fewer resources tomorrow.

Easterly and Servén are of the view that in the 1980s it had indeed become necessary to improve the health of public finances and that the state economic leadership model was exhausted. What these authors believe to be questionable is the extent to which infrastructure public expenditure should carry the burden of the cut in fiscal deficit. Latin American countries, although with important internal variations, adjust to the pattern that is also described in other studies, according to which infrastructure expenditure in less developed countries (LDCs) often suffers a disproportionate compression in times of fiscal austerity. Such compression is largely due to the myopic use of the ratio between the current budget deficit and the gross national product (GNP) as a measure of fiscal performance, while the only fiscal constraint that matters economically is the intertemporal fiscal constraint. Similar criteria of fiscal illusion were used by the EU countries on occasion of the fulfillment of the Maastricht criteria for the creation of the single currency in the 1990s, which coincided with a significant increase of privatizations in countries such as Spain or Italy.

Private provision of infrastructure is seen by these authors as a promising formula, but with few exceptions (which can be found especially in the telecommunications sector) this is a process that finds itself at a very preliminary stage. Even where private provision is feasible, transition from state to private ownership must be thought through very carefully. Opening up infrastructures to the private sector may make a great deal of sense, but cutting expenditure in infrastructures with a high rate of return and wait for the private sector to take over overnight is called by Easterly and Servén a 'leap of faith'. After an exhaustive empirical study, the authors conclude that opening up infrastructures industries to the participation of the private sector has offered mixed results and has not resolved the infrastructure problems in Latin America.

Among the empirical results that these authors present, they find that public infrastructure expenditure has a superior impact on growth when public infrastructure expenditure strongly complements private capital. Moreover, they point out that there is no reason to expect that opening up infrastructure investment to the private sector will attract as much investment as to compensate public cuts, for two reasons:

1. Private investment requires an appropriate institutional and regulatory framework.
2. Complementarity and substitutability characteristics between public and private capital may differ across industries.

In fact there is scarce evidence that the decreasing trend in public infrastructure investment has been accompanied by a symmetric increasing implication of the private sector in the provision of infrastructures. In some cases, higher private investment was accompanied by higher public investment, suggesting that public and private sectors often play complementary roles. The authors stress that the only sector where the positive expected reaction from the private sector did emerge was the telecommunications industry.

The substantial impact of infrastructures over growth documented by Easterly and Servén (2003) suggests that the fiscal adjustment of the 1980s and 1990s represented a highly inefficient way to adjust public finances.

Campos et al. (2003) quantify the macroeconomic effects of private sector participation in infrastructure for 21 Latin American countries between 1985 and 1998, although they recognize the data shortcomings of their exercise and the lack of a theoretical model on which they can base their hypotheses. They find that transport reform (airports, ports, railways, roads) has had a significant effect on per capita income, but that utilities' reform (electricity, gas, sanitation, water and telecommunications) has no significant effect.

The expenditure, price and tax implications of infrastructure industries reform extend to all activities, and therefore a general equilibrium approach would be the most appropriate method of evaluation. However, data availability and communication between economics sub-disciplines has limited the production of such studies. One exception is Benítez et al. (2003) for the case of Argentina's reform. Although these authors point out that the reform of utilities in Argentina had a positive effect on national welfare and on distribution, the gains from reform were insufficient to offset the losses from the credit shocks[9] that the Argentinean economy experienced in the 1990s. They rely on a general equilibrium model of Argentina's economy calibrated for 1993, isolating the distributional effects of utilities reform from the impact of other reforms taking place in the country during the 1990s. All agents win from increases in productivity and access, but the welfare gains are offset by financial crises.

Overall, the empirical results on the impact of reform policies on the welfare of different income groups of consumers are inconclusive. Improved access rates in some areas (although in a context of insufficient infrastructure expansion) are probably compensated with higher and more transparent charges for some groups. The cuts in public infrastructure investment, which accompanied policies to involve the private sector, have done more harm than good to the overall macroeconomic impact of reform policies in highly unequal societies.

A weakness of reform packages has been the failure to attract the support of the public opinion. According to recent surveys the percentage of respondents who disagree that privatization has been beneficial for their country has been increasing and seriously compromises the future of reforms based on selling assets to the private sector (Corporación Latinobarómetro, 2001 and 2004). This negative change in the social acceptance of privatization has been accompanied by a negative change in regulatory quality, as measured by the World Bank (this index refers to general regulatory quality, not only that related to specific regulatory agencies). The only country in the region that experienced an increase in regulatory quality, according to this index, is Mexico.

It is widely accepted among commentators and scholars that one country with high regulatory quality and high levels of success in attracting private investment and liberalizing some industry segments has been Chile. It was the first Latin American country to implement radical industry reform (liberalization and privatization) and some of its recipes have been imitated in countries that reformed later on. It is thus important to pay some detailed attention to this case, as it may shed light on the drivers of success in industry reform.

The case of Chile is, however, in a sense very unique and the overall success story masks some important inter-industry differences. Common to all industries is a reliance on private property rights that have been drastically protected since a brutal military dictatorship (1973–88) could introduce unopposed an economic model designed by local experts trained in the University of Chicago (see Shirley, 2008), and introduced a new Constitution that has been basically untouched since the restoration of democracy in 1988. This constitution not only protected private ownership, but also made legislative change very difficult in an electoral system that always give a say to the first minority in Parliament and that reserved a role for many years to non-elected, mainly conservative, legislators. This basic institutional framework, together with a historical tradition for credibility,[10] judicial independence and respect for contracts, provides the basic institutional framework that has made private investment and liberalization possible in Chile. It is doubtful, however, that the same model can be exactly applied in other countries that start reform in a democratic and open society. Moreover, as Shirley (2008) points out, Chileans show in opinion surveys that they are far from fully happy as a result of the industry reform process in general.

Across industries the experiences in Chile have gone from the very successful (water and telecommunications – although the latter was not free of controversy) to the disastrous (the bus system in Santiago), with electricity and roads somewhere in between.

The success in water (according to Foster, 2005, Stern, 2009, and Shirley, 2008) has been based on a targeted subsidy system that has made possible for the operator to keep a healthy financial position to expand access and keep high quality levels, and at the same time for consumers to have affordable access to water and sanitation, regardless of their income levels. Although the reform that ultimately made private ownership possible was designed by the military government in the 1980s, the new democratic government delayed actual full privatization for almost ten years, and implemented first the price and subsidy reforms that ultimately made privatization sustainable. Legislation sets up a three expert arbitration process in case of tariff disputes, by which the firm appoints one expert, the regulator another expert, and a third is appointed by consensus. This arbitration panel has to decide in little more than one month, providing much speedier and specialized resolution than would be achieved by courts of justice.

In Chilean telecommunications, interconnection and equal access requirements, as well as initial industry structure, facilitated a rapid transition to facilities-based competition (Spiller and Cardilli, 1997). Regulatory reforms required that the disputes among entrants and incumbents be resolved also by arbitration rather than litigation, and gave SUBTEL, the telecommunications non-independent but highly technical regulatory agency, substantial powers. SUBTEL set interconnection charges based on long-run average incremental costs, and local competition started taking hold, something that many developed countries have found very difficult to achieve. Chile replayed this scenario with long-distance competition. When ENTEL's long-distance monopoly was abolished in 1992, the local-to-long-distance inter-connection rules were specified vaguely, and it was only after much litigation that consumers were allowed to select their own long-distance carriers and that real competition began in this market. The process was highly acrimonious, as relations between the two incumbents, their prospective competitors and the regulator SUBTEL deteriorated rapidly, requiring SUBTEL to intervene on almost every issue and to endure numerous constitutional challenges to its authority.

Until mid-1994, CTC (the dominant local operator) contracted wholesale long-distance, and assigned outbound calls randomly to long-distance carriers in proportion to the size of that carrier's contract. The growing pressure on the government to let customers choose their own long-distance carrier, spurred by a Supreme Court decision, forced the regulator to take a radical approach and place all carriers on an equal basis. The mandate was that all long-distance carriers be accessible only by dialling the carrier's unique access code, thus requiring customers to select a long-distance carrier explicitly every time they made a call. The transition to equal access happened virtually overnight for the whole country, and it dramatically cut prices and raised consumption of long-distance in Chile.

Chile, on the other hand, entered the era of reformist telecommunications policy with structural separation already in place: one state-controlled company dominated local phone service, while a separate state-controlled firm dominated long-distance service, each facing limited competition from smaller private firms. In 1994, at the end of an acrimonious legal battle, CTC and ENTEL were allowed to enter each other's business – albeit with some market share restrictions – while all restrictions on entry by others into both segments of the market were essentially lifted. In Chile, eight companies were providing long-distance service at the end of the 1990s. Prices plummeted. By September

1996, average long-distance rates had fallen by more than 50 per cent since the liberalization of long-distance in 1994, and domestic long-distance traffic had risen by more than 50 per cent, according to Spiller and Cardilli (1997). To survive this rugged competition, almost all long-distance companies have adopted a strategy of becoming integrated providers, concentrating on a particular set of customers rather than a set of services. The provision of integrated services is essential in Chile today, and since there were no unbundling or resale requirements for a long time, construction of facilities was the only way to provide local service.

The electricity industry was privatized much faster than the water industry, and the shares were mainly sold to insiders (Trillas, 2001). Liberalization was much slower than in telecommunications. The problems of corporate governance that the selling method caused were made evident when a takeover for the main electricity holding, Enersis, by the Spanish firm Endesa, in the late 1990s, uncovered a dual shareholding structure, by which the executive owners' shares had a much higher price than ordinary shares. The political scandal that followed obliged the bidder to pay a much higher price for the acquisition than initially expected. In any case, private investment was sustained since the late 1980s by a very rigid legal system that guaranteed a 10 per cent return on investment to shareholders and a very high degree of vertical integration. The legal and industrial structure has been made more flexible and competitive over time, as consumer complaints increased and a mainly hydro-based system proved very vulnerable to droughts.

Private investment in roads, particularly in toll highways, has changed the quality of life to many Chileans. Today it is possible to go by car to the centre of the capital, Santiago, in a fraction of time than used to be the case a decade ago (and park in private subterranean parking lots that are also the result of foreign, mainly Spanish, investment), with tolls charged automatically by a tag system that makes it unnecessary to stop the car in toll stations as is usual in many European countries. Highway construction and maintenance has been possible thanks to a bidding system for concession contracts. The bidding system has, however, been controversial, as contract renegotiations and some corruption allegations have probably made it necessary to introduce reforms, such as going from fixed concession periods to variable periods depending on demand levels, and the introduction of a separate (independent from the authority granting the contract) regulator, in charge of monitoring contract enforcement (Engel et al., 2007).

The greatest failure has come from the reform of the highly inefficient bus system of the capital, called Transantiago (Gómez-Lobo, 2007). The reform was based on bidding for a new line system that completely changed the routes and vehicles of the old system. The result was a complete mess in terms of routes for the citizens of Santiago, with enormous difficulties to connect the new lines with the highly efficient (and publicly owned and managed) subway train system ('metro'). The delays that for many months produced the new system had a very negative impact on the quality of life for many Chileans, and caused serious difficulties for the political survival of the national government.

As the case of Chile illustrates, success is possible but not guaranteed, and it requires appropriate safeguards both for investors and consumers, as well as setting up a regulatory and institutional framework that accompanies the process, an issue towards which we turn in the next section.

REGULATORY APPROACHES IN LATIN AMERICA

Private investors would not risk their money unless there were institutional safeguards that they would receive a sufficient return. These institutional safeguards range from the basic institutions of law and order, to the specific industry regulatory agencies. There is a broad consensus among scholars that the basic regulatory governance must be credible, stable and effective, to accompany industry reform. Sound regulation must accompany, and if possible precede, a transformation based on liberalization and private investment. Liberalization and privatization in infrastructure sectors are never accompanied by complete deregulation, but by a careful regulatory reform that takes into account both the potential for market failures and regulatory imperfections. The key ingredient of regulatory reform, that must be adapted to the specific conditions of each country, is that it must make possible the intertemporal cooperation between investors, governments and consumers. Investors will only risk their resources in long lived projects where the value of assets in alternative projects is very low or zero, if governments and consumers are able to commit through regulatory institutions to remunerating them.

A number of studies, the best example of which is probably Gutiérrez (2003), show that regulatory governance, is a key ingredient of the success of reform packages in the infrastructure sectors. A full characterization of regulatory governance includes transparency, accountability and regulatory independence, as shown in Gasmi et al. (2006) and Andrés et al. (2007, 2008). We focus here on regulatory independence vis-à-vis governments, because it is the area that has attracted most academic research.

Most countries in the region embraced the independent regulatory agency model with enthusiasm, at least legally. The first country in Latin America to create a telecommunications regulatory agency (TRA) was Chile in 1977, although it would be the least (legally) independent of all agencies in the region; subsequently, it was in the 1990s that Argentina (1990), Venezuela (1991), Colombia and Peru (1994) followed. Until 2002 the remaining countries in the subcontinent created their TRA with different degrees of independence, roles and responsibilities. Among 23 countries, in 17 of them the period in office of the TRA head is legally prescribed. Of these, the law allows for re-election in 11 cases. The average age of a TRA as of 2004 was nine years. Among the countries with a prescribed time limit, the shortest period is that of Colombia (16 months) and the longest is Jamaica (seven years).

Legal independence has been measured in a number of studies by summarizing in an index such component variables as source of funding, or appointment rules. However, recent research shows that there is a lot of variation across countries in the extent to which legal provisions have translated into independence in practice, at least in the telecommunications sector. One of the most representative cases is Argentina, where there have been 11 directors of the regulatory agency CNC (previously CNT) in ten years. During the presidency of Carlos Menem there were seven different directors. The local press also reflects other cases of political pressures to force the resignation of the director, as in Bolivia, Brazil and Panama; or resignations due to differences among cabinet ministers, as in Chile; or resignations due to a change in party membership of the head of the agency, as in Nicaragua. The turnover ratio of regulators has been in general high, although there is significant cross-country variability.

In Montoya and Trillas (2009) we present case studies of the evolution of regulatory independence in practice in the telecommunications industry for 23 Latin American and Caribbean countries. Based on these studies, we construct two realistic indices of regulatory independence, which improve upon the measures of independence that have been used so far in the empirical regulation literature. We show that legal indices may give a partially distorted picture of the commitment ability of institutions. Basic illustrative econometrics suggests that the combination of *de facto* and *de jure* independence has a positive and significant impact on network penetration. Many of these agencies were given some degree of legal independence from the government to facilitate commitment not to expropriate sunk investments.

Independence does not solve the commitment problem (not to expropriate sunk investments necessary to expand service). Instead, it transforms itself into one of the government's credible commitments not to undermine the independence of the regulator, which many countries have found very difficult. Independence in practice has improved over time, especially since 2000, but some countries' governments encountered serious problems in committing to preserve regulator independence.

Of the 17 countries in the region which have a period established by law, only in four did the regulator complete the full period: Belize, Jamaica, Uruguay and Peru (which actually went beyond the prescribed period), and the remainder (with the exception of Bolivia, Colombia and Surinam) stayed in office for approximately half of the prescribed period. When measuring turnover in the head of TRAs in Latin America, we find that in 61 per cent of occasions, there is a change in the head of the agency within one month of a political change. If we take six months, in 73 per cent of the times in which there is a change in president, there is a turnover in the head of the TRA. The data confirm that more realistic measures of independence still have a positive and significant impact on network penetration. Although the coefficients measuring legal or legal and practice independence have different magnitudes, we do not want to overemphasise this difference at this stage, since we are aware that as with any index, the components and weights are, to some extent, arbitrary.

We leave an in-depth analysis of the differential impact of independence in practice for future research. We only point out here that, with this preliminary analysis (which brings the empirical literature on regulation one step closer to the more advanced empirical literature on Central Bank independence), regulator independence matters, even (or especially) when we take into account the problems that governments face when committing to respect the independence of the regulator.

On more detailed regulatory incentive systems, the study presented in Estache et al. (2005), based on 1000 contracts in Latin America, shows that there was a tendency in favour of price cap (used to regulate 56 per cent of the contracts) vis-à-vis rate or return regulation (20 per cent of the cases). However, the same authors point out that the 'initial enthusiasm for the inclusion of efficiency in the design of regulation in developing countries is however being adjusted in developments observed over the last 2–3 years' (Estache et al., 2005, p. 3). In that respect, they observe a tendency away from 'pure' price or revenue cap regimes in favour of 'hybrid' regimes which entail some pass-through of costs to consumers, most notably in activities subject to exchange risks and the ones involving negotiated long-term arrangements (for example, labour contracts). The other tendency they observe is the increasing interest in national and international performance benchmarking (yardstick competition) in LDCs, albeit with slow progress.

Also, the slowdown in the volumes of private capital in financing infrastructures in LDCs since the 1997 East Asia crisis makes the authors conclude that 'the next generation of contracts seems to be moving towards management contracts' and that 'the public sector will continue to be an important actor' (Estache et al., 2005, p. 4).

The issue of contract renegotiation with a focus on LDCs, particularly on the Latin American and Caribbean countries, is studied by Guasch et al. (2002), who combine game-theoretic analysis and empirical evidence. In a context where concessions and privatizations, rather than management contracts, were the 'salient modes of private sector participation' (ibid., pp. 77–8), contractual incompleteness and imperfect enforcement leads to wide scope for renegotiation of contracts. The inherently higher risk entailed by price cap schemes tends to exacerbate the tendency towards renegotiation. The authors find that the existence of an independent regulator reduces significantly the probability of renegotiations and increases the quality of enforcement by better commitment.

The issues related to regulation of universal service obligations (USO) are particularly prominent in the region, especially when urban areas with low cost of providing access are open to competition and cross-subsidization becomes unfeasible. Universal service is often seen as one important aspect of economic development per se, most notably as the expansion of the network has not reached rural areas in many areas. The aforementioned high cost of public funds, and therefore the high deadweight loss caused by redistribution through taxation, implies that USO policies are often seen as a sensible instrument for redistribution of income.

Laffont (2005) develops an insightful theoretical model, based on asymmetric information, of USO policies, and shows that uniform pricing policies distort incentives and therefore exacerbate the problem of the limited expansion of the network in rural areas. He points out that when competition is introduced in low-cost areas, new methods of financing access in high cost areas must be put in place, for instance via taxation on the industry or on a broader basis.[11]

One conclusion which can be widely agreed upon as emerging from the literature is that policy makers do need to take distribution concerns seriously into account, when designing effective universal service policies. Universal service programmes[12] should create special funds (from all operators that achieve a given market share, or from general taxation if the tax system is well developed enough), using the vertical chain of government to carefully target subsidies. It also appears that direct subsidies are usually preferable to cross subsidies because they do not jeopardize liberalization, but they must run in parallel with fiscal policy reform and with an improvement in the quality of government (otherwise we would be back to the previous stage, where company deficits were financed through opaque fiscal deficits).

We thus observe that optimal regulatory policies in developing countries face a number of additional constraints. Most of these additional constraints exacerbate the trade-off between rents and efficiency, and hence add distributive demands on optimal policies, generally making reform more difficult to implement or sustain.

At least since the 1960s, inequality in Latin American countries has been higher than in any other region of the world (see Ferranti et al., 2003). With the exception of countries in sub-Saharan Africa, the difference in Gini coefficients between Latin America and other regions is large. This gap narrowed in the 1970s and became wider again in the 1980s. There was no clear pattern in the 1990s, when Latin America performed better than some regions in distributional terms (for example, Eastern Europe) and worse than others (for example, South Asia). It is sometimes argued that inequality is related to the state of development in a country and comparisons should therefore be made that control for this factor. Even controlling for this factor, however, the authors that present this evidence conclude that the Latin American region suffers from high 'excess inequality'.

In 2003, the richest one-tenth of the population of Latin America and the Caribbean earned 48 per cent of total income, while the poorest tenth earned only 1.6 per cent. In industrialized countries, by contrast, the top tenth received 29.1 per cent, while the bottom tenth earned 2.5 per cent.

Inequality would not be a problem for the private sector investment in infrastructure services if markets and institutions worked perfectly (see Banerjee, 2004). Financial markets would allow the implementation of any positive net present value project and private contracts or contracts between the public sector and the private sector would be enforced by well-functioning institutions. If inequality has any impact on investment, it must be because some markets and/or institutions do not work perfectly. We contend that more attention should be placed on the fact that inequality triggers political dissatisfaction (because some markets do not distribute resources to the satisfaction of polities, and some policies do not reach enough the poor or have an impact on some sectors such as middle and lower classes which is perceived to be negative), which in turn makes commitment to reward investment difficult.

CONCLUSIONS

In this chapter we have provided a synthesis of industry reform in Latin America and the Caribbean, the region that has attracted most private investment in the infrastructure sectors in the recent past. We have shown that regulatory commitment, distribution and welfare interact in subtle ways. In the end, whether a reform is successful depends on it being able to attract investment in a sustained way, and that depends on careful institutional design and on the political process.

Current research focuses on how to design second generation reforms, so that more stability can be given to reforms (see, for example, Estache and Wren-Lewis, 2009). The objective is to make regulatory reform more robust in front of political and macroeconomic shocks, avoiding some of the mistakes that have negatively affected the welfare of citizens. The reaction of the Chilean institutions to the major fiasco of the Transantiago transport reform is illustrative (see Gómez-Lobo, 2007).

Future challenges that research in this field is starting to analyse include the role of international cooperation (for example, in the field of yardstick competition, see Estache et al., 2004) and the role of corporate governance (along the lines of Trillas, 2001, and Parisi and Yáñez, 2000) in improving the performance of privatized firms.

Engel et al. (2007) show however the limits of the involvement of the private sector. They argue that modern public–private partnerships (PPPs), a contracting form favoured by many governments because it gives the appearance of not relinquishing public sector control and at the same time using private sector management, shares more characteristics with public rather than private sector ownership.

Part of the rationale for the reform of infrastructure industries in Latin America has been predicated under the hypothesis that regulatory reform is separable from other economic policies. It seemed that industry reform could not wait for other macroeconomic dimensions (monetary policy, fiscal policy, economic growth and income redistribution) to be undertaken. However, experience has shown that reform policies are actually vulnerable to imbalances in these other fields.

The evidence so far shows that success is possible, although it may take different paths for different countries, and it requires close attention to institutional quality and to the political viability of reforms.

NOTES

1. We thank Carlos Cardoso for his excellent work as research assistant.
2. Galiani et al. (2005) for example find that in Argentina child mortality fell 8 per cent in the areas that privatized their water services and that the effect was largest (26 per cent) in the poorest areas, where the expansion of the water network was greatest.
3. See for example Krueger (2000), Basanes and Willig (2001) and Kessides (2004).
4. Gómez-Ibáñez (2003) argues that 'continuing frustration over economic reforms associated with privatization, or continuing disputes among consumers, investors, and governments about the fairness of regulated prices, may eventually reduce popular support for private infrastructure to the breaking point'. He adds that 'the perception of fairness is as important as the reality, so that regulation is as much a political as a technical act'.
5. Focusing on the electricity industry, Holburn and Spiller (2002) argue that it is a myth that economies of scale in generation limit the potential for competition in small markets. However, Jamasb (2006) argues that electricity retail competition should remain a long term objective in most developing countries, due to implementation costs. On how some small developing countries are able to develop competitive market structures in telecommunications, see Spiller and Cardilli (1997).
6. Guasch (2004) shows that price caps are a positive determinant of contract renegotiation (the same result is reported in Laffont, 2005).
7. In addition, average cost pricing has the desirable feature that it is consumer demand, and not public sector planning, the key factor in deciding which projects are built at all (this is the traditional Coasian argument in favour of average cost pricing; see Laffont, 2000).
8. The impact of infrastructure investment on growth can be high both if the investment is carried out by the public or the private sector. See for example, in the transport sector, Banister and Berechman (2000).
9. For the relationship between regulatory commitment and macroeconomic shocks, including exchange rate risk, see Harris (2003).
10. In the 2008–09 global financial crisis, Chile has been one of the countries that better resisted the recession, thanks to the ability of the Finance Minister to commit to keeping (resisting strong pressures) a budget surplus in the previous years, characterized by vigorous economic growth.
11. For references to specific universal service schemes targeted to the poor, see Estache et al. (2002).
12. Chisari et al. (2003) describe and evaluate the different universal service policies existing in Latin America.

REFERENCES

Andrés, L., Guasch, J.L., Diop, M. and López, S. (2007), 'Assessing the governance of electricity regulatory agencies in the Latin American and Caribbean region: a benchmarking analysis', mimeo.

Andrés, L., Guasch, J.L. and Azumendi, S.L. (2008), 'Regulatory governance and sector performance: methodology and evaluation for electricity distribution in Latin America', World Bank Policy Research Working Paper 4494.

Banerjee, A.V. (2004), 'Inequality and Investment', mimeo.

Banister, D. and Berechman, J. (2000), *Transport Investment and Economic Development*, London: UCL Press.

Basanes, F. and Willig, R. (eds) (2001), *Second-Generation Reforms in Infrastructure Services*, Baltimore, MD: John Hopkins University Press.
Benítez, D., Chisari, O. and Estache, A. (2003), 'Can the gains from Argentina's utilities reform offset credit shocks?', in C. Ugaz and C. Waddams-Price (eds), *Utility Privatization and Regulation. A Fair Deal for Consumers?* Cheltenham, UK and Northampton, MA, USA: Edward Elgar.
Bouille, D. and Wamukonya, M. (2003), 'Power sector reform in Latin America: a retrospective analysis', in M. Wamukonya (ed.), *Electricity Reform: Social and Environmental Challenges*, Roskilde, Denmark: United Nations Environment Programme, pp. 99–115.
Campos, J., Estache, A., Martín, N. and Trujillo, L. (2003), 'Macroeconomic effects of private sector participation in infrastructure', in W. Easterly and L. Servén (eds), *The Limits of Stabilization. Infrastructure, Public Deficits and Growth in Latin America*, Stanford, CA: Stanford University Press and the World Bank.
Chisari, O., Estache, A. and Waddams-Price, C. (2003), 'Access by the poor in Latin America's utility reform: subsidies and service obligations', in C. Ugaz and C. Waddams-Price (eds), *Utility Privatization and Regulation. A Fair Deal for Consumers?*, Cheltenham, UK and Northampton, MA, USA: Edward Elgar.
Corporación Latinobarómetro (2001), 'Latinobarómetro 2001', Santiago de Chile, at: www.latinobarometro.org (accessed 8 August 2009).
Corporación Latinobarómetro (2004), 'Latinobarómetro 2004', Santiago de Chile, at: www.latinobarometro.org (accessed 8 August 2009).
Delfino, J. and Casarin, A. (2003), 'The reform of the utilities sector in Argentina', in C. Ugaz and C. Waddams-Price (eds), *Utility Privatization and Regulation. A Fair Deal for Consumers?*, Cheltenham, UK and Northampton, MA, USA: Edward Elgar, pp. 149–74.
Easterly, W. and Servén, L. (eds) (2003), *The Limits of Stabilization. Infrastructure, Public Deficits and Growth in Latin America*, Stanford, CA: Stanford University Press and the World Bank.
Engel, E., Fischer, R. and Galetovic, A. (2003), 'Privatizing highways in Latin America: fixing what went wrong', *Economia: Journal of the Latin American and Caribbean Economic Association*, **4** (1), 129–58.
Engel, E., Fischer, R. and Galetovic, A. (2007), 'The basic public finance of public–private partnerships', mimeo, Cowler Foundation Decision Paper No. 1618.
Estache, A. and Wren-Lewis, L. (2009), 'Towards a theory of regulation for developing countries: following Jean-Jacques Laffont's lead', *Journal of Economic Literature*, **47** (3), 729–70.
Estache, A., Foster, V. and Wodon, Q. (2002), 'Accounting for poverty in infrastructure reform. Learning from Latin America's experience', Washington, DC: World Bank Institute.
Estache, A., Perelman, S. and L. Trujillo (2005), 'Infrastructure performance and reform in developing and transition economies: evidence from a survey of productivity measures', World Bank Policy Research Working Paper 3514.
Estache, A., Rossi, M.A. and Ruzzier, C. (2004), 'The case for international coordination of electricity regulation: evidence from the measurement of efficiency in South America', *Journal of Regulatory Economics*, **25** (3), 271–95.
Ferranti, D., de Perry, G., Ferreira, F. H.G. and Walton, M. (2003), 'Inequality in Latin America and the Caribbean. Breaking with History?', Washington, DC: World Bank.
Foster, V. (2005), 'Ten years of water service reform in Latin America, toward an Anglo-French model', Water Supply and Sanitation Sector Board Discussion Paper Series, 3.
Galiani, S., Gertler, P. and Schargrodsky, E. (2005), 'Water for life: the impact of the privatization of water services on child mortality, *Journal of Political Economy*, **113** (1), 83–120.
Gasmi, F., Noumba, P. and Recuero, L. (2006), 'Political accountability and regulatory performance in infrastructure industries: an empirical analysis', World Bank Policy Research Working Paper, 4101.
Gómez-Ibáñez, J.A. (2003), *Regulating Infrastructure. Monopoly, Contracts and Discretion*, Cambridge, MA: Harvard University Press.
Gómez-Lobo, A. (2007), 'Transantiago: una reforma en panne', *TIPS, 4*, Departamento de Economía, Universidad de Chile.
Guasch, J.L. (2004), 'Granting and renegotiating infrastructure concessions', *Doing It Right*, WBI Development Studies, Washington, DC: World Bank.
Guasch, J.L., Laffont, J.J. and Straub, S. (2002), 'Renegotiations of concession contracts in Latin America', World Bank Policy Research Working Paper 3011.
Gutiérrez, L. (2003), 'The effect of endogenous regulation on telecommunications expansion and efficiency in Latin America', *Journal of Regulatory Economics*, **23** (3), 257–68.
Harris, C. (2003), 'Private participation in infrastructure in developing countries', World Bank Working Paper No. 5.
Holburn, G. and Spiller, P. (2002), 'Institutional or structural: lessons from international electricity sector reforms', in E. Brousseau and J.-M. Glachant (eds), *The Economics of Contracts: Theories and Applications*, Cambridge: Cambridge University Press.

Jamasb, T. (2006), 'Between the state and market: electricity sector reform in developing countries', *Utilities Policy*, **14**, 14–30.
Kessides, I.N. (2004), 'Reforming infrastructure', *Privatization, Regulation and Competition*, Washington, DC: World Bank and Oxford: Oxford University Press.
Krueger, A. (ed.) (2000), *Economic Policy Reform: The Second Stage*, Chicago, IL: University of Chicago.
Laffont, J.-J. (2000), *Incentives and Political Economy*, Oxford: Oxford University Press.
Laffont, J.-J. (2005), *Regulation and Development*, Cambridge: Cambridge University Press.
Leipziger, Z., Estache, A. and Gómez-Lobo, A. (1999), 'Utility privatization and the needs of the poor in Latin America. Have we learned enough to get it right?', World Bank Working Paper No. 2407.
Martimort, D. and Straub, S. (2009), 'Privatization and changes in corruption patterns: the roots of public discontent', *Journal of Development Economics*, **90** (1), 69–84.
Montoya, M.A. and Trillas, F. (2009), 'The measurement of regulator independence in practice: Latin America and the Caribbean', *International Journal of Public Policy*, **4** (1/2), 113–34.
Parisi, F. and Yáñez, G. (2000), 'The deal of the century in Chile: Endesa España's takeover of Enersis', *International Review of Financial Analysis*, **9** (1), 103–16.
Parker, D. and Kirkpatrick, C. (2005), 'Privatisation in developing countries: a review of the evidence and the policy lessons', *Journal of Development Studies*, **41** (4), 513–41.
Sancho, D. and Jordana, J. (2004), 'Institutional constellations and regulatory policy', in Levy-Faur, D. and Jordana, J. (eds), *The Politics of Regulation*, Cheltenham, UK and Northampton, MA, USA: Edward Elgar, pp. 296–321.
Santhakumar, S. (2005), *Impact of the Distribution of the Cost of Reform on Social Support for Reforms: A Study of Power Sector Reforms in Indian States*, Gurgaon: India Development Foundation.
Shirley, M.M. (2008), *Institutions and Development. Advances in New Institutional Analysis*, Cheltenham, UK and Northampton, MA, USA: Edward Elgar.
Sioshansi, F.P. and Pfaffenberger, W. (2006), *Electricity Market Reform: An International Perspective*, Vancouver: Elsevier.
Spiller, P.T. and Cardilli, C.G. (1997), 'The frontier of telecommunications deregulation: small countries leading the pack', *Journal of Economic Perspectives*, **11** (4), 127–38.
Stern, J. (2009), 'The regulatory and institutional dimension of infrastructure services', CCRP Working Paper, 15, City University London.
Torero, M. and Pascó-Font, A. (2003), 'The social impact of privatization and the regulation of utilities in Perú', in C. Ugaz and C. Waddams-Price (eds), *Utility Privatization and Regulation. A Fair Deal for Consumers?*, Cheltenham, UK and Northampton, MA, USA: Edward Elgar.
Trillas, F. (2001), 'The takeover of Enersis: the control of privatized utilities', *Utilities Policy*, **10**, 25–45.
Ugaz, C. and Waddams-Price, C. (eds)(2003), *Utility Privatization and Regulation. A Fair Deal for Consumers?*, Cheltenham, UK and Northampton, MA, USA: Edward Elgar.
Victor, D.G. and Heller, T.C. (2007), *The Political Economy of Power Sector Reform: The Experiences of Five Major Developing Countries*, Cambridge: Cambridge University Press.

23 Reform of China's energy and telecommunications sectors[1]

Pun-Lee Lam and Alice Shiu

INTRODUCTION

China's rapid economic growth since its adoption of an open door policy in 1978 has brought about increasing demand for energy and telecommunications services. The development of these two sectors is of fundamental importance to the economic growth and development of China, as both sectors provide essential infrastructures to support the development of other industries. Previous studies have found a causal relationship between energy consumption and economic growth.[2] A similar relationship exists between telecommunications development and economic growth.[3] An insufficient energy and telecommunications infrastructure will therefore constrain the growth of many other industries. It will also discourage the wider use of consumer appliances, computer equipment and automobiles, and hence lower the living standard of the general population.

Despite the remarkable growth of the energy and telecommunications sectors over the past three decades, per capita energy consumption and teledensity in China are still far below the levels of developed countries. For strategic reasons, the two industries have remained under the ownership and control of the state. Even with China's accession to the World Trade Organization (WTO) in 2001, foreigners are still not allowed to be majority shareholders of telecommunications enterprises. Major energy and telecommunications enterprises are among the largest state-owned enterprises (SOEs) in China. However, the productivity of these enterprises is low compared with the productivity of their counterparts in other countries, and the problem of labour redundancy is serious. In recent years, China has occasionally experienced energy shortages that have severely disrupted the country's industrial output. In the telecommunications sector, the number of Internet users has increased substantially, but a great digital divide exists between urban and rural areas in China.

In November 2008, in response to the global financial crisis, the Chinese government announced a stimulus package of 4 trillion yuan (US$586 billion) to boost the economy over the next two years. This expenditure is equivalent to 16 per cent of China's gross domestic product (GDP), and will go mainly to new infrastructure projects, such as the construction of power grids, pipelines, and telecommunications networks. It is hoped that the new investment will not only help reduce the negative impact of the global economic slowdown, but also strengthen the infrastructure in the less developed regions of China.

This chapter examines the impact of the reform process on the development of China's energy and telecommunications sectors. First, the reform and development process of China's energy and telecommunications sectors is reviewed. Then, critical issues pertaining to infrastructure development in these two sectors over the past few decades are discussed. Finally, the prospects for the energy and telecommunications sectors are considered.

CHINA'S ENERGY AND TELECOMMUNICATIONS SECTORS

Since 1978, the Chinese government has implemented a series of economic reforms. However, for strategic reasons, the energy and telecommunications sectors remain under tight government control, and are dominated by SOEs. Productivity is low, and the problem of labour redundancy is serious. This section reviews the development of the two sectors over the past decades.

The Energy Sector

Since the Communist Party gained power in 1949, the development of China's energy sector has varied.[4] In the initial period of rehabilitation and reconstruction, major infrastructure facilities, such as electric power and transport systems, were revamped. In late 1952, the government commenced long-term economic planning, adopting the Soviet model of development. Heavy industries held a strategic position throughout the 1950s, and energy prices were kept low to speed up their development and that of other industries (Nolan and Wang, 1999). The government made great efforts to expand the electric power industry and built large generation equipment.

The havoc caused by the Great Leap Forward (1958–61) made the government realize that its strategies for accelerated growth in the heavy industry sector were impracticable, and the economy underwent a period of recovery and readjustment from 1961 to 1965. Then, China's coal industry was particularly hard hit by the political and social disturbances in the early years of the Cultural Revolution (1966–76). Production of coal dropped substantially, and the shortfall in the coal supply affected a number of industries, especially the electric power and the iron and steel industries (Thomson, 1996). Power shortages continued into the 1970s, which had an adverse effect on the outputs of many industries.

With its adoption of an open-door policy in 1978, the Chinese government encouraged foreign capital, technology and joint ventures that could assist the country's modernization. However, for strategic reasons, the energy sector remained under tight government control until recently. Since the mid-1990s, structural reforms, market incentives and decentralization policies have been introduced to attract foreign investment into the energy sector. In 1997, the government introduced a framework for the restructuring of the electric power industry that extends to 2020, and includes four steps: (1) separation of the regulatory and business functions of power enterprises, (2) vertical separation of generation from transmission, (3) national interconnection of power grids and (4) creation of a national power market.[5]

In 1998, the coal, oil and gas industries were restructured and enterprise reform was undertaken in the energy sector to improve the productivity performance of SOEs. In the Tenth Five-Year Plan (2001–05), entitled 'Revamping (or developing) the West', the government made a substantial investment in the development of energy resources (including hydropower, oil and gas) in West China and their transportation to the

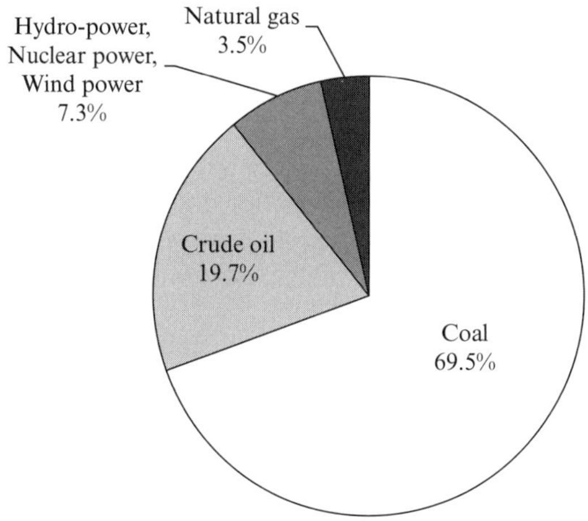

Figure 23.1 Energy consumption in China (2007)

coastal regions. In the current Eleventh Five-Year Plan (2006–10), great emphasis has been placed on energy conservation and sustainable development of China's energy resources.

Throughout China's rapid economic growth during the past three decades, coal has remained the country's dominant source of energy, and accounted for 70 per cent of total energy consumption in 2007 (see Figure 23.1).[6] A fundamental problem in the coal industry is the regional imbalance between coal production and consumption. Most of the coal reserves are in the Northern regions of China, whereas most of the consumption takes place in the coastal regions. As the coal producing and consuming regions are geographically separated, adequate provision of railway and coastal shipping facilities for coal transportation is crucial. Bottlenecks in transportation have raised the price of coal and affected the shipment of other commodities (Todd and Lei, 1994).

China's coal production can be divided into three main categories: state mines under central administration, local state mines and local non-state mines. Most of the non-state mines are small township and village enterprise (TVE) mines. Before the early 1980s, the annual output of coal from state mines was well above the output of local state or TVE mines. However, the output of TVE mines increased rapidly in the mid to late 1980s, and by the early 1990s, exceeded the output of state mines. The former mines competed with the latter for the coal markets and railway capacity for transportation (Wright, 2000). Unlike state mines, TVE mines did not bear many of the social costs of their operations or other welfare responsibilities. They produced a higher level of pollution and operated at a lower level of safety – the death rate in terms of deaths per million tonnes of output in TVE mines was higher than the figure for state mines. Finally, their low cost of operation and rapid expansion in coal production eroded the market share of state mines.

To avoid price-cutting in the coal market and improve the competitiveness of state mines, the central government attempted to restrict the activities of TVE mines by imposing safety and pollution controls. In 1996, the government's Coal Law came into effect. This law covers all aspects of coal production, including the planning for resource exploitation, approval of new mines, issuance of mining and production permits, safety standards and business management. In 1998, the State Council mandated the closure of 25 000 non-state mines and 40 state mines. Some provincial governments then prohibited power companies and other major users from purchasing coal from TVE mines. State mines were encouraged to merge with each other to increase efficiency, and some were transferred to a single mining group under the supervision of provincial governments (Andrews-Speed et al., 2000). These policies not only helped the state mines to regain a larger market share, but also resulted in the improvement of the quality and safety of coal production. Since the closure of so many coal mines, however, production has been unable to meet the growing demand, and coal prices have soared.

China's second-largest energy source is oil, which accounted for 20 per cent of total energy consumption in 2007 (see Figure 23.1). With the discovery of major oilfields in Northern China in the late 1950s, the country started the large-scale production of oil, and this production expanded rapidly during the 1960s and 1970s. In the 1980s, oil production slowed down, and China suffered a drastic drop in oil export earnings as oil prices plummeted. Despite the increase in oil demand, oil production in the country stagnated during the 1990s. As a result, China went from being an oil-exporting country to a net oil-importing country in 1993. Since then, its dependency on external sources of oil has increased. In 2007, more than half of the oil consumed in China was imported.

China is rich in natural gas resources, but produced virtually no gas before 1957. In contrast to its oil production, the country's natural gas production has increased steadily over time. Despite this continuous growth, natural gas accounted for less than 4 per cent of China's total energy consumption in 2007. However, the natural gas share of total energy consumption is expected to increase to 8 per cent by 2020. In future, new demand for natural gas will mainly come from power generation and domestic use, rather than from industrial use (Ogutcu, 1998).

The Chinese government considers the oil and gas industry to be one of the 'pillar' industries that play a leading role in the national economy. The petroleum and petrochemicals industry in China is organized into three segments: (1) onshore and offshore exploration and production, (2) manufacturing and refining of refined petroleum products and petrochemicals and (3) distribution, marketing and sales of refined petroleum products. In 1998, the government restructured the petroleum and petrochemicals industry by reorganizing the oil companies and separating the regulatory and business functions of the government. The oil and gas industry is now dominated by three major groups: China National Petroleum Corporation (CNPC), China Petrochemical Corporation (Sinopec) and China National Offshore Oil Corporation (CNOOC) (Sepulchre, 2004).

China National Petroleum Corporation and Sinopec are the only oil and gas companies engaged in all activities along the vertical production chain, which include onshore crude oil and natural gas exploration, refining, marketing and chemical operations in China. China National Petroleum Corporation is engaged in onshore crude oil and natural gas exploration and production in the Northern and Western regions, Sinopec is engaged in these activities in the Eastern, Southern, coastal and central regions in China and CNOOC is principally engaged in offshore exploration and production. All of the oil and gas reserves in the country are held by these three major groups and, among them, CNPC is the largest producer of crude oil and natural gas.

The electric power industry is the major user of coal and oil. Since 1949, the Chinese government has placed great emphasis on developing the power industry.[7] The industry expanded from the coastal to the interior regions and from the cities to the extensive countryside. In the 1980s and 1990s, following China's adoption of an open-door economic policy, structural reforms, market incentives and decentralization policies were introduced to attract local and foreign investment in the power sector, and independent power producers (IPPs) were allowed to enter the market. After the privatization of the industry in the late 1990s, capital was raised on the market to finance the expansion of China's generating capacity.

The Telecommunications Sector

For a long time, China's telecommunications sector was viewed as an instrument of government control and national defence. As such, its full development as a commercial service provider was not given priority (Xu and Pitt, 2002). Even after the adoption of an open-door policy in 1978, the telecommunications sector remained under tight control by the government and was not opened to foreign investors. The Ministry of Post and Telecommunications (MPT) and its operating arm, China Telecom, continued to control the whole post and telecommunications sector.

In the 1980s, incentive schemes and new accounting systems were introduced into the sector, but they failed to improve its operational performance. In spite of strong economic growth in China, the telecommunications sector remained underdeveloped, and the penetration rate of telephone services (measured by the number of telephone sets per 100 persons) increased only from 0.38 in 1978 to 1.11 in 1990 (see Table 23.1). The reform process did not really start until the early 1990s.

The MPT's monopoly position in telecommunications was first challenged by other ministries that manufactured equipment and owned private networks. Dissatisfied with the services provided by the MPT, the Ministry of Electronics Industry, in collaboration with the ministries of Electric Power and Railways and the People's Liberation Army, proposed the formation of the China United Communications Corporation (China Unicom) in 1992. Their proposal was approved by the State Council, and the company was formally established in 1994. The new company was allowed to utilize the existing private networks owned by other ministries to compete with China Telecom in the long-distance and mobile telephone markets. Another new company, the Ji Tong Communications Corporation (Ji Tong), was formed in 1993 to develop communications technology and provide network services for selected customers.[8] Hence, the reform of China's telecommunications sector could be considered to have started in 1994, with the entry of new players into the market.

As expected, the MPT tried to exert political influence to prevent such competition. Disputes between China Unicom and China Telecom over interconnection occurred from time to time. The MPT, which also acted as an industry regulator, was not impartial in the settlement of these disputes. Protracted negotiations over interconnection issues, high interconnection costs and delays in the delivery of services hampered the expansion of China Unicom. From 1994 to 1997, the company was still too small to challenge the incumbent. In 1997, the mobile business of China Telecom was listed in Hong Kong to capture foreign investment so as to facilitate its expansion.

As a result of the inability of the MPT to improve telecommunications services or maintain market discipline, a new round of reforms was started by the State Council

Table 23.1 Penetration rate of telephone services in China (1978–2007)

Year	Penetration rate
1978	0.38
1979	0.40
1980	0.43
1981	0.44
1982	0.46
1983	0.49
1984	0.53
1985	0.60
1986	0.67
1987	0.75
1988	0.86
1989	0.98
1990	1.11
1991	1.29
1992	1.61
1993	2.20
1994	3.20
1995	4.66
1996	6.33
1997	8.11
1998	10.53
1999	13.00
2000	20.10
2001	25.90
2002	33.60
2003	42.16
2004	50.03
2005	57.22
2006	63.40
2007	69.45

Note: Penetration rate is the number of telephone sets (including both fixed-line and mobile telephone sets) per 100 persons.

Sources: *China Statistical Yearbook*; *Yearbook of China Transportation and Communications*; Ministry of Information Industry website: http://www.miit.gov.cn.

in 1998. The reform process involved the separation of government and enterprise functions, the improvement of market competition and the ending of China Telecom's monopoly. The Ministry of Information Industry (MII) was formed to replace the MPT and charged with the responsibility of maintaining a level playing field in the telecommunications industry. China Telecom was divided into four separate groups: China Telecom, China Mobile, China Satcom (a provider of satellite services) and the Guo Xin Paging Company (see Figure 23.2).

Vertical divestiture along business lines allowed each of the four entities to concen-

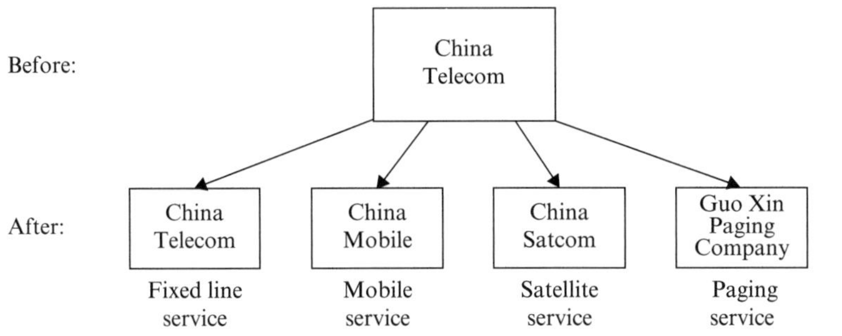

Figure 23.2 Vertical divestiture of China Telecom (1998)

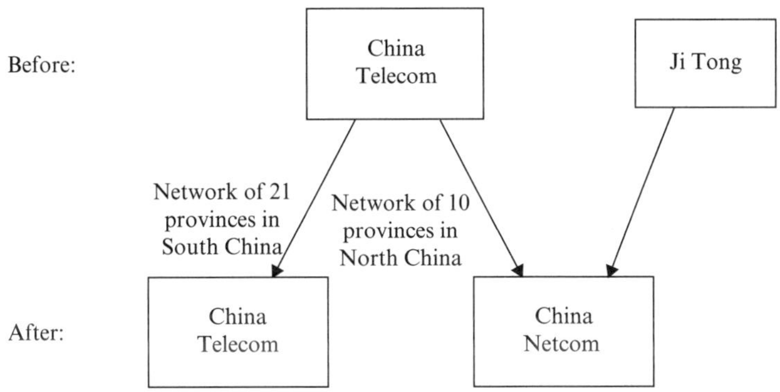

Figure 23.3 Horizontal divestiture of China Telecom (2001)

trate on its core competence without any regional barriers, as the companies were given permission to provide services across the whole country. The new China Telecom was allowed to keep the company name and concentrated on the fixed-line business after the restructuring. However, this new structure lasted for only a short period, and the company faced another divestiture a few years later.

In 2001, the State Council announced a horizontal divestiture of China Telecom (see Figure 23.3). The company was divided into north and south parts. The networks of ten provinces in Northern China were taken over by Netcom, a small Internet protocol telephone company with strong political support, to form the China Netcom Corporation (China Netcom). The assets of Ji Tong were merged with those of China Netcom, and China Telecom was entitled to retain the goodwill and control the remaining networks of 21 provinces in Southern China. The new organizational structure came into effect in May 2002. After these two rounds of divestiture, China Telecom no longer dominated the telecommunications market.

The divestiture of China Telecom and the subsequent listings of the major telecommunications companies on the stock markets helped the industry raise capital for expansion. After the vertical and horizontal divestitures, China Mobile, China Unicom, China Telecom and China Netcom became the four major players in the telecommunications market. The former two were predominantly mobile operators, whereas the latter two

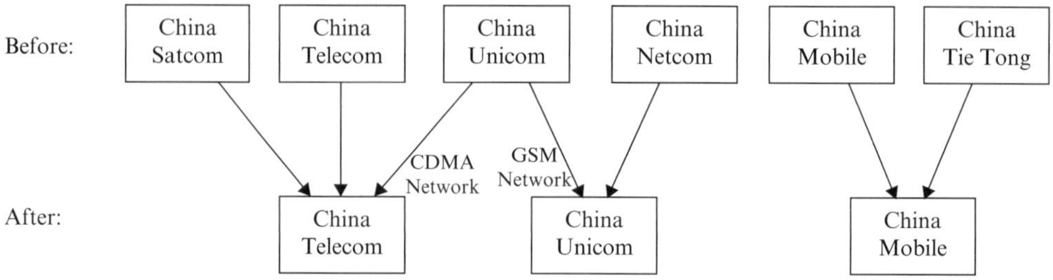

Figure 23.4 Restructuring of China's telecommunications sector (2008)

were fixed-line service providers. These four state-owned telecommunications companies were privatized one after another and listed on the stock markets. Two smaller players in the market were China TieTong and China Satcom.

In 2008, the Chinese government announced another restructuring of the telecommunications sector. A new ministry called the Ministry of Information and Information Technology (MIIT) was formed to replace the MII. The six market players were restructured again to form three new companies (see Figure 23.4). China Telecom acquired China Satcom and China Unicom's CDMA network. China Unicom's global system for mobile communications (GSM) network was incorporated into China Netcom and a new China Unicom was established. China Mobile took control of China TieTong, a small fixed-line operator in China. After the restructuring, each of the three new operators will provide both fixed-line and mobile services in China, and the government will issue 3G licences to them. It is hoped that the restructuring process will increase the competitiveness of China Telecom, China Mobile and China Unicom and improve the productivity of the country's telecommunications sector.

As discussed, the Chinese government has introduced a number of policies to reform the telecommunications sector since the 1990s. The country's entry into the WTO in 2001 hastened the reform process and stimulated the development and growth of the telecommunications sector. Since 2001, this sector has expanded significantly, and the number of subscribers has increased rapidly over the last few years. In 2003, the number of subscribers of mobile services surpassed that of fixed-line services. Table 23.1 shows that the penetration rate of telephone services has increased remarkably since the early 1990s, and reached about 70 per cent by 2007. The growth in the penetration rate was much faster than the economic growth (around 10 per cent) over the same period.

The rapid development of China's telecommunications sector in recent years can be attributed to not only the reform policies introduced by the government but also technological 'leapfrogging' (Dai, 2000, 2003; Mu and Lee, 2005). China's telecommunications infrastructure, as in other developing countries, was at a low level of development in the 1980s. However, this resulted in a latecomer's advantage when the country developed information superhighways in the 1990s. Instead of going through the costly process of re-engineering the old analogue copper wire network, China got a jump-start by deploying the latest technologies to develop a nationwide fibre-optic cable network. This 'leapfrogging' strategy allowed the country to skip the intermediate stages of copper wire and analogue telephone systems and facilitated the rapid expansion of the telecommunications sector (Dai, 2003).

REFORM OF ENERGY AND TELECOMMUNICATIONS ENTERPRISES

China's energy and telecommunications industries were under state ownership and control for a long time. As the primary objective of the enterprises in the two sectors was not profit maximization, they were very unproductive compared with their counterparts in other countries. This section reviews the approaches adopted by the Chinese government to reform state-owned energy and telecommunications enterprises. The goal of the reform process was to improve the productivity performance of the two sectors.

China's State-owned Enterprises

After the Communist Party took power in 1949, a system of SOEs was established in China. These enterprises were the production units of the country and received capital from the government for their establishment and operation. They did not have a clear legal status, and their aim was not profit maximization but rather meeting the output targets set by the authorities.[9]

Key and backbone enterprises, including those in the energy and telecommunications sectors, were under the control of the central government. A number of large SOEs were charged with the responsibility of developing the infrastructure of the whole country, and long-term development projects such as the construction of electric power systems, pipelines and telecommunications networks were funded by the central government. The provincial and local governments provided financial support to SOEs of smaller sizes.

It is argued that only SOEs with government financial support would take on projects that were not commercially viable. A state monopoly could take on universal service obligations and ensure that infrastructure facilities and services would be provided to remote rural areas. For this reason, the SOEs in the energy and telecommunications sectors were granted monopoly rights to explore and extract energy resources, construct power systems and expand the telecommunications networks throughout China.

Hence, the state sector dominated the Chinese economy for a long time. State-owned enterprises had to shoulder the social responsibility of creating employment and providing employees with various kinds of benefits and services including housing, education and medical care. As a result of this social burden, energy enterprises in China were grossly unproductive compared with their counterparts in other countries. Many of the SOEs were debt-ridden and suffered chronic problems of mounting losses, huge borrowing and low productivity. The problem of labour redundancy was very serious. To assist the development of other industries, the prices of energy and telecommunications services were kept at low levels. Because the prices were below production costs, the SOEs in the energy and telecommunications sectors suffered heavy losses and had to rely on government subsidies.

Reform of State-owned Enterprises

The reform of the state sector started in the early 1990s and reached a critical point in the late 1990s. The reform process can be divided into three different stages, as follows.

First stage: entry of new players

The reform of the energy and telecommunications sectors started with the entry of new players to challenge the incumbents. In the 1980s and 1990s, the state-owned power enterprises were slow in expanding their generating capacity and could not meet the growing demand for electricity. Power shortages existed in many regions of China, which hampered economic growth and development in these areas.

To alleviate the power shortage problem, the central government allowed IPPs to enter the generation market and local governments and power enterprises to cooperate with foreign investors to form IPP companies. In 1985, a Hong Kong company called Hopewell formed a joint venture with the local government to build a power plant in Shenzhen, one of the four special economic zones in China. In 1994, a major IPP, Huaneng Power International Company, was established to operate and manage large power plants throughout China, and challenged the monopoly power of local state-owned power enterprises in different regions of the country.

New players also entered the telecommunications arena. As mentioned, Ji Tong was formed in 1993 to develop communications technology and provide network services for target customers. In 1994, China Telecom's monopoly position in telecommunications was challenged by the establishment of China Unicom. The new company was allowed to utilize the existing private networks owned by other ministries to compete with China Telecom in the long-distance and mobile telephone markets. In 1999, with the political support of the central government, China Netcom was formed as a wholesaler for the high-speed data networks. The entry of new players hastened the reform process of China's telecommunications sector.

Second stage: industry restructuring

In the past, the Chinese government functioned as both a regulator of the economic sector and as an owner and operator of state-owned enterprises. To facilitate the entry of new competitors, the regulatory function of SOEs was separated from their enterprise or business function. The aim was to provide a level playing field for both the incumbents and new entrants. In addition, to reduce the market power of the incumbents, the energy and telecommunications sectors were restructured or reorganised through divestitures (both horizontal and vertical).

Consider the electric power industry. In 1997, the State Council formed the State Power Corporation of China (SPC). The management of state-owned power enterprises and their assets, which had formerly been undertaken by the Ministry of Electric Power (MEP), was transferred to the SPC upon its formation. The MEP continued to play a regulatory role in the industry. Most of the regional and provincial power corporations fell under the control of the SPC, and the company owned all of the transmission networks and a large proportion of the distribution facilities in China. Other IPPs sold electricity to the SPC for transmission and distribution, or directly to local distributors (Local DisCo) in urban and rural areas (see Figure 23.5).

In 2002, the State Council announced a major restructuring of China's electric power industry: the SPC would be divided into separate regional transmission companies and required to divest its generating facilities (Ma and He, 2008). Two major regional transmission companies were formed, one in the south and the other in the north of

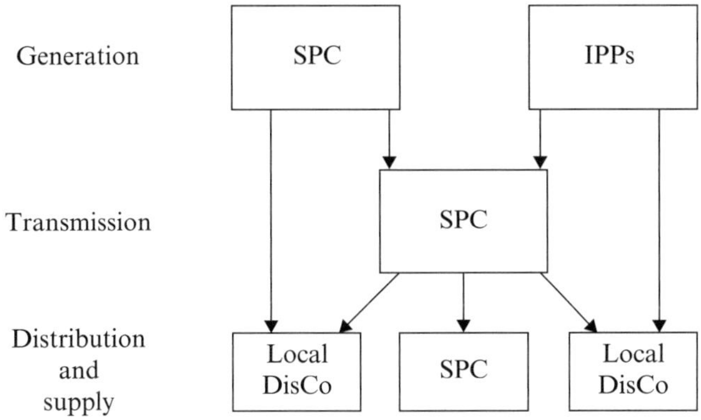

Figure 23.5 The structure of China's electric power industry under the State Power Corporation (1997)

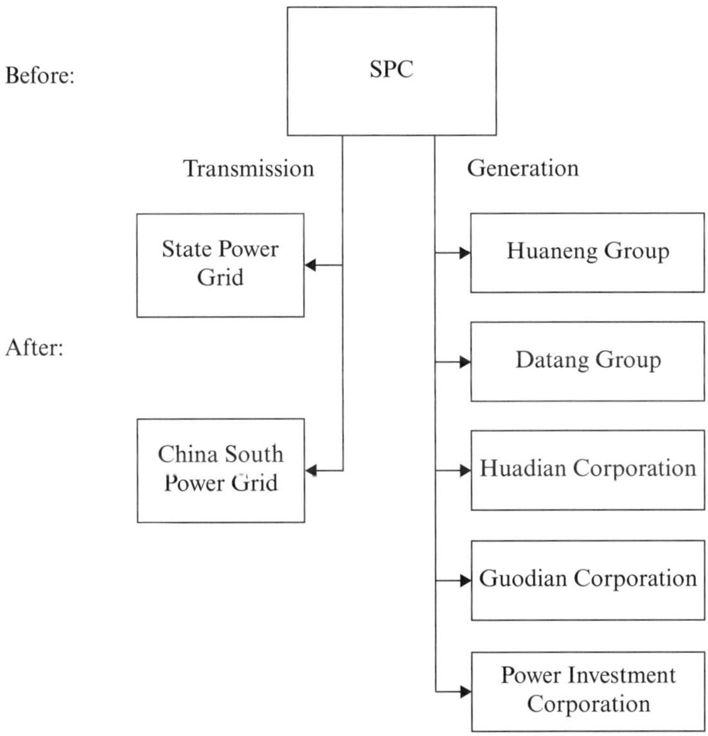

Figure 23.6 Divestiture of the State Power Corporation (2002)

China. The power generation assets of the SPC were divested and transferred to five independent generating companies (see Figure 23.6).

In the telecommunications sector, the MPT had played the dual roles of industry regulator and player for some time. Following the entry of new competitors into the market in the early 1990s, the MPT was criticized for protecting the interests of its operating arm, China Telecom. It imposed high interconnection costs on competitors and delayed their entry into the market. In 1998, the MII was formed to replace the MPT and charged with the responsibility of maintaining a level playing field in the telecommunications industry. China Telecom was divided into four separate companies along business lines. As discussed in the previous section, the telecommunications sector underwent a second and third round of restructuring in 2001 and 2008, respectively.

Third stage: corporatization and privatization

In the past two decades, the Chinese government has introduced various measures to reform the ownership structure of SOEs. In 1988, the National People's Congress enacted the Law on Industrial Enterprises Owned by the Whole People. SOEs were granted independent legal status and the right to make their own managerial decisions to protect them from any government intrusion. They were also required to take full responsibility for profits and losses and practise independent accounting.

In July 1994, the first Company Law in China came into effect. Its primary purpose was to standardize and modernize enterprise law for the state and non-state sectors (Chow, 1997; Smyth, 1999). It was designed to restructure SOEs using international capitalist models of the corporate form. SOEs were treated as independent legal entities, separated from the state, with the power to manage their assets. The Company Law contemplated several kinds of corporations, including limited liability, wholly state-owned joint stock and majority state-owned joint stock companies. A joint stock company could issue shares, which could be listed on China's two stock exchanges or foreign stock exchanges.

After the introduction of the Company Law, many of the SOEs in the energy and telecommunications sectors were corporatized and converted into 'independent corporations' that were responsible for economic management and planning in their respective regions. After restructuring, all energy and telecommunications corporations were supposed to be operated on commercial principles, and no longer received large government subsidies.

In 1995, the government introduced the 'grab the large and release the small' policy. The aim of the policy was to consolidate top and key SOEs and put small and medium-sized enterprises on the market. The central government would retain ownership and control over large and super-large SOEs, including those in the energy and telecommunications sectors, and turn them into world-class enterprises. To obtain new capital for expansion, large energy and telecommunications corporations issued shares to private shareholders, and publicly traded shares became available to domestic and foreign investors through initial public offerings (IPOs). It was hoped that corporatization and listing on the stock markets would create greater pressure on SOEs in the energy and telecommunications sectors to improve productivity.

State-owned energy and telecommunications enterprises were restructured before going public. Historical burdens such as non-performing assets (for example, schools, hospitals and restaurants) and redundant personnel were taken over by the original enterprises, whereas productive assets (including manpower) were transferred to the newly established enterprises to ensure the success of their IPOs. The restructuring process improved the financial performance of the SOEs and facilitated their listing on the stock markets.

In the telecommunications sector, as a result of a lack of capital funds, the expansion of telecommunications networks was limited to the major cities along the Eastern coast during the early period of development in the 1990s. To facilitate IPOs, the more productive telecommunications assets (that is, facilities and services) in the Eastern and coastal provinces were first injected into the listed companies. As the telecommunications systems were better developed in the Eastern and coastal regions, the productivity of enterprises in these provinces was much higher and such an arrangement was intended to attract foreign investors to buy shares of the listed companies (Lam and, Shiu 2008).

Later, additional capital was raised in the stock markets to acquire and expand telecommunications networks in other, less productive provinces in the central and Western regions. This strategy has facilitated investment in and expansion of a national telecommunications network in China over the last decade. A similar strategy has been used in the energy sector. The capital raised by listed companies in overseas stock markets has been used to acquire productive assets from their parent companies in China.

FUTURE CHALLENGES AND PROBLEMS

After experiencing a period of rapid growth and development, the energy and telecommunications sectors in China are currently under consolidation. After restructuring, the power of incumbents such as China Telecom and SPC has been greatly reduced. The priority of reform has shifted from industry restructuring to other goals, such as sustainable development and improvement of the technology and governance structure of the two sectors. This section discusses the future challenges and problems faced by China's energy and telecommunications sectors.

Sustainable Development of Energy Resources

China's great reliance on coal fuel has lowered its energy efficiency. Because of the widespread use of low-quality coal for power generation and other domestic purposes, the thermal efficiency of China's energy sector is lower than that of many developed countries. The country's biggest consumer of coal is the electric power industry. The predominance of coal-fired thermal plants in China has caused serious local and regional pollution problems. Coal combustion leads to the emission of harmful chemicals, including nitrogen oxide (NO_X), carbon dioxide (CO_2) and sulphur dioxide (SO_2), and particulates. These pollutants affect not only China but also neighbouring countries such as Korea and Japan (Byrne and Shen, 1998).

Energy shortages are another major problem affecting the future growth of China, which is currently the world's second-largest energy producer and consumer. Despite the rapid growth in energy production in the last two decades, production has been lagging behind economic growth. Since 1993, China has been a net importer of oil, and the amount of imported crude oil has increased over time. In 2007, more than half of the oil consumed was imported from other countries. China's increasing reliance on foreign sources will make it vulnerable to fluctuations in world oil markets and international political instability, and increases in the import prices of oil will adversely affect its balance-of-payment position and domestic prices.

Although China's power generation capacity has increased substantially, the problem of power shortages still exists in many of its Eastern and Southern regions. As the industrial sector consumes around 75 per cent of China's electricity, continued growth in this sector requires that the country speed up the expansion of its electric power sector. The impact of power shortages on industrial development is significant, as such shortfalls create a bottleneck in economic development, which in turn retards industrial production.

As rapid economic growth has brought about a growing demand for energy, China faces the great challenge of expanding energy resources to meet future demand. Government control of energy prices is considered to be a major cause of the country's energy shortage problem. In recent years, energy prices have been allowed to increase to levels closer to world prices. However, from time to time, the Chinese government has reduced the pace of price reform in the energy sector for fear of inflation. State-owned oil and power enterprises have complained that they are not allowed to raise the prices of oil products and electricity tariffs to reflect the increases in fuel costs. The result is the deterioration of the financial performance of these companies.

In the current Eleventh Five-Year Plan (2006–10), China places great emphasis on the building of an energy-saving and environmentally friendly society and the sustainable development of the economy. The Plan projects that by 2010, energy consumption per unit of GDP will have decreased by 20 per cent, and the total amount of major pollutants discharged will have been reduced by 10 per cent. Rigorous energy conservation measures have been proposed to meet these ambitious targets. Beyond the current Five-Year Plan, China aims at increasing renewable energy consumption from 10 per cent of total energy consumption by 2010 to 15 per cent by 2020 (State Council Information Office, 2007). In 2005, the Renewable Energy Law was passed to provide the legislative foundation and articulate the development direction for the country's renewable energy industry.[10]

As the problems of pollution and global warming affect the well-being of all people and are not confined to individual territories or countries, China needs to design incentive policies to promote energy efficiency and renewable energy development, and enhance its cooperation with the international community. The reduction in the use of low-quality coal and improvement of the energy efficiency of coal-fired plants are of vital importance to the country's energy efficiency. To promote energy conservation and sustainable development of the energy sector, the Chinese government must continue its efforts to implement price reform in the energy sector.

Development of the Information Infrastructure in Rural Areas

With the introduction of regulatory reforms and the launch of policies that have encouraged investment, China has achieved tremendous telecommunications development over the past decade. However, the market-oriented restructuring of the sector, along with the country's accession to the WTO, has resulted in a digital divide between the urban and rural areas of the nation. This digital divide should be given special attention as its widening could worsen the existing imbalance of social and economic development in China.[11]

The urban/rural digital divide is apparent in certain areas of information and com-

Table 23.2 Number of fixed-line subscribers and penetration rate in urban and rural China (five-year interval from 1965–1995, 1996–2007)

Year	Total number of urban subscribers	Total number of rural subscribers	Urban penetration rate	Rural penetration rate	Ratio of urban to rural penetration
1965	771 104	492 222	0.59	0.08	7.14
1970	784 130	527 365	0.54	0.08	7.07
1975	1 032 827	659 220	0.64	0.09	7.47
1980	1 341 715	799 036	0.70	0.10	6.98
1985	2 189 554	930 744	0.87	0.12	7.57
1990	5 384 494	1 465 809	1.78	0.17	10.24
1995	32 635 600	8 070 000	9.28	0.94	9.88
1996	42 778 000	12 169 000	11.90	1.41	8.45
1997	52 444 000	17 866 000	14.18	2.06	6.88
1998	62 598 000	24 823 000	16.50	2.86	5.77
1999	74 633 000	34 084 000	19.19	3.92	4.90
2000	93 116 000	51 713 000	20.31	6.40	3.17
2001	111 937 000	68 431 000	23.29	8.60	2.71
2002	135 791 000	78 431 000	27.04	10.02	2.70
2003	171 097 000	91 650 000	32.67	11.93	2.74
2004	210 251 000	101 505 000	38.73	13.41	2.89
2005	239 753 000	110 692 000	42.65	14.85	2.87
2006	251 329 000	116 456 000	43.55	15.79	2.76
2007	248 598 000	117 040 000	41.87	16.09	2.60

Note: Penetration rate is the number of fixed-line telephone sets per 100 persons.

Sources: *China Statistical Yearbook*; *China Posts and Telecommunications Annual Report*; website: http://china.org.cn.

munications technology (ICT), including fixed-line connections, mobile communications and Internet penetration. In 2007, the rural fixed-line penetration rate was only 16.1, whereas the urban telephone penetration rate was 41.9 (see Table 23.2).

Similarly, for mobile communications, although the penetration rate had reached 41.6 by the end of 2007, the rural penetration rate was only 12. The more developed coastal provinces have much higher mobile penetration rates than have the less developed Western provinces. The urban/rural gap in Internet penetration is even wider. The total number of Internet users in China reached 162 million by mid-2007, but the Internet penetration rate in rural areas was only 5.1 compared to 21.6 in urban areas. Once again, the more developed coastal provinces have much higher Internet penetration rates than have the less developed Western provinces.

In the past, because of the low population density and geographical disadvantages of rural areas, the expansion of telecommunication services into these regions was not a priority. In particular, when competition intensified in China's telecommunications market, industry incumbents tended to duplicate the construction of networks in developed, urban areas rather than cross-subsidizing and expanding services into the less profitable rural areas. In addition, the switch from debt financing and reliance on installation fees to equity financing prevented telecommunications operators from extending their services to the rural areas.[12]

In the current Eleventh Five-Year (2006–10) Plan, the Chinese government pays special attention to the social and economic development of rural areas, with the aim to reduce poverty, inequality and economic exclusion. As the expansion of telecommunications services into rural areas plays a critical role in promoting equitable social and economic development, rural communications has become the key focus of 'universal service'.

Universal service provision received prominent attention in China only after market-oriented restructuring had taken place. In the Chinese context, universal service means basic access to telephone services in rural and high-cost areas, or more precisely, 'telephones to every village'.[13] However, given rapid changes in technology and the markets, universal service objectives now include mobile and Internet services and other advanced ICTs. The provision of universal service has long been a challenge for China. Before 1994, the monopoly provision of telecommunication services meant that universal service obligations in the country were borne by a sole operator, China Telecom. The company was required by the government to extend services to rural areas, and was permitted to impose extra charges for other services to cross-subsidize such expansion. However, this internal cross-subsidization mechanism no longer functioned once competition was introduced.[14] In 2004, in the absence of a proper support mechanism for the provision of universal service, the MII decided to distribute the tasks of universal service provision amongst all of the operators via the Village Access Project (VAP).[15]

The VAP was initiated with the aim to fulfil the Tenth Five-Year Plan target of provision of telephone access to 95 per cent of all administrative villages.[16] All of the operators were required to share the obligations of universal access based on the revenues and geographical divisions of their networks. However, in the absence of effective enforcement, opportunistic behaviour including the suspension or delay of services occurred frequently.[17] Although the results of the VAP in recent years have been declared successful by the MII, there are still many challenges ahead in the provision of universal service to rural populations throughout China.

Geographical isolation and low population density are the main problems leading to the high cost of providing wire-line connections to rural areas. Wireless connectivity, which requires relatively less terrestrial infrastructure, may help circumvent the cost of infrastructure development and facilitate the bridging of the urban/rural digital divide in China. In particular, it is expected that the development of 3G mobile technology could help the country reach the objective of universal service. In 2008, the restructuring of the telecommunications industry paved the way for China's 3G launch.[18] 3G wireless networks have the potential to help China extend telecommunications services to rural areas as signals remain strong over long distances, with the lower user density mitigating the reduced throughput.

However, in addition to the difficulty of raising capital to roll out 3G networks, telecommunications operators need to pay attention to the cost of mobile phone ownership, access and connection fees for the Internet and pricing strategies under the new regulatory environment, as these are the crucial factors that will determine the affordability of 3G mobile technology and its success in bridging the urban/rural digital gap.

Improvement in Governance Structures

Although all major energy and telecommunications enterprises are corporatised and listed in the stock exchanges, political intervention in their management and business is very common (for example, appointment of senior personnel, imposition of price controls and monopoly profit taxes, subsidization of state-owned enterprises, divestitures and mergers). To ensure the smooth functioning of these enterprises, the Chinese government can borrow from the experience of other countries to further improve their regulatory and corporate governance structures.

In the reform of the energy sector, there has been a problem in the delineation of the responsibilities of the central and local governments. Provincial governments have gradually increased their influence over energy enterprises while the central government has lost its ability to enforce regulations and make policy decisions (Andrews-Speed et al., 2000). Regulatory staff members at the local level are usually employed by the local government but asked to enforce regulations imposed by the central government. The policy and regulatory functions of various commissions of the State Council and local government agencies are not clearly defined. Some groups are reluctant to fulfil their responsibilities and compete for resources. One example is the poor enforcement of safety regulations and closure of illegal mines owing to conflicts of interest between the local and central governments.

The corporatization of SOEs has exacerbated agency problems in companies in the energy and telecommunications sectors. China's enterprises continue to allow the government to play a role in their operation despite the efforts made to free enterprises from government intervention. At present, large-scale SOEs still lack independent corporate governance such as a board of directors that has power to appoint senior management. Appointments to senior positions are made by government officials who might have close relations with the enterprises. In the telecommunications sector, the government occasionally reshuffles the chairmen or chief executive officers among competitive enterprises. In 2008, the Chinese government instructed China Mobile to transfer a large amount of capital to assist competitors to develop their networks. In the energy sector, the practice of simultaneous employment in the government, board of directors, and enterprise groups is still common. All these practices have hampered the enterprise reforms, which are meant to separate enterprise and government functions. They have also lowered the incentive of telecommunications enterprises to improve managerial performance. Without appropriate corporate governance mechanisms, enterprise autonomy may accelerate the loss of state assets and provide more opportunities for rent-seeking activities.

Hence, it is important for the Chinese government to provide a clear regulatory framework that governs the energy sector. Andrews-Speed et al. (2000) proposed that China follows the federal system in place in the USA and Canada, in which the regulatory policies of the central government (related to safety standards and inter-provincial issues) and local agencies are clearly defined and enforced. A unified regulatory authority governing all energy matters at the state and provincial levels should also be considered (Ma and He, 2008). For the telecommunications sector, the government needs to establish a regulatory framework that provides incentive mechanisms that promote greater transparency, consistency and broader stakeholder participation in the regulatory process. This can help to limit the power of special limited groups in the reform process.[19]

It is argued that to avoid undue political influence on management, the government has to divest its shareholding in large-scale energy enterprises. The ownership and income rights of shareholders have to be defined clearly so that the contractual parties are protected and management cannot use profits from listed companies to subsidize other unproductive businesses. There is also a need to tie the compensation of managers of corporatised enterprises to their performance. One alternative is to diversify the ownership of a state enterprise into the hands of several large minority shareholders to avoid monopoly control; this will create an incentive for each of these shareholders to monitor the performance of managers.

To sum up, in the process of transforming energy and telecommunications enterprises from state into legal entities, the roles, ownership rights and relations among the state, industrial sectors and enterprises should be redefined clearly.

CONCLUSION

After 30 years of rapid economic growth, China has developed a basic energy and telecommunications infrastructure. However, the tremendous development of the two sectors has given rise to various problems. In the energy sector, major problems include energy shortages, unreliable energy supplies, bottlenecks in energy transmission, and distribution and adverse environmental impacts owing to heavy reliance on coal fuel. These problems pose serious threats to China's sustainable development. In the telecommunications sector, the major problem is the existence of an urban/rural digital divide, and the widening of this divide will worsen the existing imbalance of social and economic development.

To overcome these problems, the Chinese government should formulate appropriate policies to restore energy balance and encourage the wider use of gas and renewable resources to reduce the environmental damage created by the energy sector. Regarding the telecommunications sector, the Chinese government should speed up the licensing process and rollout of 3G networks that could provide greater access to wireless telecommunications services for rural populations and thereby bridge the digital divide. The global financial crisis provides an opportunity for China to increase investment in its energy and information infrastructure, the improvement of which would allow more of the Chinese population to enjoy the benefits of the nation's economic growth and development.

NOTES

1. The work described in this chapter was supported by a research grant from the Research Grants Council of Hong Kong (CERG PolyU 5404/05H; Project Account Code: B-Q906).
2. Refer to Shiu and Lam (2004) for a summary of previous studies of the causal relationship between energy consumption and economic growth.
3. Refer to Shiu and Lam (2008) for a summary of previous studies of the causal relationship between telecommunications development and economic growth.
4. Refer to Dorian (1998) for a thorough discussion of the historical development of the energy sector in China since 1949. In addition, the development strategies of the Chinese government in the 1950s and 1960s and their effect on the energy industry can be found in Lin et al. (1996).

5. Refer to the China Electric Power Information Centre (1997–2002) for details about the restructuring process.
6. Refer to Lam (2005) for details about the energy consumption patterns and structure of the energy sector in China.
7. Refer to Lam (2004) for a detailed discussion of the development of China's electricity industry.
8. Refer to Mueller and Tan (1997), Xu and Pitt (2002), Lu and Wong (2003) and Li and Wang (2009) for a detailed discussion.
9. Refer to Zhou (2006) for a detailed discussion of China's SOEs.
10. Refer to Ma and He (2008) for a detailed discussion of the future development of China's renewable energy sources.
11. Refer to Wang (2001) for a detailed discussion of why a growing digital divide should be cause for concern.
12. Refer to Xia and Lu (2005) and Harwit (2008) for more detailed discussions of barriers to rural access.
13. Refer to Zhao (2007) for a thorough explanation of political, economic, social and right-based arguments for universal service in China.
14. Refer to Li and Wang (2009) for a detailed discussion of the mechanism for financing a universal service.
15. Refer to MII (2006, 2007a, 2007b) for updated reports on the current development of the VAP.
16. Given the importance of the Internet, the government has also set a goal of providing Internet access to all villages by 2010. See World Bank (2007).
17. Refer to Xia and Lu (2005) for a detailed discussion of the VAP.
18. Refer to chapter 3 of World Bank (2007) for more information of the construction of 3G telecommunications networks.
19. Yu et al. (2004) proposed a new regulatory framework for the future development of China's telecommunications sector.

REFERENCES

Andrews-Speed, P., S. Dow and Z. Gao (2000), 'The ongoing reforms to China's government and state sector: the case of the energy industry', *Journal of Contemporary China*, **9** (23), 5–20.

Byrne, J. and B. Shen (1998), 'The challenge of sustainability: balancing China's energy, economic and environmental goals', *Energy Policy*, **24** (5), 455–62.

China Electric Power Information Centre (1997–2002), *Electric Power Industry in China 1997–2002*, Beijing: China Electric Power Information Centre.

China Posts and Telecommunications Annual Report, various issues, Beijing: China Posts and Telecommunications Editorial Group of MPT.

China Statistical Yearbook, various issues, Beijing: China Statistical Press.

Chow, D.C.K. (1997), 'An analysis of the political economy of China's enterprise conglomerates: a study of the reform of the electric power industry in China', *Law and Policy in International Business*, **8** (2), 383–433.

Dai, X.D. (2000), *The Digital Revolution and Governance*, Aldershot: Ashgate.

Dai, X.D. (2003), 'ICT in China's development strategy: implications for spatial development', in C.R. Hughes and G. Wacker (eds), *China and the Internet: Politics of the Digital Great Leap Forward*, London and New York: Routledge Curzon, pp. 8–29.

Dorian, J.P. (1998), *Energy in China: Poised for the 21st Century*, London: Financial Times Energy.

Harwit, E. (2008), *China's Telecommunications Revolution*, New York: Oxford University Press.

Lam, P.L. (2004), 'Pricing of electricity in China', *Energy*, **29** (2), 287–300.

Lam, P.L. (2005), 'Energy in China: development and prospects', *China Perspective*, **59** (May–June), 14–25.

Lam, P. L. and A. Shiu (2008), 'Productivity analysis of the telecommunications sector in China', *Telecommunications Policy*, **32** (8), 559–71.

Li, M. and J. Wang (2009), 'China's telecommunications industry: transformation and challenges', in P. Wilson (ed.), *Economic Policies and Social Welfare in the 21st Century: Challenges and Responses for China and Thailand*, Singapore: Cengage Learning Asia Pte Ltd, pp. 207–20.

Lin, J.Y., F. Cai and Z. Li (1996), *The China Miracle: Development Strategy and Economic Reform*, Hong Kong: Chinese University of Hong Kong Press.

Lu, D. and C. K. Wong (2003), *China's Telecommunications Market*, Cheltenham: Edward Elgar.

Ma, C. and L. He (2008), 'From state monopoly to renewable portfolio: restructuring China's electric utility', *Energy Policy*, **36** (5), 1697–711.

Ministry of Information Industry (MII) (2006), *The MII sets new goal for the VAP in 2006*, World Wide Web page: http://www.miit.gov.cn/art/2006/02/23/art_21-6906.html (in Chinese, accessed 5 October 2008).

Ministry of Information Industry (MII) (2007a), *Briefing on the VAP*, No. 1 of 2007, World Wide Web page: http://www.miit.gov.cn/attach/0/080116153853981.doc (in Chinese, accessed 5 October 2008).
Ministry of Information Industry (MII) (2007b), *Briefing on the VAP*, No. 2 of 2007, World Wide Web page: http://www.miit.gov.cn/attach/0/070612161844798.doc (in Chinese, accessed 5 October 2008).
Mu, Q. and K. Lee (2005), 'Knowledge diffusion, market segmentation and technological catch-up: the case of the telecommunication industry in China', *Research Policy*, **34** (6), 759–83.
Mueller, M. and Z. Tan (1997), 'China in the information age: telecommunications and the dilemmas of reform', *Journal of Comparative Economics*, **25** (3), 465–67.
Nolan, P. and X. Wang (1999), 'Beyond privatization: institutional innovation and growth in China's large state-owned enterprises', *World Development*, **27** (1), 169–200.
Ogutcu, M. (1998), 'China and the world energy system: new links', *The Journal of Energy and Development*, **23** (2), 281–326.
Sepulchre, A. (2004), 'Energy and globalisation: oil and gas in China', *China Perspective*, **59** (July–August), 17–28.
Shiu, A. and P.L. Lam (2004), 'Electricity consumption and economic growth in China', *Energy Policy*, **32** (1), 47–54.
Shiu, A. and P.L. Lam (2008), 'Causal relationship between telecommunications and economic growth in China and its regions', *Regional Studies*, **42** (5), 705–18.
Smyth, R. (1999), 'Enterprise bankruptcies and the restructuring of China's state-owned sector: a review of some recent problems and prospects for future success', *China Information*, **13** (4), 66–82.
State Council Information Office (2007), *China's Energy Conditions and Policies*, White Paper on Energy, Beijing: State Council Information Office.
Thomson, E. (1996), 'Reforming China's coal industry', *The China Quarterly*, **147** (September), 726–50.
Todd, D. and Z. Lei (1994), 'Ports and coal transfer: hub of China's energy supply policy', *Energy Policy*, **22** (7), 609–22.
Wang, W. (2001), 'Bridging the digital divide inside China', paper presented at the 3rd Annual Conference of the Association of Internet Researchers, 13–16 October, Maastricht, The Netherlands.
World Bank (2007), *China's Information Revolution – Managing the Economic and Social Transformation*, Washington, DC: World Bank.
Wright, T. (2000), 'Competition and complementarity: township and village mines and the state sector in China's coal industry', *China Information*, **14** (1), 113–30.
Xia, J. and T.J. Lu (2005), 'Universal service policy in China: building digital bridge for rural community', paper presented at the 33rd Conference on Communication, Information and Internet Policy, 23–25 September, George Mason University School of Law, Arlington, VA.
Xu, Y. and D. C. Pitt (2002), *Chinese Telecommunications Policy*, London: Artech House.
Year Book of China Transportation and Communications, various issues, Beijing: Yearbook House of China Transportation and Communications.
Yu, L., S. Berg and Q. Guo (2004), 'Market performance of Chinese telecommunications: new regulatory policies', *Telecommunications Policy*, **28** (1), 715–32.
Zhao, Y. (2007), '"Universal service" and China's telecommunications miracle: discourses, practices and post-WTO accession challenges', *Info*, **9** (2/3), 108–21.
Zhou, L. (2006), 'State enterprises and other players in the economy', in L. Zhou, *China Business: Environment, Momentum, Strategies, Prospects*, Singapore: Pearson Prentice Hall, pp. 143–77.

24 Infrastructure reforms in India: an analysis of successes and failures

Leena Srivastava and Shahid Hasan

INTRODUCTION

Infrastructure reforms started in the second half of the 1980s – with the opening up of the oil and gas sector to private participation followed by reforms in the telecom and electricity sectors. The design of further sectoral reforms since then, and its implementation, has undergone significant changes. Not only has there been a shift in emphasis from upstream reforms to downstream reforms but also from a focus purely on supply enhancement to one where demand management is receiving increasing attention. Another key factor that has influenced the design of reforms processes has been the challenge of providing access to services to the vast majority of India's rural and poor population.

New independent regulatory commissions have been created to provide a level playing field between incumbent public sector service providers and newer private participants and to promote competition. The Institution of the Appellate Tribunal has also been effected to circumvent the need to approach India's notoriously slow civil courts. However, the inadequacy of powers and staffing of these commissions is increasingly resulting in an erosion of their credibility and legitimacy.

This chapter analyses and compares approaches that have been taken in a few sectors of the economy and will argue the reasons for perceived and actual successes and failures in India's reform efforts.

INFRASTRUCTURE SITUATION IN INDIA: EVOLUTION SINCE THE 1980S

Infrastructure is the key in transforming a low productive economy into a fast-growing agro-industrial oriented and highly productive 'new' economy (Ahluwalia, 2003). Construction of roads, electricity facilities and telecommunication networks can provide the poor with access to a vast range of economically gainful activities and markets, which can have a dramatic impact on productivity of the economy and create opportunities at all levels.

The telecommunications sector was wholly under government ownership until 1984, and was characterized by underinvestment, outdated equipment and growth well below the potential of the market. India was among the most poorly developed telecommunication sectors in the world with an abysmally low overall penetration of 0.5 telephones per thousand population during the 1980s. This was low even when compared with other developing countries of the world. The availability of telecommunication services was mostly skewed towards urban areas. Rural teledensity was much lower than urban

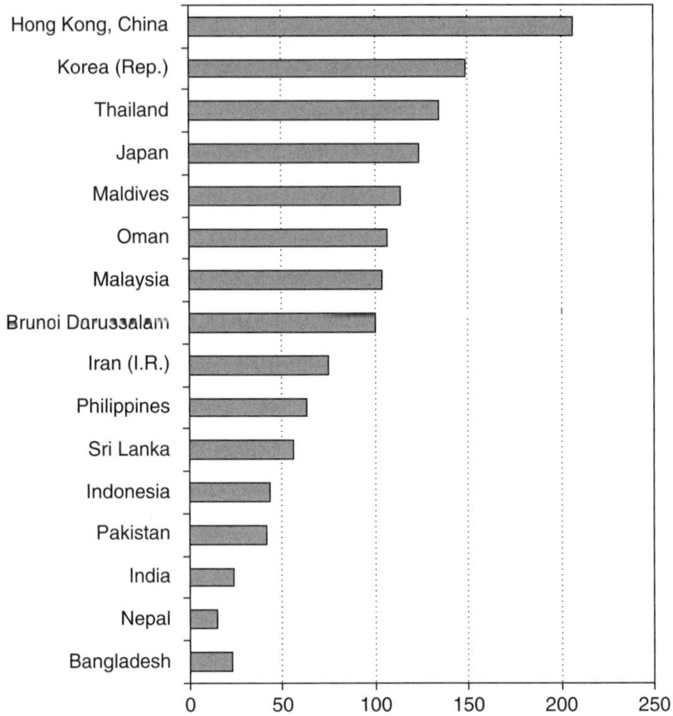

Source: International Telecommunication Union, www.itu.int/en/pages/default.aspx (accessed 5 November 2008).

Figure 24.1 Teledensity of select countries for 2007 (per 1000 inhabitants)

teledensity. However, India's vast size and as yet unserved population provided a huge potential for the expansion of telecommunication services (Figure 24.1). In terms of investments, only 12.6 per cent of total budget was devoted to the telecom sector during the Sixth and Seventh Five-Year Plan (1980–90).

Matching electricity generation, transmission and distribution infrastructure is essential for sustaining the economic growth. During the 1980s, the industry was characterized by poor quality and reliability of power supply as state electricity boards (SEBs) were not able to add requisite generation capacity owing to their poor finances. The most important reason for poor state of SEB finances was the fact that the selling price of electricity was lower than its cost of supply, although there was a great deal of variation among states and as well as consumer categories. As a result, most SEBs were losing money – particularly if the subsidies they were to receive were not taken into account. The problems caused by low average revenue were compounded by high receivables. As a result of the deteriorating state of SEB finances, the dues owed by them to public sector undertakings continued to rise. The transmission and distribution (T&D[1]) losses in India were higher than the average levels prevailing in most countries. While the government's record estimated losses to be around 21 per cent, most of the independent sources put this figure at a much higher level. High T&D losses obviously reflect sub-optimum utilization

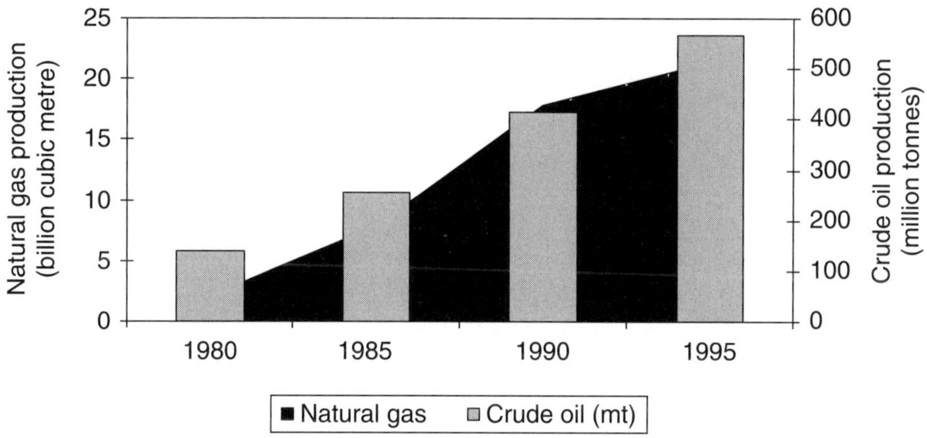

Source: http://www.indiastat.com (accessed 11 November, 2008).

Figure 24.2 Production of crude oil and gas in India

of existing assets (generation, transmission and distribution). With the exception of the Second and the Third Five-Year Plans, the share of electricity sector in the total plan outlays has been over 15 per cent. The focus in power planning and investment decisions in the sector has largely been on setting up new generation capacity. Inadequate investments in the T&D system was one of the major reasons for poor quality of supply of electricity, which is characterized by wide voltage variations, frequency fluctuations and frequent breakdowns. The annual commercial losses (without subsidies) of all the SEBs put together in the country, increased from US$0.35 billion in 1984/85 to US$0.98 billion in 1990/91 (current prices). Low plant load factor (PLF) and high T&D losses were not the only problems responsible for the poor financial performance of SEBs. The customer mix was another. Almost half of SEB sales were accounted for by sectors who pay much less than the cost of production (domestic segment) or nothing at all (agriculture).

During 1980–95, production of crude oil and natural gas witnessed steady growth (Figure 24.2) due to offshore discoveries of modest size. Production from Bombay High offshore field was responsible for the rapid growth of the country's total crude oil production in the late 1970s and throughout the 1980s. In financial year (FY) 1990, oil production peaked at 35 million tons, of which Bombay High accounted for the major share. By the end of 1980s, oil production had begun to decline, but there was a steady increase in consumption. The indigenous crude oil production was able to meet only about 35 per cent of the domestic requirement. The resource crunch in early 1990s further compounded the situation. Natural gas production was just 2 billion cubic metres in 1980, but there has been a significant increase in its production capacity over the years. However, in early 1980s, a significant share of offshore had to be flared either because of absence of compression and dehydration facilities offshore or lack of demand. This was as high as 43 per cent in 1984–85. In the second half of 1980s, natural gas was emerging as an important source of energy and feedstock. In 1984, Gas Authority of India Limited (GAIL) was set up to look after transportation, processing and marketing of natural gas and natural gas liquids. It has been instrumental in the laying of a 1700 km-long gas pipeline (HBJ pipeline) from Hazira in Gujarat to Jagdishpur in Uttar Pradesh, passing through Rajasthan and Madhya Pradesh. Refinery capacity also doubled during 1980–1990. The refinery capacity in terms of total crude throughput also increased from 25 million tonnes in 1980 to 51.6 million tonnes in 1990.

In rural areas, the water supply coverage[2] per households was only 27 per cent in 1980, which increased to 56 per cent in 1990. From 1980, 75 per cent of urban households had access to drinking water facilities, which increased to 81 per cent in 1990. On the urban front, there is also a large disparity in quantity of water supplied. The primary responsibility for providing drinking water and sanitation facilities in the country rests with the state governments and, more specifically, the local bodies in urban areas. Even though there has been a significant improvement in availability of water, there are large distributional inequities between states and cities, and per capita water consumption is still below international norms. The quality of water leaves a lot to be desired both in terms of level of treatment and regularity of supply. Overall, the sector has been suffering from poor financial health. Most water supply and sanitation (WSS) operations do not generate sufficient revenues to recover their operation and maintenance (O&M) expenditures. Water supply and sanitation tariff structures are generally skewed to benefit high- and middle-income domestic consumers. Most tariff structures include a lifeline tariff aimed at low-income households, which in fact benefits middle-income households, as the poor seldom have access to piped water. In the absence of systematic metering, bills are often established on assessed consumptions or simply on lump sum basis. The performance of Indian WSS service providers is the lowest among comparable Asian countries. In spite of the fact that water supply and sanitation were added to the national agenda during the First Five-Year planning period (1951–56), the government investment in the water and sanitation was merely 3–4 per cent of the total outlays during the Sixth and Seventh Five-Year Plan (1980–90).

Current Scenario

The past two decades have been eventful for the Indian economy, both in terms of a paradigm shift in development thinking as well as improved infrastructure (Table 24.1) as a result of economic reforms that were started in the mid-1980s and were followed by more comprehensive reforms in 1991.

Table 24.1 Investments in infrastructure (1980–2006)

	1980–81	1985–86	1990–91	2000–01	2005–06
GDP (at current prices)	31.4	60.5	123.5	444.0	874.1
Energy	0.7	1.7	3.3	9.4	5.1
Transport	0.6	1.3	2.6	5.9	8.9
Communications	0.1	0.2	0.7	7.4	0.1
Total	1.4	3.2	6.6	22.7	14.1

Note: All figures in equivalent US dollars.

Source: NCAER (1996); GoI (2008).

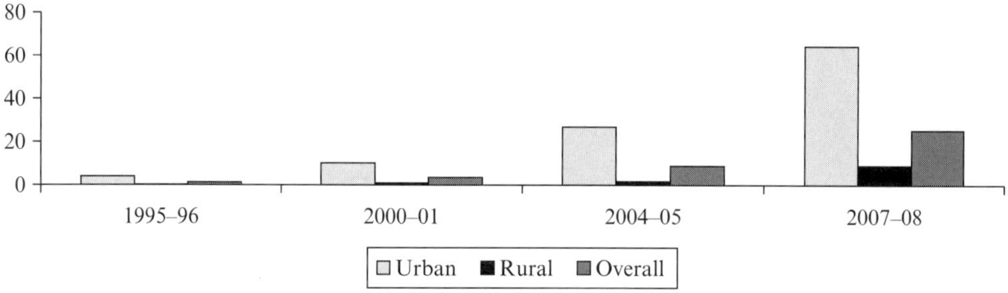

Source: TRAI and India Infrastructure Report, 1996–2008.

Figure 24.3 Teledensity (per 1000 population)

Viewed sectorally, the most significant changes have been in the telecommunications sector. Over the last one and half decades, telecommunications has seen large investments and rapid growth by private and public sector firms. From nearly 2 million telephones in 1980–81, it rose to 3 million in 1985–86 and the number of telephones nearly quadrupled in the next ten years. The total number of telephones in the country crossed the 300 million mark on 18 June 2008.[3] Growth in India's mobile sector has been astronomical. From a humble start in the mid-1990s, today India has nearly 350 million mobile subscribers (including global system for mobile communications [GSM] and code division multiple access [CDMA]). The overall teledensity has increased to 26.89 per cent in June 2008 (Figure 24.3).[4] India is also the base of some of the world's leading telecommunication handsets manufacturers, such as Nokia, Samsung, Motorola and LG, along with leading global service companies and infrastructure majors, such as Vodafone, AT&T, Ericsson, Alcatel, Singapore Telecom and Siemens. Intense competition both in manufacturing and service provisions has brought down the cost of handsets and call rates drastically. The market was growing at an annual rate of around 60 per cent and still expanding. A major push was given to take telecommunication services into the poorer and rural areas of the country. The liberalization efforts of the government are evident in the growing share of private sector in total telephone connections, which has increased from 39.2 per cent in 2004 to 72.4 per cent in December 2007.

Private sector participation in equipment manufacturing was allowed in 1991. Foreign direct investment (FDI) inflow in the telecom sector was only US$ 0.48 million in 1993. However, as can be seen from Figure 24.4, FDI inflow increased several fold in 1995, which was mainly due to enactment of the National Telecommunications Policy (NTP) 1994. However, the objectives of this policy could not be achieved due to absence of any effective regulatory regime. The NTP 1999 brought greater regulatory certainty and other effective policy changes in the subsequent years as a result of which FDI inflow again jumped several fold in the year 2001.

In the power sector, in spite of the fact that generation capacity has almost tripled since 1985, the deficit in the power supply in terms of peak and energy availability has not changed much. The peak and energy shortages were 14.8 per cent and 8.4 per cent respectively for the year 2007–08.[5] The PLF of thermal power stations has improved significantly since 1980s. State, central and private sector plants have reported a plant load factor (PLF) of 70.2 per cent, 85.4 per cent and 92.5 per cent respectively. At present, all

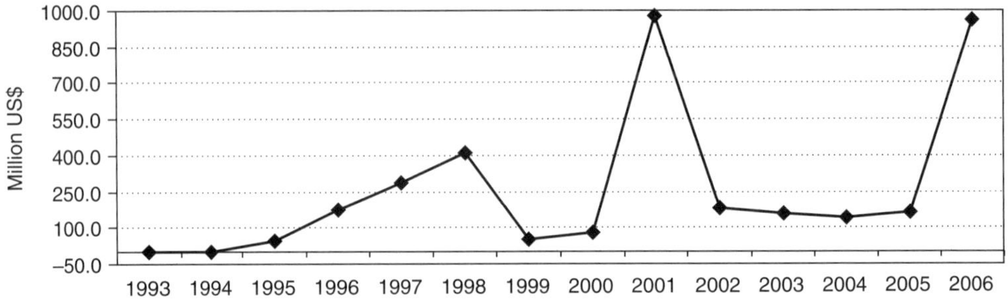

Source: Department of Telecommunications, Ministry of Communications and Information Technology, Government of India.

Figure 24.4 Growing FDI in telecom sector in India

the operating thermal power units are sub-critical units. However two ultra mega power projects based on super critical technology are under construction. Transmission and distribution losses are still alarmingly high in excess of 30 per cent.

Government of India (GoI) allowed private sector participation in generation in 1991, and offered several incentives to attract it. These incentives seemed to work during the initial years when over 250 Memorandums of Understanding (MoUs) were signed between SEBs and the private developers for a capacity addition of over 90 GW (more than the national installed capacity at that time). The government also declared eight of the most promising projects as 'fast track' and offered expedited clearance procedures, as well as provided government counter-guarantees and escrow accounts against non-payment of dues by SEBs. However, the overall performance has been very dismal. Only few of these contracts, that is, MoUs actually resulted in capacity addition.

The power-generating capacity of the private sector at June 2007 was only 18 419 MW. Out of this, 1306 MW was in hydro, 9022 MW was in thermal and 8091 MW in the wind sector. The total generating capacity in the private sector with respect to the total installed capacity in the country is only 13.6 per cent. Since 1991, only 15 503 MW[6] of generation capacity has been added in the private sector. In the Ninth Five-Year Plan (1992–97), the Government of India had the ambitious target of 17 589 MW capacity additions (17 039 thermal and 550 MW hydro) in the private sector. Against this target, the private sector was able to achieve only 5031 MW, which is less than 30 per cent of the fixed target. The Tenth Plan Working Group considered a capacity addition of 7121 MW in the private sector. Out of which 6313 MW has already been commissioned, and the revised target of capacity addition during the 10th plan is 4898 MW. The power generating capacity of the private sector as on 31 March 2003 was about 12 225 MW.[7] This implies that only 6194 MW of capacity has been added in the private sector since the enactment of the Electricity Act 2003. Overall investment in the power sector has remained in the range of 2–3 per cent of the total gross domestic product (GDP) since 1980s.

Crude oil production has been more or less static around 37 million tonnes in the absence of new finds in the country. With hardly any accretion in the reserves taking place in the last decade, the reserves to production ratio is likely to decline. The growing gap between domestic production and consumption has led to increasing oil import dependency of the country. The main thrust for acceleration of exploration activities began with the introduction of the New Exploration Licensing Policy (NELP) in 1999 under which the erstwhile nomination system favouring the national oil companies was discontinued. In order to enhance oil supplies, the government is also encouraging indigenous oil companies to acquire stake in international oil fields. The refining capacity has increased from 34 MMTPA (million metric tonne per annum) in 198–81 to 149 MMTPA in 2008. The public sector share in the total refinery capacity stood at 71 per cent.[8]

Natural gas has experienced the fastest rate of increase of any fuel in India's primary energy supply. The share of gas in the Indian energy basket has increased from a negligible contribution in the mid 1980s to 6 per cent in 2004,[9] and is expected to grow further. Several new finds have been discovered in the recent past. Most of the gas discoveries were made in the Krishna-Godavari Basin, off the east coast of India. Production of indigenous gas has increased from 8.134 BCM (billion cubic metres) to 31.747 BCM in 2007. India's natural gas reserves are currently (January 2006) estimated at about 37.22 trillion cubic feet (tcf), or about 0.59 per cent of the world total.

The spread of oil and gas pipelines network is limited and has not expanded in the recent past. Indian Oil Corporation (IOC) is the biggest player in crude oil and product pipelines network with a network spanning over 2387 km and having aggregate capacity of 25.20 MMTPA accounting for more than 50 per cent of the national network in capacity terms. The Gas Authority of India Limited (GAIL) distributes almost all the gas produced in the country, and operates over 4200 km of gas pipelines, the most important being the 2300 km HBJ pipeline with a capacity to handle 33.4 million metric standard cubic metres per day (mscmd). In addition, a few players also own city gas distribution networks. Liquefied petroleum gas (LPG) availability has increased significantly from less than 1 million tonnes in 1980s to nearly 12 million tonnes in 2007/08 (according to the Petroleum Planning and Analysis Cell, Ministry of Petroleum and Gas, Government of India).

The major problems in the water supply sector which work against its commercialization are low water tariffs, the high cost of production, high system losses from supply, poor demand management and low efficiency. Most of these problems still exist. However, the households with access to safe drinking water from the protected sources in rural areas have increased from 56 per cent in 1990 to 69 per cent in 2000 as per the Census of 2001. The same figures for urban households have increased from 81 per cent in 1990 to 93 per cent in 2000 (GoI, Planning Commission, 2002).

In terms of investments, the government has substantially increased its commitment to the water supply and sanitation from the beginning of the Sixth Five-Year Plan (1980–85). The Ninth Five-Year Plan (1997–2002) provides for US$9.11 billion as the outlay for the water supply and sanitation sector, which works out to about 4.6 per cent of the total Ninth Plan Public Sector outlay.

CHRONOLOGY OF INFRASTRUCTURE REFORMS IN INDIA

In the past, the responsibility for providing infrastructure services in India was vested with the government, which often combined operational responsibilities with policy and regulatory roles. However, this old paradigm of infrastructure being a public sector monopoly was challenged primarily by fiscal constraints. The gross fiscal deficit of India stood at US$12.36 billion in 1990–91, which was 9.4 per cent of GDP in 1990–91. The deterioration in the fiscal situation was the outcome of growing deficits on the revenue account which itself was high at 4.2 per cent of GDP in 1990–91. The mounting deficits not only restrict the capacity of the public sector to invest in the infrastructure sectors but affect the overall growth of the economy. Furthermore, the traditional organizational framework for the delivery of infrastructure has shown serious weaknesses in the planning and implementation of projects. Non-availability of funds on a timely basis and lack of proper evaluation and implementation of projects have led to major time and cost overruns.

Starting from the 1960s several East Asian countries achieved sustained high rates of growth that were unprecedented. While institutional, cultural and political factors are certainly relevant it became evident that India could also do better. As a result of this, the heavy dependence on the public sector for providing infrastructure services and the growing fiscal constraints, the government was forced to change its old development paradigm from control oriented to a competitive environment. Several initiatives were taken up in the second half of 1980s in order to accelerate economic growth. Some of these include lowering the direct tax rates, expanding the role of private sector, and liberalizing government's control trade and foreign investment. More comprehensive reforms were introduced in 1991 as a result of favourable experience of liberalization in the 1980s and looming balance of payments (BOP) crisis of 1990/91, which had greatly reduced the capacity of public finance to provide the required investment capital. The process of liberalization started with policy initiatives to mobilize and attract investments from all over the world, primarily for power generation. Subsequently, the telecommunications sector was also opened up for private investors. The GoI and several state governments since then have taken bold measures to initiate reforms in their states. A chronology of infrastructure reforms in India is presented in Box 24.1.

INITIAL REFORM DESIGN AND ITS EVOLUTION

Reforms in the Telecommunications Sector

Telecommunications reforms in India started in the 1980s by allowing private manufacturing of equipment used on customer premises. In 1985, the Department of Telecommunications (DoT) was constituted within the Ministry of Telecommunications. The DoT was to oversee the maintenance and development of telecommunications services. The objective was to focus on improving services in the telecommunications sector. To promote rapid development in all aspects of the sector including production, services, technology and finance, the Telecommunications Board was replaced by the Telecom Commission in 1989. Two large corporate entities were spun off from the Department of Telecommunications, for example, Mahangar Telephone Nigam Limited (MTNL) for Delhi and Mumbai and Videsh Sanchar Nigam Limited (VSNL) for all international services.

Later it was realized that in order to grow fast, the sector needed competition in the provision of services, and this could be achieved only by unbundling the sector. The government

BOX 24.1 CHRONOLOGY OF INFRASTRUCTURE REFORMS

Telecommunications

Year	Reform initiative
1984	Manufacturing of subscriber terminal equipment opened to private sector.
1985	Telecom was constituted into a separate department with a separate board.
1986	MTNL and VSNL created as corporations.
1988	Government introduces in-dialling scheme. PABX services only within a building, or in adjoining buildings.
1989	Telecom Commission formed.
1991	Telecom equipment manufacturing opened to private sector. Major international players like Alcatel, AT&T, Ericsson, Fujitsu, and Siemens entered equipment manufacturing market.
1992	VAS sector opened for private competition.
1993	Private networks allowed in industrial areas.
1994	Licences for radio paging (27 cities) issued.
	New Telecom Policy announced.
	Broad guidelines for private operator entry into basic services announced.
	Licences for cellular mobiles for four metros issued.
	Tenders floated for bids in cellular mobile services in 19 circles, excluding the four metros, on a duopoly basis.
1995	Tenders floated for second operator in basic services on a circle basis.
	Cellular tender bid opened.
	Basic service tender bid opened; the bids caused lot of controversy. A majority of bids were considered low.
	LOIs issued to some operators for cellular mobile operations in circles.
1996	Rebidding takes place for basic services in 13 circles. Poor response. The Telecom Regulatory Authority of India (TRAI) formed by ordinance.
	LOIs being issued for basic services.
1997	The TRAI Act passed in Parliament.
1998	Several VASs available through private operators. The first private basic service becomes operational.
1999	Announcement of National Telecom Policy.
2000	Amendment to the TRAI Act.
	Announcement of Domestic Long Distance Competition Policy.

Telecommunications

Year	Reform initiative
2000	National Frequency Allocation Plan comes into force.
2000	Telecom Disputes Settlement and Appellate Tribunal formed to adjudicate disputes and dispose of appeals with a view to protect the interests of service providers and consumers.
2001	Communication Convergence Bill approved by Cabinet.
2004	Broadband Policy announced.

Source: 3i Network (2001).

Power

Year	Reform initiative
1991	Amendment to ESA for facilitating private sector participation in generation projects of all types (except for nuclear). 100 per cent foreign investment & ownership allowed. Several other incentives also offered.
1995–96	First state level regulatory commission was established in Orissa under the Orissa Electricity Reform Act 1996. Electricity distribution privatized.
1996	Common Minimum Action Plan passed by the Chief Ministers that recommended creation of CERC and SERCs and greater autonomy to the SEBs, which are to be restructured and corporatised and run on commercial basis.
1997	CEA Clearance exempted for projects under 1000 MW but state government environment clearance required up to 250–500 MW. Liquid fuel policy announced.
1998	Mega-Power Policy: special incentives for the construction and operation of hydro-electric power plants of at least 500 MW and thermal plants of at least 1000 MW.
	Electricity Regulatory Commissions Act by the Government of India, which led to creation of CERC and SERCs in a number of States. Reform legislation enacted by a few states also paved the way for restructuring of the SEBs.
2001	Energy Conservation Act, which came into effect in March 2002, has laid down concrete measures to ensure the efficient use of energy and its conservation.
2003	Electricity Act 2003 by the Government of India, which provides for independent regulation, private sector participation, redefining the role of different entities, and so on.

Power

Year	Reform initiative
2005	National Electricity Policy announced that provides guidelines for overall development of the power sector.
2006	Tariff Policy for providing framework for determination of tariffs.
	Rural Electrification Policy announced.
2008	National Hydro Policy announced.
	First Energy Exchange established.

Source: Ministry of Power, Government of India.

Oil and gas

Year	Reform initiative
1994	The Directorate General of Hydrocarbons (DGH) established to regulate the upstream petroleum industry.
1996	First refinery in JV sector at Mangalore established with minority share holding of PSU.
1993	Parallel marketing of LPG/SKO started.
1998	New Exploration Licensing Policy (NELP) announced for attracting foreign oil companies in exploration.
1998	Retention pricing for refineries abolished.
	Refining sector de-licensed.
1998	Cost plus formula for indigenous crude oil abolished and linked to international price effective.
1998–2002	APM dismantled during.
1998	Petronet LNG formed.
1999	Bids for NELP 1 invited in January 1999. So far VI rounds of bidding completed and 162 blocks awarded.
1999	Fully private sector refinery at Jamnagar commissioned.
2001	Bids for CBM blocks invited.
2002	Dismantling of APL completed.
2004	LNG imports commenced.
2005	First LNG terminal Commissioned.
2007	Petroleum and Natural Gas Regulatory Board set up.

Source: Compiled by the authors.

Water supply and sanitation

Year	Reform initiative
1987	First National Water Policy announced.
1993–94	Accelerated Urban Water Supply and Sanitation Programme
1999	Total Sanitation Campaign Programme was launched.
1999	Sector Reforms Projects for Water Supply in rural areas
2002	National Water Policy revised.
2004	Maharashtra Water Resources Regulatory Authority established.

also recognized the need to separate the operations, management and regulatory responsibilities of the DoT. The second phase of reforms commenced with the general liberalization of the economy in the early 1990s and announcement of a new economic policy (NEP) in 1991. Telecommunications equipment manufacturing was delicensed in 1991 and value-added services were declared open to the private sector in 1992, following which radio paging, cellular mobile and other value-added services were opened to the private sector. The National Telecom Policy was announced in 1994, with a major thrust on universal service and qualitative improvement in telecommunications services and the start of private sector participation in basic telephone services. A crucial aspect of the institutional reform of the Indian telecommunications sector was the setting up of an independent regulatory body in 1997. The Telecom Regulatory Authority of India (TRAI) was established to assure investors that the sector would be regulated in a balanced and fair manner. The TRAI was vested with powers to ensure its independence from the government. The government itself has retained the licensing function. Further changes in the regulatory system took place with the TRAI Act 2000 that aimed to restore functional clarity and improve regulatory quality. The TRAI can frame regulations and levy fees and charges for telecommunications services as deemed necessary. To fairly adjudicate any dispute between licensor and licensee, between service provider, between service provider and a group of consumers, a separate disputes settlement body was set up, called Telecom Disputes Settlement and Appellate Tribunal (TDSAT), in 2000. Since then the sector has been witnessing tremendous growth with the presence of private sector provision of telecommunications services in the country (Ministry of Communications and Information Technology, 2008–09).

Reforms in Power Sector

The first generation of reforms in the power sector was influenced by the looming power crises in the late 1980s. Given the huge investment needs in the power sector, the role of private sector in enhancing system capacity became imperative. Recognizing the need, the GoI allowed private sector participation in generation in 1991, and offered several incentives to attract them. These incentives seemed to work during the initial years when over 250 MoUs were signed between SEBs and the private developers for additional capacity of over 90 GW (more than the national installed capacity at that time). The government also declared eight of the most promising projects as 'fast track' and offered expedited clearance procedures, as well as provided government counter-guarantees and escrow accounts against non-payment of dues by SEBs. However, the overall performance has been very dismal. Only a few of these contracts, that is. MoUs actually resulted in additional capacity. Lack of creditworthiness of the SEBs, inadequate procurement and payment guarantee mechanisms, delays in finalization of contracts (Power Purchasing Agreement [PPA], Fuel Supply Agreement and Fuel Transportation Agreement) and inadequate financial strength of SGs to cover large number of projects were some of the critical factors for under-achievement. Moreover, in many cases the private power was more expensive than SEBs' own power.

By 1995–96 it was becoming clear that liberalizing investment in generation was unlikely to solve the problems of the power sector. Despite the enthusiastic response to the private power policy, the result has not been encouraging. Reforms also did not take place in the regulation of the industry. With most electricity boards pricing power for certain groups of consumers, such as agricultural consumers and some high-tension industries, with varying levels of subsidies, their financial health and accountability remained seriously eroded. Because of the failure of the fast-track projects and the slow progress of state-level reforms, the central government acknowledged the need for more comprehensive reforms at the national level. As a result, in the second generation of reforms, structural and regulatory changes were proposed to address some of the issues and problems of the first generation of reforms. Independent regulatory commissions at the central and state levels have been established for rationalizing bulk and retail tariffs, regulating investments and improving quality of supply and services. Moving from the single-buyer model that dominated initially, it was recommended to restructure SEBs through corporatization and introduction of competition in the wholesale market. It was also desired that utilities should commercialise their operations and professionalize their management. In the third generation of reforms, a new Electricity Act 2003 was introduced. It consolidated the laws relating to electricity and aims to ensure availability and affordability of power by promoting competition and inducing states to reform. The Electricity Act 2003 has introduced significant changes in industry structure by moving from single-buyer market to a multi-buyer multi-seller system. It allows private sector participation in generation, transmission and distribution, trading of electricity, calls for competitive selection of new generation capacity in future, removes barriers in harnessing captive power, establishment of the Appellate Tribunal, and so on. The Act envisages a competitive regime in all the segments of the power sector, especially in the distribution sector. Creation of a direct linkage between the generator and the consumer thereby bypassing the state electricity boards, is a key feature of this Act. Open access provisions enable a consumer to source electricity from a generator or source of his or her choice by availing the open access provisions in the Act. It creates a competition by removing the monopoly of the local utility and enables competition between the generators.

Reforms in the Oil and Gas Sector

Prior to the mid-1990s, the oil and gas sector was largely dominated by the National Oil Company operating in a controlled regulatory environment. Most of the segments were not open to foreign or private sector participation. Given the rather depressing scenario on the fiscal and crude output fronts, the government initiated reforms in petroleum sector in 1990. As a first step towards deregulation of the oil and gas industry, the Administered Pricing Mechanism (APM) was recommended to be abolished in a phased manner (1998–2002) rather than in one shot. The objective of dismantling the APM was to allow freedom to oil companies in fixing retail prices with a view to ensuring better margins and better returns for the refining and marketing companies. In other words, petroleum prices were expected to reflect market prices and the subsidies on kerosene and LPG were to come down.

India was one of the least explored countries, and government did not have adequate resources to put in. Therefore, in order to increase exploration activities, it allowed the foreign and private sector participation in the upstream sector by announcing the New Exploration Licensing Policy (NELP) in March 1997. The main objective was not only to mobilize resources from the private sector but to level the playing field in the upstream sector between private and public sector companies in all fiscal, financial and contractual matters. The move was seen as inevitable if India had to attract funds and technology from abroad into its petroleum sector.

The private sector was also allowed into the refining sector, but government kept the majority control over marketing petroleum products. Only those who have invested US$0.46 billion in refineries or are producing at least 3 MT (million tonnes) of crude oil annually were eligible to have marketing outlets for controlled products (motor spirit and high-speed diesel). It was felt that a greater deregulation in the marketing sector is essential for a fully competitive market to operate. The next important development was the creation of a regulator for the oil and gas industry in the country. It was long felt that ensuring a level playing field for the public and private sectors, encouraging investment in India's burgeoning natural gas sector and introducing competition are crucial, and hence, presence of an independent regulator would greatly help in ushering the competition in the industry that has seen much growth in the natural gas sector and an increase in the number of players in the fuel retailing segment. Overall, the objectives of the deregulation of oil and gas industry were to (1) increase competition in the industry by allowing the entry of more players, (2) attract private capital, and (3) remove constraints on economic pricing of products.

Reforms in the WSS Sector

Traditionally, and as per the Indian Constitution, the primary responsibility of providing drinking water and sanitation in the country rests with the local governments. Over the years, larger allocations have been made by the centre and state governments, however, the funds remain inadequate to meet the requirements of growing population in the country. The major change in the strategy came through the Seventy-third Constitutional Amendment, which provided greater role and involvement of local bodies in drinking water and sanitation. This led to the inception of the Sector Reforms programme, which was based on more decentralized approaches of delivering by involving the people rather than centralized and government controlled system of implementation. This paradigm shift transformed the government's role from being service provider to facilitator. The National Water Policy was revised to incorporate the new approach in order to improve the sustainability of resources and system. The main features of the policy include (1) adoption of demand responsive approaches based on empowerment, to ensure full participation in decision-making, control and management by communities, (2) shifting role of governments from direct service delivery to that of planning, policy formulation, monitoring and evaluation, and partial support, and (3) partial capital cost sharing and full responsibility of O&M by users.[10] Under the new reform programme, the government allowed communities and local bodies/institutions to choose the appropriate institutional model for service delivery suited to their own specific local context. It was expected that new thinking and a new approach would help in achieving environment, institutional and financial sustainability. The Sector Reforms Projects (in 65 districts) are being implemented by the government without any external assistance in several districts spreading across 25 states covering a population of over 70 million.[11] Reforms in urban water and sanitation are also in their early stages. Urban water Supply and Sanitation (UWSS) services are also characterized by inefficiency and poor service quality. Most cities provide only intermittent water supply. A greater role by urban local bodies (ULBs) and privatization of services through build – operate – transfer (BOT) and build – own – operate – transfer (BOOT) routes are encouraged. Overall, the reforms both in rural and urban areas are guided by four principles, and these include (1) more transparency in operations, (2) greater community participation, (3) increased awareness generation, and (iv) social auditing.

KEY ACHIEVEMENTS AND FURTHER CHALLENGES

In the telecommunications sector, the most noteworthy achievement has been the increase in the number of fixed and cellular telephone connections. The most important landmark in telecommunications reforms, however, came with the New Telecom policy in 1999 (NTP'99). This allowed the existing operators to more from a fixed licence fee regime to a revenue-sharing regime. The NTP'99 opened up the national and long-distance markets to private sector, corporatization of telecommunication services, and so on. These policy measures have resulted in significant reduction in long-distance tariffs due to competitive pressures. The intense competition resulted in price slash by players several times. Creation of an independent institutional regulatory mechanism in the shape of the Telecom Regulatory Authority of India (TRAI) and the Telecom Disputes Settlement and Appellate Tribunal (TDSAT) further helped to gain the confidence of the investor's community. It can be seen that the growth pattern has been synchronous with the telecommunications reforms. The telecom sector is a major recipient of FDI flows.

As is evident from the teledensity, the growth of telecommunication sector has been skewed towards urban areas. Therefore, extending the telecommunications facilities, including broadband access, into the rural and inaccessible areas should be given priority. Since, providing services in rural areas are not expected to cover the actual cost, the main issue was, who should 'subsidize' operations in rural areas? The Universal Service Obligation (USO) Fund was created in 2002 by the Ministry of Communications and Information Technology, to provide telecommunications services in the rural areas at reasonable rates. Telecommunications firms have been contributing 5 per cent of their revenue to this fund. So far US$3.46 billion has been collected.[12] However, as per the latest report of the Comptroller and Auditor General of India tabled in Parliament, only one-third (US$1.15 billion) of the fund was utilized during 2002–03 to 2006–07. The fund is administered by the Department of Telecommunication and, initially, was being used only to support rural fixed line telephones, but this was later amended to include mobile services as well. The biggest challenge lies in devising an effective USO mechanism for encouraging rural connectivity.

In the power sector, the most commendable achievement has been the creation of independent regulatory authorities at both central and state level. Twenty-six states out of 29 have notified the creation of the SERC. However, one of the key challenges being faced by the regulatory commissions is to attract and retain well-qualified staff. Currently, the trend in most commissions is that the support staff consists of engineers and competent senior employees deputed from the erstwhile SEBs. Though they are adept at providing excellent technical support, there remains a possibility that undue influencing of decisions and very often bias to their parent organization poses a hindrance to the fair and just analysis of the petitions of the utility under consideration. Another concern is that the tenure of some of the regulators is too short. By the time they start understanding the complexities of the sector, their tenure is over. The hunt for new regulators starts, and sometimes there are deliberately long delays in filling these posts. This seriously affects the smooth and effective functioning of the commission. The task of appointing a competent person to this post is again is a big challenge in itself.

Rationalization of tariffs was accorded top priority by almost all the regulatory commissions in the inception period, which still continues. One or more tariff order has been issued in 25 states. Improving the quality of supply and services is also gaining the attention of regulators. In this regard, several SERCs have framed standards of performance to be followed by the utilities. However, monitoring, reporting and evaluation are still poor. One key achievement of reforms in this sector is the extent of transparency that has emerged in regulatory processes as well as consumer awareness.

Thirteen states have completed the structural unbundling of SEBs. In two states, distribution has been privatized. The outcome has been mixed. Financial viability of the sector is still the most dominant reason hampering private sector participation in generation. The sector has to be made financially strong internally to attract private investments. Transmission and distribution losses and the demand–supply (both peak and energy) gap still remain high. In spite of the rationalization of retail tariffs, the tariffs do not reflect cost of supply. Government still subsidizes tariffs for certain class of consumers, mainly domestic and agriculture as part of its social commitments.

The Indian hydrocarbon sector has undergone significant transformation since the start of deregulation of the sector. The key achievements in the sector have been:

- Dismantling of the APM – earlier, functioning of the oil industry was governed by the APM. A ballooning oil pool deficit forced the government to bring domestic prices in line with the international prices, and subsequently APM pricing was dismantled in a phased manner spanning over four years (1998–2002). The dismantling was expected to introduce a new era in the Indian hydrocarbon sector, where all petroleum pricing prices were expected to move to market prices and the subsidies on kerosene and LPG were expected to decline. However, even after six years of dismantling, the pricing of petroleum products (diesel and petrol) still does not reflect the true market prices, and the subsidy bill on LPG and kerosene has been increasing steadily.

- New Exploration and Licensing Policy – the main thrust of the NELP was to accelerate exploration activities by providing a level playing field to the operators. Under this, the erstwhile nomination system favouring the national oil companies was discontinued. Private investors were allowed to conduct exploration activities and to market oil and gas produced in domestic markets. Seven rounds of NELP have taken place so far, and the response has been satisfactory. Through, this policy, the government is making efforts to expeditiously explore the inadequately explored and unexplored areas of the country's sedimentary basins. Steps have also been taken to reappraise the sedimentary basins of the country with latest data acquisition and interpretation technology in order to make these areas more attractive for foreign exploration and production (E&P) companies.
- Private participation in the refining sector – with the opening up of the refining sector for private investments, the refining capacity has increased significantly. This has greatly helped the country in becoming a net exporter of petroleum products from being a net importer earlier.
- Independent regulator – the establishment of the Petroleum and Natural Gas Regulatory Board (PNGRB) in the country is expected to provide an impetus to the competitive environment in various activities in the downstream oil and gas sector, safeguard interests of the consumer and help in developing the natural gas market, especially city gas distribution.
- Increasing the competition in the industry – the government is also playing an important part in encouraging competition in the oil industry. Various restrictive policies on the marketing side, including the establishment of new outlets, appointment of dealers, and so on, have been abolished. According to a government notification, subsequent to dismantling of the APM in the country, any company which has invested or committed to invest US$0.46 billion rupees in the Indian oil and gas sector over a period of ten years, could undertake the marketing of transportation fuels. Deregulation has also brought in customer-focus in the working of the oil companies. The companies are competing to establish their 'brand image' in the market. The international format of a new generation outlet was introduced by oil companies.

In the water sector, despite significantly increased funding and higher levels of coverage, access to quality water remains a dominant issue both in rural and urban areas. In addition, the proper upkeep of water supply schemes has been a problem. As part of Sector Reforms Projects in rural areas, 65 projects at an estimated cost of US$0.47 billion are being implemented. The government has already spent about Rs11.51 billion out of its total share of US$0.44 billion. Also, US$0.03 billion have been mobilized from the community. On the urban front, several cities have attempted private sector participation in variety of forms, such as BOT and build finance and transfer (BFT), for providing water services.

Recognizing the rapid rate of urbanization and the poor conditions of UWSS, low tariffs, poor recovery rates, poor quality of services, the central government has initiated a number of programmes to support reform and upgrading of the quality of urban water supply. The 'City Challenge Fund' has been established to fund part of the transition costs for a city government or public service provider to advance systematically towards creditworthiness and a sustained improvement in public services. The 'Urban Reform Incentive Fund' has been established to enhance the creation of efficient urban land markets and develop the legal and tariff framework to advance urban governance and utility reforms. The Pooled-Finance Mechanism has been established to enhance the ability of small towns to access capital markets in a cost-effective manner and on better terms than if each tried to access the market individually. In accordance with the National Water Policy 2002, some states have come up with their own State Water Policy with an integrated water resource management as a core strategy for achieving sustainable development of the sector. Some of the critical challenges that the sector faces are (1) declining availability of water, (2) poor reliability and quality of water supply, (3) inequities in distribution, (4) achieving financial and environmental sustainability, and (5) unclear roles and responsibilities of institutions, and so on.

SUGGESTIONS FOR FURTHER ACTION

There is no doubt that India has made noticeable progress both in terms of expanding infrastructure facilities and structural and regulatory changes aimed at furthering the competitive environment in procuring and delivering the services. There are areas where India has done remarkably well compared with other countries who had introduced economic reforms. However, a lot needs to be done in order to achieve the perceived objectives.

Telecommunications

In the telecommunication sector, increasing the penetration of telecommunications services (fixed and mobile) in rural areas should be given priority. It is been observed that the USO Fund is not effectively utilized. One of the reasons is the complicated and time-consuming procedures for claiming this subsidy amount as it first goes to the Consolidated Fund of India. The procedures for withdrawal of funds need to be simple, so that it encourages operators in rural areas to make necessary investments. Further, bidding, evaluation, contract and settlement procedures for utilizing the USO Fund should be simplified. Infrastructure-sharing in rural areas should be encouraged as it will not only bring down the cost of providing the services, but operators would be able to provide benefit to large number of consumers through effective utilization of the available subsidy.

Power

In the power sector, development of a positive public image of regulatory bodies is very important. Independent regulation is new in India and therefore, the public's image of regulators is still being formed. Public respect for regulators is essential for independent regulation to succeed. On this issue, the government has a very important role to play in establishing the legitimacy of the regulatory bodies. Therefore, the government must be careful in its public statements regarding the regulatory body, even when the regulatory body takes a decision that is not entirely to the government's liking. Many governments have supported the regulatory bodies even when the regulatory bodies made decisions that did not enjoy whole-hearted public support. However, there have also been some states where there has been conflict between the government and the SERC. As a result of this conflict, the government has sometimes resorted to criticizing the SERC in the media. Such situations could result in severe media backlash, thereby creating a bad image among the consumers. It is also most important to improve the selection process. In spite of the requirements in the statutes to fill regulatory bodies with people of varied experience, in India, the selection of regulators shows a significant bias with a very large majority of the regulators being retired bureaucrats or retired judges. In the case of the SERCs, some regulators are also retired or retiring engineers of the Central Electricity Authority (CEA) or the SEBs. There is a feeling among stakeholders and others interested in the infrastructure sectors, that retiring officers are trying to make regulatory appointments an avenue for continued employment. Clearly, the selection process is not attracting a variety of talented people. There should be a period of succession in retirement of regulators. This would provide an opportunity to the newly appointed regulators to acquire an understanding of the sector from their other colleagues. Therefore, the large majority of the regulators should not retire at the same time, which otherwise would create a void of sector experts and may cause delays in the review and award of orders. We propose anybody appointed should have service tenure of at least two or three years before reaching the retirement age of 65 years, as per the Electricity Act 2003.

The pace of reduction of cross-subsidy should be speeded up. The issue of fuel security is a major barrier to private players as is the uncertainty regarding pricing and procurement of input fuels. The government needs to have a clear and integrated policy defining the approach to energy provision and pricing for the next several years. In a similar vein, the environmental clearance procedures should be streamlined and should not reflect biases against private sector projects. The Electricity Act 2003 provides the requisite framework for encouraging competition as well as private sector participation in the electricity sector. It is extremely important for the future development of the sector that the Act is implemented in the right spirit. In particular, regulations relating to open-access, third-party sales must be drafted to promote competition.

Oil and Gas

The proposed reforms related to phasing out subsidies on domestic LPG and public distribution system (PDS) kerosene is appreciated. But, in the meantime, efforts should be made to effectively target and deliver the subsidies to the deserving section of the population until subsidies are prevalent in the system. At the same time, innovative technological solutions such as smart cards[13] may be taken up at the earliest opportunity for targeted subsidy delivery. This would also help in tracking the pattern and extent of subsidy utilization by the targeted users. The government should undertake a pilot project for the introduction of smart cards for delivery of energy subsidies on a priority basis. Internationally, there are several successful examples of the use of smart cards for efficient delivery of a range of services.

With the burgeoning oil demand in the country, oil import dependency occupies central stage in India's emerging energy policies. This dependency is projected to rise further in the future. In this regard, it is important to maximize domestic production, diversify sources, increase share of term contracts, acquire equity oil abroad and build strategic oil reserves.

Water Supply and Sanitation

One of the main reasons for the poor state of water utilities in India is the lack of accountability and transparency in operations. Moreover, these are loosely regulated. An equally important factor is that local and national governments determine the price of water, that is, set the water tariffs. As a result, water tariffs are frequently based on political considerations and vested interests instead of the costs that are involved in producing and delivering water. Despite significantly increased funding and higher levels of coverage, access to urban water services in Indian cities remains skewed and the quality of services inadequate. There is growing recognition of the institutional weaknesses in the delivery of urban water supply and sanitation services. Functions of policy-making, oversight and service provision are not clearly delineated and lead to conflicting objectives, political interference and lack of incentives and accountability. Therefore, there is an urgent need to accelerate institutional reforms to bring in more clarity in the role definition.

It has been observed that a variety of government institutions at the centre (Central Water Commission, Central Ground Water Board, National Rivers Conservation Directorate) have a role, either direct or indirect, in the management of the drinking water supply. It is important to have an effective interface and a high degree of functional coordination between the development policy-making and planning and implementation institutions. Equally important is for rural water supply and sanitation (RWSS) and UWSS agencies to operate hand-in-hand with their health and education counterparts to jointly monitor indicators of RWSS/UWSS, health, education, poverty and equity in order to make significant headway in the respective sectors. Separation of roles for policy, regulation and operations are badly needed. Accountability should be the guiding principle for institutional restructuring WSS institutions. The community should play an enhanced role in service delivery. Currently, no regulatory reforms have been carried out in the sector at central level. No central-level regulatory body exists. Also state-level regulatory boards are not in place. Creation of regulatory bodies can help (1) improve the sustainability of the sector through cost recovery/adequate return on capital, (2) provide a greater degree of fairness to all users including the poor, and (3) improve the production and allocation efficiency in WSS services provision.

NOTES

1. It comprises of technical and non-technical losses (which also called 'commercial losses'). Theft and pilferage account for significant share in the non-technical losses.
2. Water Supply and Sanitation, A WHO-UNICEF Sponsored Study, Planning Commission, Government of India, 2008.
3. India Brand Equity Foundation website, last updated July–September 2008.
4. The Indian Telecom Services Performance Indicators, Telecom Regulatory Authority of India (TRAI), April–June 2009.

5. Annual Power Supply Position, Central Electricity Authority, Ministry of Power, Government of India, 2008–09.
6. Various annual reports of the Ministry of Power, Government of India.
7. *All India Electricity Statistics: General Review 2005*, Central Electricity Authority, Government of India; and *Annual Report 2004-05*, Ministry of Power, Government of India.
8. Petroleum statistics 2007–08, Ministry of Petroleum and Natural Gas, Government of India.
9. World Energy Outlook, 2006, OECD/International Energy Agency.
10. India – Assessment 2002, Water supply and sanitation, Planning Commission, Government of India.
11. Sector Reforms Projects in Rural Drinking Water Supply – All India Evaluation Study, CMI Social Research Centre, New Delhi, 2005.
12. Report No. PA 1 of 2008 – Union Government – Civil and Postal Departments; the Comptroller and Auditor General of India, 2007.
13. A smart card is a plastic card of the size of a credit card with an integrated circuit built into it. Information in the microchip can intravenously verify the cardholder's identity and any other privileges to which the cardholder may be entitled. These can be used as an effective tool for efficient delivery mechanism in a range of applications including effective targeting of energy subsidies. Such a card would maintain the personal identity and details about the entitlements of energy subsidies for the cardholder. It will also record the transaction history, pattern and extent of subsidy utilization. The card could be renewed after a specified time.

REFERENCES

Ahluwalia, M.S. (2003), 'India's economic reforms'. www.planningcommission.nic.in. (accessed October 2008).

Government of India (GoI), Planning Commission (2002), 'India assessment 2002: water supply and sanitation', report of the Planning Commission, Government of India

Government of India (GoI) (2008), *Economic Survey 2007–08*, New Delhi: Ministry of Finance, Economic Division, and Oxford University Press.

Ministry of Communications and Information Technology (2008–09), *Annual Report*, Department of Telecommunications, Ministry of Communications and Information Technology, Government of India, New Delhi.

National Council of Applied Economic Research (NCAER) (1996), *India Infrastructure Report 1996*, vol. 3.

Telecom Regulatory Authority of India. (2009), *The Indian Telecom Services Performance Indicators, April–June 2009*, New Delhi: Telecom Regulatory Authority of India.

3i Network (2001), *India Infrastructure Report 2001*, New Delhi: Oxford University Press.

25 Liberalization of infrastructures in Russia[1]

Hella Engerer

INTRODUCTION

Russia is a country which has gone through a complex transition from the planned economy to a market economy. Similarly to other eastern European countries, the liberalization of infrastructures which took place in Russia has gone hand in hand with a restructuring of state-owned enterprises. In addition, the liberalization and restructuring of the so-called 'strategic' sectors such as energy or transportation has served as a sign of the political willingness to reform the economy and privatize formerly state-owned enterprises. Given the importance of a developed infrastructure for economic growth, the progress of the infrastructural sector reforms also mirrors the progress of transition.

In Russia, the specific approaches to reform and restructuring of various infrastructure sectors are shaped by a range of internal factors, such as resource endowment, inherited capital stock and vast geographical distances. Therefore, liberalization and restructuring processes in Russia may appear to be of a somewhat special character ('Russian style'). Nevertheless, the Russian reforms of infrastructures are still, at least in part, influenced by the ongoing debate on theoretical approaches as well as other countries' experiences, in particular in Western Europe. This analysis of the liberalization of infrastructure in Russia will highlight key internal factors, as well as external influences.

This chapter provides an overview of the transition of the Russian economy and infrastructures. First, it analyses the development of the Russian economy during the transition period, focusing specifically on the transition's main economic challenges. Then a snapshot of infrastructure reforms in Russia will be presented. Finally, the liberalization of infrastructures in Russia will be compared with the progress of liberalization of infrastructures in other Eastern European countries as well as in Western Europe. So far, the progress of infrastructural reform in Russia has been limited as compared with other European countries. Thus, I will focus on liberalization and restructuring of various infrastructure sectors. Within the Russian context, privatization and ownership changes have been important factors which therefore deserve to be addressed at some length. The liberalization of market entry (for example, for foreign investments) as well as the issues of competition and concentration will also be discussed. In addition, I will examine whether the liberalization and restructuring of a certain sector was based on a (theoretical) model and which kind of regulation is likely to be applied in the future. I find that the strategies used by Russia to liberalize and to regulate various infrastructure sectors have differed from industry to industry. For instance, much attention and resources have been devoted to the reform of the energy infrastructure, whereas changes in the infrastructure in other sectors have been much less pronounced. Thus, liberalization and restructuring differ across the various sectors. However, as generally, infrastructure reforms in Russia tend to comprise several stages, of which the later are frequently postponed or extended. I summarize the Russian approaches to infrastructure liberalization. Across the various sectors progress of telecommunications liberalization and deregulation is uneven. In all sectors, however, there is a lack of modernization. Therefore, additional financial resources must be mobilized in the future, either domestically or from abroad.

OVERVIEW OF THE RUSSIAN ECONOMY AND INFRASTRUCTURE

Over the past several years, the Russian economy has been undergoing a process of recovery from the so-called transition crisis of the 1990s. During the 1990s gross domestic output fell sharply, and main sectors of the economy were affected by a massive decline in production and investment.

At the end of the 1990s. a severe financial crisis struck the Russian economy, upsetting the incipient upward trends. It was only after year 2000 that Russia has been able to report stable growth rates for its main economic indicators, such as the gross domestic product (GDP), industrial production and investment.

At first glance, at the beginning of 2008 Russia seemed to be a rapidly growing, thriving economy, with high GDP growth rates, as well as an impressive foreign trade surplus. The latter is mainly due to Russia's extensive earnings from energy exports, whose share in overall export earnings of goods has been steadily growing. Such an upward trend is the result of two underlying factors: the increase in energy export volumes accompanied by massive increases in international oil prices, which, combined, have generated windfall profits for the Russian economy.[2] Therefore, at least in financial terms, Russia has had the necessary means to restructure and reorganize its economy which has relied predominantly on heavy industry. However, energy still dominates Russian exports and export earnings, and energy production has an increasing share within the industrial sector. Instead of undertaking a profound restructuring of its industry, Russia has been lagging behind in what has to do with the production and export of high-technology products. This places Russia at a disadvantage in the global competition.

The recovery of the Russian economy and the windfall profits from external trade have reduced the pressure for further (industrial) restructuring as well as for the implementation of institutional reforms. This holds also true with regard to the reform and development of infrastructure, where some progress can be observed, yet much remains to be done.

Against this background, the global financial crisis and the decline of oil prices has strong negative effects on the Russian economy and its potential for infrastructure reforms. Statistically, GDP growth in 2008 only has reduced to 5.6 per cent (2007: 8.1 per cent). However, according to the International Monetary Fund, the Russian economy will decline by 2 per cent in 2009 and will only slowly recover thereafter. In addition, the huge private capital inflows as well as the windfalls profits from oil exports have reduced substantially thus limiting the potential for future investment outlays in infrastructures.

SNAPSHOT OF THE RUSSIAN INFRASTRUCTURE REFORMS

The term 'infrastructure' can be defined in a number of different ways. In general, one can distinguish among three meanings (see http://dictionary.reference.com/browse/infrastructure): (1) the basic, underlying framework or features of a system or organization, (2) the fundamental facilities and systems serving a country, city, or area (for example, transportation and communication systems, power plants and schools) and (3) the military installations of a country. Most economic analyses of infrastructures rely on either the first or the second definition.[3] A further distinction can be made with regard to different types of infrastructure facilities and systems:[4]

- technical infrastructure (transport systems and networks, pipelines)
- social infrastructure (hospitals and schools)
- institutional infrastructure (legal norms and procedures)
- personal infrastructure (intellectual and entrepreneurial skills).

In the following we use the term infrastructure as referring to the fundamental facilities or the technical infrastructure. The latter can further be divided into:

- (public) utilities: energy utilities (gas, electricity and heat) and other utilities (water and sewerage)
- transport (rail and air transport, roads and maritime ports)
- communications (telecommunications, Internet and others).

These kinds of technical infrastructure differ widely with regard to technical installations and investment needs. In addition, technical infrastructure may vary depending on whether providers are public or private, as well as on the number of users. Therefore, different approaches to liberalization and deregulation have been used across Europe (including Eastern Europe); these models will be discussed in the relevant subchapter. Before analysing infrastructure sectors, we will present the European Bank for Reconstruction and Development's (EBRD's) composite index on infrastructure reforms in Eastern Europe.

RUSSIAN INFRASTRUCTURE REFORM AS MIRRORED BY THE EBRD INFRASTRUCTURE INDEX

The European Bank for Reconstruction and Development has developed an index for infrastructure reforms which is a composite score for the progress of reform in such sectors as electricity, railways, roads, telecommunications and water and waste water. It measures the progress with commercialization, tariff reform, quality of regulation and private sector involvement (EBRD, 2004, p. 4). By doing so, it can provide a first, rough impression of infrastructure reforms in Europe.

According to the EBRD index, Russia made some progress with infrastructure reforms during the transition. In 2007, Russian infrastructure reform was assigned a score of '3+' (where 0 indicates no reform at all and 4+ indicates the market economy). This score

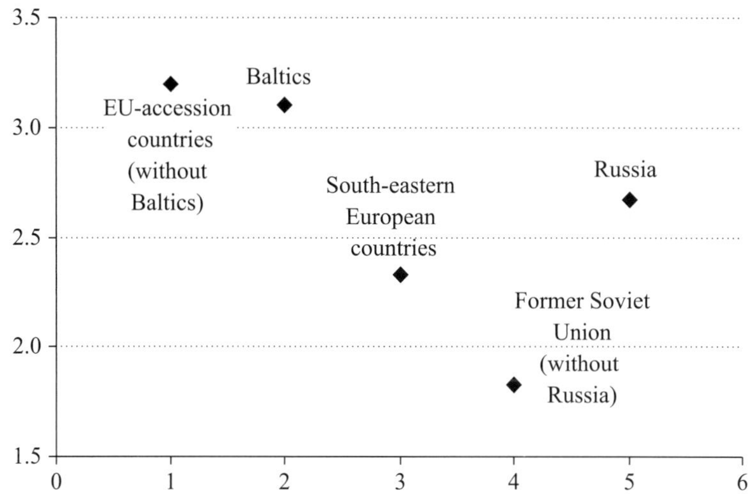

Source: EBRD (2007), own calculations.

Figure 25.1 Infrastructure reforms in Eastern Europe

Table 25.1 Russian infrastructure reforms by sectors, 2000–07

	2000	2001	2002	2003	2004	2005	2006	2007	2008
EBRD index of infrastructure reform	2.3	2.3	2.3	2.3	2.7	2.7	2.7	2.7	2.7
Electric power	2.0	2.0	2.3	3.0	3.0	3.0	3.0	3.0	3.3
Railways	2.3	2.3	2.3	2.3	2.7	2.7	2.7	3.0	3.0
Roads	2.0	2.0	2.3	2.3	2.3	2.3	2.3	2.3	2.3
Telecommunications	3.0	3.0	3.0	3.0	3.0	3.0	3.0	3.0	3.3
Water and waste water	2.3	2.3	2.3	2.3	2.3	2.3	2.3	2.3	2.7

Source: EBRD, institutional and structural change indicators, http://www.ebrd.com/country/sector/econo/stats/index.htm (accessed summer 2009).

places Russia in the medium performance range in Eastern Europe (see Figure 25.1). In this region, countries have proceeded quite differently with infrastructure reforms in recent years. Countries which have subsequently joined the European Union (Eastern European accession countries and the Baltic States) have made remarkable progress, whereas Southeastern European countries and former Soviet republics have been slow in implementing the reforms (see Figure 25.1).

The EBRD infrastructure index encompasses reforms in the following sectors: electric power, railways, roads, telecommunications and water. Russia has made some progress with reforms in the electric power sector, telecommunications and railways, whereas reforms in roads and water lag behind (see Table 25.1). The indicator also shows that progress over the past years has been limited.

According to the infrastructure index, Russia has not yet made sufficient progress in the area of infrastructure reforms. While this may be true for western observers and the measurements they use, such an evaluation neglects the fact that Russia has had a difficult task of completely reorganizing its entire system from a planned economy toward a market economy since 1992. Liberalization and restructuring of the Russian economy go far beyond infrastructural reform, and even within infrastructure sectors the reforms needed were greater than in Western European market economies. This will be taken into account when reforms of various infrastructure sectors are analysed in the next chapter.

LIBERALIZATION AND RESTRUCTURING OF INFRASTRUCTURES IN RUSSIA

Russian infrastructures have special characteristics because of their socialist legacies. At least at the beginning of transition, infrastructure facilities were technically outdated and in poor condition. In addition, infrastructures were constructed to mainly serve the needs of heavy industry (for example, railways), instead of the needs of private businesses or individual households, such as access to telephones. This misbalance became all the more obvious when market activities gained momentum. Apart from the common legacies of the planned economy, in Russia – as elsewhere – different infrastructure sectors have different characteristics. In some sectors reforms have been based on western reform models or influenced by western experiences. The following section highlights the similarities and differences of infrastructure reforms in different sectors, starting with energy, which is by far the biggest sector of the Russian economy.

Energy

During the socialism era, the Russian energy sectors – as other sectors of the economy – were owned by the state. Prices were regulated and deliveries to Russian and foreign customers were centrally planned. The transition challenge was twofold: first, to (fully) liberalize and privatize those parts of the energy sector (for example energy production, distribution to end-users) which can be conducted by 'normal' private business and which typically had been liberalized in western economies (for example, oil production and gasoline retail trade). With regard to the Russian oil industry, the liberalization and privatization were accomplished during the 1990s; however, the government continues to impose special duties and apply special regulations (such as export duties and barriers for market entry).[5] The second part of the challenge was to liberalize and deregulate the gas and electricity sector. Thereby, the competitive part (production and supply) was to be separated from those parts which are supposed to be a natural monopoly. The state-owned companies have to be unbundled and privatized. In addition, the regulatory schemes and rules for the access to and usage of transport lines had to be introduced. These objectives were achieved with different models and varying success in different energy sectors. The following section will take a closer look at the electricity and gas sector reforms.

Electricity

Russian electricity generation drastically shrank during the 1990s, reflecting the massive industrial production downturn caused by the transition crisis. The decrease in the amount of energy generated was most noticeable for thermal power plants. At the end of 1990s, signs of recovery can be observed. Since then, electricity generation has increased; however, it still has not regained previous levels. As a consequence of the global financial crisis and shrinking production the electricity consumption is expected to decrease by 4.5 per cent in 2009.

Up until recently, the Russian electricity sector was headed by the Unified Energy System of Russia Limited (RAO UES of Russia) which was created in 1992. RAO UES owns 72.5 per cent of installed capacity of Russian power plants and generates 69.5 per cent of Russian electricity production.[6] In addition, it controls 96 per cent of the total length of Russia's trunk transmission lines. The nuclear power sector is outside the structure of RAO UES.

During the period from 1992 to 2008, a reorganization and subsequently (2003–08) a restructuring of the RAO UES took place. In addition to the restructuring of the company, the whole electricity sector was completely reorganized. During this process, some of the previous plans and decisions were altered and the reform programme revised in several aspects. Thus, the reform process was extremely complex. In what follows we present the rough guidelines for reforms and simplified structures of their outcomes.

The RAO UES was included in the voucher privatization scheme which took place in Russia from 1992 to 1994. During that period 20 per cent of the shares of the RAO UES were distributed to the public and another 5 per cent were given to the company's employees. Later, within the framework of sales privatization in the second half of the 1990s the Russian government sold shares to private business firms and commercial banks but retained its majority shareholder position. In 2000, the state owned about 52 per cent of shares in the RAO UES.

In 1992 energy companies (so-called AO energos) were created at the regional level to operate smaller power plants and the low voltage grid. The RAO UES was the main single owner of these energos, holding at least 49 per cent of the energos' equity. Only four energos (Irkutskenergo, Tatenergo, Bashkirenergo, and Novosibirskenergo) have remained independent. In 2000, the RAO UES held equity stakes in 72 energos, in addition to owning 36 federal power plants (see Figure 25.2).[7] Thus, during the 1990s the electricity sector remained a vertically integrated monopoly.

In 2000/01 both Russian and foreign advisers presented a range of proposals to the reform and deregulation of the sector. According to Engoian (2006), two competing models have survived: (1) the RAO UES–Arthur Andersen–Minekonomrazvitia (Ministry for Economic Development) model, which was based on asset privatization and total liberalization,[8] and (2) the Gossovet Group model, which guaranteed the interests of the regional authorities, the energos and minority shareholders (thus being closer in nature to the German model). In the end, a combination of these two models was chosen with an emphasis on privatization and liberalization (see Resolution no. 526 'On reforming the electricity sector in the Russian Federation' of 11 July 2001). According to the chosen reform model the main elements of electricity reforms would be (1) breaking up of the vertically integrated sector and separating those parts where competition

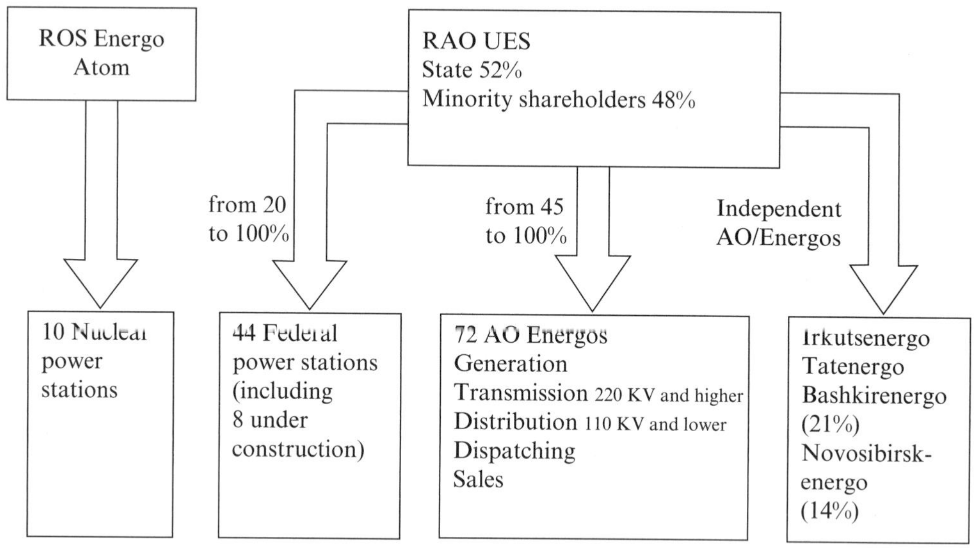

Source: RAO UES Rossii Power industry reform background; own illustration.

Figure 25.2 Structure of the Russian electricity sector in 2000

is feasible (production and supply) from transmission and distribution which are characterized by natural monopoly; (2) fostering competition, and (3) creating a regulatory framework for transmission and distribution activities.

The reforms started with a preparatory stage (for example, inventory and accounting). In fact, it was only in spring 2003 that key laws were adopted (including the 'Law on the Electric Power Industry' – Russian Federation, 2003) which would provide the legal basis for reforms of the electricity sector. In addition, the RAO UES itself presented a plan for restructuring the holding, called the "5+5 Plan" (referring to the five years needed to restructure the holding and the five years afterward). The deadline for the completion of the reforms and restructuring of the electricity sector was set for 2008/2009. The period from 2003 to 2009 was divided into three phases (see IEA, 2005; OECD, 2005):

Phase 1 (2003–04), which envisaged physical unbundling of vertically integrated entities. In addition, the following reform issues were on the agenda: establishment of wholesale generating companies and the national system operator, unbundling of four energos in a pilot project, implementation of a transitional wholesale market, development of the regulatory framework for competition and of the scope of the regulatory authority. Not all of the objectives could be reached on time. With regard to physical unbundling this phase was later extended to 1 April 2006.

Phase 2 (2005–06): the extension and functioning of the wholesale market were a core task during this phase, featuring plans for the creation of wholesale generation companies and the restructuring of the energos. In addition, the RAO UES, the Federal Grid Company and the national system operator were to be separated. Privatization of thermal power plants was to begin.

Phase 3 (2006–09): in this phase the competitive segment of the wholesale market will be extended and existing price controls abolished. In addition, the regulatory environ-

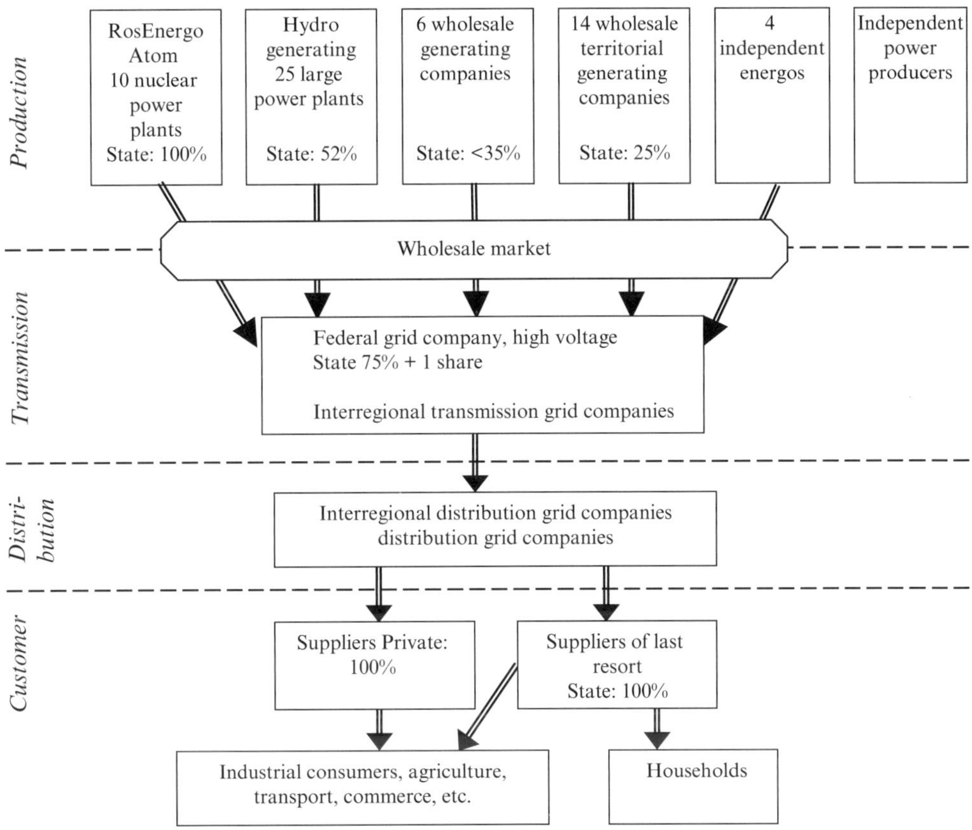

Source: RAO UES of Russia (2008), Tompson (2004); own research.

Figure 25.3 Target structure of the Russian electricity sector, 2008–09

ment for the wholesale market will be completed and the guidelines for the retail market will be developed (see the target structure in Figure 25.3).

Indeed, the electricity reforms have roughly followed these guidelines. However, the reform repeatedly fell behind schedule. The creation of most of the wholesale generation companies (so called OGCs) was completed in autumn 2006, while the restructuring of the energos and the establishment of 14 territorial generating companies (TGCs) as well as the independent grid companies (IDGC) were completed only in late 2007.[9] In 2007 and 2008 the process of selling shares of OGCs and TGCs has been continued. As of July 2008, foreign companies have been able to acquire shares in three TGCs and three OGCs. But not only foreigners have acquired shares in power generating companies. Gazprom, too, bought four important companies which supply Moscow and St Petersburg. The new established power companies have announced ambitious investment programmes to modernize generation facilities. However, the implementation of these programmes are highly questionable or will be postponed as long as electricity consumption declines as a result of the financial crisis.

A core element of the electricity sector reform is the creation of a wholesale market.

Free trade on the wholesale market in the European part of Russia and the Urals region commenced in November 2003. Gradually trade has been extended to other regions further east as well. The share of the regulated sector within the total trade volume in the wholesale market amounted to 85 per cent and was set to be reduced in the following years, compensated by the growth of the competitive sector. However, in autumn 2006, new rules were introduced on the wholesale market. At that time, the regulated sector and the free trade sector were replaced by a system of regulated contracts between buyer and supplier of electricity; contract duration is for one year or more and prices are initially regulated by the Federal Tariff Service. Volumes traded at regulated prices in the wholesale market will be steadily reduced and at the beginning of 2011 electricity will be exclusively sold at free prices (Resolution no. 205, 7 April 2001 'On amending certain resolutions of the Russian Federation Government related to the calculation of electricity volumes sold at free (competitive) prices'. A gradual approach will also be used to liberalize the retail market.

Transmission and distribution of electricity have also been reorganized. In 2002 the Federal Grid Company (FGC) was established. In the course of the reorganization of the RAO UES, grid companies (trunk grid companies as well as distribution grid companies) were spun-off from the energos. In addition, interregional transmission companies (ITC) were formed. In December 2007 the decision was made to merge the 56 trunk grid companies and the seven ITC with the FGC. The distribution grid system was also reorganized. In spring 2007, 11 interregional distribution grid companies existed; this number was reduced to eight by mergers with corresponding distribution companies in autumn of the same year. In early 2008 decisions were made for further reorganization again in the form of a merger. In sum, the organization of the grid system has been centralized in recent years. It has been announced that in a pilot project tariffs will be regulated by a model based on the rate of return on invested capital starting on 1 July 2008.

Energy sales companies have also been created during the unbundling of the energos. On principle, these sales companies can either operate as competitive sales companies or as lenders of last resort which are obliged to supply energy to end consumers, either households or industrial customers. Until end-user prices are not fully liberalized, in practice consumers will buy electricity from their regional supplier of last resort (see Doeh et al., 2008). Sales companies will receive the status of a supplier of last resort by an open tender which is organized by regional authorities. Thus, energy supply remains under state influence. So far, the overall functioning and the legal basis for a liberalized retail market remains an open question.

In Russia, different state authorities are responsible for regulation, tariff setting and fostering competition in the electricity sector. Since 2004, a federal executive agency called the Federal Tariff Service[10] – which is the successor of the Federal Energy Commission – has been responsible for the regulation of natural monopolies, such as electricity and gas, oil pipelines, railways as well as ports and airports. The tasks of the Federal Tariff Service encompass the price (tariff) setting and the control of price (tariff) determination and application.[11] Thus, it is not really an independent authority but can serve the government in its double interests as the regulator of the electricity market and a market participant (as the state still holds shares in various energy companies.). Regional energy commissions set tariffs for the retail segment of the electricity sector. According to Tompson (2004), regional energy commissions can set different prices for different customers. This enables them to conduct a kind of industrial regional policy.[12] The Federal Antimonopoly Service is responsible for fostering competition and preventing abuse of monopoly power. In 2006 a new competition law was enacted[13] that has changed some provisions of the old law and added new definitions (for example, the threshold for the 'dominant position' has been reduced from 65 per cent to 50 per cent of the market; the definition of 'monopoly prices' has also been changed).

In sum, the reorganization of generation, transmission, distribution and sale of electricity has resulted in the break-up of the former state monopoly. The reforms created new structures but not necessarily a better profile of the electricity sector with regard to competition and efficiency. It remains at least doubtful whether the electricity reform has actually created competitive structures on the supply side and in the sales sectors of the economy; in these sectors, companies which might exert regional market power still exist.[14] In the transmission segment, mergers of interregional or regional transmission companies have increased concentration and centralization. At the same time, the Federal Tariff Commission is not truly independent in its capacity as a regulatory agency. It will depend on the regulatory practice and the concrete competition policy whether the potentially competitive segments of the electricity sector can survive in the years to come. For the time being, it remains an open question, whether the reforms will actually result in a long-lasting liberalization and deregulation of the electricity sector. The liberalization is hampered by current problems due to the financial crisis. The financial crisis has re-created non-payments of energy bills, which were a widespread phenomena of the transition era. In almost all sectors of the power industry, investment outlays had been reduced in 2009 and postponed. Thus, the modernization of the sector is further delayed.

Natural Gas

Russia is the world's largest producer of natural gas. Production of natural gas decreased in the 1990s, the trend that was reversed around 2000/01. Since then, production increases have been registered. In 2008 about 664 billion m^3 have been produced. In the first half of 2009 production shrank by more than 20 per cent as a result of the financial crisis. (see also IEA, 2006.)

In the future, any significant gas production increases are unlikely. One reason is that production from large gas fields is decreasing, while the exploration and production from new (smaller) fields is unable to cover the resulting deficit. At the same time, gross domestic demand is increasing, which, coupled with supply-side constraints, has led analysts like Sagers (2007) to warn against an impending 'gas crisis' in Russia.

Within Russia, Gazprom (and its affiliates) dominates the domestic gas sector in almost all its segments from the exploration of natural gas fields, to pipeline transmission and Russian gas exports. Gazprom is licensed to develop 60 per cent of Russian gas reserves. In addition, Gazprom is the only owner of the high-pressure interregional gas pipelines[15] and the sole operator of the 24 gas storages. It controls all gas exports to western countries. In 2008 Gazprom produced about 83 per cent of the Russian gas production. Thus, only a small – albeit growing – share of natural gas is produced by other independent companies.[16]

The reorganization of Gazprom can be divided into at least two phases. During 1992–98 it seemed that the reorganization and privatization of Gazprom would eventually lead to a mixed shareholding model with the government being an important but not the sole shareholder. In summer 1992 Gazprom was first transformed into a '*kontsern* (corporate group) of the Russian state' and later into a state-owned joint stock company. The state retained 40 per cent of the shares. The remaining 60 per cent of the shares were included in the voucher privatization programme of 1994. These shares were distributed to the broad population (34 per cent), workers and management (15 per cent), Rozgasifikasija (1 per cent) and Gazprom (10 per cent). It was planned that 9 per cent of the shares owned by Gazprom would be sold to foreign investors. In 1996 foreigners were first allowed to hold shares as American Depositary Receipts (ADR). In 1998 the first sale to a foreign company occurred when German Ruhrgas acquired 2.5 per cent of the shares.[17] At the same time, the government held about 39 per cent of shares. Since then, the ownership structure has significantly changed in favour of state ownership; by 2005 the government has regained the majority shareholding in Gazprom. Overall, in the last decade the reorganization of Gazprom served to strengthen the role of the state, rather than to reduce it. See also, on deregulation of the Russian natural gas sector and gas supply, Grigoryev (2007).

According to Ahrend and Tompson (2004), the Russian Ministry for Economic Development and Trade elaborated several proposals on gas sector reforms including the restructuring of Gazprom which were successfully resisted by Gazprom in recent years. Gazprom has even thwarted relatively modest attempts at an internal reorganization of its subsidiaries proposed by the ministry as an alternative to breaking up the whole company. This reform resistance coincided with Gazprom's further expansion into activities of other energy sectors (for example, electricity) and other sectors of the economy (such as banking, insurance and mass media) by establishing new companies or buying shares in existing enterprises.

Gazprom has been said to be a state within the Russian state. However, it does not wield unlimited control over the Russian gas sector policy. As it has been mentioned above, the state has been able to increase its direct influence on Gazprom by means of increasing ownership participation in recent years. Indirectly, Gazprom's business activities are also constrained by governmental regulations in the area of tariffs and prices. Gazprom is required to supply the Russian gas market with an annually negotiated volume of natural gas at regulated prices which do not even cover production costs.[18] In 2007 the regulated sector was to account for 72 per cent of domestic gas consumption. The remaining natural gas for the domestic market was supplied by Gazprom (4 per cent) and independent producers (about 24 per cent), with the majority of this volume delivered at so-called free prices. In spring 2007 price ceilings and minimal prices were introduced in the regulated sector.[19] In addition, wholesale prices in the regulated sector are set to increase annually over the period from 2007 to 2010. It is planned that domestic gas prices in the regulated sector will cover production costs by the end of 2010, and that by 2011, the separation between the regulated sector and the free price sector will cease to exist. On average, the Russian domestic price level for natural gas remains much lower compared to the prices that Western European consumers are charged. In addition, Russian gas price policy stands in contrast to that of western countries where large industrial consumers benefit from price reductions as compared to households. In what can be seen as the influence of the socialist past, prices for households are still lower than for industrial consumers.[20] Gas bills for households are often calculated not on the basis of actual consumption, but rather on the basis of a flat rate per square metre or per person living in the household. Logically, as long as the price level remains low and payments are not tied to actual consumption, there are hardly any incentives on the part of domestic consumers to save energy.

The separation between the domestic gas sector with low regulated prices and the foreign gas market where international market prices are used and real revenues can be collected, will be maintained at least until 2011. Gazprom has a strong interest in exporting expensive gas to western European countries as compared to CIS countries, which buy gas at a cheaper price. Gazprom's objective is to raise CIS export prices to the Western European price level in the coming years.

Gazprom – and more precisely its subsidiary Gazpromexport – controls the export of natural gas to foreign customers. In theory this is no impediment if all gas exporting companies are granted equal access to the pipeline network. In Russia, Third Party Access (TPA) to the pipeline network is guaranteed by law. The Federal Tariff Service sets the tariffs for pipeline use by independent companies. Since 1 October 2005 the tariff has been fixed at 23.84 RUR per 100 cu m per 100 km.[21] However, it remains unclear whether independent companies get fair access in practice. Some legal provisions allow Gazprom to circumvent the original spirit and purpose of the law. For instance, Gazprom has a right to deny access because of capacity problems. In this situation, however, outsiders (including the regulatory authority) find it extremely difficult to obtain the needed information in order to verify whether there are indeed capacity constraints. Independent gas producers are also separated from foreign markets and have restricted access to profitable business activities abroad, while Gazexport remains the sole exporter to Western European countries. Deliveries to European countries make up the bulk of Russian gas exports, as lower export prices for CIS countries make exports to these countries less profitable.

Although independent gas producers have become more visible on the supply side of the Russian gas sector, Gazprom maintains a key function in transmission and exports. Because of information asymmetries, Gazprom also has a stronger position vis-á-vis the regulatory authority. The increasing share of state ownership then might indicate that the government is trying to regain control and regulatory power over Gazprom and thus to put a 'fence' around the 'state within the state'.

Transport

The Russian transport system has to be considered from the background of the country's vast territory and its extreme climatic conditions. In addition, the transformation crisis in the first half of the 1990s caused a sharp reduction in the volumes of freight transport as well as passenger transport. Freight transport recovered during the second half of the 1990s whereas passenger transport is still in decline.

With regard to the type of transport rail accounts for more than 55 per cent of overall freight transport as measured in tons per kilometre (t-km). In the eastern part of Russia its role is even more important. Shipping plays a minor role as freight transport medium. Inland shipping is constrained because rivers flow in a north–south direction, whereas major traffic flows go from east to west. In addition, in winter rivers are frozen and non-navigable. Compared to rail, the importance of roads in freight transport is still low but has been increasing in recent years.

With regard to passenger transport, the use of all kinds of transport vehicles has decreased during the period 1990–2007. Buses are still the dominant means of transportation followed by metro, trolleybuses and tramways. In inter-city connections rail accounted for two-thirds of passenger kilometres in 2005, by air transport which amounted to 22 per cent of passengers kilometres. These shares have changed in favour of air transport during 2000–05. Prior to the era of high oil prices such a development was viewed by many as a sign of the growing importance of the airline industry in long-distance passenger transport – a prediction which appears less likely to hold true with the oil prices setting ever higher records and the airline industry going through a crisis. In addition, it is possible that ground cargo transportation will increase its share in the overall volume of freight transport. Nevertheless, railway will remain the most important transport medium in the Russian Federation.

In Russia, the railways network, airports and (maritime) ports are listed in the registry for natural monopolies.[22] The Federal Tariff Service is responsible for tariff setting for railways, airports and (maritime) ports.[23] Proposals to reform various transport sectors differ in extent. Whereas the privatization of airports is still under discussion,[24] some progress can be observed with the restructuring of railways.

Railways

The Russian rail system is one of the largest in the world. The combined length of all routes stands at 85245 km, with annual passenger traffic 1.3 billion people and cargo traffic of about 1281 billion tons. The Russian railway system employs over 1 million people. See also Tsuij and Sergachev (n.d.).

According to Friebel et al. (2007), the preconditions for Russian railway reforms significantly differ from those in Western European countries. In Russia, low-value, bulk commodities (for example, coal and construction materials) dominate freight transport. Producers heavily rely on railways, especially in the eastern part of the country where the roads are in poor condition or are even nonexistent. In contrast, in Western Europe railway reforms are initiated while rail transport faces heavy competition by road transport for both freight and passenger transport. In Western Europe high value commodities are also carried by motor vehicles. In addition, in Western Europe the countries and the distances to overcome are small compared to Russia. Finally, the Russian railway system is old and in high need of repair. Therefore, Russia had to opt for a reform model that takes into account these large investment needs.

Friebel et al. (2007) distinguishes between two railway models: the North American model and the European model. The North American model is characterized by two independent vertically integrated railway companies which are privately owned. Those structures have their historical origins in small private railway enterprises. In this setting the reform objective is to create a kind of 'geographic' or 'source' competition: it might be the case that there is only one railroad company offering a demanded route, but other companies do offer alternative routes and destinations. Therefore, this model does not provide perfect competition and does not always avoid natural monopoly positions. However, in offering alternative suppliers and routes, it set limits to monopoly power. In contrast, railways in the European Union (EU) have other traditional characteristics. The EU reform model is based on privatizing the railway sector apart from the grid. The latter constitutes a natural monopoly and remains regulated (and is often state owned). Railway companies should have equal access to the infrastructure. Across Western European countries differences exist with regard to access, the extent of privatization and the extent of vertical separation of the competitive and the monopoly sector.

Russian railways reform has generally followed the European reform model. However, the reforms are characterized by Russian peculiarities. In 2003 the Federal Law on Railway Transport came into force.[25] The law created a legal basis for the separation of infrastructure and train operations. Thereby, open and fair access to the infrastructure should be guaranteed legally. In October 2003 the Russian railroad company, Rossijskie Zhelesnye Dorogi (RZhD), was created as a joint stock company. The company is state owned and vertically integrated.[26] The creation of the railroad company was the central part of the first stage of the envisaged railway reform (see European Conference of Ministers of Transport, 2004). During this stage, the former Ministry of Railways was renamed the Federal Railway Transport Agency (FTRA) as the regulator of railway transport. In the second stage (2003–05) it was envisaged that the funding for loss-making passenger transport would be divided between federal, regional and local budgets. The Russian railroad company would maintain its major function in freight transport and would operate the infrastructures. However, independent companies would also have the opportunity to enter the market. In addition, subsidiaries of the railroad company would be created (among others, suburban passengers transport, freight transit operations). In the third stage (2006–10) the subsidiaries will be privatized. In addition, the role of independent companies in freight transport and thus competition will subsequently increase.

Whereas in the first stage reforms were implemented quickly and almost completely, the reform measures of the second stage have been implemented only partially (for example, the central government and regional and local government could not agree upon funding for passenger transport, see Pittman, 2007).[27] However, the major shortcoming of reforms during the second and, in particular, the third stage is the still prevalent lack of competition. It was only in 2007 that stock companies were created, passenger services were separated and more private companies engaged in freight transport (EBRD, 2007). As a prominent investor the EBRD has agreed to acquire a minority stake in TransContainer, Russia's largest provider of rail container services which handles more than half of all container traffic in the country. In addition, the creation and subsequent privatization of other ancillary activities are under way.[28]

The Russian railway strategy for 2030 envisages huge investments, for example, in acquiring new rolling stock and constructing new railways. The financial crisis made the implementation of the modernization strategy more difficult. In summer 2009 the EBRD has provided the Russian railroad company with a long term loan of US$500 in order to continue its reform programme.

Telecommunications

Telecommunication is a fast growing sector in Russia. Growth in key indicators such as the number of telephones and the number of calls made serves as proof of this positive development. Within the period 2000–06 the number of long distance calls tripled. In particular, the number of mobile phones has rapidly increased in recent years. In 2006 the saturation of the market even surpassed 100 per cent (which means that there were more mobile phones in the country than the number of population).[29] In the first quarter of 2008, the number of mobile telephony units reached 177 million, an 11 per cent increase from the first quarter of 2007. In accordance with increasing number of telephone calls and telephones, revenues from telecommunications services have sharply grown in recent years. Again, the most prospering segment is the markets for mobile phones.

Reforms of telecommunications have a signalling function in former socialist countries. During the Soviet times all means of communication were controlled by the state and used for propaganda and control. During the 1990s, the Russian telecommunication sector was reorganized and partially privatized. In 1993, the Rostelecom became the single national network operator, while local network operators were privatized. At that time each region had its local operator. In 1995 state-owned shares of the local operators were consolidated into a holding company, Svyazinvest. The initial plan to (partially) privatize the holding could not be totally implemented. Today, the Russian state still owns 75 per cent minus one share of Szyazinvest. Other shareholders are JSC Comstar-UTS (17.31 per cent plus one share) and MGTS FINANCE S.A. (7.69 per cent).[30] In 2002 a reorganization of the regional level had taken place and seven regional operators had been established. Svyazinest holds at least 51 per cent of the shares in these regional companies. In addition, Svyazinvest incorporates Giprosvyaz and Rostelecom. The latter functions as a domestic long-distance and international operator.

The Russian telecommunication sector is regulated by the Ministry of Communication and the Federal Communication Agency which is part of the ministry. The Federal Anti-monopoly Service and local administrations are also involved in local tariff setting. The legal basis for the operation of telecommunications is the Federal Law 'On Communication' which was adopted in summer 2003.[31] One of the key objectives of the law was to stimulate competition in the sector, in particular, in long-distance telephony.

In 2006 Rostelecom's monopoly in long-distance telephony was abolished. Today, about 30 companies are licensed to provide long-distance services in Russia. Aside from Rostelecom, important players are MTT, Transtelekom, Golden Telecom, and Equant (Orange). However, Rostelecom remains the dominant player; in 2007 it controlled about 56 per cent of the long-distance market, MTT accounted for 7 per cent (with the exception of Moscow, where MTT controls 17 per cent of the market) and Golden Telecom controlled 17 per cent. The rest – about 20 per cent – belonged to Transtelekom and other companies, including IP telephony operators.

In contrast to long-distance telephony, the market for mobile telephony is dominated by three leading private companies, Vimpelcom, MegaFon and MTS. The market share of Vimpelcom as well as MegaFon is estimated at about 35 per cent; MTS holds another 25 per cent. Other companies on the Russian market are Uralsvyazinform (which is owned by 41 per cent by Svyazinvest) and Tele2. Because the market has reached a certain level of maturity most major providers have chosen a new business strategy – namely, to increase the average revenue per user (APRU), which has been traditionally low in Russia.

Postal services were liberalized during transition. In 1996 state monopoly on special postal services was abolished. Since then, commercial postal enterprises have been established, with over 80 organizations obtaining licences for postal services.[32] However, they often provide niche products. Thus, the Russian Post remains an important organization for traditional postal services. It is required to provide postal services even in remote areas. The efficiency and financial performance of the Russian Post are low and its technical infrastructure is outdated. The federal budget provides financial resources to improve the technical infrastructure. Some proposals currently under discussion offer to provide the Russian Post with additional financial resources from the stabilization fund (see Bundesagentur für Außenwirtschaft, 2007). However, a comprehensive development programme for the Russian post is still outstanding.

Water and Wastewater

The Russian water network encompasses about 528 000 km, sanitation about 176 000 km. About 78 per cent of the population is connected to the centralized water supply and about 62 per cent of the population to the wastewater collection (see OECD, EAP Task Force, 2007). There are differences between urban and rural areas: in 2005, the coverage of population in urban areas amounted to 97 per cent, whereas in rural areas water pipelines were available only in 38 per cent of population centres. An even greater difference exists with regard to wastewater. Sewerage systems are installed in 76 per cent of urban areas, as compared with only 4 per cent of rural population centres (data for 2003; OECD, 2003).

Water use decreased during the period 2000–06. However, consumption levels are still high. Whereas in European cities daily per capita water consumption ranges between 129 litres in Germany and 265 litres in Spain, water consumption in Russian cities amounts to 300 litres a day. One reason for this are the still low water prices (on average about 10.30 RUR/cu m fresh water; 7.9 RUR/cu m waste water)[33] which do not cover supply costs. In addition, water meters are not widely used in Russia (see OECD, EAP Task Force 2007). Again, as a result of legacy of the Soviet past, tariffs for households are lower than for commercial consumers. The underpayment by the population is covered by federal subsidies. However, in recent years the level of household prices has increased, cost coverage has improved and cross subsidization has decreased (see OECD, EAP Task Force 2007.

Water supply is mainly operated by municipal enterprises. Private enterprises play a minor, albeit increasing, role in Russian water supply, according to OECD, EAP Task Force (2007). In contrast to this conclusion, Hall and Popov (2005) have observed a rapid and risky privatization of Russian water supply. However, Hall and Popov make it clear that private operators are usually not the owners, but control assets by means of a lease, rent or concession agreement for periods ranging from 25 to 49 years.[34] For these contracts a competitive tendering usually does not take place. Water companies are often owned by holdings, for example, Rossiiskie Kommunalnye Sistemy (Russian Utility Systems) is owned by the IES (Integrated Energy Systems) holding, and Rosvodokanal belongs to the Alpha Group. Another main supplier is Evraziiskoe vodnoye partnerstvo (Eurasian Water Partnership).

Until 2006 water tariff regulation depended on ownership type (see OECD, EAP Task Force, 2007). That means that the prices of municipal enterprises were regulated by local governments, while other enterprises' tariffs were regulated by federal authorities. This differentiation between ownership types has been abolished by the new Federal Law 'On principles for tariff regulation of municipal organizations' of 30 December 2004, No. 210, which came into force on 1 January 2006. In addition, the law stipulates that tariffs have to meet the financial needs of municipal organizations and help in the implementation of their investment programs for the modernization of the infrastructure. The latter is extremely important, because the Russian water infrastructure is in poor condition with a high amount of leakage (see Bundesagentur für Außenwirtschaft, 2008).[35] According to OECD, EAP Task Force (2007) the share of water supply networks that require replacement increased from 29.5 to 35 per cent during 2000–05. The accident rate – measured per kilometre of water supply – ranges from 0.38 to 0.42. For sewerage network the accident rate increased from 0.18 to 0.22 during the period 2000–05.

In sum, reforms of water supply and wastewater were not always conducted in an orderly manner. Obviously, there was no coherent reform model for ownership changes and privatization as well as tariff regulation (since 2006). However, water tariffs which allow for investment outlays and strategic investors or owners are preconditions for the modernization of the outdated water infrastructure.

Until mid-2009 there has been no comprehensive programme for the further development of water supply and distribution. The so-called 'Water Strategy' is still under consideration. Nevertheless, on the regional level the modernization of water infrastructure is under way.

FINAL REMARKS AND FUTURE PERSPECTIVES

In Russia, the transition from central planning toward a market economy has broadly affected the infrastructural sectors. Given the difficult initial conditions and the lack of experience with regard to economic liberalization and regulation, Russia has indeed made some good progress in the field of infrastructure reforms. During the socialist past, infrastructures were seen as important, in particular in their capacity as key sectors of the economy which allowed to control energy and transport flows, as well as telecommunications and the media. Throughout the transition, the energy sector kept its so-called 'strategic sector' status, which entailed special legal regulation. In general, reforms in most infrastructure sectors have not reached their final stage. In addition, there are a number of differences among different infrastructure sectors with respect to reform concepts, implementation and progress.

The electricity sector differed from other infrastructure-related sectors in that its reform programme was the most elaborate of all, although the scope and timing of the reform were changed in the course of implementation. Still, the ambitious reform programme has been implemented only partially. The reform involved the breakdown and reorganization of the former state monopoly, the RAO UES; as a result, foreign investors have been allowed to acquire shares in wholesale generation companies. However, it remains doubtful whether these measures have actually fostered the creation of competitive structures, in particular with regard to regional markets. At the same time, mergers of grid companies have led to a recentralisation of the grid system. In addition, price liberalization is far from being complete. Therefore, liberalization and deregulation would have to be continued consistently during the next years. However, at the moment there appears to be no strong actors who might encourage further reforms.

The Russian natural gas sector is dominated by Gazprom, which also has a monopoly on gas exports to western countries. During the 1990s a (semi)privatization of Gazprom was envisaged in order to reduce the state's role in the gas sector. Despite this initial privatization plan, Gazprom's ownership structure has actually changed in favour of the government. Other gas companies have gained some importance with regard to gas production in recent years. However, foreigners still face difficulties to enter the gas market and engage in long-term business activities, as it has been the case with Royal Dutch Shell/Mitsubishi/Mitsui which had to give up their majority stake in Sakhalin II in favour of Gazprom. In addition, reforms of the domestic Russian natural gas market have been slow. Wholesale prices are being liberalized only gradually, a process which will last at least until 2011. Until then, a free sector and a regulated sector will have to coexist. The gradual nature of price increases is officially justified by social aspects. For economic reasons, social issues can be better tackled by direct transfers to poor households. In the end, low natural gas prices do not provide the right incentives for energy savings. However, a decrease in (or a lower rate) of domestic gas demand is necessary to maintain high natural gas exports in the next decade when natural gas production from smaller fields or in remote areas will become more difficult. Thus, the liberalization of domestic gas prices cannot be delayed endlessly.

In a fast-growing economy transportation is a key issue. In Russia, the railways remain the most important means of transportation. The railway reform programme comprises the deregulation and creation of competition in three stages up to the year 2010. Whereas the creation of the Russian railroad company as a joint stock company (which was the central part of the first stage of the reform in 2003) was implemented very quickly, the reform measures of the second stage (2003–05) were implemented only partially. In addition, so far the creation of competition which is the central task of the third stage of the reform (2006–10), has been slow. The EBRD has agreed to acquire a minority stake in TransContainer,which is a welcome development, since foreign investment can have positive consequences for the privatisation process, helping it to regain its momentum.

The Russian telecommunication sector is a very dynamic sector with regard to the density of the telephone networks and the number of calls made annually. During the 1990s the reorganization and partial privatization of the sector resulted in the market entry of foreign telecommunication companies. In 2006 the monopoly of Rostelecom in long-distance telephony was abolished. However, Rostelecom remains the dominant player in the market. In mobile telephony, however, it is three private companies which hold the largest shares of this dynamically developing and already highly mature market. Liberalization has also formally been conducted in postal services. However, market entry for new service suppliers has been difficult. While new entrants may provide niche products, the Russian Post remains the key supplier of traditional postal services. In sum, the telecommunication sector is an example that state firms have succeeded in exerting their market power and impeding the entry of new suppliers to the markets.

In the area of water supply and waste water, no comprehensive reform programme has been followed. Water supply is mainly operated by municipal enterprises. The procedure of privatization or other means of private control (for example, lease, concession) are not always transparent. Until 2006, water tariff regulation was differentiated depending on ownership types. The new tariff system tries to introduce tariffs which would cover costs and allow for the required modernization of the outdated water infrastructure.

A comparison of the reforms in various infrastructure sectors shows that the results of the reforms have so far been quite mixed: the electricity sector seems to be the sector that has progressed the furthest. Even there, however, the reforms have only resulted in the creation of a kind of wholesale market, and it remains unclear whether the retail market liberalization will actually succeed. In sectors where reform strategies did exist (for example, railways), the latter usually envisaged gradual implementation, thus entailing the risk that the subsequent reform stages would be postponed or even abandoned altogether. The structural and ownership changes in the telecommunications sector as well as in water supply seem to have been driven by ad hoc measures and took place rather spontaneously. Nevertheless, the telecommunications sector has experienced a dynamic development. Up until now, it has been the gas sector that has successfully resisted most attempts at reform.

To summarize, the modernization of infrastructures remains one of the main challenges Russia is facing today. So far, the existing infrastructure reforms have emphasized the reorganization and creation of regulatory authorities, but they have neglected the problem of technically outdated infrastructure. Therefore, additional financial resources must be mobilized in the future, either domestically or from abroad. However, foreign investors still face difficulties in their attempts to conduct business in Russia, as the example of Sakhalin II clearly shows. In addition, as a consequence of the global financial crises foreigners are reluctant to invest in Russia. In general, then, Russian enterprises must rely on their own resources.

The conditions for such resource mobilization seem to be the best in the gas sector, due to its high export revenues. In addition, Gazprom has increased its activities and investment in foreign gas markets. This will not only result in increased influence on foreign markets but it may also contribute to a transfer of valuable know-how. In electricity, the privatization of spun-off enterprises has only recently attracted foreign investors, and so far, little is known concerning their future business strategies in Russia. Again, owing to the financial crisis they will likely wait with huge investments. Within the telecommunications, the mobile telephony sector has attracted foreigners, but this development will most likely come to a standstill as the market is nearing saturation. Other sectors are still closed to outsiders or are only of local importance, as is the case with water supply where tariff increases are the financial preconditions for cost coverage and modernization. In general, Russia has to provide better conditions for foreign investment and for domestic resource mobilization. Nevertheless, the reliability of Russian infrastructure services must be improved. Unless this is achieved, infrastructural weaknesses may come to be one of the chief limitations on the future growth and development of the Russian economy.

NOTES

1. I wish to thank for technical assistance Wolfgang Härle and Yana Zabanova. This analysis of the Russian economy was mainly written before the whole consequences of the financial crisis became obvious and a recovery took place (for example, affecting oil prices which have rapidly increased since then). Therefore changes which have occurred since the last update of the chapter (affecting infrastructure developments in 2010/11) are not included.
2. In 2004, Russia established the so-called stabilization fund, which accumulates the revenues from oil exports sold above a crude oil price of \$20 per barrel. Because of this, the volatility of world market prices on raw materials has a reduced impact on the domestic economy. In addition, another objective of the fund is to absorb excessive liquidity. In rouble terms the amount of the stabilization fund grew from

106 billion RUR at the beginning of 2004 to 2346 billion RUR at the beginning of 2007. At the beginning of August 2007 the amount stood at 3263 billion RUR (US$127.5 billion). From 1 February 2008 the Stabilization Fund will be split into a Reserve Fund and a National Prosperity Fund. The Reserve Fund will invest in low risk securities and will serve as security fund in case of a fall in energy prices. Its amount will be about 10 per cent of GDP. Revenues above this limit will be allocated to the National Prosperity Fund. See Engerer and Kemfert (2008).
3. Studies on military installations are a special feature; the defence industry and its installations are seldom analysed within infrastructure economics.
4. See, for definition, http://lexikon.meyers.de/meyers/Infrastruktur (accessed summer/autumn 2009).
5. In addition, pipeline transport for oil and oil products is seen as a natural monopoly; third-party access to pipelines is guaranteed by law and tariffs for pipeline use are regulated. Since spring 2008 pipeline tariffs have been fixed at 176.55 RUR per tonne (without VAT), see Prikaz No. 214 of 30 April 2008.
6. See http://www.rao-ees.ru/en/info/about/main_facts/show.cgi?pozition.htm (accessed summer/autumn 2009).
7. According to Tompson (2004), energos are controlled by regional administrations rather than by the RAO UES. Regional energy commissions also play a central role in the regional electricity sector as they set tariffs for the different types of electricity consumers.
8. The project was first elaborated by the RAO UES itself in collaboration with Arthur Andersen. The project was supported by the Ministry for Economic Development (see Engoian, 2006). According to this project, the energos would be spun off into structurally distinct segments, competition would be introduced in electricity generation and supply, whereas transmission and distribution as natural monopoly would be regulated adequately
9. There is a special time schedule for the reorganization of hydropower plants. The first stage of consolidation of the hydro wholesale generation companies was completed at the end of 2007/beginning 2008, (http://www.rao-ees.ru/en/reforming/reason/show.cgi?background.htm, accessed summer/autumn 2009) The Far East region will be exempt from reforms and will remain regulated.
10. See Decree of the Russian Federation Government, No. 294, 9 April 2004.
11. See http://www.fstrf.ru/eng (accessed summer/autumn 2009).
12. According to Yudashkina und Pobochy (2005) the price tariffs are calculated on the basis of the 'cost plus profit' method. Thereby generation and transmission costs are calculated. Federal as well as regional components are considered for retail prices. In addition, the government determines minimum and maximum tariffs for each Russian region once a year. This provides a corridor for regional regulators.
13. Federal Law, No. 135-FZ of 26 July 2006, 'On Protection of Competition'.
14. The IEA (2005) has calculated for the proposed structure of the wholesale market the so-called Herfindahl-Hirshman Index which is a measurement for concentration. It turns out that from a global perspective the reforms will result in higher competition of the wholesale market. From a regional perspective, however, the concentration ratio remains high.
15. An exemption has been the low-pressure gas distribution network. In the 1990s Rosgazifikatsiya operated the low-pressure distribution network. In 1992 the local offices became independent and most of them were transferred to joint stock companies. However, many of the companies were highly indebted. By shares-for-debt swaps Gazprom bought indebted companies and obtained access to the gas distribution segment.
16. These companies are often subdivided into other independent gas companies (for example, Norilskgazprom) and vertically integrated oil companies (such as Lukoil) producing natural gas.
17. Foreign participation was first limited to 20 per cent of Gazprom's capital. This restriction was abolished in 2005.
18. According to Ahrend and Tompson (2004), the government and Gazprom negotiate a 'gas balance' for the domestic supply at the end of the year for the year ahead. By this, the gas to be supplied to the domestic market at regulated prices is determined. Gazprom informs the industrial consumers on their quotas and also corrects their bids if necessary. Because of its position as a gas producer, pipeline operator and gas exporter Gazprom benefits from the existence of information asymmetries. In sum, this kind of rationing system is – according to Ahrend and Tompson – completely opaque.
19. See Postanovlenie No. 33/2007.
20. There are different price zones in Russia. Until 2005, seven pricing zones existed. Since 2005, 11 pricing zones have been established. In addition, wholesale prices are adjusted for seasonal differences in consumption. Price zones and seasonal adjustments are only examples for the complicated price formula which is applied for domestic price determination.
21. See http://www.fstrf.ru/otrasli/gas/viewdoc?url=/fst/dep/gas/intrest/6 (accessed summer 2009). The tariff has not changed in the meantime. Gazprom has complained that the tariff does not cover its costs to maintain the pipeline network.
22. See Federal Tariff Service, 'Registry of subjects of natural monopoly in transport'.

23. For legal regulations and tariffs, see http://www.fstrf.ru/otrasli/terminal/?start=11&topics=terminal&topics2=t (accessed summer 2009).
24. At the end of 2006 Sochi airport was auctioned to Oleg Deripaska's Basic Element group for 5.5 billion RUR. See http://en.wikipedia.org/wiki/Sochi_International_Airport (accessed summer 2009).
25. See Federal'nyi zakon, No 18, 10.01.2003, 'Ustav zheloznodorozhnogo transporta Rossijkoj Federacii'.
26. At the beginning of 2008 the railroad company had 17 territorial branches,
27. Passenger transport is partially loss-making. There exist four classes of accommodation. The Russian railroad company can fix the fares for first and second class accommodation which is profitable. Fares in third and fourth class are regulated and make a loss. In 2008 the government will pay about 16.5 billion roubles in subsidies, rising to 30 billion the next year (see Anonymous 2008). Freight tariffs are also a problematic issue. According to Tsuji and Sergachev the Russian tariff policy is rather complex. Freight tariffs, among others, differ according to the type of cargo, as well as distance and weight. In spring 2008 a regulation announces that new transport tariffs (including airports and maritime ports) will be established by the end of 2009 (see Postanovlnie Pravitel'stva Rossiskoj Federacii No. 293/2008).
28. For instance, in March 2007 the Russian Railways Trading Company was created as subsidiary.
29. This number usually is calculated on the basis of sold SIM cards and not on the SIM cards actually in use. There might be persons, who have contracts with different mobile providers and do not actually use all of them.
30. See http://eng.svyazinvest.ru/about/ (accessed summer 2009).
31. See Federal Law of June 25, 2003 No. 126 'On Communication'.
32. See http://www.russianpost.ru/portal/ru/home/normative/developmentconcept/dc2 (accessed summer 2009).
33. See bfai http://www.bfai.de/ext/Export-Einzelsicht/DE/Content/__SharedDocs/Links-Einzeldokumente-Datenbanken/fachdokument,templateId=renderPrint/MKT200804238010.pdf (accessed 4 April 2011).
34. In summer 2005 the law 'On Concession Agreements' came into force. See for a comment Russian Law, 'New concessions law in Russia: infrastructure projects facilitated', http://www.russianlaws.com/newsdetail.aspx?news=6677 (accessed 4 April 2011).
35. According to OECD, EAP Task Force (2007) 35 per cent of water supply networks and 29 per cent of wastewater collection networks used up their resources and needed replacement.

REFERENCES

Ahrend, Rudiger and William Tompson (2004), 'Russia's gas sector: the endless wait for reform?', OECD Economics Department, Working Paper No. 402.

Anonymous (2008), 'Economic surge fuels Russian rail revival', *International Railway Journal*, June, 16–18, http://www.railjournal.com/A/xfeature1.html (accessed April 2011).

Bundesagentur für Außenwirtschaft (2007), 'Russlands Post braucht Milliardeninvestitionen', https://www.bfai.de/DE/Content/__SharedDocs/Links-Einzeldokumente-Datenbanken/fachdokument,templateId=renderSE.html?fIdent=MKT200712138004&source=Google&sourcetype=SE (accessed 4 April 2011).

Bundesagentur für Außenwirtschaft (2008), 'Russlands Wasserwirtschaft steht vor großen Investitionen', http://www.bfai.de/ext/Export-Einzelsicht/DE/Content/__SharedDocs/Links-Einzeldokumente-Datenbanken/fachdokument,templateId=renderPrint/MKT200804238010.pdf (accessed 4 April 2011).

Decree of the Russian Federation Government, No. 294 of April 9, 2004, http://www.fstrf.ru/eng/about/law/17 (accessed summer/autumn 2009).

Doeh, Doran, Charles Wood, Alexander Popov, Serguei Fominykh, Natalia Mouratova and Denton Wilde Sapte (2008), 'Reform of the Russian electricity sector', *EBRD Law in Transition*, October 2008.

European Bank for Reconstruction and Development (EBRD) (2004), *Transition Report 2004: Infrastructure*, London: EBRD.

Engerer, Hella (2002), 'The Russian gas sector: survival of the planned economy or evolution of market mechanism?', in Maarten J. Arentsen and Rolf W. Künneke (eds), *National Reforms in European Gas*, Amsterdam and Boston, MA: Elsevier, pp. 133–62.

Engerer, Hella and Claudia Kemfert (2008), 'The Russian energy sector 1990–2005 and climate policy: special emphasis on energy production and external trade', Deutsches Institut für Wirtschaftsforschung Berlin: Politikberatung kompakt 37, http://www.diw.de/documents/dokumentenarchiv/17/78006/diwkompakt_2008-037.pdf.

Engoian, Alda (2006), 'Industrial and institutional restructuring of the Russian electricity sector: status and issues', *Energy Policy*, **34**, 3233–44.

European Bank for Reconstruction and Development (EBRD) (2007), *Transition Report 2007. People in Transition*, London: EBRD.

European Conference of Ministers of Transport (2004), *Regulatory Reform of Railways in Russia*, Paris: OECD.
Federal Law of the Russian Federation, No. 147 of August 17, 1995, 'On natural monopolies', http://www.fas.gov.ru/english/legislation/5414.shtml.
Federal Law of June 25, 2003 No. 126 'On Communications', http://www.medialaw.ru/e_pages/laws/russian/communications.htm.
Federal Law of December 30, 2004, No. 210 'On principles for the regulation of tariff of minicipal organizations', http://www.fstrf.ru/eng/about/law/21.
Federal Law, No. 135-FZ of July 26, 2006, 'On protection of competition', http://www.fas.gov.ru/english/legislation/8955.shtml.
Federal Tariff Service, 'Registry of subjects of natural monopoly in transport' http://www.fstrf.ru/otrasli/terminal/viewdoc?url=/fst/dep/law/doc/26 (accessed summer 2009).
Federal'nyi zakon, No 18, 10.01.2003, 'Ustav zheloznodorozhnogo transporta Rossijkoj Federacii', http://www.glin.gov/view.action?glinID=177558 (accessed 4 April 2011).
Friebel, Guido, Sergei Guriev, Russell Pittman, Elizavaeta Shevyakhova and Anno Tomová (2007), 'Railroad restructuring in Russia and central and eastern Europe: one solution for all problems?', *Transport Reviews*, **27**, 251–71.
Goskomstat Rossii, *Statistical Yearbook*, various issues (including Russian versions).
Grigoryev, Yuli (2007), 'Today or not today: deregulation of the Russian gas sector', *Energy Policy*, **35**, 3036–45.
Hall, David and Vladimir Popov (2005), 'Privatization and restructuring of water supply in Russia and Ukraine', Public Services Research Unit, http://209.85.135.104/search?q=cache:XZyxdqqvQDoJ:www.psiru.org/reports/2005-01-W-RussiaUkrainewater.doc+Hall+Popov+Privatisation&hl=de&ct=clnk&cd=2&gl=de (accessed 4 April 2011).
International Energy Agency (IEA) (2005), *Russian Electricity Reform. Emerging Challenge and Opportunities*, Paris: IEA.
International Energy Agency (IEA) (2006), *Optimising Russian Natural Gas. Reform and Climate Policy*, Paris: IEA.
Organisation for Economic Co-operation and Development (OECD) (2003), *Urban Water Reform in Eastern Europe, Caucasus and central Asia. Progress since the Almaty Ministerial Conference*, Paris: OECD.
Organisation for Economic Co-operation and Development (OECD) (2005), *Russia: Building Rules for the Market*, Paris: OECD.
Organisation for Economic Co-operation and Development (OECD), EAP Task Force (2007), 'Monitoring of the system of water supply and wastewater in the Russian Federation', Document 2, http://www.oecd.org/dataoecd/8/49/38936803.pdf (accessed 4 April 2011).
Pittman, Russell, Oana Diaconu, Emanual Šip, Anna Tomová and Jerzy Wronka (2007), 'Competition in freight railways: "Above-the-rail" operators in central and eastern Europe', *Journal of Competition Law and Economics*, **3** (4), 673–87.
Pittmann, Russell (2007), 'Restructuring the Russian electricity sector: re-creating California?', *Energy Policy*, **35** (3), 1872–83.
Postanovlenie No. 33/2007, 'O sovershenstvovaniye gosudarstvennogo regulirovanija cen na gaz', http://www.fstrf.ru/otrasli/gas/viewdoc?url=/fst/dep/gas/doc/196 (accessed summer 2009).
Postanovlenie Pravitel'stva Rossiskoj Federacii No. 293/2008, "O gosudarstvennom regulirovanii i kontrole cen (tarifov, sborov) na uslugi subjectov estestvennych monopolii i transportnych terminalach, portach, aeroportach i uslugi po ispol'sovoniju infrastructurnye vnutrennich vodnych pute', http://www.fstrf.ru/otrasli/terminal/viewdoc?url=/fst/dep/transport/doc/68 (accessed summer 2009).
Prikaz No. 214 of 30 April 2008, http://www.fstrf.ru/otrasli/oil/viewdoc?url=/fst/dep/gas/doc/222 (accessed summer 2009).
RAO UES of Russia (2008), 'Information Bulletin, Restructuring of RAO "UES of Russia" in 2007', http://www.rao-ees.ru/en/reforming/info_bull_pdf/year_2007_eng.pdf.
RAO UES Rossii, 'Power industry reform background', http://www.rao-ees.ru/en/reforming/reason/show.cgi?background.htm (accessed 4 April 2011).
Russian Federation (2003), 'Law on Electric Power Industry', approved by the Federation Council on March 12, 2003. http://www.rao-ees.ru/en/reforming/laws/show.cgi?flaws.htm.
Russian Federation Government (2001) Resolution no. 526, 'On reforming the electricity sector in the Russian Federation' of 11 July 2001, available at: http://www.rao-ees.ru/en/info/about/str_comp/show.cgi?content.htm.
Russian Federation Government (2007) Resolution no. 205 of 4 April 2007, 'On amending certain resolutions of the Russian Federation Government related to the calculation of electricity volumes sold at free (competitive) prices', available at: http://www.rao-ees.ru/en/reforming/laws/show.cgi?flaws.htm (accessed 4 April 2011).

Tompson, William (2004), 'Restructuring Russia's electricity sector: towards effective competition of faux liberalisation', OECD Economics Department, Working Paper No. 403.

Sagers, Matthew J. (2007), 'Developments in Russian gas production since 1998: Russia's evolving gas supply strategy', *Eurasian Geography and Economics*, **48** (6), 651–98.

Tsuji, Hisako and Dmitry L. Sergachev (n.d.), 'An overview of Russian railways: current reforms and expansion in the far eastern region (summary)', http://www.erina.or.jp/eu/Research/db/pdf2004/04032.pdf (accessed 4 April 2011).

World Bank, 'Railways Databases Update 2007', http://web.worldbank.org/WBSITE/EXTERNAL/TOPICS/EXTTRANSPORT/EXTRAILWAYS/0,,contentMDK:21416523~isCURL:Y~menuPK:1309984~pagePK:148956~piPK:216618~theSitePK:515245,00.html (accessed 4 April 2011).

Yudashkina, Galina and Serguey Pobochy (2005), 'Regional aspects of electricity sector regulations in Russia', Economics Education and Research Consortium, Working Paper Series, No. 05/15.

26 Recent Australian infrastructure liberalization

Gary Madden, Jeffrey Petchey and Aaron Morey

BACKGROUND

Australia is a relatively small country in population and gross domestic product (GDP), with trade critically important to growth and prosperity. Additionally, the nation is geographically isolated from major trading partners the USA, Japan, China and EU member countries. Further, trade is increasingly reliant on mining export growth. In particular, total mining industry gross value added almost doubled between 1985–86 and 2005–06, with these exports comprising 37 per cent of total export value (ABS, 2008a). Tourism is another substantial export component that contributes A$38935 million (in 2006–07). Also significant is the supply of international education services, contributing A$9981 million in 2005–06 (ABS, 2008c). Clearly, well-functioning integrated airport, road, rail, ports and telecommunications infrastructure are required to facilitate export activity. This infrastructure must also support the coastal concentration of the population. At 2005, 60 per cent of the population reside in urban areas, with most communities located in coastal cities. By contrast, the national population density is only 3 persons per square kilometre (World Bank, 2008). This sparse pattern of settlement and regional business isolation impose significant demands on infrastructure placement. For example, regional rail, road and air infrastructure is required to efficiently transport goods and services from geographically remote areas. For example, stresses on infrastructure through increased commodity demand and need for an integrated supply chain is illustrated by an infrastructure bottleneck that occurred at Dalrymple Bay Terminal (North Queensland) in 2004–05. In this instance, rapid increase in throughput resulted in ship queues that lasted an average 21 days, and are estimated to have cost the coal industry A$2 million per day (Pettitt, 2007). Additionally, telecommunications infrastructure is critical in facilitation of international and trans-Australian commerce. This chapter focuses on infrastructure challenges that Australia is facing. The chapter reviews infrastructure policy reforms since the 1980s, and currently used infrastructure funding mechanisms. Next, industry-level and economy-wide impacts of policy reform are presented. Finally, issues and infrastructure requirements are considered.

POLICY REFORM

In the 1980s an infrastructure reform agenda emerged through policy that attempted to introduce international competition into traditionally protected utility industries. Policies included tariff and import quota reductions, and the float of the Australian dollar. Anti-protectionist policies are not universal, with many industries such as transport and energy sheltered from competition (Fels, 2002). A more coherent infrastructure policy reform was introduced in 1995 with the implementation of the National Competition

Policy (NCP) agenda. The NCP introduced competition into energy, airport and rail industries through the removal of competitive restrictions, structural reform of public monopolies and price regulation in provision of infrastructure (Samuel, 2006). A key underpinning agreement of the NCP was the Competition Policy Agreement (CPA). Under the CPA, state and federal governments were obliged to implement competitive neutrality principles to ensure public and private companies were operating under identical regulatory conditions. Further, the CPA required government to structurally reform public monopolies. In particular, governments were committed, under the CPA, to corporatize and consider unbundling vertically integrated public infrastructure monopolies into natural monopoly networks, and introduce competition into contestable segments of facilities and services, prior to privatization (McInerney et al., 2007). Government entities also became subject to provisions relating to competition in the Trade Practices Act (TPA) 1974. Hence, provision of critical infrastructure services such as telecommunications, rail and energy were open to potential private provision (McInerney et al., 2007). Additionally, an independent body, the National Competition Council (NCC), was formed to oversee implementation of, and monitor and report on, NCP reform.

In 2006 the Council of Australian Governments (COAG)[1] announced the implementation of the National Reform Agenda (NRA) to extend the process of reform that began with the NCP. The NRA's brief is broader than improving infrastructure, with stated goals to improve productivity and workforce participation. The NRA's streams are: (1) competition which aims to increase efficiency and competition in the electricity and transport sectors, as well as simplify and make consistent the approach to infrastructure regulation; (2) regulatory reform with the general aim of implementing best-practice regulation, and (3) human capital which aims to improve the health and basic knowledge of the workforce (Productivity Commission, 2006c). With regard to the impact on infrastructure, the NRA aims to introduce structural reform to land transport and energy providers so that increased competition will ultimately drive prices closer to marginal supply costs. Further, the NRA seeks to align the regulatory approach to infrastructure through the Competition and Infrastructure Regulation Agreement (CIRA). The agreement aims to minimize regulatory and bureaucratic burdens imposed on potential investors and users of cross-jurisdictional infrastructure such as rail and road networks from having to comply with several regulatory systems. There are no provisions for water and telecommunications reforms under the NRA (McInerney et al., 2007). More recently the commonwealth government's Infrastructure Australia Act 2008 established the statutory advisory council Infrastructure Australia (IA) to achieve, in coordination with the other Australian governments, a strategic blueprint for Australia's future infrastructure needs. Specifically, IA seeks to identify any infrastructure gaps and bottlenecks, and recommend investment, policy and regulatory reform priorities. The inaugural general infrastructure assessment is to be completed by December 2008 (ALP, 2008). In addition to this national approach, specific reforms in key infrastructure industries are outlined in the remainder of this section.

Rail Reform

The federal government AusLink Green Paper (AusLink, 2002) recommended integration of road and rail regulation, and general policy formation. The need to synthesize the sectors is recognized in the formation of the Australian Rail Track Corporation (ARTC) in 1997. The ARTC aims to simplify access for rail users via deregulation and streamlining of jurisdiction access regimes (Everett, 2006). The most significant rail reform arising from the NCP is the vertical separation of above-and-below state rail infrastructure monopolies. This reform is to enable separation of the rail industry into contestable and non-contestable segments. The non-contestable rail industry segments of tracks and signalling equipment, due to their natural monopoly characteristics, remain owned by the state government. The contestable, above-rail segment, consists of carriages and other rolling stock, when open to competition allow prices in the segment to move toward the long-run marginal cost of provision. However, with state governments retaining ownership of track infrastructure, and being service providers under vertical separation, it was necessary to introduce access regimes through legislative changes to the TPA 1974, which required the granting of third-party access to services deemed of national significance (Healey, 1996). The access regimes, in addition to regulatory and pricing mechanisms, are State determined and in direct conflict with the ARTC's aim to make consistent inter-jurisdictional access regimes, and the CIRA agreement, where COAG agreed to employ a national system of rail access regulation. The NRA's competition policy transport stream, adopted in 2005, seeks to integrate and reform rail and road regulation and provide nationally consistent procedures to evaluate proposed rail infrastructure projects (Productivity Commission, 2006c). The NRA's infrastructure planning stream principal strategy is the AusLink National Network (ANN). The ANN, as outlined in the AusLink (National Land Transport) Act 2005, contains detailed reforms regarding road and rail infrastructure. The main aim of the Act is to increase the competitiveness and efficiency of freight rail. Generally, the Act seeks to integrate road, rail and port infrastructure and recognizes the complementary roles of commonwealth, state and local governments, as well as the private sector, in providing adequate infrastructure (Ridout, 2004).

Road Reform

Congestion in capital cities and the growing importance of transporting natural resources for exportation means that road infrastructure is critical to the functioning of the Australian economy (COAG, 2007). Initial road infrastructure reform predated the NCP under the National Road Transport Commission (NRTC). These reforms concern inter-jurisdictional consistency of regulation, and the alignment of heavy vehicle charges more closely with road usage costs. Alignment is attained through a template legislation framework whereby the commonwealth enacted legislation to the Australian Capital Territory (ACT) and the States then apply the legislation to their jurisdictions (Productivity Commission, 2005). In total, 31 NRTC initiatives were developed by 2001. Of these initiatives, 25 were endorsed by COAG for implementation and assessment under the NCP. Despite implementation of this legislation by 2004, industry complained of inconsistency and shortcomings in the process. In response, an NCC review of the reform agenda (NCC, 2003a) concluded that the implemented initiatives are not sufficient to achieve a nationally consistent regulatory regime (Productivity Commission, 2005). Under the NRA, new reform commitments are outlined as part of the competition policy transport stream. The commitments include harmonization and reform of road and rail regulation, and adoption a consistent national framework to evaluate new public road infrastructure projects (Productivity Commission, 2006c). The NRA can also impact on the road sector via the infrastructure regulation stream. In particular, this area of reform aims to, through CIRA, reduce regulatory uncertainty for national infrastructure stakeholders to foster and encourage third-party access to significant infrastructure on commercial terms and conditions (Productivity Commission, 2006c). Finally, under the NRA infrastructure planning stream the ANN is committed for completion. With regard to road infrastructure, the AusLink (National Land Transport) Act 2005 aims to: define a national network of road infrastructure links; implement the National Land Transport Plan to improve and integrate the national network, and create a single funding regime for the national network.

Telecommunications Reform

Prior to NCP implementation scant telecommunications industry reform occurred. With state governments responsible for many infrastructure assets, the telecommunications industry is responsible to the federal government through the Trade Practices Act. Partial deregulation and restructuring, and tentative privatization initiatives prior to 1990 were largely ineffectual (Productivity Commission, 2005). Typically, labour union and community group power prevented any serious effort to privatize the incumbent operator Telstra during this period (McInerney et al., 2007). Domestic telecommunications infrastructure regulation and policy was significantly impacted by the Telecommunications Act 1991. The Act established the initial right to interconnect to Telstra's network. The regulator also developed a charging method based on incremental costs and introduced a universal service obligation (USO) regime. Unbundling, however, was not required in an extensive form (Spiller and Cardilli, 1997). Despite the reform agenda, the federal government kept Telstra vertically integrated. However, a fully operational wire-line competitor, Optus, is authorized to compete with Telstra from 1991. Mobile operators, Optus and Vodafone, were introduced as competitors. No new entrants are allowed in either wire-line or mobile markets until 1997, when the Telecommunications Act was implemented (Spiller and Cardilli, 1997). The Telecommunications Act 1997 aims to achieve efficiency and international competitiveness in the telecommunications industry and serve the long-term interests of industry consumers.

The present regulatory conditions in the telecommunications sector are derived from this Act, which introduces full competition to the sector by providing for open entry, an anti-competitive conduct framework and specific access regime. The aim of the reforms is to achieve technological neutrality to allow investors to select technologies based on specific market needs (COAG, 2007). In 1997 the Telstra (Dilution of Public Ownership) Bill 1996 received royal assent for the privatization of a third of Telstra. The government then passed (at a second attempt in 1999) the Telstra (Transition to Full Private Ownership) Bill 1998 to sell another 16 per cent of the company. Finally, in 2005 the Australian government passed the Telstra (Transition to Full Private Ownership) Act 2005 to allow full privatization of Telstra without further parliamentary approval. Currently, 83 per cent of Telstra is owned by private institutions (Global Competition Review, 2008). Under the NCP, government enterprises are subject to the competition provisions of the Trade Practices Act 1974. This treatment enables privately owned firms to compete in infrastructure markets. However, Telstra retains dominance through local loop ownership. Accordingly, the federal government introduced telecommunications-specific regulatory measures. First, an anti-competitive conduct regime placed the onus of proof on the service provider, rather than the Australian Competition and Consumer Commission (ACCC), to demonstrate the provider is not engaging in anti-competitive behaviour. The second regulatory measure is a framework to allow providers access to non-contestable network facilities. Additionally, since 2003 the government requires Telstra to produce separate wholesale and retail accounts (Productivity Commission, 2005). The NRA does not include provision for telecommunication sector reforms, which the federal government deals with separately (COAG, 2007). However, in 2008 the federal government announced acceptance of tenders to roll-out a national fibre broadband network to reach 98 per cent of the population at a minimum speed of 12 Mbps. The government intends to subsidise the winning bid by A$4.7 billion (DBCDE, 2008).

Port Reform

Commodity exporters have substantially increased throughput at Australian ports. For instance, in 2005–06 the combined volume of shipments from Port Hedland and Dampier is 45 per cent higher than in 2000–01. Newcastle and Hay Point also experienced significant increases in shipments. Furthermore, Australian containerized trade is forecast to increase on average by 5.4 per cent per annum from 2006 to 2016 (COAG, 2007). Prior to the NCP, most ports were state government-owned monopolies with regulatory responsibility residing with the relevant minister. While some commercialization of ports occurred in the 1980s, however, there was minimal meaningful policy reform (Everett, 2008). After implementing the NCP framework government-owned ports were brought under the jurisdiction of the Trade Practices Act 1974. This treatment, when added to the NCP reforms (such as promoting competitive neutrality, structural reform of public monopolies and third-party access to infrastructure services), provided significant influences on the sector (Productivity Commission, 2005). In particular, the CPA's competitive neutrality principles required that government-owned ports adopt corporatized business models which commercially focus port facility operations, and general independence from government business (NCC, 2001).

Most port authorities are corporatized and adopt a landlord business model, whereby the authority provides core activities such as channel maintenance and navigation. More contestable services such as harbour towing, stevedoring and dredging are provided by contractors. The approach is not consistent across jurisdictions, with South Australian ports operated by the private sector (Productivity Commission, 2006c). Under the NRA, CIRA outlines initiatives to increase competition and efficiency at nationally significant ports. Accordingly, jurisdictions agree to transparently review significant ports and authorities to determine whether entry prevention by competitors occurs. Generally, CIRA provides a consistent national approach to regulation (McInerney et al., 2007). In particular, the NRA through the competition policy infrastructure regulation stream provides necessary competitive and regulatory conditions for further investment in port infrastructure via pricing and investment signals. Specific initiatives include: ensuring ports are only subject to regulation when competition must be promoted; facilitating entry of new suppliers; and ensuring third-party access is competitively neutral (COAG, 2006b).

Airport Reform

The Two-Airline policy (TAP) restricts competition so that supply consists of a public and private airline only. Following the 1990 withdrawal of the policy, air services are assessed under a regulatory framework that treats airlines as economic entities guided by market forces. This change in policy stance is guided by the principle that the domestic air service market is contestable due to low entry and exit costs on a route-by-route basis (Weller, 2007). However, only with the reinforcement of the Workplace Relations Act 1996 is the dominance of incumbent airlines challenged. Under the Act, primary responsibility for industrial relations and awards is with employees and employers at workplace and enterprise levels. This treatment enables highly competitive, low-cost carriers unconstrained by pre-liberalization industrial agreements to provide services. For instance, Virgin Blue entered the market and adopted a business model centred on labour-saving technology, policies and practices that enable greater scheduling flexibility and lower labour costs (Weller, 2007). With the removal of the TAP, federal national aviation infrastructure funding was withdrawn, and a user-pays framework introduced with operational responsibilities under the jurisdiction of state governments and private firms. For example, in 1998 the Federal Airports Corporation (FAC), that operated and advised federal airports, was dismantled as increasing numbers of international airports were privatized (Weller, 2007). The privatization of 17 major Australian airports occurred in 1997–98, with 50-year leases sold to private operators. Further, a restriction of cross-ownership of certain airport pairs and airline ownership of 5 per cent was imposed. Price regulation of aeronautical services under this structure was via price caps at major airports[2] (Productivity Commission, 2002). This price regulation was later replaced in 2001 by an *ex post* framework of price monitoring. Further, all airports are required to formulate 20-year master plans, which are revised every five years. This requirement ensures that airport lessees maintain long-term strategic views of future infrastructure requirements (COAG, 2007). To boost tourism as part of the federal government's internationalization strategy, in 1992 international carriers were granted limited access to Australian airports. This strategy promotes the Single Aviation Market (SAM) between Australia and New Zealand (Weller, 2007). Privatization of Qantas occurred only after British Airways purchased a 25 per cent share in the carrier in the early-1990s. With Qantas fully privatized by 1995, limits on foreign ownership remain (Weller, 2007).

Electricity Reform

Restructuring the electricity sector gained momentum in 1995 with NCP initiatives to create a fully competitive market via vertical de-integration of generation, transmission and distribution elements; corporatization of government electricity business; establishment of an interstate transmission network with third-party access; abolishing entry barriers to the generation and retail sectors; and transmission pricing based on costs (Chester, 2007). In 1998, the newly established National Electricity Market (NEM) allowed the trading of electricity supplied by the generation sector across states.[3] The NEMS's regime specifies a compulsory market, maximum spot price and jurisdiction-specific derogation (Chester, 2007). Other reforms through the 1990s included the separation and disaggregation of generation, transmission, distribution and retail functions across the sector, with the exception of Western Australia. Another reform was the fostering of conditions for private sector entry into ownership and operation, and the creation of new, nationally consistent regulatory arrangement (Fels, 2002). Vertical and horizontal de-integration also occured with 34 government-owned electricity companies in 1990, falling to 43 in number by 2006; half of these companies are privatized (Chester, 2007).

Gas Reform

Under the NCP, regulatory and legislative barriers to interstate gas trade were removed. Further, government-owned gas utilities were corporatized and structural separation of transmission and distribution elements occurred. A national approach to third-party gas transmission and distribution pipeline access, and full retail contestability to allow customer choice between suppliers were implemented (Productivity Commission, 2005). Furthermore, in 2004 the Ministerial Council on Energy (MCE) developed a framework to consolidate governance arrangements for the economic regulation of gas (which was eventually incorporated into the competition stream of the NRA). A key policy objective was to reduce regulatory risk and compliance costs. Additionally, the MCE constructed principles aimed at promoting investment, entry and transparency in gas markets, with the aim of strengthening competition, and developing reliable and secure household, commercial and industrial gas supply (Productivity Commission, 2006c).

Water Reform

In 1994, COAG agreed to implement a nationally comprehensive reform framework for the water industry, with most of the framework adopted later by the NCP (COAG, 1994). Integral to the framework are institutional arrangements requiring commercialization and corporatization of state government water service companies by 1998. The reforms are intended to vertically separate the water industry and introduce commercial discipline to the market (Productivity Commission, 2005). Comprehensive pricing reform also implemented under NCP initiatives include the removal of cross-subsidies, reservation of funds for future asset refurbishment and infrastructure upgrade, and the introduction of consumption-based and cost recovery pricing. Further, investment reforms were also introduced to ensure investment in water supply schemes only occurs for economically feasible and ecologically sustainable projects (Productivity Commission, 2005). In 2004, COAG developed the National Water Initiative (NWI)[4] to extend and progress NCP-based reform. A component of the NWI is to encourage water trading by removing trade-restricting institutional barriers and introducing more efficient trading arrangements through the elimination of cross-subsidies. The NWI maintains full-cost recovery pricing yet requires urban water business tariffs move towards upper-bound pricing by 2008 to encourage investment.

FUNDING MECHANISMS

Any consideration of infrastructure funding in Australia must occur within the context of an understanding of the federal system. In 1901 the colonies federated to create the Commonwealth of Australia. The financial provisions of this federation granted much of the taxation power to the new commonwealth government, principally through Section 90 of the Constitution, which centralized the excise and tariff powers. Subsequent High Court interpretations of Section 90 excluded state access to the consumption tax base. Further concentration of taxation power occurred during and after World War II when the income tax power was relinquished to the commonwealth. Income taxation has since remained the preserve of the central government. Thus, unlike many federations, where central and sub-national governments have access to income and consumption tax bases, Australian states rely on a small number of relatively inefficient taxes on payroll, various licences and fees, and land taxes, for their own source revenue. In contrast, spending responsibilities, for example, for health, education and infrastructure provision (namely, electricity, water, road, and rail and port facilities) remained predominantly with the states. This means that most Australian public infrastructure spending is undertaken by the states, and local governments for local roads, rather than the commonwealth, with the exception of defence, telecommunications and infrastructure related to the information economy. Provisions of such services are almost exclusively the responsibility of the commonwealth. The discussion on funding mechanisms here concentrates on infrastructure funding for states. A consequence of highly centralized financial powers combined with decentralized expenditure responsibilities is that there exists a substantial 'fiscal gap' at the national level, that is, an excess of revenue collected over central government spending requirements. The fiscal gap (surplus) is allocated to states annually to enable the funding of services, including infrastructure provision via: (1) unconditional general revenue assistance; and (2) conditional grants (or specific purpose payments).

General Revenue Assistance

The Australian fiscal equalization system is among the most comprehensive in federated countries (Petchey and Wells, 2004), with equalization making explicit account of individual state infrastructure requirements. That is, general revenue assistance provided to states is subject to fiscal equalization, but otherwise funding is unconditional (that is, once received expenditures are discretionary). The fiscal equalization model, developed by the Commonwealth Grants Commission (CGC) – the national body responsible for making recommendations to the Commonwealth on federal–state financial relations – allocates funds on a per capita basis, after adjusting those shares by adding expenditure and revenue 'needs'. The CGC calculates expenditure 'disabilities' or service-specific parameters normalized around unity. When a state expenditure disability exceeds unity for a particular service, the commission views that the state faces higher than average costs in providing that service (relative to the state average). This assessment can be due to population dispersion (for example, Western Australia is geographically four times the size of France, with a population of less than 2 million), age-health profile of the population (particularly relevant for health services) or economies of scale.

Alternatively, when a state has an assessed expenditure disability (less than unity) for a particular service it is deemed to have a less than average cost of provision. Once service- and state-specific expenditure disabilities are calculated they are used to estimate state-specific aggregate expenditure needs – the value is positive or negative depending on the sign and magnitude of service expenditure disabilities. This aggregate expenditure need is added to the notional state grant (based on equal per capita share of the general revenue assistance grant pool). When an estimated expenditure need is negative, implying that (on average) a state has lower costs of service provision relative to the average the grant available is set below the equal per capita share. However, when the assessed need is positive, implying that, on average, a state has relatively high costs then the state's share of the grant pool is set above the equal per capita share. Thus, relatively high cost states are compensated through equalization with additional grant funding beyond that implied by an equal per capita allocation. This additional funding is sufficient to allow the provision of average levels of service while applying an average tax effort.[5]

Equalization and Infrastructure

The CGC distinguishes between needs arising from recurrent spending (for example, labour and materials) and that arising from infrastructure provision. Again, the commission considers that states potentially face differential infrastructure costs due to differences in population dispersion, asset longevity (owing to natural hazards and climate), and economies of scale and population growth.[6] The CGC argues that such disability results in differential rates of depreciation of state- and service-specific capital stocks (and hence infrastructure maintenance costs). Such differentials are reflected in the CGC methodology used to estimate aggregate expenditure need. The commission also argues that states face interest rate differentials on infrastructure borrowing (debt charges) owing to population differences. In particular, the smaller states face relatively high borrowing costs. Accordingly, an assessment of state aggregate expenditure need includes a component that recognizes differences in debt charges and costs associated with depreciation of infrastructure. This procedure is currently under review (CGC, 2007), with changes to be implemented from 2010. As noted above, the equalization model also adjusts states' per capita share of the general revenue assistance pool with an assessment of revenue need. This proceeds in much the same way as for expenditure needs in that it calculates revenue 'disabilities' for each state tax base (for example, payroll tax). A revenue disability of less than unity for a particular base and state means the state is required to employ a lower than average tax effort to raise a given revenue (compared with the all-state average). This is likely to occur in relatively 'rich' states. The reverse applies to relatively 'low income' states, that is, revenue disabilities exceeding unity imply that a greater than average tax effort must be employed to raise a given revenue. Once the state per capita share is adjusted using the assessed expenditure and revenue needs, the CGC calculates the actual share of the general revenue assistance pool.[7] Whether an individual state receives more or less than its per capita share depends on the sign and magnitude of own revenue and expenditure needs.

The (equalized) general revenue assistance received from the commonwealth augments state revenue from payroll tax, stamp duties, licence fees, gambling taxes and land taxes.[8] The augmented revenue base is disbursed on both recurrent spending and infrastructure provision, with the split determined by the state. In particular, the division and the distribution of state capital works are determined on an annual cycle based on budgets which are presented to state parliaments mid-year. The proportion of the revenues devoted to infrastructure is based on macroeconomic and long run growth considerations, as well as, voter preference over service provision. Most states have best practice project evaluation guidelines in place to assess the 'economic worth' projects (Government of Western Australia, 2005). As concerns are raised concerning the adequacy of state infrastructure investment, procedures are in place at state level, as well as through COAG, to increase the rate of investment. In particular, several states have developed longer-run infrastructure strategies to remove bottlenecks and enhance productivity (Government of Western Australia, 2006). Nevertheless, there is evidence, provided by a recent study on state capital stocks (Petchey and Fernandez-Dodsley, 2008), that the stock of state infrastructure is declining relative to the stock of national infrastructure for 15 general government services (including health, transportation and education).

Specific Purpose Payments

The remainder of the fiscal gap is returned to states as tied specific purpose payments (SPPs). For 2008–09 this amounts to A\$33.1 billion or 11.3 per cent of total commonwealth expenditure. Of this amount, A\$3.492 billion (set to increase to A\$4.275 billion in 2011–12) is in direct support of infrastructure provision at the state and local government levels. Most infrastructures spending, which occurs through SPPs, is related to road provision at the state, but more particularly, local government levels. Such payments occur under the auspices of the AusLink programme (Commonwealth Budget Paper No. 3, 2008, Table 3.1, p. 32). Petchey and Shapiro (1997) are critical of the efficiency of SPPs as a vehicle to fund recurrent spending and infrastructure. They argue SPPs overlap and duplicate services as they facilitate cost shifting between levels of government. There were about 90 different SPP grants across the Australian federation, with each having a recurrent and capital component depending on the type of grant. In January 2009, COAG implemented a new framework for federal–state relations which consolidated the number of SPPs, reducing them to five or six. The SPP conditions were also relaxed, allowing states greater discretionary power. For instance, as of January 2009 certain payments that previously were made as SPPs are included in general revenue assistance, with states only required to spend allocated funds within a certain sector (for example, hospitals) but retain flexibility to determine the distribution of funding within the sector (Commonwealth Budget Paper No. 3, 2008, p. 29).

State Borrowing

In addition to funding infrastructure investment from recurrent revenues states also borrow in financial markets. Overall state borrowing is subject to oversight and regulation by the Australian Loan Council. There is no explicit commonwealth guarantee given to state borrowings. Accordingly, investors lend directly to states as sovereign entities, although there is still an implicit commonwealth guarantee. Nonetheless, there are differential state borrowing costs that reflect state-specific risk assessments determined by financial markets. There is also substantial monitoring and regulation of state borrowing by financial markets. Such monitoring is recently assisted by the adoption of standardized accrual accounting across states. Major international credit rating agencies also provide credit ratings for state governments. Despite this oversight, market regulation of states is considered inadequate, and that Loan Council monitoring of state debt is needed. This view is based on the belief that: (1) in federations there is a moral hazard problem arising from implicit central government guarantee of sub-national debt; and (2) governments generally responding inefficiently to financial market signals (Petchey and Wells, 2004).

COAG Reform Fund

Questions about the adequacy of investment in public infrastructure, particularly at state level, resulted in the development through COAG of a new 'COAG Reform Fund' which will receive additional monies from the commonwealth's budget (not from revenue currently allocated to the states). From this fund three intended 'nation building' funds are to be established: the Building Australia Fund, the Education Investment Fund and the Health and Hospitals Fund to correct for perceived key infrastructure shortages (Commonwealth Budget Paper No. 3, 2008, p. 18). The allocation mechanism for the funds is as yet to be decided. Currently, governance arrangements are only progressed for the Building Australia Fund. The development of this fund is the current focus of COAG meetings in partnership with Infrastructure Australia, the advisory council set up under the Infrastructure Australia Act (2008). As noted earlier, the current focus of attention is the undertaking of an audit of national infrastructure that is to be finished by the end of 2008. The Australian Loan Council is also to advise on whether the spending from the COAG Reform Fund, together with existing spending on infrastructure, is consistent with macroeconomic goals.

Summary

Most public infrastructure provision in Australia occurs at state level, and for local roads, at local government level. Because of the fiscal dependence of the states, the funding for infrastructure at the state level is traditionally derived from general revenue assistance grants, which are subject to fiscal equalization, a component of which takes explicit account of infrastructure needs, specific purpose payments, borrowings and own source revenues. However, funding mechanisms for infrastructure are undergoing a metamorphosis with the development of the COAG Reform Fund. When this fund is substantial it has the potential to dominate traditional forms of funding for infrastructure.

IMPACT OF LIBERALIZATION

In 2005 the federal government commissioned the Productivity Commission[9] to assess the NCP reforms. The commission's core finding is that NCP (and related) reforms implemented through the 1990s increased GDP 2.5 per cent (A$20 billion) by 2005. This estimate ignores dynamic efficiency gain attributed to improved service quality and reliability, and an expanded range of goods arising from more competitive markets

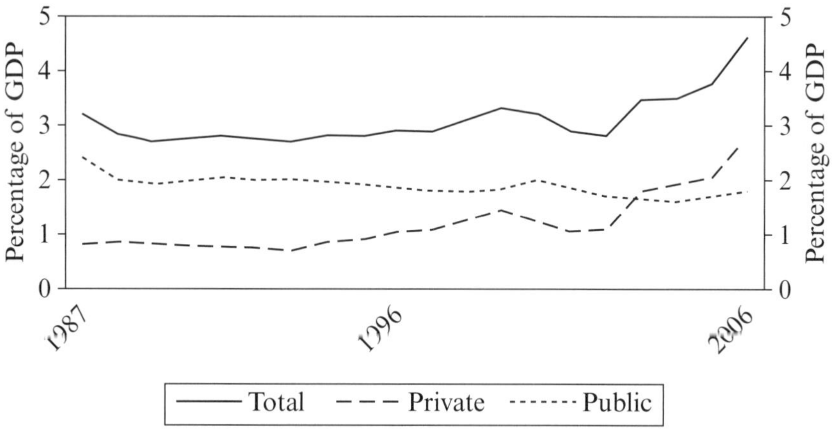

Source: Coombs and Roberts (2007).

Figure 26.1 Share of public/private investment in economic infrastructure

(Productivity Commission, 2005). Also, the reforms are identified by the Organisation for Economic Co-operation and Development (OECD, 2006) as critical to Australian economic performance. In particular, NCP (and other microeconomic) reforms substantially impacted on the structure and infrastructure investment spending magnitudes. For instance, the ratio of total Australian infrastructure investment to GDP rose from an average 3 per cent for 1987–2000 to 4.5 per cent by 2006 (Coombs and Roberts, 2007). The structure of infrastructure spending, proxied by the ratio of private investment to public investment, increased from 0.3 in 1987 to 1.5 in 2006 (Coombs and Roberts, 2007). Importantly, the decline in public infrastructure investment since the mid-1990s is more than offset by the rise in private sector investment (see Figure 26.1).

Interestingly, the effects of liberalization on infrastructure price vary by industry. In real terms, the price of some infrastructure services fell since the implementation of the NCP. However, such price declines are not necessarily efficient when investment incentives are diminished to the extent that long-term service provision is compromised. This issue is considered further in the following sector-specific analyses. For example, for 1990–2004, household electricity prices increased 4 per cent whereas business prices fell by 27 per cent. Further, household gas price rose 13 per cent for the period, while the corresponding business price fell by a similar magnitude.[10] Finally, the analysis of price change is incomplete without considering service quality. Overall, service quality generally improved in infrastructure sectors, while a wider range of services is made available. However, offsetting transitional employment and income costs incurred in certain regional communities must also be acknowledged (Productivity Commission, 2005). An examination of sector-specific changes concludes the section.

Rail Outcomes

The current rail infrastructure is inadequate for forecast export and corresponding inter-state freight demand increases. In particular, the Engineers Australia organization reports widespread delay and uncertainty in new investment, and that improved infrastructure is necessary to support the transfer of interstate freight from road to rail (Engineers Australia, 2005). The ARTC's aim, to create a single, national rail regulatory body for rail operators, is unrealized, with certain jurisdictions resistant to relinquishing control of essential rail infrastructure (and developed independent access regimes). In this situation, when access is sought via state government access providers access to the national track is problematic (and inefficient). Further, in Western Australia and Victoria while railway tracks are transferred from the public sector to private operators the sectors remain effectively vertically integrated (Everett, 2006). Also, vertical separation does not rule out subadditivity[11] or double marginalization (Wills-Johnson, 2008). However, some sector outcomes are positive, for example, the Productivity Commission (2005) estimates that sector total factor productivity (TFP) increased on average 8 per cent per year for 1989–98. Further, freight rates declined by 18 per cent in real terms for 1990–97. This outcome is associated with an increase in operator numbers, from 8 to 29, during the period. However, determining whether competition caused freight rate reductions is complicated by the confidentiality of operator contractual arrangements (Everett, 2006).

Road Outcomes

The Productivity Commission (2006a) reports the absence of 'clear connection' between road charging and investment decisions. For example, in 2004–05, national public expenditure on roads is A$10.4 billion, while heavy vehicles are required to pay only A$1.6 billion in registration fees and diesel fuel excise. The commission noted that should an independent jurisdictional road fund create a direct link between road charges and efficient investment then significant efficiency gains are realizable. In particular, current road pricing prevents competitive neutrality between road and rail, for example, the effective annual registration is reduced by more intense vehicle use. Further, diesel excise need not reflect marginal road use costs per se, implying an inefficient supply of road transport. The quality of infrastructure is also a barrier to competitive neutrality. Specifically, poor rail quality along transport corridors increases price and transit time of rail relative to road, reflecting an investment disparity between road and rail infrastructure. Jointly these factors contribute to an oversupply of road freight relative to rail (Productivity Commission, 2005). Further, persistent inter-jurisdictional regulatory inconsistency affects regulations governing higher mass limit truck access to arterial road networks, multiple rail access regimes and the presence of seven rail safety regulators. The Productivity Commission (2005) argues this outcome resulted from the manner in which road reform is committed to under the NCP. For example, while reform measures for the energy sector is both specific and detailed, road reform measures are of a more general nature. This general specification resulted in regulatory agencies continually clarifying their roles and responsibilities, as well as requiring measures, which delay and blunt reform initiatives. In response, the NCC admitted that the reform agenda to 2001 did not include all initiatives required for a nationally consistent framework (NCC, 2003b).

Telecommunications Outcomes

With telecommunications market liberalization the incumbent provider Telstra retained market dominance through continued ownership of the local loop. Accordingly, rivals had to negotiate access with Telstra to provide services. Although interconnection is mandatory, and the sector open to full competition (from 1997), Telstra maintains dominance in the local call market (COAG, 2007). The ACCC (2005) claims that the current regulatory framework, despite attempts to introduce account separation, is not sufficiently monitoring and controlling Telstra's ability to price rivals 'out of the market' through high wholesale interconnection charges. This strategy is manifest in Telstra's wholesale and retail broadband price margins. Also raised are concerns that Telstra's continued ownership of copper and cable networks limits the deployment of new competitive networks (Productivity Commission, 2005). Despite such concerns, the Allen Consulting Group (ACG, 2002) estimates the total impact of the telecommunications reform at A$10 billion (in real GDP). Further, the reforms estimated to increase employment by 100,000 additional positions, while real wages increased. Further, McInerney et al. (2007) asserts that output per worker doubled during the 1990s, while average telecommunications charges fell substantially (see their table 26.1). Telstra report fault repair, service availability and fault free services (99.9 per cent and 99.1 per cent, respectively, in 2003) performance significantly improved with the implementation of reform (ACA, 2003). However, with privatization, fulfilling USOs is the responsibility of the private sector. Accordingly, public USO funding fell 30 per cent in real terms between 1998–99 and 2003–04 (Productivity Commission, 2005). McInerney et al. (2007) argue that the legislation regulating USOs only provides an indirect means to encourage connection of 'uneconomic' services, set standards and introduce technology. Finally, existing barriers limit infrastructure diffusion to rural and regional areas, for example, inadequate telecommunications infrastructure; lack of awareness of ICT benefits; high prices; relatively less educated community; and higher training costs (Madden and Coble-Neal, 2003).

Port Outcomes

Key motivations to reform of the port sector remain unrealized, in particular inter-jurisdictional consistent regulation. That is, with nine economic regulators applying distinct legislation increases the risk of investment proposals. Further, privatization of selected ports created problems associated with third-party access. Prior to privatization there are no incentives to prevent access; however, this is not the case after the transfer from public to private monopoly that responds to the applicable regulatory framework, not to market forces (Everett, 2008). Also, doubts remain as to whether Australian ports are capable of responding to forecast increases in international demand for coal and iron ore. Indeed, many ports struggle with current capacity constraints, with significant infrastructure bottlenecks projected to develop through port channels without adequate policy response (Exports and Infrastructure Taskforce, 2005). However, port reforms have realized service price declines of 52 per cent from 1990–91 to 2000–01, and substantial productivity improvement in rail port services. Isolating the impact of sector corporatisation on these outcomes is difficult due to coincident changes to waterfront enterprise agreements (Productivity Commission, 2005).

Airport Outcomes

The abolition of the TAP opened the Australian airport industry to competition. Virgin Blue entered the national trunk market and instigated intense price competition on selected routes. Eventually, Ansett Airlines collapsed in response to severe price competition, increased costs and falling demand. Since the Ansett collapse the trans-national market adopted the low-cost business model of Virgin Blue. For example, Qantas downgraded economy service and launched a discount subsidiary, Jetstar (Weller, 2007). Post-liberalization mandated USOs are no longer in place to ensure regional area service delivery, and internal cross-subsidizations are removed. When the national market is unable to support commercial airport service, local and state authorities are expected to intervene. Ultimately, rather than compete inefficiently, airlines are to establish route-based monopolies that serve particular regions. However, ubiquitous access remains an issue (Weller, 2007). Price caps prior to 2001 kept aeronautical charges at Australian airports low by international standards. With transition from the price-cap to price-monitoring systems list prices increased. Airport operators claim such increases are necessitated because of lower traffic volumes and increased security costs. Ongoing regulation is argued as necessary given the continued presence of market power for major airports in core aeronautical services due to barriers to entry, economies of scale and low price elasticity of demand (Productivity Commission, 2002). However, many operators argue that the application of regulation is inconsistent, inflexible and of questionable relevance (BTRE, 2005).

Electricity Outcomes

While a national electricity market is established conditions for effective market operation are not in place. That is, generators possess significant market power from insufficient competition between intra- and inter-state firms (McInerney et al., 2007). Indeed, in some jurisdictions the continued ownership of electricity generation assets hinders competition (COAG, 2006a). Absence of competitive pressure is identified by COAG (2002) as exacerbating electricity market price volatility. Market power clearly affects the volatility of NEM spot prices, for example, for 1998–2006 approximately 4 per cent of half-hourly NEM spot prices exceed A$200 per MWh, however, only once is the maximum spot price (in given year) at peak annual demand (Chester, 2006).

Factors that restrict competition in NEM markets include dominance of generation capacity by a few firms, with government ownership accounting for two-thirds of capacity. Hence, suppliers affect market price and quantity by withholding capacity. Another factor is that the NEM is not a national market per se. That is, because of limited connection of capacity across regions, there are higher wholesale prices than without capacity constraints. Therefore, a principal NEM aim, the export of electricity to regions where demand is unmet by local generation, is not realized.[12] Finally, electricity industry market structure is becoming more concentrated as a result of consolidation in generation and retail sub-sectors, and re-integration of generation in retail markets. Re-integration occurred for nearly 45 per cent of total generation capacity under the NEM, and is expected to increase (Chester, 2007). Also, industry is concerned about non-commercial and excessive investment by publicly owned utilities. Specifically, investment at lower rates of return to sub-commercial levels discourages new private investment in base-load power. The Energy Reform Implementation Group (ERIG), established by COAG to recommend proposals for electricity reform, suggests that government-owned generators be vertically disaggregated to strengthen competitive neutrality between public and private electricity business (McInerney et al., 2007). Conversely, labour productivity more than doubled during 1993–2002 as a result of the reforms (Productivity Commission, 2005). Further, average electricity tariffs remain low by international comparison, and fell in real terms. However, this price decline is concentrated in business markets, with consumer prices remaining at best constant (Chester, 2007). Finally, supply interruption, as measured by system minutes supplied, fell from 13.53 in 2000–01 to 4.43 in 2004–05 (DITR, 2006).

Gas Reform Outcomes

In assessing implementation of NCP gas industry reforms, the NCC conclude that despite delays conditions for national free trade in gas markets are established. However, COAG argues that much regulatory inconsistency remains because of overlapping state and national regulation (COAG, 2002). Accordingly, in 2005 COAG established the national regulator – Australian Energy Regulator (AER) – responsible for economic regulation and ensuring compliance with electricity and gas rules nationally (OECD, 2006). Subsequent industry performance includes substantially increased capital expenditure on new gas pipelines (ESAA, 2004).[13] However, COAG (2002) finds removal of restrictions on the interstate trade in gas and provision of pipeline and customer access encouraged new pipeline development and gas reserve exploration. Further, Australian markets exhibit industrial gas prices and unplanned supply disruption among OECD member countries (Productivity Commission, 2005). Finally, labour productivity rose more than six-fold between 1991 and 2000 (AGA, 2003).

Water Reform Outcomes

Sibly and Tooth (2008) argue that, despite full implementation of NCP reform, water authorities and regulators remain insufficiently independent from government, and that governments are unwilling to price efficiently. Further, investment decisions are distorted and delayed due to institutional arrangements that allow government to select infrastructure projects based on political, rather than economic criteria, exacerbated by mandated third-party access to government business not consistently met (Sibly and Tooth, 2008). Absence of truly competitive markets results in an inefficient and unsustainable allocation of urban water, with prices not reflecting the full scarcity costs of supply (Productivity Commission, 2005). The water trading initiative introduced under the NWI is also unsuccessful. At 2006 only 4 per cent of permanent water entitlements are traded (ABS, 2006). The Productivity Commission (2006b) in attempting to identify impediments to planned development of rural water markets find that restricted trade between irrigation districts and the introduction of exit fees are principal inhibitors. Exit fees charged per mega litre for trade intra-district increase entitlement prices for importing regions and reduce entitlement prices to sellers in exporting regions. This treatment contributes a reduced quantity of water being traded.

UNRESOLVED ISSUES

Australian regulators recently made substantial progress in infrastructure provision policy however, there remains scope for further reform. Principal challenges concern the avoidance of future capacity constraints and infrastructure bottlenecks. Also, there is scope to improve the inter-jurisdictional alignment of regulation across sectors (Exports and Infrastructure Taskforce, 2005). Of particular importance is the need to synthesize road, rail and port infrastructures to prevent bottlenecks such as occurred at Dalrymple Bay Terminal in 2005. Maintaining international trading-partner confidence by exporting in a timely fashion via efficient infrastructure operation is also a crucial performance requirement (COAG, 2007). In particular, current arrangements whereby road is favoured over rail needs to be neutralized to ensure efficient supply. As to telecommunications, evolving technology and regulatory trends suggests incumbent ability to limit competition through core network ownership will gradually erode. Other sector challenges include ensuring that the path to next-generation networks and ubiquity of access to telecommunications services are achieved. Final proposals for the construction of the National Broadband Network (NBN) are due for submission to the Federal government on November 26, 2008. The contract award is subsidized by A$4.7bn from the federal government (Bingemann, 2008). The fibre network is intended to ensure that Australia is internationally competitive in Internet access. However, reviews of ownership and regulatory arrangements are needed to ensure that network roll-out and long-term supply is efficient. Further, a forecast national population of between 30.9 and 42.5 million by 2056 will generate substantial additional demand for energy and water services (ABS, 2008b). To meet these requirements additional investment in infrastructure is needed.

Policy and other regulatory arrangements must also to be conducive to private investment (COAG, 2007). In particular, current price cap arrangements and government investment are likely to attenuate electricity and water sector investment (McInerney et al., 2007). Further, the NEM must become more competitive to completely realize potential benefits, while rural and urban water reform must be better integrated to achieve efficient allocations (Productivity Commission, 2005). Finally, new private airport operators should consider the expected strong passenger arrival growth at major Australian airports when making investment decisions (COAG, 2007).

NOTES

1. COAG is an inter-governmental forum comprised of the Prime Minister, State Premiers, Territory Chief Ministers and the President of the Local Government Association.
2. Sydney, Melbourne, Brisbane and Perth are considered by the Productivity Commission to be major airports.
3. The NEM is a wholesale market for electricity supply in New South Wales, Victoria, Queensland, South Australia, the Australian Capital Territory and Tasmania.
4. Western Australia and Tasmania are not party to the NWI.
5. As might be expected the states to game against such a formula by deliberately focussing on high-cost services or inflating cost structures in order to influence their share of the pool. Petchey (2008) explores this behaviour in a game-theoretic model. The commission claims to have methods in place to impede such gaming.
6. This is partly like the Swiss equalization model that estimates the expenditure needs of localities related

to height above sea level. The argument is that it is more costly to build roads in mountainous areas than on flat country. Localities with more mountains therefore receive additional funding.
7. The CGC formula adjusts the per capita share for differences in the budgetary position of the states. However, this procedure is ignored in the current discussion.
8. Other sources include discretionary taxes, that is, states are able to select the base and tax rate, although both choices indirectly affect the general revenue assistance received via equalization.
9. The Productivity Commission is the principal, independent review and advisory body on microeconomic and regulatory policy to the Australian government.
10. The introduction of the GST in 2000 potentially exaggerates these differences, as business is exempt from GST on these services (Productivity Commission, 2005).
11. When a cost function is subadditive then a given output is produced at least costs by a single firm.
12. This outcome requires substantial investment in high-voltage and high-capacity transmission lines.
13. This response is argued due to higher demand rather than the economic regulatory regime.

REFERENCES

AusLink (2002), Green Paper, http://www.auslink.gov.au/publications/historical/pdf/Green_Paper_5_Nov.pdf (accessed 23 August 2008).

Australian Bureau of Statistics (ABS) (2008a), *Year Book Australia 2008*, cat. no. 1301.0, Canberra: ABS.

Australian Bureau of Statistics (ABS) (2008b), *Population Projections, Australia, 2006 to 2101*, cat. no. 3222.0, Canberra: ABS.

Australian Bureau of Statistics (ABS) (2008c), *Tourism Satellite Accounts*, cat. no. 5249.0, Canberra: ABS.

Australian Bureau of Statistics (ABS) (2006), *Water Account, 2004–2005*, cat. no. 4610.0, Canberra: ABS.

Australian Competition and Consumer Commission (ACCC) (2005), 'Review of price control arrangements', February, available at: http://intranet.accc.gov.au/content/item.phtml?itemId=670116&nodeId=87d2b6e9b5f93ed6ede6c61dd8ad1a3d&fn=Final%20report%E2%80%942004%20review%20of%20Telstra%20price%20control%20arrangements%20(Feb%2005).pdf (accessed 4 April 2011).

Australian Communications Authority (ACA) (2003), *Telecommunications Performance Report 2002–03*, Canberra: ACA.

Allen Consulting Group (ACG) (2002), 'Benefits resulting from changes in telecommunications services', http://www.allenconsult.com.au/resources/ACA_Benefits_Final.pdf (accessed 21 August 2008).

Australian Gas Association (AGA) (2003), *Gas Statistics Australia 2003*, Canberra: AGA.

Australian Labor Party (ALP) (2008), 'Infrastructure Australia', http://www.infrastructureaustralia.gov.au/index.aspx (accessed 23 August 2008).

Bingemann, M. (2008), 'Broadband players battle for big prize', *Australian IT*, 4 September, http://www.australianit.news.com.au/story/0,24897,24291825-5013041,00.html (accessed 24 August 2008).

Bureau of Transport and Regional Economics (BTRE) (2005), *General Aviation: An Industry Overview*, Report 111, Canberra: BTRE.

Chester, L. (2006), 'The conundrums facing Australia's national electricity market', *Economic Papers*, **25**, 362–77.

Chester, L. (2007), 'Who benefits from the restructuring of the Australian electricity sector?', *Journal of Economic Issues*, **41**, 981–1001.

Commonwealth Budget Paper No. 3 (2008), *Australia's Federal Financial Relations*, Commonwealth Government, Canberra.

Commonwealth Grants Commission (CGC) (2007), 'Building blocks for the assessment of capital related expenses in the 2010 review', Staff Discussion Paper, CGC 2007/22-S, Australian Government, Canberra.

Coombs, G. and Roberts, C. (2007), 'Trends in infrastructure', *Economic Roundup*, Summer, 1–16.

Council of Australian Governments (COAG), (1994), *Report of the Working Group on Water Resource Policy to the Council of Australian Governments*, Canberra: COAG.

Council of Australian Governments (COAG), (2002), *Towards a Truly National and Efficient Energy Market*, Energy Market Review, Canberra: COAG.

Council of Australian Governments (COAG), (2006a), 'Market structures: efficiency via competition', Energy Reform Implementation Group, discussion paper, COAG, Canberra.

Council of Australian Governments (COAG), (2006b), 'COAG background paper: COAG national competition policy review', http://www.COAG.gov.au/COAG_meeting_outcomes/2006-0210/docs/national_competition_policy_report.pdf (accessed 30 August 2008).

Council of Australian Governments (COAG), (2007), 'Australia's infrastructure: national overview report', http://www.COAG.gov.au/COAG_meeting_outcomes/2007-04-13/docs/infrastructure_reports_nat.pdf (accessed 30 August 2008).

Department of Broadband, Communications and the Digital Economy (DBCDE) (2008), 'National broadband network', http://www.dbcde.gov.au/communications_for_business/funding_programs__and__support/national_broadband_network (accessed 31 August 2008).

Department of Industry, Tourism and Resources (DITR) (2006), *Energy in Australia, 2006*, Canberra: ABARE.

Energy Supply Association of Australia (ESAA) (2004), *Submission to the Productivity Commission Review of National Competition Policy Arrangements*, Canberra: ESAA.

Engineers Australia (2005), '2005 Australian infrastructure report card', http://www.infrastructurereportcard.org.au/ (accessed 3 September 2008).

Everett, S. (2006), 'Deregulation and reform of rail in Australia: some emerging constraints', *Transport Policy*, **13**, 74–84.

Everett, S. (2008), 'Public sector deregulation: a paradigm in conflict', *Maritime Policy and Management*, **35**, 341–51.

Exports and Infrastructure Taskforce (2005), 'Australia's export infrastructure', report to the Prime Minister, ABARE, Canberra.

Fels, A. (2002), *National Infrastructure Summit, August 14, 2002: Reforming Australia's Infrastructure*, Melbourne: ACCC.

Global Competition Review (2008), *Telecoms and Media 2008*, London: Law Business Research.

Government of Western Australia (2005), *Project Evaluation Guidelines*, August, Perth: Department of Treasury and Finance.

Government of Western Australia (2006), *Framework for the State Infrastructure Strategy: Green Paper*, September, Perth.

Healey, D. (1996), 'Access regimes in practice – part 111A', National Competition Policy Reforms/Trade Practices Update Seminar, 9 July.

Madden, G. and Coble-Neal, G. (2003), 'Internet use in rural and remote Western Australia', *Telecommunications Policy*, **27**, 253–66.

McInerney, L., Nadarajah, C. and Perkins, F. (2007), *Australia's Infrastructure Policy and the COAG National Reform Agenda*, Canberra: Australian Treasury Department.

National Competition Council (NCC) (2003b), *Annual Report 2002–03*, Canberra: Ausinfo.

National Competition Council (NCC) (2003a), *Assessment of Governments' Progress in Implementing the National Competition Policy and Related Reforms: 2003, Volume One – Overview of the National Competition Policy and Related Reforms*, Canberra: Ausinfo.

National Competition Council (NCC) (2001), *2001 Assessment of Governments' Progress in Implementing National Competition Policy and Related Reforms*, Canberra: Ausinfo.

Organisation for Economic Cooperation and Development (OECD) (2006), 'Economic survey of Australia 2006', http://www.oecd.org/document/28/0,3343,en_2649_33733_37147228_1_1_1_1,00.html# (accessed 7 September 2008).

Petchey, J. (2008), 'Theoretical analysis of equalization and spatial location efficiency', *Regional Studies*, **42**, 1–16.

Petchey, J. and Fernandez-Dodsley, D. (2008), 'Australian state and service specific (public) capital stock estimates', joint report with the WA Department of Industry and Resources, Perth.

Petchey, J. and Shapiro, P. (1997), 'One people one destiny: centralization and conflicts of interest in Australian federalism', in D. Wildasin, (ed.), *Fiscal Aspects of Evolving Federations*, Cambridge: Cambridge University Press.

Petchey, J. and Wells, G. (2004), 'Australia's federal experience', in M. Baimbridge and P. Whyman (eds), *Fiscal Federalism and European Economic Integration*, London: Routledge.

Pettitt, T. (2007), 'Will a national port regulator solve congestion problems in Australian ports?', *Maritime Policy and Management*, **34**, 121–30.

Productivity Commission (2002), *Price Regulation of Airport Services*, Report no. 19, Canberra: Productivity Commission.

Productivity Commission (2005), *Review of National Competition Policy Reforms*, Report no. 33, Canberra: Productivity Commission.

Productivity Commission (2006a), *Road and Rail Freight Infrastructure Pricing*, Productivity Commission Inquiry Report No. 41, Canberra: Productivity Commission.

Productivity Commission (2006b), *Rural Water Use and the Environment: The Role of Market Mechanisms*, research report, Melbourne: Productivity Commission.

Productivity Commission (2006c), *Potential Benefits of the National Reform Agenda*, report to the Council of Australian Governments, Canberra: Productivity Commission.

Ridout, H. (2004), 'Gov takes the long road to gain transport benefits', *Manufacturers' Monthly*, **5–5**, July.

Samuel, G. (2006), *Australian Infrastructure Reform: Where to from Here?*, May, Adelaide: South Australian Centre for Economics.
Sibly, H. and Tooth, R. (2008), 'Bringing competition to urban water supply', *Australian Journal of Agricultural and Resource Economics*, **52**, 217–33.
Spiller, P. and Cardilli, C. (1997), 'The frontier of telecommunications deregulation: small countries leading the pack', *Journal of Economic Perspectives*, **11**, 127–38.
Watanabe, B. and Mueller, E. (1984), 'A poverty profile for rural Botswana', *World Development*, **12** (2), 115–27.
Weller, S. (2007), 'Strategy and the contested politics of scale: air transportation in Australia', *Economic Geography*, **83**, 137–58.
Wills-Johnson, N. (2008), 'Separability and subadditivity in Australian railways', *Economic Record*, **84**, 95–108.
World Bank (2008), *World Development Indicators 2008*, Washington, DC: Development Data Group.

Legislation

AusLink (National Land Transport) Act 2005 (Cwlth)
Infrastructure Australia Act 2008 (Cwlth)
Telecommunications Act 1997 (Cwlth)
Telstra (Dilution of Public Ownership) Bill 1996 (Cwlth)
Telstra (Transition to Full Private Ownership) Bill 1998 (Cwlth)
Telstra (Transition to Full Private Ownership) Act 2005 (Cwlth)
Trade Practices Act 1974 (Cwlth)
Workplace Relations Act 1996 (Cwlth).

27 Infrastructure liberalization: challenges to the new economic paradigm in the context of developing countries

Catarina Figueira and David Parker

INTRODUCTION

In all economies the provision of efficient, reliable and affordable infrastructure services is an essential prerequisite of economic growth and sustainable development. Key infrastructure services include the provision of water and sanitation, power, transport and telecommunications. Without an adequate infrastructure, market development and output growth is likely to be seriously slowed (World Bank, 1994). Moreover, affordable infrastructure services are critical to the improvement of household welfare and the reduction of poverty. It is a sobering fact that today around 1 billion people lack access to adequate, clean water supplies and a further 1.2 billion to adequate sanitation. The result is disease and high mortality especially amongst infants.

In the twentieth century, generally, infrastructure provision was the exclusive province of the public sector, with large, state-owned enterprises (SOEs) being responsible for investment and service delivery. However, in recent years the inadequacies of state enterprises have been revealed with sometimes costly and inefficient services and insufficient capital investment. When the water and sewerage industry was privatized in England and Wales in 1989, this followed years of under-funding. By 2010 the industry will have invested £69 billion since privatization in improving drinking water and raising environmental standards. The result is cleaner and more reliable supplies and less polluted rivers and beaches. But it has come at a price. Water and sewerage tariffs in England and Wales rose by 39.1 per cent in real terms between privatization in 1989 and 2006 (Ofwat, 2007). Meanwhile, through the private finance initiative and public private partnerships the private sector has been brought more directly into the building and running of roads, schools, government buildings, defence and hospitals. In Britain over £40 billion has been spent on private finance initiative (PFI) schemes.

Internationally, public–private partnerships (PPP) are mushrooming with France transferring its toll roads to the private sector as a recent example. At the same time, private capital markets have responded with a number of companies specializing in these sorts of investments. Exchange traded funds (ETF) have been established with stock-listed and tradeable-like equities to invest specifically in infrastructure assets.[1] Barclays Global Investors manages one such fund. The Australian bank, Macquarie, is also very actively involved in this market. Infrastructure assets appeal to private investors because of the expectation of steady income flows and non-competitive pricing of services to consumers. Also, returns, like those from government bonds, are effectively backed by the taxpayer. Infrastructure investments can provide big returns. Standard and Poor's infrastructure index, including 75 of the world's biggest operators in the energy, water and transport sectors, returned 24.3 per cent a year over the five years to 2007, compared with 11.3 per cent from a portfolio of global equities (*The Business*, 2007).

The result is that since the early 1990s a new dominant economic paradigm has emerged that infrastructure development requires market liberalization and privatization. Since the mid-1980s, governments around the world have pursued policies to involve the private sector in the delivery and financing of infrastructure services. The growth in foreign direct investment (FDI) accelerated in the 1990s, rising to US$331 billion in 1995 and to US$1.3 trillion in 2000 (UNCTAD, 2002). A principal feature of the growth in FDI has been the rise in foreign investment in services, which is now the dominant sector in global FDI (Kirkpatrick et al., 2006a). A significant part of the increase in foreign investment in the services sector has been the growth in private capital flows for infrastructure development. For example, private sector participation in infrastructure projects in developing countries rose dramatically after 1990 with annual investment commitments reaching a peak of US$128 billion in 1997. In contrast, there was a sharp decline in donor support for infrastructure projects during the 1990s, with aggregate flows of official development assistance falling by around a half during the decade (Willoughby, 2002). According to the World Bank's Private Participation in Infrastructure (PPI) database, 26 countries awarded only 72 infrastructure projects with private participation in the period 1984–89. In the 1990s this grew sharply with the new interest in privatization in donor agencies and governments. A total of 132 low- and middle-income countries pursued private participation in infrastructure – 57 of them in three or all four of the sectors covered in the database (transport, energy, telecommunications, and water and sewerage), albeit sometimes in response to 'conditionality' terms attached to foreign loans and grants. In 1990 to 2006 developing countries transferred to the private sector the operating risk for over 4000 infrastructure projects, attracting investment commitments of more than $1100 billion (World Bank, 2008). After dipping following the Asian financial crisis in 1997, since 2001 private financing has revived.

There are clear attractions to introducing private capital into infrastructure delivery for policy-makers, but at the same time a number of difficulties. This chapter reviews the difficulties, particularly in the context of developing economies and their need for sustainable economic development and poverty reduction. It provides a summary of the literature on the relationship between the quality of the regulatory framework and investment in infrastructure in middle- and lower-income developing countries. The evidence confirms that investment in infrastructure responds positively to an effective domestic regulatory context. In other words, the provision of adequate infrastructure is not simply a matter of seeking out the optimal ownership arrangement, it is critically dependent upon having 'sound government'.

THE DIFFICULTIES OF PRIVATE SECTOR INFRASTRUCTURE PROVISION

Before concentrating specifically on the problems facing infrastructure provision in developing countries, it is useful to say a few words about some of the general problems that face private sector infrastructure delivery. Encouraged by international organizations such as the World Bank, privatization has been a major component of the economic reform programmes pursued by many developing countries over the past two decades (Parker and Kirkpatrick, 2005). Privatization has been advanced because it is expected to promote more efficient operations, reduce the financial burden on government budgets, expand service delivery, and increase the level of foreign and domestic private investment (World Bank, 1995). However, because infrastructure services typically have some natural monopoly characteristics arising from pervasive economies of scale and scope, competition is unlikely to develop or quickly becomes uneconomic because of the duplication of assets. In some sectors technological advances, notably in telecommunications and electricity generation, have reduced the economic advantages of monopoly provision, but natural monopoly remains pervasive in the water and sewerage industry, electricity and gas distribution and transmission, and fixed-line telecommunications (except, perhaps, in dense user areas such as major cities). There is also limited scope for new competition in trunk roads (even if planning permission could be obtained to build new highways), bridges and airports. In health and education competition may be economically feasible but tends to be politically and socially divisive. As a consequence, privatization of these industries, in whole or in part, is problematic. Where it does occur it requires either an effective state regulatory system, or competition 'for the market' rather than 'in the market' through concession and management agreements and the like. Concession and management contracts with the private sector involve a form of continuing state regulation, from the writing of the initial contract specification to the *ex post* contract monitoring.

The railway network in many parts of the world was developed by private capital and private firms operated 'turnpike' roads in Britain in the eighteenth and early nineteenth centuries. So complete private provision of infrastructure is feasible – the problem is it tends not to last! From 1844, in Britain, the state became more and more involved in regulating the railways and their charges because of public pressure and concerns about monopoly pricing. The turnpike roads became uneconomic, because of competition from rail, and fell under government control during the nineteenth century. The gas, electricity and water industries in Britain developed as a mixture of public and private sector provision from the Industrial Revolution. The result was inconsistent service quality, a lack of economies of scale and sometimes inefficient duplication of facilities. In all cases, government became involved in regulating the private operators to prevent monopoly abuse. In the twentieth century these sectors came under almost complete state control. For example, in 1926 the state-owned Central Electricity Board was set up to rationalize electricity transmission and develop a high-voltage national grid, which it did very successfully. The rest of the electricity industry was nationalized in 1947. The gas industry was nationalized in the following year. In England and Wales the water industry was rationalized into mainly ten state-owned regional water and sewerage utilities in 1973 to take advantage of river basin water management schemes.[2] Overall, the historical evidence suggests that private sector involvement in infrastructure provision has been inadequate over time and transitory (for example, the turnpike roads). Nor is there, it has been suggested, any overwhelming evidence that the private sector provided infrastructure services more efficiently and effectively than the state enterprises that took them over (Foreman-Peck and Millward, 1994).

That private investment in infrastructure is problematic and may fail to generate the economic benefits that result from private over state ownership in competitive markets,

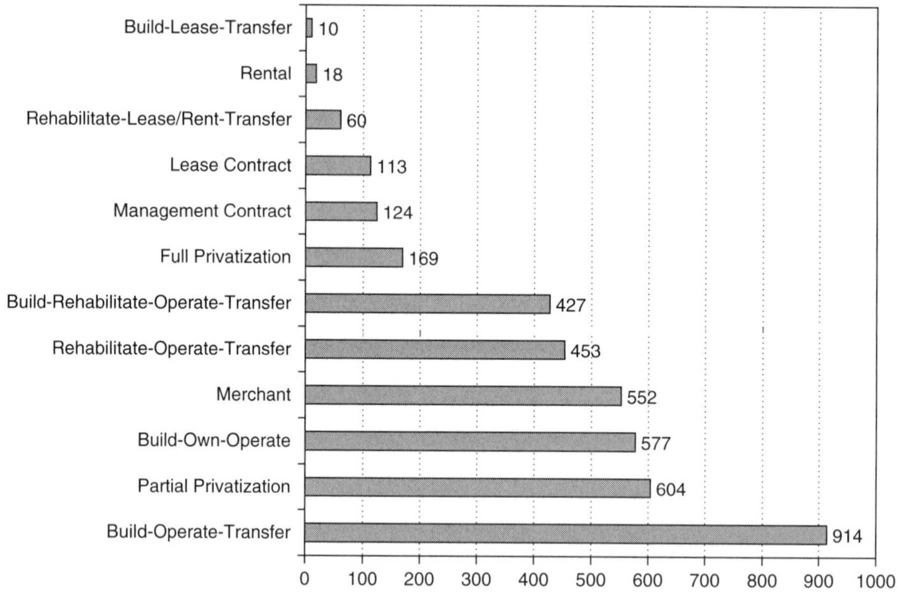

Figure 27.1 Number of projects by type of contract, 1990–2007

such as manufacturing, relates to the issue of natural monopoly. Infrastructure provision is often a one-off provision where providing a competing structure would be highly uneconomic – so much so that the private sector would not supply it. The reasons for private sector failure relate to the technology of the provision, the scale of the investment, and the resulting economies of scale and sometimes of scope. Private ownership is most effective where there is competition in supply. Competition acts as the incentive to management to hold down costs and maximize consumer benefits. The private sector monopolist may settle for 'an easy life', just like its public sector counterpart.

In infrastructure delivery rivalry under private management is mainly in the form of 'competition for the market' or competition to win the contract or concession agreement (World Bank, 2004, ch. 5). Private infrastructure investment has taken a number of forms internationally, involving differing degrees of investment risk (as summarized in Figure 27.1). There has been some divestment of state assets to private ownership, mirroring the infrastructure privatizations in Britain. There have also been some 'greenfield' projects or completely new investment by the private sector in infrastructure schemes. But some form of contracting for services with the private sector is prevalent. Management and lease contracts typically involve a private entity taking over the management of the state-owned enterprise for a given period, although the facility continues to be owned by the public sector, and the public sector remains wholly or largely responsible for new investment. Under a concession agreement a private firm takes over the management and assumes more risk by agreeing to undertake investment. At the end of the contract period the arrangement is renewed through a further round of competitive bidding or the activity reverts to state control. However, problems can and do arise. These problems relate essentially to the existence of information asymmetries and pervasive transaction costs in infrastructure markets. Infrastructure contracting is subject to the costs of negotiating, monitoring and enforcing the contracts in a world where the contracting parties inevitably face uncertainty about future unforeseen and perhaps unforeseeable events (Williamson, 1985). The government and the winning firm can be expected to have different levels of information about such matters as costs, revenues and demand. The government enterprise may hold the information that a firm needs to bid optimally and the government must establish bidding rules and processes to being forward an adequate number of bidders. Following the contracting, it is highly unlikely that government, now as regulator or contract enforcer, will receive all of the information required to regulate to maximize social welfare. The results of contracting in terms of outputs and prices are therefore likely to be 'second best' to those of a competitive market.

One particular outcome of competitive bidding for infrastructure is the 'winner's curse', where a contract is won but the successful bidder subsequently discovers that it is unable to perform the contract to specification at the prices agreed and make a profit. This leads to contract reneging, where the contractor walks away from providing the service, often leaving a gap in service provision and the government to step back in, or contract renegotiation. In Mexico privately built toll roads proved unprofitable and were taken into state control. One of the two water concessions in Manila in the Philippines quickly failed (Cuaresma, 2006). The much lauded privatization of the water supply in Buenos Aires in Argentina turned sour after 2000 owing to a change in the economic environment and disappointing profits. Where contract renegotiation occurs either the company or the government could be the loser, depending upon the results of the renegotiation. Guasch (2004, p. 13) concludes that a staggering 74 per cent of water concession contracts in Latin America were renegotiated significantly within a few years of being signed – in Buenos Aires prices were raised within months of the start of the water concession (Alcazar et al., 2000). Studying cancelled concession contracts in developing countries, Harris et al. (2003) find that water and sewerage concessions have the second highest incidence of cancellation after toll roads.

The explanation is that the transaction costs in infrastructure contracting can be high and revenues and costs difficult to forecast when contracts are negotiated. The failure of the Mexico toll road scheme occurred because of disappointing traffic flows when drivers chose at an unexpectedly high rate to use the 'free' public roads. Moreover, where the private sector is expected to absorb the costs of bidding and may have to make appreciable investments in the service, the contractor is open to 'hold up'. Privatization requires investors to sink funds into fixed assets that are specific to the venture. Where the investor fears that the government will later attempt to exploit this situation and renegotiate the contract to the contractor's disadvantage, investors will be deterred from committing to the investment (Hart and Moore, 1988). For example, water provision is a politically sensitive issue and liable to government intervention in pricing and investment responding to public pressure (Kirkpatrick et al., 2006b). Alternatively, investors may require a front-end loading of returns or sovereign guarantees from the state or international agencies. In turn such guarantees reduce the net economic benefits of attracting private capital by reducing managerial incentives to control costs. There is the separate but significant difficulty of maintaining incentives for the private contractor to invest and maintain the quality of the infrastructure as a contract gets closer to its completion date. Moreover, there are usually appreciable sunk costs to participating in the bidding process, to which must be added the costs of fixed asset investments once the contract begins. The existence of the 'hold up' threat is an important reason why since the Asian financial crisis, when a number of governments reneged on contracts, privatization of the water industry has been almost entirely in the form of management contracts where the state remains responsible for the fixed assets.

The problems facing infrastructure privatization through contracting are compounded by the fact that the contracts are typically for long periods of time to justify the sunk costs involved in taking on the contract, especially when large fixed asset investments are required. Inevitably the contract will be incomplete, in terms of specifying all of the contingencies that may trigger contract adjustments and the forms the renegotiation might take. This places a large emphasis on the skills of both government regulators and companies when operating concessions, to ensure as far as possible that the outcome remains mutually beneficial. Having government remain responsible for fixed asset investment is one possible solution. But the incentive to privatize infrastructure on the part of cash-strapped administrations lies in bringing forward new capital investment by the private sector.

In addition, bidding for infrastructure schemes may be subject to a small numbers problem, undermining the benefits of 'competing for the market'. In some countries only two or three firms have been serious bidders for infrastructure contracts and in some cases there has been only one bidder. In France a lack of competition led the government to change the rules on bidding for water contracts. Low numbers bidding reduces the competitiveness of the contract and can occur because of over-restrictive pre-qualification criteria for contracts, as well as the conclusion by firms that the risks of winning the contract do not justify the costs of bidding. However, it can also occur because of the relatively small number of large multinational companies that pursue infrastructure concessions and management contracts internationally (albeit often in partnership with local firms) (McIntosh, 2003, p. 2). This is not assisted by mergers and acquisitions affecting infrastructure companies. Since 1998 there have been reportedly 3165 mergers and acquisitions of infrastructure companies at a cost of over US$500 billion. The peak year was 2006 when 400 deals involving investments of almost US$148.4 billion were arranged (*The Business*, 2007). Most of this has involved smaller, regional companies. But in so far as infrastructure provision becomes more concentrated, the scope for competitive bidding recedes. Also, the smaller the number of bidders the greater is the scope for either actual or tacit collusion between bidders. The result can be both adverse selection and moral hazard. Adverse selection takes the form of suboptimal contracts at the outset, resulting from one of the contracting parties acting opportunistically to arrange especially favourable terms; while moral hazard occurs when one of the contracting parties renegotiates the terms of the contract in their favour during its lifetime. The risk of the latter grows when the existing contracting firm knows there is really no alternative supplier to which the government can quickly turn.

THE IMPORTANCE OF GOVERNANCE IN INFRASTRUCTURE DELIVERY

Transaction costs inherent in infrastructure delivery using the private sector are reinforced by weaknesses in government (Spiller and Savedoff, 1999, pp. 1–2). The decline in private sector infrastructure investments in developing countries after the Asian financial crisis in 1997 was consistent with growing concerns amongst investors about regulatory capacity (Harris, 2003). Some countries lack the administrative and institutional capacity to regulate effectively, whether through formal regulatory offices or concession and management contracts.[3] All methods of price, profit and quality of service regulation are demanding in terms of their information needs. The design of contracts and post-privatization regulatory measures becomes more complex when the need to allow for the impact on different income groups and especially the poor is an issue, as it should be in developing economies. Ensuring that the poor are not disadvantaged by private sector involvement requires an institutional capacity to regulate private sector activities effectively to ensure affordability, especially where there is an inadequate or non-existent social welfare policy (Clarke and Wallsten, 2002). However, in developing countries, in particular, there is frequently a lack of regulatory capacity, including a limited availability of skilled staff to bring about the desired result. Shortage of regulatory skills constrains the regulatory authorities' ability to formulate and design appropriate regulatory measures, including writing and monitoring contracts certain to advance the interests of the poor. In other words, the necessary skills of government regulators to negotiate and monitor private investments are often weak.

Constraints on regulatory capacity increase the degree of regulatory risk and transaction costs (Parker, 2002). Many developing countries lack the necessary trained personnel to sustain regulatory commitment and credibility. Regulatory offices in developing countries tend to be small, under-manned for the job they face, and more expensive to run in relation to GDP than in developed countries (Domah et al., 2003). Familiarity with regulatory models and methods of regulatory policy analysis is often limited (Kirkpatrick et al., 2005). The other main difficulties found in many developing countries relate to broader governance problems (Minogue, 2002; Stern and Holder, 1999) or to the legal powers and responsibilities of regulators, including their effective independence from regulatory (including political) capture. Regulatory risk is the outcome of uncertainty and inconsistency in a regulatory regime, which leaves private agents fearful of current and future regulatory decisions. Where regulatory risk is appreciable, investors either will not invest or will seek compensation in the form of a larger expected return, leading to a higher cost of capital. A higher cost of capital will mean higher consumer charges (or state subsidies) or a lower rate of investment (Guasch and Hahn, 1999; Hahn, 1998).

State regulation including writing contracts attempts to 'mimic' the economic welfare results of competition. But it can do so only in a 'second best' way because competitive markets generate superior knowledge of consumer demands and producer supply costs (Sidak and Spulber, 1997). Indeed, government involvement introduces important economic distortions into market economies: 'regulation . . . is far from being a full substitute for competition, it can create systematic distortions, it generally faces a trade-off between promoting one type of efficiency at the expense of another, and it is likely to generate significant costs, in terms of both direct implementation and exacerbation of inefficiency' (Hay and Morris, 1991, pp. 636–7). The difficulties in designing an effective and efficient regulatory framework acquire an additional degree of complexity in the context of developing countries where significant capacity and resource constraints often arise. The impact of infrastructure regulation on market incentives, and on investment behaviour in particular, is therefore uncertain and difficult to predict a priori.

The aim of infrastructure regulation should be to establish a policy environment that sustains market incentives and investor confidence. For this to be achieved, the regulator needs to be shielded from political interference, and government needs to support a regulatory environment that is transparent, consistent and accountable (Parker, 1999). This implies that the capacity of the state to provide effective and efficient regulatory institutions will be an important determinant of infrastructure provision. The challenge of providing infrastructure regulation which establishes credibility with the private sector and at the same time ensures efficient economic performance on the part of the regulated enterprises is not easily achieved. There is an extensive literature on the distorting effects of state regulation even when conducted by dedicated regulatory bodies (Armstrong et al., 1994; Guasch and Hahn, 1999). This leads to 'credibility' and 'commitment' considerations: *credibility* on the part of investors that the regulatory rules will bring about the intended outcome; and *commitment* of government to the regulatory rules, so that post-privatization or post-concession award, the regulator does not act opportunistically to reduce the prices and profits of the private regulated businesses. Regulatory regimes are also prone to 'capture', where the regulatory process becomes biased in favour of particular interest groups and notably the regulated companies in an attempt to acquire economic rents (Laffont, 1999; Peltzman, 1976; Stigler, 1971).

What is clear is that the capability of firms to influence public policy is an important source of comparative advantage (Shaffer, 1995). Balanced against the risks of regulatory capture, however, is the possibility that regulators might develop a culture of arrogant independence, bordering on vexatious regulation. This creates some uncertainty about the desirable degree of regulatory independence that is desirable. The 'independent agency' is normally favoured by western advisors, who draw from the experience of regulation in the UK and the USA. However, regulatory independence and an impartial judicial review of due process may not be credible in some institutional structures.

An important policy implication that follows from the regulatory capacity constraints existing in many countries is that the optimal regulatory measures to be adopted need to be developed in the specific context of these institutional constraints. Recent research relating to utilities' privatization in developing countries has shown that the sequencing of privatization and regulation reforms has a significant impact on the economic outcomes. In particular, the establishment of an effective, independent regulator before embarking on privatization is associated with more favourable outcomes in terms of capacity expansion, service penetration and productivity (Wallsten, 2001; Zhang et al., 2005, 2007). More generally, the experience of private investment in the former communist countries of Central and Eastern Europe, and in Latin American economies with 'Washington Consensus' market liberalization reforms, has highlighted the importance of first establishing an institutional infrastructure including appropriate regulatory systems (Kuczynski and Williamson, 2003). As Stiglitz (2002, p. 18) comments in relation to economic development: 'successful economic programs require extreme care in *sequencing* – the order in which reforms occur' (emphasis in the original).

The importance of the regulatory environment is reflected in the data on private sector investment in infrastructure. As Figure 27.2 shows, Latin America and the Caribbean accounted for 39 per cent of the cumulative investment in infrastructure between 1990 and 2006. Also the region dominated in terms of divestitures and concessions and was second to East Asia in terms of greenfield projects (see Figure 27.3). In Latin America

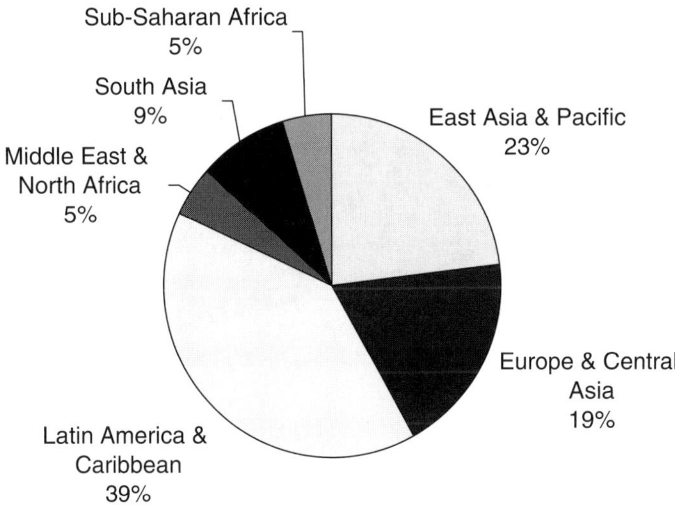

Figure 27.2 *Cumulative investment in infrastructure projects with private participation by region, 1990–2006*

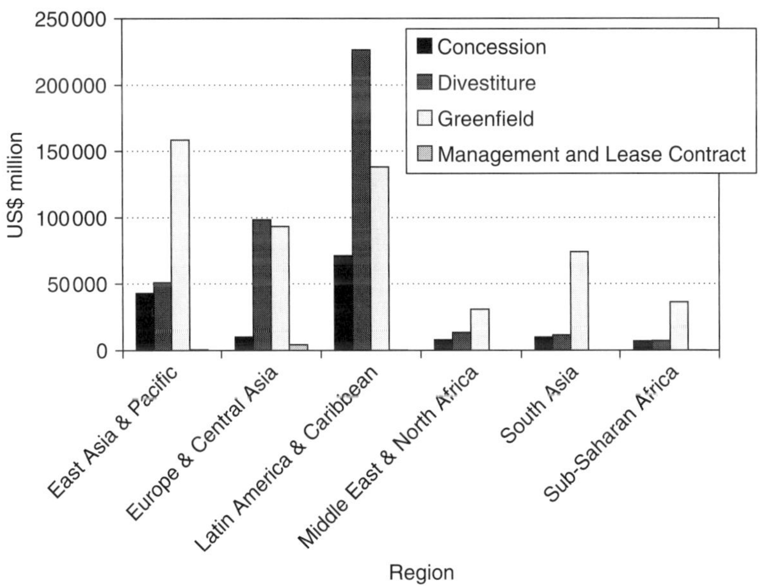

Figure 27.3 *Cumulative investment in infrastructure projects with private participation by type and region, 1990–2006*

and the Caribbean, private participation in infrastructure was often part of a broader sectoral reform programme, aimed at enhancing performance through private operation and competition and generating the financial resources needed to improve service coverage and quality through tariff adjustments (World Bank, 2003, pp. 2–3).

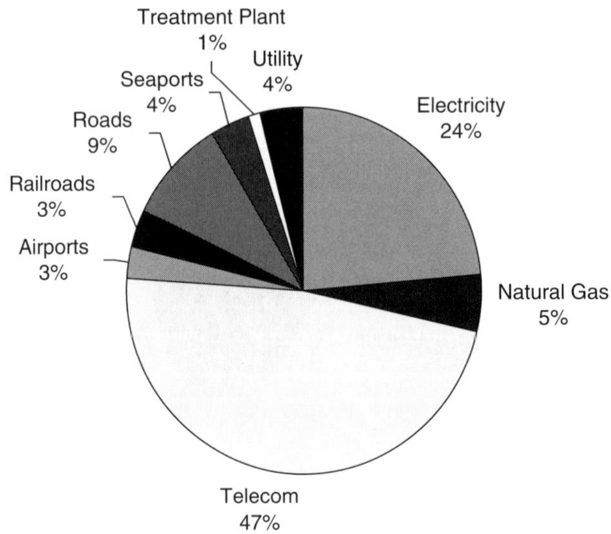

Note: The term 'utility' (as denoted by the World Bank) includes sewerage collection and treatment, as well as water utility both with and without sewerage.

Figure 27.4 Cumulative investment in infrastructure projects with private participation in developing countries by sector, 1990–2006

The East Asia and Pacific region has been the second largest recipient of private investment in infrastructure. Over the period 1990–2006 it accounted for 23 per cent of cumulative private participation in infrastructure in developing countries. However, in South Asia, the Middle East and Africa, where the regulatory resources are in particularly short supply and governance regimes are often poor, private investment in infrastructure has been very limited. In both the Middle East and North Africa as well as sub-Saharan Africa, private investment has been negligible – yet the need for infrastructure investment is nowhere more pressing.

Also, consistent with the expectations from transaction cost theory, private participation in infrastructure in developing countries has been concentrated in the telecommunications sector (Figure 27.4). In this industry competition 'in the market' is feasible, especially due to wireless technology. Therefore, long-term contracts between the government and the private sector either may not be necessary or need to be less detailed. Telecommunications accounted for 47 per cent of the cumulative investment in 1990–2006. Energy, which includes electricity and the transmission and distribution of natural gas, attracted the second largest share of capital, accounting for 29 per cent of the cumulative investment in private infrastructure projects over the same period. Electricity generation is particularly suitable for competition, except where there is a dependence on hydro or nuclear schemes. In contrast, private participation in the water and sewerage sector has been very limited, accounting for 5 per cent of cumulative investments. The limited amount of private involvement in water utilities reflects the inherent difficulties that face privatization in this sector, in terms of the technology of water provision and the nature of the product and transaction costs, alongside regulatory weaknesses (Kirkpatrick et al., 2006b).

It is increasingly recognized that differences across countries in economic conditions provide only a partial explanation of the location choices of multinational enterprises and that the quality of a country's institutional framework can have a significant impact on the perceived investment environment. Institutions have been defined in a variety of ways. Following North's (1990, 1991) widely cited definition, the institutional framework has been represented as the set of informal and formal 'rules of the game', which constrain political, economic and social interactions. Similarly, Groenewegen and Künneke (2005) explain that institutions both reflect people's mental maps and constrain them. From this perspective, a 'good' institutional environment is one that establishes an incentive structure that reduces uncertainty and promotes efficiency, thereby contributing to stronger economic performance. Included in this institutional structure are the laws and political and social norms and conventions that are the basis for successful market production and exchange. This broad concept of institutions has been incorporated into empirical studies of FDI using a range of indicators. It is now common, for example, to include a variable to control for inter-country differences in the broad political environment (Altomonte, 2000; Morisset, 2000), although as noted by Dawson (1998) the results have been mixed. A measure of inter-country differences in corruption has also been shown in several studies to have a significant impact on private investment (IFC, 2002; Wei, 2000). The extent of legal protection of private property and how well such laws are enforced, is an additional factor that has also been shown to have a significant effect on foreign investors' location decision. A parallel stream of research has focused on perceptions and assessments of the quality of public institutions – especially on how well they function and what impact they have on private sector behavior (IMF, 2003).

The term 'governance' has been adopted in the literature to cover different dimensions of the quality of public institutions, including government effectiveness and efficiency. Recent empirical evidence has confirmed that cross-country differences in growth and productivity are related to differences in the quality of governance (IMF, 2003; Jalilian et al., 2007; Rodrik, 2000). This approach has been extended recently to consider the impact of governance on cross-country differences in FDI flows. Globerman and Shapiro (2002) use the six governance indicators estimated by Kaufmann et al. (2003) to assess the impact of governance quality on both FDI inflows and outflows for a broad sample of developed and developing countries over the period 1995–97. The Kaufmann indices describe various aspects of the governance structures, including measures of political instability, rule of law, graft, regulatory burden, voice and political freedom, and government effectiveness, and therefore encompass many of the individual institutional variables used in earlier studies. The Kaufmann governance variables are combined with measures of physical, human and environmental capital to explain FDI flows, and the results indicate that the quality of governance infrastructure is an important determinant of both FDI inflows and outflows (Globerman and Shapiro, 2002, pp. 1908–14).

The study by Stein and Daude (2001) uses the gravity model approach to test for the role played by institutional quality on FDI location in Latin American countries during the period 1997–99. A group of four alternative measures of institutional quality is combined with two other sets of variables and tested as potential determinants of FDI flows. The first consists of variables which are typically used in gravity models of trade, such as GDP, per capita income and distance between the source and host countries (Greenaway and Milner, 2002). The second group consists of variables, other than the institutional ones, which can affect the attractiveness of a country as a location for FDI, such as the level of taxes on foreign investment activities, human capital and infrastructure quality. The results show that the governance variables are almost always statistically significant, confirming that the quality of institutions has a positive impact on FDI. The results are shown to be robust to the use of a wide range of institutional variables, to different model specifications and to different estimation techniques. It is increasingly clear that privatization and the introduction of regulatory agencies modelled on western lines, at best, may have only partial effects on access, affordability and the quality of infrastructure services in developing countries. Meanwhile poor governance, reflected particularly in high levels of corruption and cronyism in a country, endemic in some developing economies, removes the efficiency incentives resulting from privatization (Estache et al., 2007).

INFRASTRUCTURE PROVISION AND SUSTAINABLE DEVELOPMENT

The Millennium Declaration of the United Nations General Assembly has emphasized the importance of infrastructure services in enabling the improvement of health, education and in achieving the income poverty objectives set out in the Millennium Declaration. Indeed, an adequate infrastructure of basic services can reduce the number of people suffering from water-related illnesses, which damage productivity in a country and increase the economic costs associated with hospitalization, malnutrition and death. Moreover, the accessibility to energy, such as electricity, also contributes to the preservation of medicines, vaccines and foods and ultimately reduces health-related problems. These arguments are widely documented in studies by Esrey et al. (1991), Listorti (1996) and Galiani et al. (2005).

Generally, the availability of water, sanitation and electricity provide scope for improvement in productivity. As suggested by de la Fuente and Estache (2004), the expansion and improvement of utility services also facilitates a decrease in the overall costs of businesses. It opens new market opportunities for businesses and increases productivity and the investment that drives economic growth. Estache et al. (2002) provide evidence for this in Latin America. They show that under-investment in basic services has impacted negatively on growth by 1 per cent to 3 per cent. In other words, the formulation of appropriate regulatory measures that promote the provision of vital infrastructure services and make the services affordable to the poorest is vital to ensure sustainable development.

Governments can design regulatory mechanisms and methods that improve provision for the poor and can attempt to achieve certain social objectives, such as providing services in remote areas and reducing the risks to public health and safety. Regulators should also be aware of changes in industrial structure brought about by deregulation and increased competition and the effects on incomes and employment among the poor. Finally, regulators in developing countries, as elsewhere, need to remain acutely aware of the threat from special interest groups leading to 'regulatory capture', which can lead to regulation having a regressive effect on income distribution. In the discussion below the concern is with the direct and indirect effects of regulation on infrastructure services and poverty reduction.

Regulatory reform in developing countries needs to address the design of pricing and access policies which balance economic efficiency and social equity (Kuyvenhoven, 2004). In order to do this successfully, Demery and Squire (1996) and Paris (2000) suggest that there is an urgent need to improve both technical and economic expertise and to design appropriate regulatory models and tools to achieve these two sets of goals, perhaps based on simplified means of intervention which are more appropriate to developing economies than exist in developed countries (Wegelin and Borgman, 1995). Moreover, mechanisms need to be in place to identify the critical circumstances where regulatory intervention is needed so as to avoid over-regulating, as argued by Kessides (2005a).

The poor are not, of course, a homogeneous grouping and may have different needs, entitlement relations and resource endowments, which complicates regulatory policy (for a discussion see Bailey, 2004; Curry, 1989; Dash, 2003; Islam, 1992; Santibanez, 2005). In particular, the poor are both consumers and owners of assets (sometimes physical assets such as land but otherwise human resources or provision of labour). Each of these roles of the poor in an economy needs to be understood by regulators if effective action is to be taken to address their poverty, as suggested by Islam (1990) and Rakodi (1995). In a study of several Asian countries, including China and Sri Lanka, Islam (1990) concluded that generally economic policies do not successfully integrate poverty alleviation measures, particularly they do not provide adequate support to the needs of certain poor, such as those who live in rural locations. A regulator in a developing country will need to understand the needs of the poor, their location and the real barriers to their access to adequate infrastructure services. For instance, research by Taylor (2002) in Botswana provides evidence as to how people used to diverse ways of life value material subsistence strategies differently. Evidence of programmes which were not well targeted have been described by Haddad and Adato (2002) and Adato and Haddad (2002), for South Africa, and Lloyd-Sherlock (2000), for some countries in Latin America. The main argument relates to the fact that the beneficiaries to these programmes were often poor, but not the poorest.

The regulator will also need to understand the different ways in which the interests of the poorest might be best advanced (Suryanarayana, 2001); for example, the promotion of local electricity supplies using small generators or solar power compared with expanding the national grid, or the provision of communal water and sanitation schemes. Information is therefore a key prerequisite of pro-poor regulation. To this end, the participation of the community is very important in achieving efficacy with the interventions carried out to alleviate poverty. As suggested by Hoddinott (2002), the engagement of communities in collective actions can even reduce the cost of implementing certain pro-poor policies. In Ghana, for example, regulators normally consult with industry, government and consumers and have public hearings (Botchway, 2001). In Jamaica, the regulator attempts to discover views through local churches and in Bolivia town hall meetings are held (Smith, 2000). In the Agra district, in India, the anti-poverty programmes have been decentralized to an extent that the decision-making process tends to be done at close to a village level, so that the programmes can deal more effectively with the demands of the local community (Sanwal, 1987). In Colombia, regulators hold public hearings involving operators and consumers and invite consumer organizations to make representations; the credibility of these organizations has been boosted by their partnerships with some of Colombia's universities (Cannock, 2002).

However, in the case of many other countries and regions, it is not clear whether and how infrastructure regulators attempt to gauge the views of the poor, and there must be concern that regulators receive most of their information from the regulated firms, politicians and higher income groups or elites (Echeverringent, 1992). For instance, in the Philippines the regulatory agencies stand accused of a lack of transparency in decision-making: 'Many complaints against regulators cite their closed-door decisions, their less-than-public hearings, their tilt toward the industry's representatives, even their ability to communicate simple information to their buyers and consumers' (Cariño, 2005). Adato and Haddad (2002) also claim that the efficacy of some of the public works programmes based in South Africa have been compromised by conflicting priorities among competing 'voices' within the local society and the inability of local authorities, in some instances, to deal with such difficulties (Nel and Binns, 2003).

It is generally recognized in the literature that the lack of access to infrastructure services is primarily a problem for the poorest people in developing countries. This is usually due to the fact that the poor do not have the monetary means to pay for basic services, but there are also a number of other factors which impede the poorest from accessing services such as water, sanitation or electricity. The geographical area where they live is usually important, because access to these services does vary across regions and even neighbourhoods. For instance, urban areas or neighbourhoods where predominantly middle-income households live tend to have broader access to utilities services, while neighbourhoods which are sparsely populated or where the majority of people are poor have a higher proportion of people not connected to basic services. A particularly acute example of this is 'unofficial' settlements or 'shanty towns'.

Regulators in some developing countries have been committed to the adoption of pro-poor policies. Examples of these policies are provided by Garg et al. (2003) who provide evidence that in India, village public telephones and public call offices have been promoted in both urban and rural areas to make telecommunications more accessible to the poor. The Brazilian Agency of Electrical Energy has decided that consumers should no longer pay for electricity connection charges so as to promote universal service (ANEEL, 2003). Other examples include a recent law passed in South Africa on water services, which states that every household has the right to a certain amount of free water per day, and the introduction in Buenos Aires of a bimonthly charge to spread the cost of new water connections over five years, interest free. This was specifically introduced to make water services more accessible to the poor (although it may have had the side-effect of curbing the expansion of the water network; Alcazar et al., 2000). Further measures were introduced in 2002 including social tariffs which benefit pensioners and the poor in specific areas. Also, in South Africa the government is committed to introducing a monthly 50 kWh of free electricity in a further attempt to curb poverty (Howells et al., 2006).[4]

However, even when poor people have access to water supply or sanitation, not always are the standards of these services acceptable: it is often the case that the quality of the service is substandard (that is the water in the pipes may not be safe to drink) and/or the services are not reliable (that is access to water supply but no water in the pipe; access to electricity but sporadic availability of the service). So, how can access to basic services be improved among poor people in developing countries? This is where regulation can play a key role, in terms of alleviating the monetary pressures which deter the poorest from accessing these services and also by helping removing other non-price obstacles (World Bank, 2006).

One frequent regulatory measure in developing countries is the modelling of the tariff structure of basic services in such a way that the access and consumption of basic services by the poor, who cannot pay market rates, is cross-subsidized by the non-poor through higher charges or directly by taxpayers. However, this calls for care in targeting the beneficiaries, because a badly conceived tariff structure may end up benefiting other groups more than the poorest. For example, subsidies provided to households that already use the service will not promote an increase in access to the service. In Uganda, US$500 million a year was spent on subsidies for electricity provision, but only 6 per cent of the population had access to electricity. Similarly, Foster et al. (2001) claim that in India in the late 1990s, only one quarter of the subsidies for water services (which amounted to around 0.5 per cent of GDP) benefited the poor. In Guatemala, the introduction of a 'social tariff' on electricity did little to benefit the poor households because it benefited only those already connected to the electricity network. As a result, 60 per cent of poor households remain without electricity (Foster and Araujo, 2004).

It is therefore the case that the poorest might benefit more from a different form of service delivery than subsidizing existing services, such as the adoption of alternative technologies (Kebede, 2006; Manning, 2001). Reiche et al. (2000) show that the use of photovoltaic systems to provide electricity in rural areas may decrease costs by 75 per cent relative to the extension of current electricity networks. Foster (2002) discusses condominial systems that have successfully been applied in 'favelas' in Brazil and more recently in Bolivia and Peru. These systems sometimes use community labour and they are a cheaper method of obtaining access to water and sanitation by installing pipes under and along sidewalks or even along backyards. In Brazil, cost savings with this method amount to around 20 per cent, while in Bolivia they reached 40 per cent for water and 25 per cent for sanitation. In Bolivia in 2001, the Institute for Technical Norms and Standards approved changes in quality standards to allow for low-cost infrastructure services to reach poor settlements by making use of a cheaper condominial technology for water and sewerage (PPIAF and WSP, 2001).

Claims have been made that any adverse effects on the poor in developing countries resulting from privatization and market liberalization programmes have been greatly exaggerated. Typical is the following conclusion:

> There is no evidence that such reforms hurt poor or rural consumers – at least in terms of access to services. Even when service prices increase, the share of poor and rural households with connections does not decrease. And in many cases coverage increases, possibly because connection fees fall once service is no longer rationed. Indeed, case studies show that allowing entry and competition in infrastructure services can dramatically increase services for poor people. (Kessides, 2005b, p. 27)

However, while it does seem that privatization and the arrival of competition has often brought about widespread benefits to all consumer groups in telecommunications, the evidence relating to other infrastructure industries is less compelling. These industries are less conducive to cost-reducing technological change. Notable in this respect is the water sector where competition in the market is ruled out by the economics of water supply and sewerage services, particularly the high costs of building supply facilities and the costs of pumping water and treating sewerage (Morris and Cabrera, 2003).

Also, there has been a tendency to extrapolate from the experiences of one country and region (much of the published research relates to Latin America) to developing economies generally. A conclusion that all income groups in Argentina benefited from efficiency, quality of service and access improvements following the privatization of utilities (Chisari et al., 1999) and that the poorest groups may have benefited most (Benitez et al., 2003), for example, does not necessarily mean that this result will apply universally. It is to be expected that the result of reforms depend upon the nature and form that they take and especially the local economic and regulatory environment (see, for instance, Foster and Araujo (2004) for a discussion of the effects of a 'social tariff' policy applied in the electricity sector). This turns attention from simply looking at privatization and market liberalization as sufficient reforms in themselves to concern with the quality of state regulatory regimes.

Moreover, because the poor are consumers and suppliers of labour, the effects of reforms need to consider not just prices and outputs but employment levels, working conditions and wages (Winters et al., 2004). A recent study of four Latin American economies suggested that privatization had no clear effect on prices, but that there had been adverse distributive effects on the poor because of redundancies in the privatized utilities. For example, Sheahan (1994) discusses how the Peruvian programme of liberalization in the early 1990s resulted in increased unemployment.

MODELLING REGULATION AND FDI IN INFRASTRUCTURE IN DEVELOPING COUNTRIES

It seems clear from the discussion so far that the conditions for the provision of adequate infrastructure in developing countries are much more than a matter of substituting private for state financing and more a matter of developing adequate regulatory capacity and providing the right governance environment to promote sufficient investment. As a result, a large number of countries have introduced new, dedicated regulatory offices with independence from day-to-day ministerial intervention to supervise the activities of their privatized utilities. Evidence on the impact of utilities regulation in developing countries is still limited; but studies for telecommunications and electricity confirm that privatization brings greater benefits when it is accompanied by an effective regulatory regime (Wallsten, 2001; Zhang et al., 2005, 2007).

Research involving one of the authors of this chapter was recently undertaken into FDI and infrastructure provision to see whether regulation had influenced the flow of foreign direct investment to the infrastructure sector, specifically in developing countries (Kirkpatrick et al., 2006a). More precisely, the research examined whether the perceived quality of the regulation framework had an impact on the locational choice of multinational enterprises when investing in infrastructure projects in developing countries. Where the regulatory regime is successful in establishing credibility with investors, regulation may be expected to have a benign influence on investment commitments. But where the regulatory institutions are perceived to lack independence of government and to be vulnerable to political interference, investors may be deterred from committing to large-scale, sunk cost capital investments. The research was therefore based on the hypothesis that the quality of regulation matters for investment and that there would be a positive relationship, other things being equal, between the quality of infrastructure regulation and the inflow of FDI to the infrastructure sector.

The relationship between FDI in infrastructure and regulation was empirically tested using a model which regressed the FDI data for a sample of 67 low- and middle-income countries on a measure of regulatory institutional quality and a set of control variables, for the period 1990–2002. Data on foreign (private) direct investment were obtained from the database on Private Participation in Infrastructure (PPI), made available by the World Bank (World Bank, 2003). The control variables that might be expected to influence FDI flows included GDP, to reflect the level of income and demand in the economy, and the inflation rate, the exchange rate and the level of taxation. The inflation measure and the exchange rate were included as macroeconomic stability measures, with the expectation that greater economic volatility acts as a disincentive to inward investment. The average tax burden was expected to have a negative impact on FDI. Other control variables included were the degree of trade openness (measured as the ratio of imports and exports to GDP), a labour force quality measure (secondary school enrolment) and a variable to reflect the level of a country's financial development. The final control related to the quality of the infrastructure stock in the sample countries. The investment decision was expected to be influenced by the need for additional infrastructure provision (Fay and Yepes, 2003; Leipziger *et al.*, 2003). Countries with greater infrastructure needs might be expected to attract foreign investment in infrastructure.

Panel data estimation methods were employed and both fixed and random effects models were tested. In almost all cases the control variables displayed the expected sign and a number of the results were statistically significant. This added to confidence that the modelling was sound. The main focus of the research was on the effect that the regulatory institutional framework may have on foreign investors' decision to commit resources to infrastructure projects in developing countries. Two variables were used as measures of the quality of the regulatory environment for the infrastructure sector. The first was taken from the set of governance-related variables estimated by Kaufmann et al. (2003). These indices describe six aspects of the governance structures for a broad cross-section of countries: voice and accountability, political instability, regulatory quality, rule of law, control of corruption and government effectiveness. A limitation of this measure is that it relates to regulatory effectiveness at the level of the economy as a whole, rather than the infrastructure sector specifically. In the light of this limitation, an alternative measure of the regulatory environment was used, in the form of a dummy variable to indicate whether independent regulators were established in the telecommunications and electric power sectors.

Considering the results for the regulation variables, the sign for each of the three regulation measures confirmed that FDI in infrastructure is positively influenced by the quality of the regulatory framework.[5] The general measure of regulatory quality was statistically significant, confirming that the overall quality of the governance environment attracts inward FDI in infrastructure. The alternative regulatory measure, a specific measure of infrastructure regulation based on the existence of an independent regulatory agency in the telecommunications and electricity sectors, was also statistically significant. However, when the independent utility regulation variable and the measure for overall governance were both included in the same equation, the former became insignificant. The research was therefore unable to detect a strong influence for independent utility regulation separate from the quality of a country's overall governance. This is consistent with the conclusion that investors in infrastructure are more likely to be influenced in their location decision by the overall governance environment in a country than by the existence of an independent utility regulatory authority. Alternatively, the existence of an independent utility regulatory authority is not credible to investors in the absence of a sound governance environment.

CONCLUSIONS

Since the early 1990s a dominant economic paradigm has emerged that infrastructure development in developing countries requires market liberalization and privatization. A number of countries have privatized infrastructure services and there has been an unprecedented increase in private foreign investment in infrastructure projects. Much of this investment has been in the telecommunications and electricity sectors, but there has also been private investment in other areas, such as roads, ports, airports, and water and sanitation. However, economic analysis suggests that private sector infrastructure investment is associated with significant investor risk linked to the long-term, sunk cost characteristics of infrastructure projects. Moreover, the involvement of the private sector in 'natural monopolies' raises challenges when creating regulatory capacity, including establishing effective regulatory agencies and/or contract negotiating and monitoring skills within government, both to control monopolistic behaviour and attract investors to the infrastructure sector.

This chapter has assessed the possible economic barriers to private sector infrastructure investment in developing economies arising from the pervasive information asymmetries and transaction costs that surround regulation and contract negotiation. It has particularly flagged the special difficulties for developing countries facing the dual challenges of achieving sustainable development and poverty reduction. A number of important problems facing private sector investment in infrastructure schemes in developing countries have been highlighted and the results of research using a dataset on private participation in infrastructure projects in developing countries for the period 1990 to 2002 have been summarized. The research confirms that foreign investment in infrastructure responds positively to the existence of good governance.

By implication, where state regulatory institutions are unpredictable and untested in developing countries, foreign investors will be reluctant to make a major commitment to large-scale infrastructure projects. The main policy implication is the need to support capacity building and institutional strengthening in developing countries. In other words, infrastructure development is not simply about liberalization and privatization and less government per se in these countries, but rather about building better government.

NOTES

1. Exchange traded funds track indices of property and construction businesses without the investor having to own the physical assets.
2. Scotland and Northern Ireland have their own, separate, state-owned water and sewerage provision.
3. Harris (2003) suggests that weaknesses in regulatory contract design have contributed to the high level of renegotiation of private infrastructure contracts.
4. On pro-poor policies aimed at making water available in South Africa, see Naraghi and Kebothalez (2004).
5. For full results and further analysis see Kirkpatrick et al. (2006a).

REFERENCES

Adato, M. and Haddad, L. (2002), 'Targeting poverty through community-based public works programmes: experience from South Africa', *Journal of Development Studies*, **38**, 1–36.

Alcazar, L., Abdala, M. and Shirley, M. (2000), 'The Buenos Aires water concession', *World Bank Research Working Paper 2311*, World Bank: Washington, DC.

Altomonte C. (2000), 'Economic determinants and institutional frameworks: FDI in economies in transition', *Transnational Corporations*, **9** (2), 75–106.

Agencia Nacional de Energia Eletrica (ANEEL) (2003), 'Destaques do Governo – Universalização. Metas de universalização', http://www.aneel.gov.br/224.htm.

Armstrong, M., Cowan, S. and Vickers, J. (1994), *Regulatory Reform: Economic Analysis and British Experience*, Cambridge, MA: MIT Press.

Bailey, J.P. (2004), 'Asset development for the poor (conservative and liberal approaches to poverty reduction within Catholic social thought)', *Journal of the Society of Christian Ethics*, **24**, 51–72.

Benitez, D., Chisari, O. and Estache, A. (2003), 'Can the gains from Argentina's utilities reform offset credit shocks?', in C. Ugaz and C. Waddams (eds), *Utility Privatization and Regulation: A Fair Deal for Consumers?*, Cheltenham, UK and Northampton, MA, USA: Edward Elgar.

Botchway, K. (2001), 'Paradox of empowerment: reflections on a case study from Northern Ghana', *World Development*, **29** (1), 135–53.

Cannock, G. (2002), 'Feedback to regulators from consumers', APOYO Consultoria report presented at the International Telecommunication Union (ITU) Global Symposium for Regulators, Hong Kong, 7–8 December.

Cariño, L.V. (2005), 'Regulatory governance in the Philippines: lessons for policy and institutional reform', mimeo, National College of Public Administration and Governance, University of the Philippines.

Chisari, O., Estache, A. and Romero, C. (1999), 'Winners and losers from the privatization and regulation of utilities: lessons from a general equilibrium model of Argentina', *The World Bank Economic Review*, **13** (2), 357 78.

Clarke, R.G. and Wallsten, S.J. (2002), 'Universal(ly bad) service: providing infrastructure services to rural and poor urban consumers', *Policy Research Paper 2868*, Washington, DC: World Bank.

Cuaresma, J.C. (2006), 'Pro-poor water services in Metro Manila: in search of greater equity', in M. Minogue and L. Carino (eds), *Regulatory Governance in Developing Countries*, Cheltenham, UK and Northampton, MA, USA: Edward Elgar.

Curry, R.L. Jr. (1989), 'The basic needs strategy, the congressional mandate, and U.S. foreign aid policy', *Journal of Economic Issues*, **23** (4), 1085–96.

Dash, A. (2003), 'Strategies for poverty alleviation in India: Cysd's holistic approach to empowerment through the self-help group model', *Ids Bulletin – Institute of Development Studies*, **34**, 133–42.

Dawson, J.W. (1998), 'Institutions, investment and growth: new cross-country and panel data evidence', *Economic Enquiry*, **36**, 603–19.

De la Fuente, A. and Estache, A. (2004) Infrastructure productivity and growth: a quick survey, mimeo, Washington, DC, WBIGF.

Demery, L. and Squire, L. (1996), 'Macroeconomic adjustment and poverty in Africa: an emerging picture', *The World Bank Research Observer*, **11** (1), 39–59.

Domah, P., Pollitt, M.G. and Stern, J. (2003), 'Modelling the costs of electricity regulation: evidence of human resource constraints in developing countries', mimeo, Risk Regulation, Accountability and Development Workshop, 26 and 27 June, University of Manchester.

Echeverrigent, J. (1992), 'Public-participation and poverty alleviation – the experience of reform communists in India West-Bengal', *World Development*, **20**, 1401–22.

Esrey, S.A., Potash, J.B., Roberts, L. and Shiff, C. (1991), 'Effects of improved water supply and sanitation on ascariasis, diarrhoea, dracunculiasis, hookworm infection, schistosomiasis and tracoma', *Bulletin WHO*, **69**, 609–21.

Estache, A., Goicoechea, A. and Trujillo, L. (2007), 'Utilities reforms and corruption in developing countries', City University Department of Economics Discussion Paper Series no. 07/07, City University, London.

Estache, S., Foster, V. and Wodon, Q. (2002), *Accounting for Poverty in Infrastructure Reform: Learning from Latin America's Experience*, Washington, DC: World Bank.

Fay, M. and Yepes, T. (2003), 'Investing in infrastructure: what is needed from 2000 to 2010', *World Bank Research Working Paper 3102*, Washington, DC: World Bank.

Foreman-Peck, J, and Millward, R. (1994), *Public and Private Ownership of British Industry 1820–1990*, Oxford: Clarendon Press.

Foster, V. (2002), *Condominial Water and Sewerage Systems: Costs of Implementation of the Model. Water and Sanitation Program*, Washington, DC: World Bank.

Foster, V. and Araujo, M.C. (2004), 'Does infrastructure reform work for the poor? A case study from Guatemala', *World Bank Research Working Paper 3185*, Washington, DC: World Bank.

Foster, V., Tre, J.P. and Wodon, Q. (2001), 'Energy prices, energy efficiency, and fuel poverty', mimeo, Washington, DC: World Bank.

Galiani, S., Gertler, P. and Schargodsky, E. (2005), 'Water for life: the impact of the privatization of water services on child mortality', *Journal of Political Economy*, **11** (1), 83–120.

Garg, A., Kabra, M. and Gulaty, S. (2003), 'Regulation in India: an overview and assessment of impact on economic growth and poverty reduction', mimeo, paper presented to the International Conference – Challenges to Development: Innovation and Change in Regulation and Competition, 14 October, Manila.

Globerman S. and Shapiro D. (2002), 'Global foreign direct investment flows: the role of governance infrastructure', *World Development*, **30** (11), 1899–920.

Greenaway D. and Milner C. (2002), 'Regionalism and gravity', *Scottish Journal of Political Economy*, **49** (5), 574–85.

Groenewegen, J.P.M. and Künneke, R.W. (2005), 'Process and outcomes of the infrastructure reform: an evolutionary perspective', in R.W. Künneke, A.F.Correljé and J.P.M. Groenewegen (eds), *Institutional Reform, Regulation and Privatization: Process and Outcomes in Infrastructure Industries*, Cheltenham, UK and Northampton, MA, USA: Edward Elgar.

Guasch, J.L. (2004), *Granting and Renegotiating Infrastructure Concessions: Doing it Right*, WBI Development Studies, Washington DC: World Bank.

Guasch, J.L. and Hahn, R.W. (1999), 'The costs and benefits of regulation: implications for developing countries', *The World Bank Research Observer*, **14** (1), 137–58.

Haddad, L. and Adato, M. (2002), 'Maximizing benefit transfers to the poor: evidence from South African employment programmes', *International Labour Review*, **141**, 203–33.

Hahn, R.W. (1998), 'Policy watch: analysis of the benefits and costs of regulation', *Journal of Economic Perspectives*, **12** (4), 201–10.

Harris, C. (2003), 'Private participation in infrastructure in developing countries: trends, impact, and policy lessons', *World Bank Working Paper No. 5*, Washington, DC: World Bank.

Harris, C., Hodges, J., Schur, M. and Shukla, P. (2003), 'Infrastructure projects: a review of cancelled private projects', *Public Policy for the Private Sector*, **252**, Washington, DC: World Bank.

Hart, O. and Moore, J. (1988), 'Incomplete contracts and renegotiation', *Econometrica*, **56**, 755–85.

Hay, D.A. and Morris, D.J. (1991), *Industrial Economics and Organization: Theory and Evidence*, Oxford: Oxford University Press.

Hoddinott, J. (2002), 'Participation and poverty reduction: an analytical framework and overview of the issues', *Journal of African Economies*, **11**, 146–68.

Howells, M., Victor, D.G., Gaunt, T., Elias, R.J. and Alfstad, T. (2006), 'Beyond free electricity: the costs of electric cooking in poor households and a market-friendly alternative', *Energy Policy*, **34** (17), 3351–58.

International Finance Corporation (IFC) (2002), 'Trends in private investment in developing countries, 1970–2000', Discussion Paper no. 44, Washington DC: World Bank.

International Monetary Fund (IMF) (2003), *World Economic Outlook: Growth and Institutions*, Washington, DC: IMF.

Islam, N. (1992), 'Poverty in South Asia: approaches to its alleviation', *Food Policy*, **17** (2), 109–28.

Islam, R. (1990), 'Rural poverty, growth and macroeconomic policies: the Asian experience', *International Labour Review*, **129** (6), 693–714.

Jalilian, H., C. Kirkpatrick and D. Parker (2007), 'The impact of regulation on economic growth in developing countries: a cross-country analysis', *World Development*, **35** (1), 87–103.

Kaufmann, D., Kraay, A. and Mastruzzi, M. (2003), 'Governance matters III: governance indicators for 1996–2002', *World Bank Policy Research Working Paper 3106*, Washington DC: World Bank.

Kebede, B. (2006), 'Energy subsidies and costs in urban Ethiopia: the cases of kerosene and electricity', *Renewable Energy*, **31**, 2140–51.
Kessides, I.N. (2005a), 'Infrastructure privatization and regulation: promises and perils', *The World Bank Research Observer*, **20** (1), 81–108.
Kessides, I.N. (2005b), 'Infrastructure privatization: gone too far? Or too early to tell?', *The PB Newsletter*, www.privatizationbarometer.net/newsletter (accessed in 2005).
Kirkpatrick, C., Parker, D. and Zhang, Y.-F. (2005), 'Price and profit regulation in developing and transition economies: a survey of the regulators', *Public Money & Management*, **25** (2), 99–106.
Kirkpatrick, C., Parker, D. and Zhang, Y.-F. (2006a), 'Foreign direct investment in infrastructure in developing countries: does regulation make a difference?', *Transnational Corporations*, **15** (1), 143–71.
Kirkpatrick, C., Parker, D. and Zhang, Y-F. (2006b), 'State versus private sector provision of water services in Africa: an empirical analysis', *World Bank Economic Review*, **20**, 143–63.
Kuczynski, P.P. and Williamson, J. (eds) (2003), *After the Washington Consensus: Restarting Growth and Reform in Latin America*, Washington, DC: Institute for International Economies.
Kuyvenhoven, A. (2004), 'Creating an enabling environment: policy conditions for less-favored areas', *Food Policy*, **29** (4), 407–29.
Laffont, J.-J. (1999), *Incentives and the Political Economy of Regulation*, Oxford: Oxford University Press.
Leipziger, D., Fay, M., Wodon, Q. and Yepes, T. (2003), 'Achieving the Millennium Development Goals: the role of infrastructure', *World Bank Policy Research Working Paper 3163*, Washington, DC: World Bank.
Listorti, J (1996), 'Bridging environmental health gaps', *AFTES Working Paper 20*, Washington, DC: World Bank.
Lloyd-Sherlock, P. (2000), 'Failing the needy: public social spending in Latin America', *Journal of International Development*, **12** (1), 101–19.
Manning, T.E. (2001), 'RE microfinance: new horizons for RE technologies in poverty alleviation projects', *Refocus*, **2** (8), 22–25.
McIntosh, A. (2003) *Asian Water Supplies: Reaching the Urban Poor*, Manila: Asian Development Bank.
Minogue, M. (2002), 'Governance-based analysis of regulation', *Annals of Public and Cooperative Economics*, **73** (4), 649–66.
Morisset, J. (2000), 'FDI in Africa: policies also matter', *Transnational Corporations*, **9** (2), 107–26.
Morris, L. and Cabrera, L.F.G. (2003), 'The involvement of the private sector in water servicing: effects on the urban poor in the case of Aguascalientes, Mexico', *Greener Management International*, **42**, 35–46.
Naraghi, S. and Kebothalez, T. (2004), 'Water as a key to socio-economic development and poverty Eradication in North-West Province, as projected for the entire South Africa', *Water Sanitation*, **30**, 681–4.
Nel, E. and Binns, T. (2003), 'Decentralising development in South Africa', *Geography*, **88**, 108–16.
North, D.C. (1990), *Institutions, Institutional Change and Economic Performance*, Cambridge: Cambridge University Press.
North, D.C. (1991), 'Institutions', *Journal of Economic Perspectives*, **5** (Winter), 97–112.
Ofwat (2007), http://www.ofwat.gov.uk/aptrix/ofwat/publish.nsf/AttachmentsByTitle/water_regfacts_figs.pdf/$FILE/water_regfacts_figs.pdf (accessed 2007).
Paris, R. (2000), 'Making development sustainable', *Organisation for Economic Cooperation and Development. The OECD Observer*, **223**, 33–35.
Parker, D. (1999), 'Regulating public utilities: what other countries can learn from the UK experience', *Public Management*, **1** (1), 93–120.
Parker, D. (2002), 'Economic regulation: a review of issues', *Annals of Public and Cooperative Economics*, **73** (4), 493–519.
Parker, D. and Kirkpatrick, C. (2005), 'The impact of privatization in developing countries: a review of the evidence and the policy lessons', *Journal of Development Studies*, **41** (4), 513–41.
Peltzman, S. (1976), 'Toward a more general theory of regulation', *Journal of Law and Economics*, **14** (August), 109–48.
Public-Private Infrastructure Advisory Facility (PPIAF) and Water and Sanitation Program (WSP) (2001), *New Designs for Water and Sanitation Transactions: Making Private Sector Participation Work for the Poor*, Washington, DC: World Bank.
Rakodi, C. (1995), 'Poverty lines or household strategies? A review of conceptual issues in the study of urban poverty', *Habitat International*, **19** (4), 407–26.
Reiche, K., Covarrubias, A. and Martinot, E. (2000), 'Expanding electricity access to remote areas: off-grid rural electrification in developing countries', in G. Isherwood (ed.), *World Power 2000*, London: Isherwood Production, pp. 52–60.
Rodrik, D. (2000), 'Institutions for high-quality growth: what they are and how to acquire them', *Studies in Comparative International Development*, **35** (3), 3–31.
Santibanez, C. (2005), 'The information basis of poverty measurement: using the capability approach to improve the CAS proxy tool', *The European Journal of Development Research*, **17** (1), 89–110.

Sanwal, M. (1987), 'The implementation of decentralization: a case study of district planning in Agra District, Uttar Pradesh, India', *Public Administration & Development (1986–1998)*, **7** (4), 383–97.
Shaffer, B. (1995), 'Firm level responses to government regulation: theoretical and research approaches', *Journal of Management*, **21** (3), 495–515.
Sheahan, J. (1994), 'Peru's return toward an open economy: macroeconomic complications and structural questions', *World Development*, **22** (6), 911–23.
Sidak, J.G. and Spulber, D.F. (1997), *Deregulatory Takings and the Regulatory Contract: The Competitive Transformation of Network Industries in the United States*, Cambridge: Cambridge University Press.
Smith, W. (2000), 'Regulating infrastructure for the poor: perspectives on regulatory design', mimeo, Infrastructure for Development: Private Solutions and the Poor Conference, 31 May–2 June, London.
Spiller, P. and Savedoff, W. (1999) 'Commitment and governance in infrastructure', in F. Basanes, E. Uribe and R.D. Willig (eds), *Can Privatization Delivery Infrastructure for Latin America?*, Baltimore, MD: Johns Hopkins University Press.
Stein, E. and Daude, C. (2001), 'Institutions, integration and the location of foreign direct investment', mimeo, Inter-American Development Bank: Washington, DC.
Stern, J. and Holder, S. (1999), 'Regulatory governance: criteria for assessing the performance of regulatory systems: an application to infrastructure industries in the developing countries of Asia', *Utilities Policy*, **8**, 33–50.
Stigler, G. (1971), 'The theory of economic regulation', *Bell Journal of Economics and Management*, **2** (Spring), 3–21.
Stiglitz, J. (2002), *Globalisation and its Discontents*, Harmondsworth: Penguin.
Suryanarayana, M.H. (2001), 'Economic reform versus food security: Kerala's Gordian knot', *Journal of International Development*, **13** (2), 239–53.
Taylor, M. (2002), 'The shaping of San livelihood strategies: government policy and popular values', *Development and Change*, **33**, 467–88.
The Business (2007), 'Taxpayers' Olympic burden is a chance to strike gold in infrastructure works', 14 April, p.45.
United Nations Conference on Trade and Development (UNCTAD) (2002), *Trade and Development Report*, Geneva: UNCTAD.
Wallsten, S. (2001), 'An econometric analysis of telecom competition, privatization, and regulation in Africa and Latin America', *Journal of Industrial Economics*, **49** (1), 1–20.
Wegelin, E.A. and Borgman, K.M. (1995), 'Options for municipal interventions in urban poverty alleviation', *Environment and Urbanization*, **7**, 131–51.
Wei, S-J. (2000), 'How taxing is corruption to international investors?', *Review of Economics and Statistics*, **83** (1), 1–11.
Williamson, O.E. (1985), *The Economic Institutions of Capitalism: Firms, Markets, and History*, Cambridge: Cambridge University Press.
Willoughby, C. (2002), 'Infrastructure and pro-poor growth: implications of recent research', report prepared for the UK Department for International Development, mimeo.
Winters, L.A., McCulloch, N. and McKay, A. (2004), 'Trade liberalization and poverty: the evidence so far', *Journal of Economic Literature*, **42** (1), 72–115.
World Bank (1994), 'Infrastructure for development', *World Development Report*, Washington, DC: World Bank.
World Bank (1995), *Bureaucrats in Business: The Economics and Politics of Government Ownership*, Oxford and Washington, DC: Oxford University Press and World Bank.
World Bank (2003), *Private Participation in Infrastructure: Trends in Developing Countries, 1990–2001*, Washington, DC: World Bank and Public Private Infrastructure Advisory Facility.
World Bank (2004), *Reforming Infrastructure: Privatisation Regulation and Competition*, Washington, DC: World Bank.
World Bank (2006), *Poverty Reduction and Growth: Virtuous and Vicious Cycles*, Washington, DC: World Bank and Oxford University Press.
World Bank (2008), *Private Participation in Infrastructure Database*, Washington, DC: World Bank.
Zhang, Y.-F., Parker, D. and Kirkpatrick, C. (2005), 'Competition, regulation and privatisation of electricity generation in developing countries: does the sequencing of the reforms matter?', *The Quarterly Review of Economics and Finance*, **45**, 358–79.
Zhang, Y.-F., Parker, D. and Kirkpatrick, C. (2007), 'Electricity sector reform in developing countries: an econometric assessment of the effects of privatisation, competition and regulation', *Journal of Regulatory Economics*, **33** (2), 159–78.

Index